Northern Horizon

Eastern Horizon

Western Horizon

Southern Horizon

The Night Sky in Summer

Second Edition

EXPLORATIONS
AN INTRODUCTION TO ASTRONOMY

Thomas T. Arny

Professor of Physics and Astronomy
University of Massachusetts, Amherst

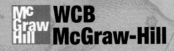

Boston, Massachusetts   Burr Ridge, Illinois   Dubuque, Iowa
Madison, Wisconsin   New York, New York   San Francisco, California   St. Louis, Missouri

WCB/McGraw-Hill

A Division of The **McGraw·Hill** *Companies*

EXPLORATIONS: AN INTRODUCTION TO ASTRONOMY, SECOND EDITION

3 4 5 6 7 8 9 0 QPD/QPD 9 0 9 8

ISBN 0-8151-2023-0

Publisher *Kevin Kane/Jim Smith*
Sponsoring Editor *Lloyd Black*
Developmental Editor *Donata Dettbarn*
Marketing Manager *Lisa Gottschalk*
Project Manager *Marilyn Rothenberger*
Production Supervisor *Cheryl Horch*
Designer *K. Wayne Harms*
Cover designer *K. Wayne Harms*
Cover Photograph *Jonathan Alpert SCRATCHworks*
Photo research coordinator *Donata Dettbarn*
Art Editor *Donata Dettbarn*
CD-ROM Development *Video Publishing Group, Inc.*
Illustration *Artscribe* and *Jay Hoagland*
Multimedia Coordination *David G. Shaw*
Multimedia Animations *Artscribe*
Compositor *Betsy Manini/Interactive Composition Corporation*
Typeface *10/12 Garamond*
Printer *Quebecor*

Library of Congress Cataloging-in-Publication Data

Arny, Thomas.
 Explorations: an introduction to astronomy / Thomas T. Arny. –
1996 version.
  p.   cm.
  Includes Index.
  ISBN 0-8151-2023-0
  1. Astronomy.   I. Titl.e
QB45.A76   1997
520—DC21

97-20191
CIP

www.mhhe.com/physsci/astronomy/arny

Contents

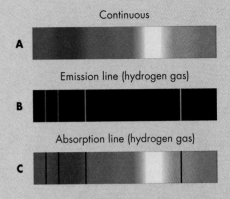

Continuous

A

Emission line (hydrogen gas)

B

Absorption line (hydrogen gas)

C

CHAPTER 4
The Earth   123

CHAPTER 7
Survey of the Solar System   217

CHAPTER 8
The Terrestrial Planets   239

CHAPTER 9
The Outer Planets   271

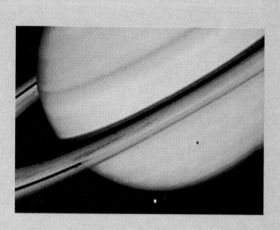

CHAPTER 10
Meteors, Asteroids, and Comets   297

CHAPTER 11
The Sun, Our Star   325

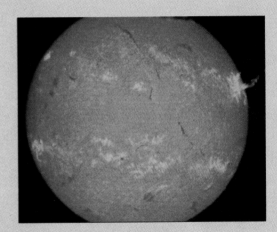

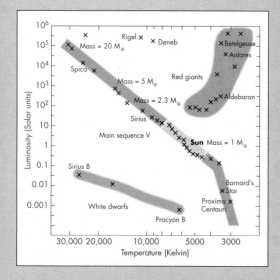

CHAPTER 14
Stellar Remnants: White Dwarfs, Neutron Stars, and Black Holes   413

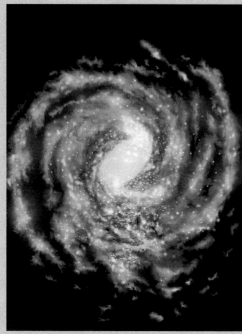

CHAPTER 17
Cosmology   505

ESSAY 3   Life in the Universe   529

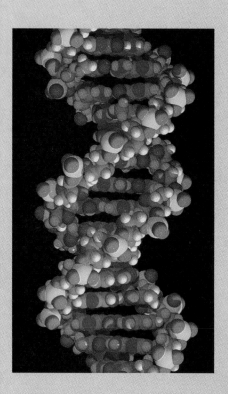

Preface

When I began writing *Explorations: An Introduction to Astronomy,* many people asked me why I was writing an astronomy book. Much of my motivation comes from wanting to share my own sense of wonderment about the Universe. I hope that in an astronomy course students can get some sense of where they fit in the astronomical Universe—a sense of location in the cosmic landscape. I also hope that students will come away from such a course with a sense of the richness of the Universe. When we look around us on our own planet, we see incredible biodiversity. So too when we look at the heavens, we see incredible astrodiversity. Stars, moons, and planets are as strange, colorful, and wonderful as tropical butterflies. Finally, I hope that students will gain some appreciation of the methods by which such tiny beings as we are have learned so much about the Universe. Not just laboratory techniques, those methods are far more important in the process of learning: the steps by which we go from observation to hypothesis and then on to what we hope is understanding.

But why write your own astronomy book when so many already exist? Most of the current books have so much material that they are impossible to get through in a single semester, and much material is omitted. I therefore decided that my first goal was to make a book that was short. However, as I worked at it, I kept finding things that I didn't want to leave out, material such as calendars and history of astronomy. But how could I write a short book and still include such topics? The solution was to organize the book so that instructors and students could omit the unwanted sections without interrupting the flow of ideas. Thus, I placed a number of topics such as time keeping and exo-biology into Essays that may be easily skipped. I also tried to make the book short by limiting its scope. Rather than covering everything, I have tried to focus on only what at the time seemed to me the most important ideas.

Another goal I set myself was to give simple explanations of why things happen. Such explanations generally involve physical principles that are unfamiliar to non—science students. However many even very complicated physical ideas can be appreciated, if not fully understood, by their appeal to analogy or to similarities with everyday phenomena. For example, diffraction effects can be seen by looking at a bright light through a lock of your hair pulled over your eyes or through glasses that you have fogged with your breath. By tying physical principles to everyday observations, many of the more abstract and remote ideas become more familiar. Thus I have used analogies heavily throughout the book, and I have designed the illustrations to make those analogies more concrete.

Another goal I set myself was to explain *how* astronomers know the many curious things they have learned about our Universe. Such explanations often require mathematics, and so I have included it wherever it is crucial to understanding a method of measurement, as in the use of the modified form of Kepler's third law to determine a star's mass or in Wien's law to measure its temperature. However, because math is so intimidating to so many students, I have tried to begin these discussions by introducing the essence of the calculations in everyday language. Thus if the student or instructor chooses to omit the math, it will not prevent an understanding of the basic idea involved. For example, Wien's law relates the temperature of a hot object to its color by a mathematical law. However, the consequences of the law can be seen in everyday life when we estimate how hot an electric stove burner is by the color it glows. Similarly, I have tried to work through the math problems step by step, explaining that terms must be cross-multiplied, and so forth.

As a final goal, I have set many of the modern discoveries in their historical context. I want to demonstrate that science is a dynamic process and that it is subject to controversy. Ideas are often not immediately accepted, and to appreciate those that scientists finally settle on, it helps to understand the arguments for the against as well as the train of reasoning that leads to the "accepted" answer. On this point I must digress and reveal my own amazement (and naiveté) at how many widely accepted ideas have such flimsy underpinnings and how many widely quoted values for astronomical quantities are very imperfectly known.

NEW TO THE SECOND EDITION

During the past two years astronomers have made important and exciting discoveries: distant, young blue galaxies; a gamma-ray source apparently located in a dim galaxy; evidence that meteorites from Mars may contain fossil evidence of ancient life on that planet; and evidence of perhaps a dozen or so planets around distant stars. Even though many of these claims and discoveries remain controversial, I think students should have the chance to learn a little about them and the evidence that supports what in some cases are truly amazing claims. For that reason a second edition of *Explorations* seems appropriate. Even now, as this edition goes to press, *further* exciting discoveries are making news: oceans beneath Europa's crust? Consensus on the value of the Hubble Constant? a reconciliation between the age of the oldest stars and the age of the Universe? In any case, a new edition was needed.

In putting this second edition together I have made relatively few changes in the book's overall structure, scope, or approach. The two most obvious changes are (1) the merging of the Essays on telescopes and light in the atmosphere into a chapter on telescopes, and (2) the addition of six "Overviews" that preview the material in related chapter blocks. My goal in the overviews is to help students see how the material they will encounter in the following few chapters fits together, so that they can see the broader context of what they are learning. The overviews have relatively little text and lots of pictures. Through this visual material, students may better remember and understand the material.

In addition to these changes, I have updated ideas in several chapters. For example, I've emphasized what I believe to be a growing belief that comets and asteroidal material have delivered much of the gas that makes up the atmospheres of the inner planets. As another example, I've discussed the observations with the Hubble Space Telescope that have begun to alter our view of how galaxies form and evolve. Moreover, observations with Hubble and Hipparchos may at last be reconciling the discrepancy between the age of the Universe and the ages of its oldest stars. All these changes may well be premature, but I think they deserve mention in the new edition. In making changes, I have tried to preserve what students and professors have told me they liked about the first edition. Please let me know your reactions to these changes, so that this book can continue to be as useful as possible.

A final comment about the new edition involves the accompanying CD-ROM. I suspect that still relatively few students have access to this technology. However, enough do to persuade the editors to include simple animations of some of the ideas that are difficult to visualize in still art. These animations were made by Greg Holt and Carolyn Duffy at ArtScribe, and I want to thank them for their usual flair for bright, clear art. I also want to thank David Shaw of McGraw-Hill Higher Education, who designed and executed the CD-ROM. He did a fine job of a very difficult task in turning a book into a multimedia product.

HOW TO STUDY WITH THIS BOOK

Learning anything requires a certain amount of work. You certainly don't expect to be able to pick up a guitar and play it without practice, nor do you expect to be able to jog 5 miles without working out regularly. Learning astronomy also requires some work. The steps below may help you learn material better and more easily.

In reading any assignment, begin by looking at the pictures. Turn the pages of the chapter and familiarize yourself with what the objects you will be reading about look like. Then read the introduction. Next, jump to the summary. Finally, start again and read the assigned material through. As you read, make notes of things you don't understand. For example, if you are puzzled about why eclipses don't happen every month, make a note. I would urge you *not* to highlight as you read. Making a few short notes is much more effective than highlighting whole paragraphs.

Look carefully at the pictures and diagrams. If the figure caption has a question in it, try to answer it. Make your own sketch of diagrams to be sure you understand what they represent.

In a first reading of a chapter, I'd suggest that if you are troubled by math you should simply skip it for the time being. Be sure however to read the material leading into the math so you at least understand what is being dealt with. When you encounter a mathematical expression of a physical law, put in words what the law relates. For example, the law of gravity relates the force of gravity to the mass of the objects and their distance from each other.

If you encounter words or terms as you read that you don't know, look them up in the glossary or index. You are just wasting your time if you read a description of some object and you don't know what it is.

When you finish the assignment, try to answer the review questions. They are short and are designed to show you whether you have assimilated the basic factual material of the assignment. Try to do this without looking back into the chapter, but if you can't remember, look it up rather than skip over the question. You might find it helpful to get a pile of scratch paper and actually write out short answers to the questions.

Having read the material once, go back and try to work through the math parts. Then try a practice problem to see if you can work through the material on your own.

If you get stuck at any point, see your teaching assistant or professor for help. Don't be shy about asking questions. I wish someone had beat this into my head earlier. Learning is a thousand times easier if you ask questions when you get stuck.

Throughout all the book, I have also tried to convey some of my own enthusiasm for astronomy. Many astronomical objects are strikingly beautiful. Others conjure up a sense of amazement. To me it is the ultimate wonder that within the Universe life has formed that can contemplate the Universe and ask what it is about. Seeing a clear night sky spangled with stars is for me a nearly religious experience. And yet the beauty that I see and my sense of wonder is enriched even more by an appreciation of the complex processes that make the Universe work. I hope this book will similarly increase your appreciation of our Universe's wonders.

If while using this book you find mistakes or if you have suggestions about how to make it better, *please* let me know. Write me at the Astronomy Program, University of Massachusetts, Amherst, MA 01003-4525, USA. If you have access to e-mail, please let me know that way. My address is arny@daisy.phast.umass.edu. I really want your feedback.

ASTRONOMY ON THE INTERNET

Over the last few years, many teachers and students have gained access to an exciting new astronomy resource: the Internet. Hundreds of scientists around the world have created picture galleries and accompanying explanations that can be read by any user with access to the Internet. All are available "free" at the click of a mouse. Moreover, no special computer expertise is needed.

To use this resource, get a computer account at your school on a machine that runs a "browser" such as Netscape or Microsoft Explorer. Check with your local computer guru about how to connect your personal computer to the network or go to your library or computer room and use one of the computers there that is already set up. Typically you'll need about 10 minutes to learn how to use the system. It is time well spent, because the Internet has become one of the best ways to find out the latest news about astronomical discoveries (similar systems exist for many other disciplines and hobbies). It would be impossible to list all the sites: they number in the thousands and change daily. Using one of the search systems such as Yahoo or Altavista will allow you to type in a few words that describe what you are interested in, and in a minute or so you may have literally hundreds of suggested addresses that you can go to by simply clicking your computer mouse.

To help you navigate the labyrinth of the internet, I've listed sites at the end of many chapters that cover the material in that chapter. To access them, type in the full address listed. Then, depending on your browser, hit the return key or click on a "go get" button.

For those not familiar with the net, a quick note on abbreviations and terms:

browser = a computer program (often free) that allows you to "browse" the web for information.
url = uniform resource location
http = hypertext transfer protocol
www = worldwide web
html = hypertext markup language
page = the computer screen display that appears when you call a given address
link = just that: a link from one page to another, perhaps to a computer on the other side of the world.

When you try to connect to a side, you need to type all the characters *as shown,* being careful about upper- and lowercase letters.

To help you get started, I've listed below a few addresses of general interest.

"Yahoo." An amazing collection of special interest pages on many subjects, including academic fields, travel, and entertainment. It also has a very good list of astronomical sources, including the beginnings of an on-line introductory astronomy text.
www.yahoo.com/Science/Astronomy/

"Welcome to the Planets." Many pretty pictures, fact sheets on the planets and the other Solar System objects, and a glossary. Prepared by the Jet Propulsion Laboratory.
www.pds.jpl.nasa.gov/planets

"Views of the Solar System," by Calvin Hamilton. Excellent pictures, fact sheets, and glossary. Lots of links to other sites.
www.bang.lanl.gov/solarsys/

"The Nine Planets," by Bill Arnett. "A Multimedia Tour . . ." One of the best astronomy sites. Lots of excellent links, pictures, text, and glossary.
www.seds.org/billa/top/

"The Solar System Live." Another good site for pictures and information about the Solar System.
fourmilab.ch/solar/solar.html

Note: Most of these addresses will suggest links to many additional sources, thus creating a web of information hence the name, "worldwide web."

ACKNOWLEDGMENTS

I owe thanks to many people for their help in this book, both for the first edition and for this new one. Help came in the form of advice, pictures, information, encouragement, and improvements to my own understanding of things. I have pestered all of my colleagues in the Five College Astronomy Department and many of them in the Department of Physics and Astronomy. Mike Skrutskie and Martin Weinberg, as neighbors down the hall, bore a disproportionate share of questions, and I owe them special thanks. Gene Golowich read over an early draft on inflationary cosmology and made valuable suggestions. I profited from many conversations with Ted Harrison, the late Ed Phinney (of the Classics Department), Peter Schloerb, and David Van Blerkom. Other people who contributed were Bill Bates and Rick Newton, who helped with setting up and taking pictures, and Linda Ray Arny, who helped me locate many references. I also want to thank Amy Lovell for her careful proofreading of the first edition.

Many readers have been kind enough to take the time to send me suggestions for ways to improve the text or to point out errors. They include Bill Dent, Bill Irvine, Daniel Jaffe, Susan Kleinmann, Lauren Likkel, Mesgun Sebahu, Ron Snell, Mark Stuckey, Gene Tademaru, and Steve Schneider. I particularly want to thank the following people for very detailed critiques of several sections: Eric Feigelson (who read and commented extensively on the Telescope chapter), Wei Lee, Rainer Mauersberger, James O'Connell, Joel Weisberg, Richard White, and Ben Zellner.

I also wish to thank William R. Luebke, who revised the test bank that he had done for the second printing of the first edition and improved it dramatically.

Many people at McGraw-Hill have helped immensely. I am very grateful to Jim Smith, who began the project and read the entire manuscript in its first draft. Mary Hill, Carl Masthay, and Arny Wastalu all helped in the first edition. I owe a special debt of gratitude to Judy Hauck, who was Developmental Editor for the first edition and turned it into such a fine and attractive book.

I am also deeply appreciative of the lovely work done by Carolyn Duffy and Greg Holt of ArtScribe who did the color figures for the book. They listened patiently to my suggestions and turned my ill-drawn scrawls into bright, clear drawings. Likewise, I want to thank Jay Hoagland for the margin sketches.

Others at McGraw-Hill who made major contributions include Lloyd Black, who offered many helpful comments and ideas for the 1996 update; John Murdzek, the developmental editor for the update, who suggested many improvements; and Donata Dettbarn who searched diligently and creatively for the many beautiful new pictures. For the second edition, I want to thank Nick Murray for both his careful copyediting and for a number of important suggestions and clarifications. I am very grateful to Marla Irion and Marilyn Rothenberger for their help in turning the manuscript into a book and for her encouragement and willingness to make last-minute changes to allow us to make the book as up-to-date as possible.

I want especially to thank Donata Dettbarn who took over as developmental editor in the midst of many confusing changes. She deserves a medal for her patience and encouragement (as well as general editorial skills) for continuing to find lovely new art and for bringing the second edition to its conclusion.

Lisa Gottshalk in the sales office also deserves a special thank you for her enthusiasm. Finally, I want to thank the many sales reps who sent me comments from adopters and potential users about points they felt needed fixing. Thank you all.

REVIEWERS

The following people have reviewed this book at various stages of its development. I very much appreciate their help, suggestions, and corrections. Any errors that remain are not their fault, but mine.

Bruce Balick, University of Washington, Seattle

Tom Balonek, Colgate University

Tai L. Chow, California State University, Stanislaus

Bruce de Mayo, West Georgia College

Alexander Dickson, Seminole Community College

Jess Dowdy, Abilene Christian University
Eric Feigelson, The Pennsylvania State University
Donald Foster, Wichita State University
Aaron Galonsky, Michigan State University
Bruno Gruber, Southern Illinois University
Heidi Hammell, Massachusetts Institute of Technology
John Greg Hoessel, University of Wisconsin
Thomas Hockey, University of Northern Iowa
Terry Jay Jones, University of Minnesota
Yong H. Kim, Saddleback College
Jeffrey Kuhn, Michigan State University
John K. Lawrence, California State University, Northridge
Carolyn Mallory, Moorpark College
J. Scott Shaw, University of Georgia
Norman Sperling, Chabot Observatory and Science Center
Michael Stewart, San Antonia College
Walter Wesley, Moorehead State University
Dan Wilkins, University of Nebraska
W.C. Woods, Glassboro State College
Jon K. Wooley, Eastern Michigan University

I am especially grateful for the very detailed and thoughtful comments and suggestions of Thomas Hockey, John Greg Hoessel, Terry Jay Jones, Norman Sperling, and Dan Wilkins, and I want particularly to thank Eric Feigelson for his helpful and thoughtful suggestions for the first edition and his careful reading of Chapter 5 for this new edition.

NEW ANCILLARY PACKAGE

The second edition of *Explorations* has a new set of ancillary materials. They include the following:

A demonstration interactive version of the text on CD-ROM.

A full interactive version of the text on CD-ROM that includes some twenty animations of figures such as eclipses, retrograde motion, and phases of the Moon.

A Visual Resources Library CD-ROM.

An Instructor's Resource Manual that includes references to additional articles and books, lecture outlines, and syllabi for a variety of course formats.

A Computer Test Bank expanded from the version prepared by Professor William R. Luebke of Modesto Junior College for the 1996 prtinting.

A set of transparency acetates.

INTRODUCTION

THE COSMIC LANDSCAPE

God taking the measure of the Universe in the etching "Ancient of Days," by William Blake (c. 1794). (Courtesy of the Rosenbach Museum and Library)

Remote galaxies as seen by the Hubble Space Telescope (Courtesy of HST)

Astronomy is the study of the heavens, the realm extending from beyond the Earth's atmosphere to the most distant reaches of the Universe. Within this vast space we discover an amazing diversity of planets, stars, and galaxies. That creatures as tiny as ourselves cannot only contemplate but also understand such diversity and immensity is amazing. But even more amazing are the objects themselves: planets with dead volcanos whose summits dwarf Mt. Everest, stars a hundred times the size of the Sun, and galaxies—slowly whirling clouds of stars—so vast that they make the Earth seem a grain of sand in comparison. All this is the cosmic landscape in which we live, a landscape we will briefly explore now to familiarize ourselves with its features and to gain an appreciation of its vast scale.

I.1 THE EARTH, OUR HOME

We begin with the Earth, our home **planet** (fig. I.1). This spinning sphere of rock circling the Sun is huge by human standards, but it is one of the smallest bodies in the cosmic landscape. Nevertheless, it is an appropriate place to start because, as the base from which we view the Universe, it influences what we can see. We cannot travel from object to object in our quest to understand the Universe. Instead, we are like children who know their neighborhood well but for whom the larger world is still a mystery, known only from books and television at second hand.

FIGURE I.1
The planet Earth, our home, with blue oceans, white clouds, and multihued continents. (Courtesy NASA.)

But just as children use knowledge of their neighborhood to build their image of the world, so astronomers use their knowledge of Earth as a guide to more exotic worlds. For example, volcanos that spew molten lava and geysers that shoot hot water into the air suggest that the center of the planet we live on seethes with heat. This heat stirs molten rock within the Earth, creating slow but powerful currents that shift our planet's crust, building mountains, heaving up volcanos, and generating a magnetic field. Looking outward to our planetary neighbors, we find landscapes on Venus and Mars that bear evidence of many of the same processes that sculpt our planet and create its diversity. Likewise, when we look at the atmospheres of other planets we see many of the same features that occur in our atmosphere. For example, winds in the thin envelope of gas that shelters us swirl around our planet much as similar winds sweep the alien landscapes of Venus and Mars.

1.2 THE MOON

The Moon is our nearest neighbor in space, a **satellite** that orbits the Earth some quarter million miles (384,000 km) away. Held in tow by the Earth's gravity, the Moon is much smaller than Earth (only about one quarter our planet's diameter).

With the naked eye, and certainly with a pair of binoculars, we can clearly see that its surface is totally unlike Earth's. Instead of white whirling clouds, green covered hills, and blue oceans, we see an airless, pitted ball of rock that shows us the same face night after night (fig. I.2).

Why are the Earth and Moon so different? Their different surfaces arise in large part from their great difference in mass. The Moon's mass is only about 1/80 the Earth's, and its smaller bulk was therefore less able to retain heat. Without that strong internal heat, no crustal motions—so important in shaping Earth—have altered the Moon's surface. In fact, the Moon has changed so little over the past 3 billion years that its surface may hold important clues to what planets may have been like billions of years ago. In addition to this scientific importance, the Moon has symbolic significance for us—it marks the present limit of our direct exploration of space.

FIGURE I.2
The Moon, as we see it with unaided eyes.
(Courtesy Stephen E. Strom.)

1.3 THE PLANETS

Beyond the Moon, circling the Sun as the Earth does, are eight other planets, sister bodies of Earth. In order of their average distance from the Sun and working out, the nine planets are Mercury, Venus, Earth, Mars, Jupiter, Saturn, Uranus, Neptune, and Pluto. These worlds have dramatically different sizes and landscapes. For example:

- Craters scar the airless surface of Mercury.
- Dense clouds of sulfuric acid droplets completely shroud Venus.
- White clouds, blue oceans, green jungles, and red deserts tint Earth.
- Huge canyons and deserts spread across the ruddy face of Mars.
- Immense atmospheric storms with lightning sweep across Jupiter.
- Trillions of icy fragments orbit Saturn, forming its bright rings.
- Dark rings girdle Uranus, its spin tipped by some cosmic catastrophe.
- Choking methane clouds whirl in the deep blue atmosphere of Neptune.
- Perpetual ice glazes dim Pluto.

To the unaided eye the other planets are mere points of light whose positions shift slowly from night to night. But by observing them, first with Earth-based telescopes, then ultimately by remotely piloted spacecraft, we have learned that they are truly other worlds. Figure I.3 shows pictures of these nine distinctive bodies and reveals something of their relative size and appearance. Some are far smaller and others vastly larger than Earth, but all are dwarfed by the Sun, whose immense gravity holds them in orbit.

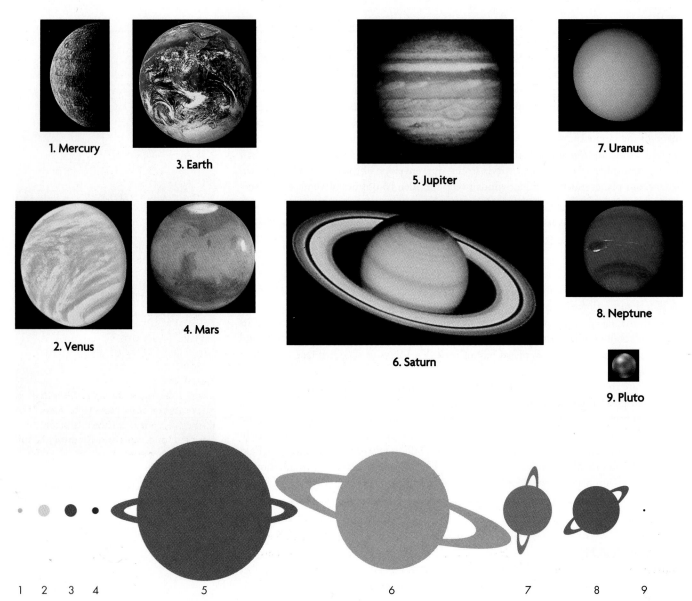

1. Mercury

3. Earth

5. Jupiter

7. Uranus

2. Venus

4. Mars

6. Saturn

8. Neptune

9. Pluto

1   2   3   4   5   6   7   8   9

FIGURE I.3
Portraits of the nine planets along with
silhouettes showing their correct relative
size. All except Pluto's picture were taken
by spacecraft. (Pictures of planets
courtesy NASA/JPL.)

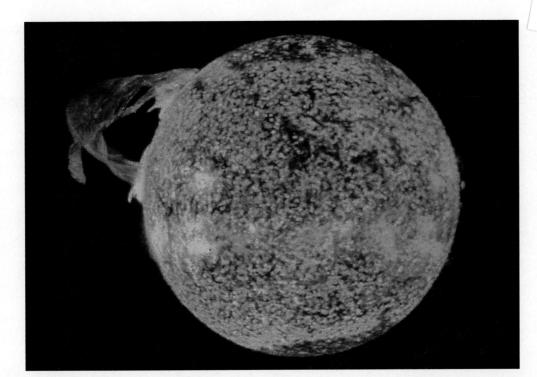

Jupiter

Earth

FIGURE I.4
The Sun as viewed through a filter that allows its hot outer gases to be seen. The Earth and Jupiter are shown to scale beside it for comparison. (Courtesy Naval Research Laboratory.)

I.4 THE SUN

The Sun is a **star,** a huge ball of gas over 100 times the diameter of the Earth and over 300,000 times more massive: if the Sun were a volleyball, the Earth would be about the size of a pinhead, and Jupiter roughly the size of a nickel. The Sun of course differs from the planets in more than just size: it generates energy in its core by nuclear reactions that convert hydrogen into helium. From the core, the energy flows to the surface, and from there it pours into space to illuminate and warm the planets (fig. I.4).

I.5 THE SOLAR SYSTEM

The Sun and its nine planets form the **Solar System.** But smaller bodies—satellites (moons) orbiting the planets, and asteroids and comets orbiting the Sun—also populate the Solar System (fig. I.5).

I.6 A SENSE OF SCALE

If the paths that the planets follow around the Sun were visible, we would see that the Solar System is like a huge set of nested elliptical rings, centered approximately on the Sun and extending about 4 billion miles outward to Pluto's orbit. It is hard to imagine such immense distances measured in miles. In fact, it is as foolish to use miles to measure the size of the Solar System as it is to use inches to measure the distance between New York and Tokyo. Whenever possible, we try to use units appropriate to the scale of what we seek to measure. For example, in earlier times people used units that were quite literally at hand, such as finger widths or the spread of a hand to measure a piece of cloth; paces to

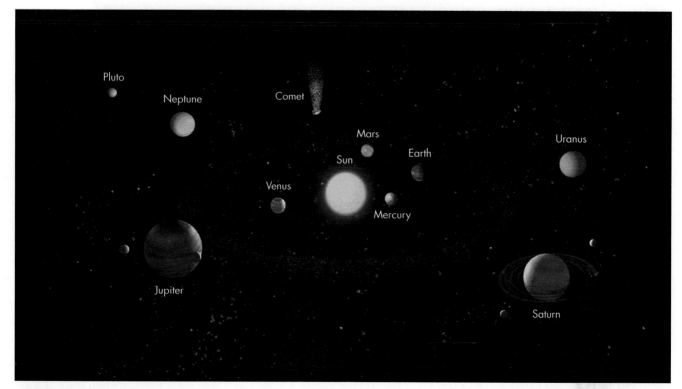

FIGURE I.5
An Artist's view of the Solar System
showing the Earth and its eight sister
planets circling the Sun. Note the
asteroids between Mars and Jupiter and
the comet moving along its elongated
orbit. (In this diagram and many others
throughout the book, distances and sizes
of astronomical bodies are exaggerated
for clarity.)

measure the size of a field. In the same tradition, although on a different scale, astronomers use distance scales related to familiar objects, such as the Earth. As we shall see in later chapters, the Earth's radius makes a convenient unit for measuring the size of the other planets. Likewise, the Earth's distance from the Sun makes a good unit for measuring the scale of the Solar System.

I.7 THE ASTRONOMICAL UNIT

The **astronomical unit,** abbreviated as AU, is defined by the distance from the Earth to the Sun.* This translates into about 93 million miles (150 million kilometers). If we use the AU to measure the scale of the Solar System, Mercury turns out to be 0.4 AU from the Sun, while Pluto is about 40 AU.

The Solar System remains the limit to our exploration of the Universe with spacecraft; our probes have penetrated only slightly beyond Pluto. But telescopes extend our view far beyond the Solar System to reveal that just as the Earth is but one of many planets orbiting the Sun, so too the Sun is but one of a vast swarm of stars orbiting the center of our galaxy, the Milky Way.

I.8 THE MILKY WAY GALAXY

The **Milky Way galaxy** is a cloud of several hundred billion stars with a flattened shape like the Solar System (fig. I.6). The Sun and other stars orbit the Milky Way at some 140 miles per second (220 kilometers per second), but so vast is our galaxy that it still takes the Sun about 240 million years to complete one trip around this immense disk. The Milky Way's

---

*Because the Earth's orbit is actually an ellipse, the AU is technically defined slightly differently, a point we will discuss further when we consider planetary orbits.

FIGURE I.6
The Milky Way Galaxy: side and top views, respectively. (Courtesy Alex Baum.)

myriads of stars come in many varieties, some hundreds of times larger than the Sun, others hundreds of times smaller. Some stars are much hotter than the Sun and shine a dazzling blue-white, while others are cooler and glow a deep red. Still others emit no light at all because their intense gravity prevents any light from escaping.

In the Milky Way, as in many other galaxies, stars intermingle with immense clouds of gas and dust. These mark the sites of stellar birth and death. Within cold, dark clouds, gravity may draw the gas into dense clumps that eventually turn into new stars, lighting the gas and dust around them. Stars eventually burn themselves out and explode, spraying matter outward to mix with the surrounding clouds. This matter from exploded stars is ultimately recycled into new stars (fig. I.7).

In this huge swarm of stars and clouds, the Solar System is all but lost—a single grain of sand on a vast beach—forcing us again to grapple with the problem of scale. Stars are almost unimaginably remote: even the nearest one to the Sun is over 25 trillion miles away. Such distances are so immense that analogy is often the only way to grasp them. For example, if we think of the Sun as a pinhead, the nearest star would be another pinhead about 35 miles away and the space between them would be nearly empty. In fact, distances between stars are so immense that even astronomical units are inappropriately small and so we again choose a new unit of length—the light-year.

FIGURE 1.7
Photographs of interstellar clouds in the Milky Way taken with an Earthbound telescope. On the scale of these pictures, the Solar System out to pluto is about 1000 times smaller than the period ending this sentence. (A) A cold, dark cloud and a star cluster beside it. Dust in the cloud blocks our view of the stars behind it. (Courtesy David Malin, Anglo-Australian Telescope Board.) (B) A group of clouds heated by young stars. Glowing hydrogen in the clouds creates their red color. (Courtesy Anglo-Australian Telescope Board.)

1.9 THE LIGHT-YEAR

Measuring a distance in terms of a time may at first sound peculiar, but we do it often. We may say, for example, that our town is a 2-hour drive from the city, or our dorm is a 5-minute walk from the library, but expressing a distance in this fashion implies that we have a standard velocity.

Astronomers are fortunate to have a superb velocity standard: the speed of light in empty space, which is a constant of nature and equal to 299,792,458 meters per second (about 186,000 miles per second). Moving at this constant and universal speed, light in 1 year travels a distance defined to be 1 **light-year,** abbreviated as ly. As we will show below, this works out to be about 6 trillion miles (10 trillion kilometers). To demonstrate that, however, is cumbersome because it involves multiplying a series of large numbers. We therefore will use a more concise way to write them called powers-of-ten notation.

In **powers-of-ten notation** (also called **scientific notation**) we write numbers using ten to an exponent, or power. Thus we write $100 = 10 \times 10 = 10^2$ and 1 million (1,000,000) as $10 \times 10 \times 10 \times 10 \times 10 \times 10 = 10^6$. Instead of writing out all the zeros, therefore, we use the exponent to tell us the number of zeros. A number like the speed of light (186,000 miles per second) may also be written in powers-of-ten notation, becoming 1.86×10^5 miles per second. Likewise, the astronomical unit (150 million kilometers) can be written as 1.5×10^8 km.

One reason to use powers-of-ten notation is that multiplying and dividing becomes enormously easier. For example, to multiply two powers of ten we just add the exponents, and to divide we subtract them. Thus $10^2 \times 10^5 = 10^7$, and $10^8/10^3 = 10^5$. More details on using powers-of-ten notation are given in the appendix, but as an illustration of its usefulness, let us now calculate the number of miles in a light-year.

To find how far light travels in a year, we multiply its speed by the travel time. One year is about 31,600,000 (about 3.16×10^7) seconds.* Multiplying this by the speed of light we get 3.16×10^7 seconds $\times 1.86 \times 10^5$ miles/second $= 3.16 \times 1.86 \times 10^{12}$ seconds $\times$ miles/second $= 5.88 \times 10^{12}$ miles, or about 6 trillion miles (10^{13} kilometers),

---

*Numbers here and elsewhere are rounded off.

as we previously claimed. In these units, the star nearest the Sun is 4.3 light years away. Although we achieve a major convenience in adopting such a huge distance for our scale unit, we should not lose sight of how truly immense such distance are. For example, if we were to count off the miles in a light-year, one every second, it would take us about 185,000 years!

We can now use the light-year for setting the scale of the Milky Way galaxy. In light-years, our galaxy is about 80,000 light-years across, with the Sun orbiting roughly 30,000 light-years from the center. Within the Milky Way's disk, stars are separated by a few light-years. The diffuse gas clouds scattered across its disk can be hundreds of light-years across.

1.10 GALAXY CLUSTERS AND THE UNIVERSE

Having gained some sense of scale for the Solar System and the Milky Way, we now resume our exploration of the cosmic landscape, pushing out to the realm of other galaxies. Here we find that just as stars are grouped into galaxies, so galaxies are themselves grouped into **galaxy clusters.**

The cluster of galaxies to which the Milky Way belongs is called the **Local Group.** It is "local," of course, because it is the one we inhabit. The Local Group is small as galaxy clusters go, containing only about 30 galaxies as members, but it is still about 3 million light-years in diameter. Yet despite such vast dimensions, the Local Group is itself part of a still larger assemblage of galaxies known as the **Local Supercluster.** Figure I.8 puts this in perspective for us.

Our supercluster consists of a few dozen member galaxy clusters, spreading over 100 million light-years, but it is perhaps itself part of an even larger structure known as the Great Attractor, a cluster of superclusters, possibly as much as 200 million light-years across. Structures of such vast size are about the largest objects we can see before we take the final jump in scale to the **Universe** itself.

The visible Universe is the largest astronomical structure of which we have any knowledge. What we can know of it is limited to what we can observe using sophisticated Earth-based and space-based telescopes and instruments. Even its size and shape are only roughly known. For example, according to some theories the Universe extends limitlessly, while according to others it spans perhaps 15 billion light-years, gradually and imperceptibly curving back on itself to form a closed system much like the surface of the Earth. But regardless of our uncertainty about the known Universe's shape and size, we can observe that its structure is surprisingly well ordered. Small objects are clustered into larger systems, which are themselves clustered: planets around stars, stars in galaxies, galaxies in

FIGURE I.8
(A) The Milky Way, which is at right in the Local Group of galaxies and (B) the Local Supercluster. The patches are individual galaxy clusters. Our knowledge of these structures is still very incomplete, so this is an artist's interpretation of data gathered to date.

Local Group

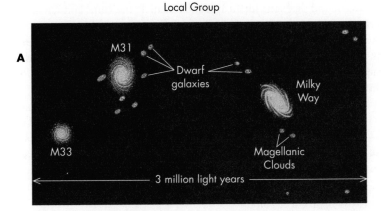

Local Supercluster

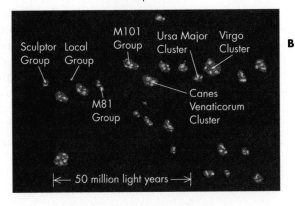

clusters, clusters in superclusters, and superclusters into even larger groups. Although astronomers do not yet understand completely how this orderly structure originated, they do know that gravity plays a crucial role.

I.11 GRAVITY

Gravity gives the Universe structure because it creates an attraction between *all* objects. You can see that attraction even in everyday life. For example, if you drop a book, the Earth's gravitational force makes the book fall. Moreover, that same force spans the vast distance between the Earth and the Moon to hold our satellite in its orbit. Similarly, gravity holds our planet in its orbit around the Sun and the Sun in its orbit around the Milky Way.

Although gravity dominates the large-scale structure of the Universe, other forces dominate on smaller scales. To understand these forces, we need to look briefly at the small-scale structure of matter.

I.12 ATOMS AND OTHER FORCES

Matter is composed of submicroscopic particles called **atoms.** Atoms are incredibly small. For example, a hydrogen atom is about one ten-billionth of a meter (10^{-10} m) in diameter. Ten million hydrogen atoms could be put in a line across the diameter of the period at the end of this sentence. But despite this tiny size, atoms themselves have structure. Every atom has a central core called the **nucleus** that is orbited by a swarm of smaller particles called **electrons** (fig. I.9). The nucleus is in turn composed of two other kinds of particles called **protons** and **neutrons.**

Although the particles in an atom exert a gravitational attraction on one another, atoms are not held together by gravity. Instead, an electric force gives them their structure. That force arises because protons and electrons have a property called **electric charge.** A proton has a positive electric charge, and an electron has a negative electric charge. A neutron, as its name suggests, has no charge.

The **electric force** can either attract or repel, depending on the charges. Opposite charges attract, and like charges repel. Thus, two electrons (both negative) repel each other, while an electron and a proton (positive and negative) attract each other. That attraction is what holds the electrons in their orbits around the nucleus of an atom.

You can see the electric force at work in many ways. For example, the electric charges generated when a clothes drier tumbles your laundry creates an attraction that may make clothes cling together. The crackling sound you hear as you pull fuzzy socks away from a shirt is the electric charges jumping and making tiny sparks.

The electric force is closely linked with the magnetic force that makes a compass work or holds the little magnets to the door of your refrigerator. In fact, the theory of relativity demonstrates that electric and magnetic forces are fundamentally the same, and scientists generally refer to them jointly as the electromagnetic force.*

Yet another force plays a critical role in atoms. This force is called the **strong** or **nuclear force** and it holds the protons and neutrons together in an atom's nucleus. Although the effects of the strong force cannot be seen directly in everyday life, without it, the nuclei of atoms would disintegrate and with them, our familiar world.

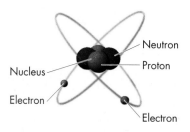

FIGURE I.9
An atom consists of a nucleus around which electrons orbit. The nucleus is itself composed of particles called protons and neutrons.

---

*The electromagnetic force is linked with yet another fundamental force responsible for the radioactive decay of atoms. These combined forces are today known technically as the electroweak force.

I.13 THE SCIENTIFIC METHOD

Our understanding of the Universe has not come easily. It has grown out of enormous intellectual work of thousands of men and women over thousands of years, through the rigorous testing of ideas. That testing is part of what we generally call the **scientific method.**

In its simplest form the scientific method is the procedure by which scientists construct their view of the Universe and its contents, regardless of whether those ideas concern stars, planets, living things, or matter itself. In the scientific method a scientist proposes an idea about some property of the Universe—a hypothesis—and then tests the hypothesis by experiment. Ideally the experiment either confirms the hypothesis or refutes it. If refuted, the hypothesis is rejected.

This same procedure may be applied to any scientific idea. In fact, whether an idea is "scientific" depends to some extent on whether it can be verified by either a real or imagined experiment. Astronomers, however, face a special difficulty in applying the scientific method because they cannot experiment with their subject matter directly: In virtually all cases, they can only passively observe. Nevertheless, they try—like all scientists—to use the scientific method.

Application of the scientific method is no guarantee that its results will be believed. For example, we will see in the next chapter that even before 300 B.C., the Greek philosopher Aristotle taught that the Earth is a sphere. Yet despite his "proofs" supporting this hypothesis, many people continued to believe the Earth to be flat. Even today, some scientific hypotheses may be rejected in the face of their experimental verification, and other may be accepted though untrue. For example, astronomer A may find evidence supporting hypothesis B, but astronomer C may claim the experiment was done incorrectly, or the data were analyzed improperly. In fact, some historians of science have argued that old ideas are discarded not so much through proof of their incorrectness but rather by the death of their proponents.

We need therefore to keep in mind throughout this book that when we discuss ideas, they are not all "proved" or universally accepted. This is especially true of ideas at the frontiers of our knowledge; for example, those dealing with the origin and structure of the Universe or those dealing with black holes. However, the tentativeness of such ideas does not always stop astronomers from being very positive about them, leading the Soviet physicist Lev Landau to state that astrophysicists are "often in error, but never in doubt." Therefore, keep in mind that some of the ideas we discuss in this book will be proved wrong in the future. That is not a failing of science, however. It is its strength.

SUMMARY

The Earth is one of nine planets orbiting the Sun, and the Sun is one of some 100 billion stars that make up the Milky Way galaxy. The Milky Way and about 30 other galaxies comprise the Local Group, which in turn is part of the Local Supercluster of galaxies. Superclusters seem to be grouped into even larger systems, all of which fit within the visible Universe. We can speak with some certainty about the size and properties of objects in our immediate neighborhood, but the farther we move from Earth, the less certain we become.

Astronomers use the astronomical unit (AU) and light-year (ly) to measure the immense size and distances of astronomical systems. The AU is defined by the average distance between the Earth and the Sun, and the light-year is defined as the distance light

FIGURE I.10
How the Earth fits into the Universe.

Table I.1   The Scale of the Universe

Object	Approximate radius
Earth	6400 km (4000 miles)
Sun	700,000 km (100 times radius of the Earth)
Earth's orbit	150 million km (200 times radius of Sun) $=$ 1 AU
Solar System to Pluto	40 AU (about 8600 times radius of the Sun)
Milky Way Galaxy	40,000 ly (about 10^8 times radius of the Solar System)
Local Group	1.5 million ly (about 40 times radius of the Milky Way)
Local Supercluster	40 million ly (about 30 times radius of the Local Group)
Visible Universe	15 billion ly (about 300 times radius of the Local Supercluster)

travels in a year, which is about 6 trillion miles. Using these units, we can see the vast scale of the Universe in figure I.10 and table I.1.

Matter is made up of atoms in which charged particles called electrons orbit a nucleus. The nucleus is itself composed of smaller particles called protons and neutrons.

Three forces give the Universe its structure: gravity on the cosmic scale and the strong and electromagnetic forces on the scale of atoms.

QUESTIONS FOR REVIEW

1. To what systems, in increasing order of size, does the Earth belong?
2. What force holds the different astronomical systems described in this section together?
3. About how much bigger in radius is the Sun than the Earth?
4. Roughly how big across is the Milky Way galaxy?

5. How big is an astronomical unit?
6. How is a light-year defined?
7. What particles make up an atom?
8. What force holds the electrons to an atom's nucleus?
9. What is meant by the scientific method?

THOUGHT QUESTION

Propose a hypothesis about something you can experiment with in everyday life and try to verify or disprove the hypothesis. For example, what kind of surfaces will the little magnetic note holders people use on refrigerators stick to? Any smooth surface? Any metal surface?

PROBLEMS

1. The radius of the Sun is 7×10^5 kilometers, and that of the Earth is about 6.4×10^3 kilometers. Use powers of ten to show that the Sun's radius is about 100 times the Earth's radius.
2. Given that an astronomical unit is 1.5×10^8 kilometers, and a light-year is about 10^{13} kilometers, how many AU are in a light-year?
3. If the Earth were the size of a B-B (about 0.1 of a centimeter radius), how big would the Sun be? How big would the Milky Way be?
4. Imagine building a model of the Solar System on your campus. Work out the diameter and spacings of the planets in millimeters and meters respectively.

5. If the Milky Way were the size of a nickel (about 2 centimeters), how big would the Local Group be? How big would the Local Supercluster be? How big would the Visible Universe be? The data in table I.1 may help you here.
6. Using powers of ten, evaluate $(4 \times 10^8)^3/(5 \times 10^{-6})^2$.
7. Calculate approximately how long it takes light to travel from the Sun to Pluto.
8. A typical bacterium has a diameter of about 10^{-6} meters. A hydrogen atom has a diameter of about 10^{-10} meters. How many times smaller than a bacterium is a hydrogen atom?

FURTHER EXPLORATIONS

Bronowski, Jacob. *A Sense of the Future.* Cambridge, Mass.: MIT Press, 1977.

Boorstin, Daniel J. *The Discoverers.* New York: Random House, 1983.

Morrison, Philip, and Phylis Morrison. *Powers of Ten.* New York: W. H. Freeman, 1982.

Pine, Ronald C. *Science and the Human Prospect.* Belmont, Calif.: Wadsworth Publishing, 1989.

Sagan, Carl. *Cosmos.* New York: Random House, 1980.

VIDEO

Powers of Ten. Pyramid Film and Video, Santa Monica, California, 1989.

TEST YOURSELF

1. The light-year is a unit of
   (a) time.
   (b) distance.
   (c) speed.
   (d) weight.
   (e) age.

2. You write your home address in the order of street, town, state, and so on. Suppose you were writing your cosmic address in a similar manner. Which of the following is the correct order?
   (a) Earth, Milky Way, Solar System, Local Group.
   (b) Earth, Solar System, Local Group, Milky Way.
   (c) Earth, Solar System, Milky Way, Local Group.
   (d) Solar System, Earth, Local Group, Milky Way.
   (e) Solar System, Local Group, Milky Way, Earth.

3. Which of the following astronomical system is/are held together by gravity?
   (a) The Sun.
   (b) The Solar System.
   (c) The Milky Way.
   (d) The Local Group.
   (e) All of them are.

4. From the lower part of figure I.3, about how much larger is Jupiter's diameter than the Earth's?
   (a) 2 times.
   (b) 5 times.
   (c) 10 times.
   (d) 25 times.
   (e) 100 times.

5. Which of the following statements can be tested for correctness using the scientific method?
   (a) The Sun's diameter is about 100 times larger than the Earth's diameter.
   (b) The sky is sometimes blue.
   (c) An astronaut cannot survive on the Moon without life-support systems.
   (d) The Moon is an uglier place than the Earth.
   (e) All of the above except (d).

KEY TERMS

astronomical unit, 6
atom, 10
electric charge, 10
electric force, 10
electron, 10
galaxy cluster, 9
gravity, 10
light-year, 8

Local Group, 9
Local Supercluster, 9
Milky Way galaxy, 6
neutron, 10
nuclear force, 10
nucleus, 10
planet, 2
powers-of-ten notation, 8

proton, 10
satellite, 3
scientific method, 11
scientific notation, 8
Solar System, 5
strong force, 10
star, 5
Universe, 9

The Night Sky

The stars of the night sky have inspired people of all times and cultures. For example, the ancient people of the Mediterranean knew the group of bright stars shown in figure OV1.1 as Orion the Hunter. Such starry patterns are called constellations, and their outlines remain essentially the same for thousands of years. We see Orion the way it looked to the ancient Greeks.

FIGURE OV1.1
Photograph of the Constellation Orion.
(Roger Ressmeyer/Corbis.)

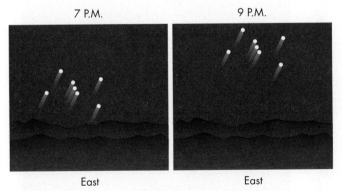

7 P.M.   9 P.M.

East   East

FIGURE OV1.2
Stars rise each evening in the east and move across the sky to set in the west.

To the people of long ago, the constellations were not only the subject of myths. They also served as markers of the passage of time, and although today many of us have lost contact with the night sky and its lore, the daily and yearly rhythms of the stars continue. For example, if we watch Orion on a midwinter night, we see him rise in the east (fig. OV1.2) and sweep overhead to set in the west.

The motion of Orion and the other stars across the night sky is caused by the Earth's rotation. As our planet spins, it carries us beneath the sky so that the stars at night move across the sky from east to west just as the Sun moves during the day (fig. OV1.3).

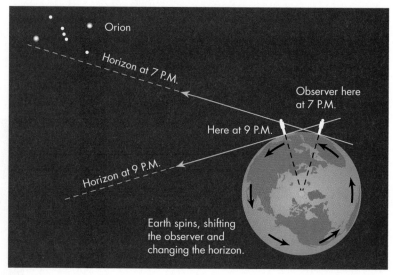

FIGURE OV1.3
The stars rise and set because the Earth spins on its rotation axis.

The daily cycle of rising and setting is not the only cycle in the sky. If we watch the sky over the course of a year, we will notice that each season has its own constellations. Orion dominates the evening sky in January, but Leo the Lion replaces him by March. By June evenings, Leo is in turn replaced by Scorpius (fig. OV1.4). Thus, the constellations visible on a given night tell us the time of year.

January

March

FIGURE OV1.4
We see different constellations at each season of the year.

FIGURE OV1.4
Continued

June

The changing constellations from season to season result from our planet's orbital motion around the Sun. As we circle the Sun, different parts of the sky move into and out of its glare— old constellations disappearing at sunset and previously hidden ones becoming visible at sunrise (fig. OV1.5).

Ancient people knew of other objects in the night sky in addition to the Sun and stars. All early people recognized the Moon and its monthly cycle of phases from crescent to full and back to crescent. In fact, markings on a 30,000 year old bone may be a record of those changes (fig. OV1.6). Many ancient cultures also discovered five objects that looked like bright stars but that moved along well-defined paths through the constellations from one night to the next. The ancient Babylonians named them for their Gods, names whose Roman versions we still use today: Mercury, Venus, Mars, Jupiter, and Saturn. To these original five planets astronomers have added three more— Uranus, Neptune, and Pluto—discovered with the help of the telescope.

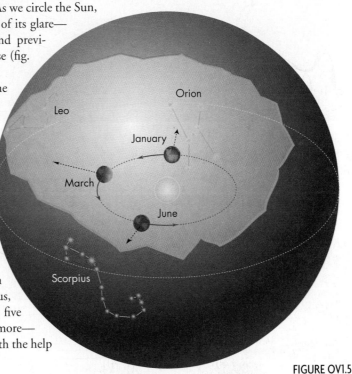

FIGURE OV1.5
We see different constellations as the seasons change because the Earth moves around the Sun, revealing parts of the sky previously hidden in the Sun's glare.

FIGURE OV1.6
Markings made on a bone fragment about 30,000 years ago may represent a record of the phases of the Moon.

The daily, yearly, and other cycles of star and planetary motion led the people of the ancient civilizations surrounding the Mediterranean to picture the heavenly objects as pivoting around the Earth on giant crystalline spheres. The spheres were aligned so that the planets always moved within a narrow band on the sky that they called the Zodiac (fig. OV1.7). Today we know that the planets follow the Zodiac because their orbits all lie in nearly the same plane as the Earth's orbit (fig. OV1.8).

Careful observation of the planets shows a curious pattern. Because of their orbital motion around the Sun, planets move across the sky at a slightly different speed than the stars. Although the planets and stars *always* rise in the

FIGURE OV1.7
The motion of the planets through the Zodiac.

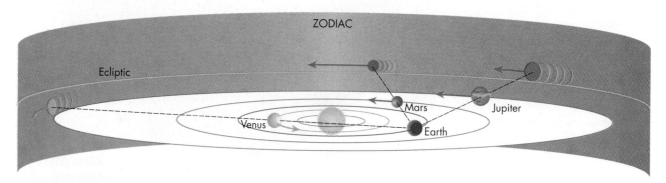

FIGURE OV1.8
The planets move across the sky within the narrow band of the Zodiac because their orbits all lie in nearly the same plane.

east and set in the west, the orbital motion of the planets makes them usually drift eastward with respect to the stars (fig. OV1.9). That is, if we focus our attention on a planet and a star near it in the sky and watch over the course of several nights, the planet moves slightly slower across the sky than the stars do. As a result, the planet will move eastward with respect to the star, even though both are moving from east to west across the night sky.

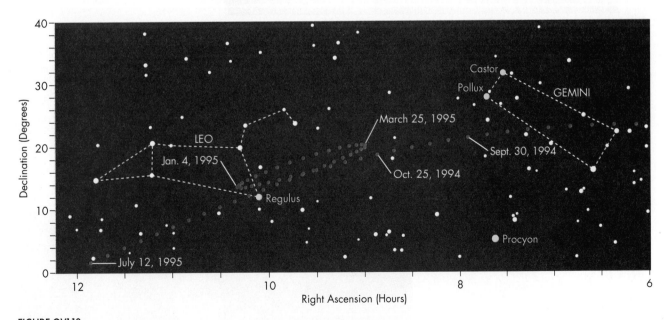

FIGURE OV1.9
From one night to another, the planets' orbital motion around the Sun normally carries them eastward against the background stars.

This normal drift of the planets may sometimes reverse, so that a given planet shifts westward through the constellations (fig. OV1.10). This is not the result of the planet changing its direction in its orbit; rather, it is caused by the Earth's motion. Just as a car may briefly look like it is moving backward when you pass it, so too a planet appears to reverse its direction as the Earth swings by it in our yearly journey around the Sun.

The regular and predictable cycles of the Sun, Moon, stars, and planets are marked at rare intervals by eclipses. Astronomers of antiquity understood the cause of eclipses, and thousands of years ago, Greek and Chinese astronomers

FIGURE OV1.10
Planets occasionally stop their eastward drift through the stars and for a brief period shift westward, undergoing retrograde motion. This "backward" drift is caused by the Earth passing the planet and does not mean the planet has reversed its orbital motion.

were able to predict these spectacular events. A lunar eclipse occurs when the Earth's shadow strikes the Moon, and a solar eclipse occurs when the Moon's shadow strikes the Earth (fig. OV1.11).

Although the Moon orbits the Earth once a month, the tilt of the Moon's orbit is such that in most months, the Moon's shadow falls above or below the Earth. Similarly, the Moon's orbital tilt causes the Earth's shadow to generally fall above or below the Moon. Under these circumstances, no eclipse occurs. However, twice each year, on the average, the Moon's orbit is tilted in just the proper way to permit the Moon's shadow to fall upon the Earth, and we then have an eclipse of the Sun. At such times, either two weeks before or two weeks after the eclipse of the Sun, we have an eclipse of the Moon.

FIGURE OV1.11
When the Moon passes directly between us and the Sun, we see a solar eclipse. When the Earth passes directly between the Sun and the Moon, we see a lunar eclipse.

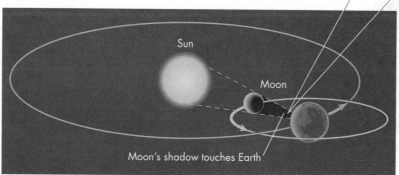

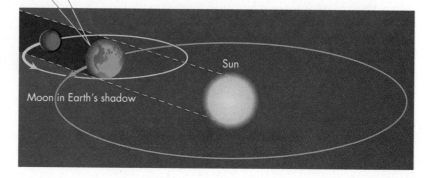

CHAPTER ONE

HISTORY OF ASTRONOMY

Stonehenge (NY/English Heritage)

A Mayan building oriented to the summer sunset (Werner Forman/Art Resource)

The Hubble Space Telescope (STSCI)

Our picture of the Universe has been assembled bit by bit from many separate discoveries—discoveries made by scientists from many parts of the world, at many times in the past, and in many disciplines. How those discoveries led to our current knowledge is the subject of this chapter. For convenience in our discussion, we will divide the history of Western astronomy into four main periods which can be characterized as follows:

Prehistoric (before 500 B.C.):* In the prehistoric era, people observed the daily and seasonal motions of the Sun, Moon, and stars and learned to use their cyclic motions for keeping time and determining direction.

Classical (500 B.C.–A.D. 1400): In the classical period, scientist-philosophers began to make measurements of the heavens and, with their knowledge of geometry, constructed idealized models that could account for the motion of heavenly bodies.

Renaissance (1400–1650): In the Renaissance period, those geometrical models were reassessed and found wanting. Astronomers therefore devised new models that took into account a much greater body of data based on observational records accumulated over centuries. Astronomers also benefited from a technological advance that allowed them to observe even more—the telescope.

Modern (1650 to the present): Finally, in the modern period scientists began the search for the physical laws (such as the law of gravity) that underlie the observed movements in the heavens. Other important contributions to our understanding of the Universe came from technological advances (for example, in optics, electronics, and computers) and better mathematical techniques (such as calculus). Such factors continue to be important today.

1.1 PREHISTORIC ASTRONOMY

We do not know when people of antiquity first began studying the heavens, but it was certainly many thousands of years ago. Astronomical observations are part of virtually every culture and include the obvious events that anyone who watches the sky can see without the need of any equipment, such as the rising of the Sun in the eastern sky and its setting toward the west; the changing appearance of the Moon at different times of the month from new moon to crescent to full and back again; and eclipses. Other discoveries, such as the one that recognizes the planets as a class of heavenly bodies different from the stars, require more careful observation but were also probably made by prehistoric people.

For many prehistoric people, observations of the heavens had more than just curiosity value. Because so many astronomical phenomena are cyclic—that is, they repeat day after day and year after year—they can serve as time keepers. For example, when is it safe to set out on a sea voyage? When is it time to harvest crops? When will an eclipse occur? Moreover, the cyclic behavior of the heavens implies that many events seen in the sky are predictable. Thus some of the impetus for studying the heavens probably came from the desire to foretell future events there, and it may have motivated early cultures to build monumental stone structures such as Stonehenge (fig. 1.1).

*A few written records of astronomical observations made by astronomers in ancient China, India, Korea, and Japan pre-date 500 B.C.

Ironically, many of the astronomical phenomena well known to ancient people are not nearly so familiar to people living today because the smog and bright lights of cities make it hard to see the sky and its rhythms. Perhaps more importantly, we no longer rely upon direct astronomical observations to tell us what season it is, when to plant, and so on. Therefore, if we are to appreciate the growth of astronomical ideas, we need to first understand what our distant ancestors knew and what we ourselves can learn by watching the sky over the course of a year.

One of nature's spectacles is the night sky seen from a clear, dark location with the stars scattered across the vault of the heavens. The night sky is a particularly appropriate place to begin our study of the history of astronomy because its array of stars has changed very little over the last several thousand years. The night sky affords us a direct link with what our remote ancestors viewed as they tried to understand the nature of the heavens. Thus our first goal is to familiarize ourselves with some general aspects of the sky at night.

In the following discussion you might want to imagine yourself as a shepherd in the Middle East, a hunter-gatherer on the African plains, a trader sailing along the coast of the Mediterranean, or even a ship's navigator in the early twentieth century. Whichever role you choose to assume, try to get out and actually look at the sky.

FIGURE 1.1
Stonehenge, a stone monument built by the ancient Britons on Salisbury Plain, England. Its orientation marks the seasonal rising and setting points of the Sun. (Courtesy Tony Stone/Rob Talbot.)

The Celestial Sphere

The thousands of stars visible on a clear night are at vastly different distances from us. For example, the nearest star is about 4 light-years away, but others are hundreds of times more distant. Such huge distances prevent us from getting any sense of their true three-dimensional arrangement in space. For purposes of naked-eye observations, we can therefore treat all stars as if they are at the same distance from the Earth, imagining that they lie on the inside of a gigantic dome that stretches overhead. Astronomers envisage this dome as part of the **celestial sphere,** and picture it completely surrounding the Earth, with the Earth at its center, as depicted in figure 1.2.*

The celestial sphere has no physical reality, but it serves as a **model** of the heavens—an easy way to visualize the arrangement and motions of celestial bodies. We use the term *model* to mean a representation of some aspect of the Universe. That is, the celestial sphere represents a way of thinking about or viewing the location and motions of stars and planets. The celestial sphere is the first of many models we will encounter that humans have used to describe the known Universe. In later chapters we will use models to enhance our

Planets?

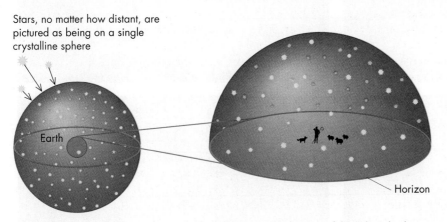

Stars, no matter how distant, are pictured as being on a single crystalline sphere

Earth

Horizon

Model: The celestial sphere

The human experience of the celestial sphere

FIGURE 1.2
The sphere of the sky surrounds the Earth and is called the "celestial sphere."

*In figure 1.2 and many others throughout the book, distances and sizes of astronomical bodies are exaggerated for clarity.

understanding whenever the size or other properties of what we study fall outside the range of our everyday experience. Thus we will speak of models of atoms, models of stars, and models of the Universe itself.

Constellations

As human beings, we seek order in what we look at. Thus, when ancient people looked at the stars, they saw in their arrangement on the celestial sphere fixed patterns, what we today call **constellations.** Sometimes these constellations resemble animals if we use a little imagination. For example, the pattern of stars in Leo looks a little like a lion, whereas that of Cygnus looks like a swan in flight, as depicted in figure 1.3. As you will realize when you learn to identify the constellations, many have shapes that bear little resemblance to their namesakes. Also, keep in mind that stars in a constellation generally have no physical relation to one another. They simply happen to be in more or less the same direction in the sky.

All stars move, but as seen from Earth, their positions change very slowly, taking tens of thousands of years to make any noticeable shift. Thus we see today virtually the same pattern of stars that was seen by ancient peoples. A shepherd who lived 5000 years ago in the Middle East would have no trouble recognizing the star patterns of the night sky we see and might even call them by the same names.

We do not know how the constellation names were chosen, but most of them date back thousands of years. In fact we don't know when the names were first given to the constellations, or why, although there is some evidence they served as mnemonic devices for keeping track of the seasons and for navigating. For example, the beginning of the stormy winter months, when sailing was dangerous and ships were often wrecked, was foretold by

FIGURE 1.3
The two constellations Leo, (A), and Cygnus, (B), with figures sketched in to help you visualize the animals they represent. (Photo (A) from Roger Ressmeyer, digitally enhanced by Jon Alpert. Photo (B) courtesy Eugene Lauria.)

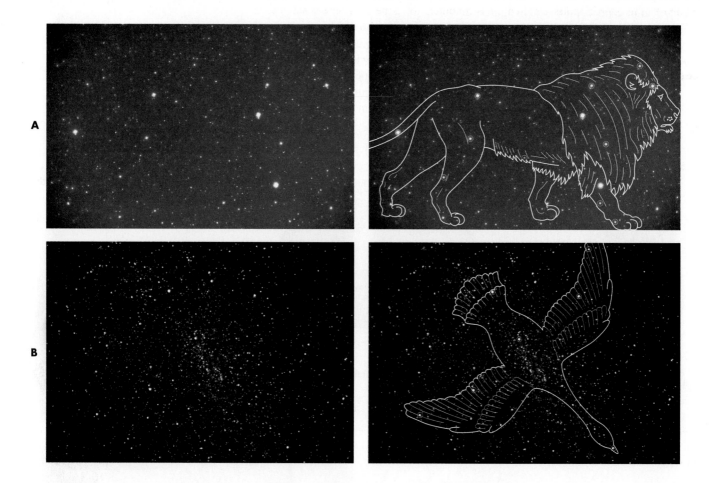

the Sun's appearance in the constellations Pisces and Aquarius, the water constellations. Likewise, the harvest time was indicated by the Sun's appearance in Virgo, a constellation often depicted as the goddess Proserpine, holding a sheaf of grain.

Motions of the Sun and Stars

Daily or diurnal motion.   Take a look at the night sky, and you will see stars rise in the east, move across the sky, and set in the west, exactly as the Sun does. You can verify this by watching the night sky for as little as 10 minutes. A star seen just above the eastern horizon will have risen noticeably higher and stars near the western horizon will have sunk lower or disappeared. Likewise, if you look at a constellation, you see its stars rise as a fixed pattern in the east, move across the sky, and set in the west. In terms of our model of the heavens based on the celestial sphere, we therefore explain the rising and setting of stars as rotation of the celestial sphere around us (fig. 1.4). For ancient people, that rotation was far easier to believe than that the Earth moved. Thus they attributed all celestial motion—that of the Sun, Moon, stars, and planets—to a vast sphere slowly turning overhead. Today, of course, we know that it is the Earth's rotation that makes the Sun, Moon, and stars rise in the east and move westward across the sky. It is not the celestial sphere that spins but the Earth.

If you look at the celestial sphere turning overhead, two points on it do not move, as you can see in figure 1.4. These points are defined as the north and south **celestial poles.** The celestial poles lie exactly above the north and south poles of the Earth, and just as our planet turns about a line running from its north to south poles, so the celestial sphere rotates around the celestial poles. Because it lies directly above the Earth's North Pole, the north celestial pole always marks the direction of true north. Therefore the north celestial pole is an important aid to travelers on land and sea and was widely used by early peoples for this purpose. Even today the celestial poles serve as reference marks for navigators. Moreover the celestial poles are used by astronomers to locate stars on the sky just as mapmakers use the Earth's north and south poles as reference marks for locating places on our planet.

Another useful sky marker frequently used by astronomers is the **celestial equator.** The celestial equator lies directly above the Earth's equator, just as the celestial poles lie above the Earth's poles, as figure 1.4 shows.

At the same time that the Earth's spin causes the apparent daily motion of the Sun and stars across the sky, the Earth's orbital motion around the Sun also causes changes in the

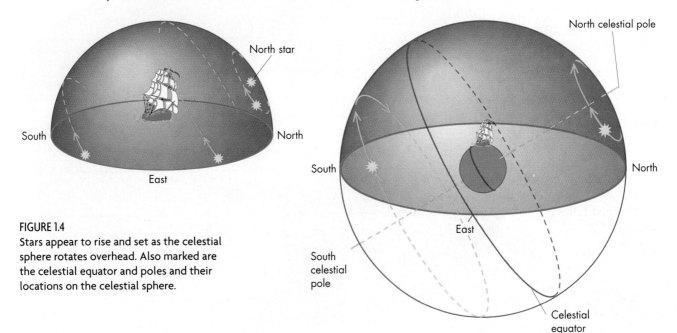

FIGURE 1.4
Stars appear to rise and set as the celestial sphere rotates overhead. Also marked are the celestial equator and poles and their locations on the celestial sphere.

sky. However, this motion is much slower and harder to observe without making records of star positions. Nevertheless, astronomers today think that many ancient peoples built monuments to keep track of such motions. Because these movements repeat on a yearly cycle, they are called *annual motions*.

Annual motion. If you watch the sky over a period of time, say, several months, you will discover that new constellations appear in the eastern sky, and old ones disappear from the western sky. For example, on an early July evening across most of North America, Europe, and Asia, the constellation Scorpius will be visible in the southern half of the sky. On December evenings Scorpius is not visible. Instead you will see the brilliant constellation Orion, the hunter.

The discovery that different stars are visible at different times of the year was extremely important to early people because it provided a way to measure the passage of time other than by simply counting days. Moreover, the stars demonstrated that many celestial events are predictable and that they may be used to order our lives on Earth. For example, using the constellations mentioned above, if the evening sky shows Leo, it will soon be time to plant. If the sky shows Scorpius, it must be summer, and if it shows Orion, it must be winter. Determining the season by using the stars may sound strange to people living in climates with strong temperature differences between summer and winter. But for the an-

FIGURE 1.5
The Sun hides from our view stars that lie beyond it. As we move around the Sun, those stars become visible, and the ones previously seen are hidden. Thus the constellations change with the seasons.

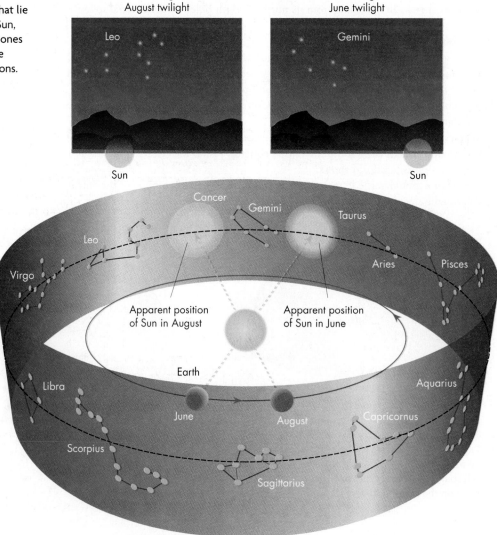

cient civilizations that were based in semitropical climates, such temperature differences are much smaller.

This change of the constellations with the seasons is caused by the Earth's motion around the Sun. As the Earth moves around the Sun, the Sun's glare blocks our view of the part of the celestial sphere that lies toward the Sun, making the stars that lie beyond the Sun invisible, as figure 1.5 shows. For example, in early June, a line from the Earth to the Sun points toward the constellation Taurus, and so its stars are lost in the Sun's glare. After sunset, however, we can see the neighboring constellation, Gemini, just above the western horizon. By early August, the Earth has moved to a new position in its orbit. At that time of year, the Sun lies in the direction of Cancer, causing this constellation to disappear in the Sun's glare. Looking to the west just after sunset, we now see Leo just above the horizon.

Month by month, the Sun covers one constellation after another. It is like sitting around a campfire and not being able to see the faces of the people on the far side, but if we get up and walk around the fire, we can see faces that were previously hidden. Similarly the Earth's motion allows us to see stars previously hidden in the Sun's glare and makes a given star rise 3 minutes and 56 seconds earlier each night. Adding up that delay over a full year gives 24 hours. Thus a year later, when the Earth returns to the same spot in its orbit, the sky looks the same again.

The ecliptic. If we could mark on the celestial sphere the path traced by the Sun as it moves through the constellations, we would see a line that runs around the celestial sphere, as illustrated in figure 1.6. Astronomers call the line that the Sun traces across the celestial sphere the **ecliptic.** The name *ecliptic* arises because only when the new or full moon crosses this line can an eclipse occur, as you will learn later in this chapter. If you look at figure 1.6, you can see that the ecliptic is the extension of the Earth's orbit onto the celestial sphere, just as the celestial equator is the extension of the Earth's equator onto the celestial sphere.

It is not easy to track the Sun's position relative to the stars because the Sun's glare makes it impossible to see the stars. However, if you look just before sunrise or just after

The Bighorn medicine wheel. Built by the Native Americans of the High Plains. The diameter of the wheel is about 30 meters (90 ft), with its spokes marking the azimuths of sunrise at winter and summer solstices. (© Tony Stone)

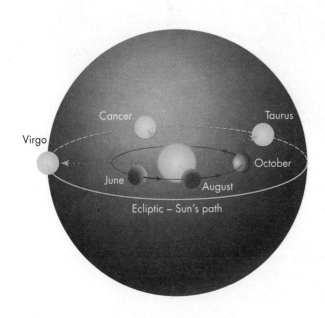

FIGURE 1.6
The Sun's path across the background stars is called the *ecliptic*. The Sun appears to lie in Taurus in June, in Cancer during August, in Virgo during October, and so forth. Note that the ecliptic is also where the Earth's orbital plane cuts the celestial sphere.

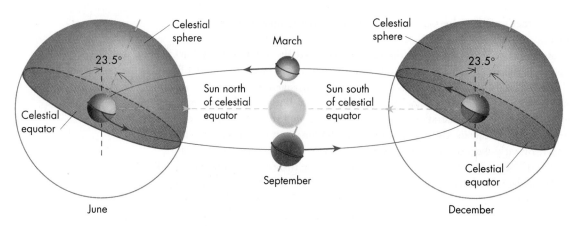

FIGURE 1.7

The Earth's rotation axis is tilted by 23.5° with respect to its orbit. The direction of the tilt remains the same as the Earth moves around the Sun. Thus for part of the year the Sun lies north of the celestial equator, whereas for another part it lies south of the celestial equator.

FIGURE 1.8

These five diagrams show the Sun's position as the sky changes with the seasons. Although the Earth moves around the Sun, it looks to us on the Earth as if the Sun moves around us. Notice that because the Earth's spin axis is tilted, the Sun is north of the celestial equator half of the year (late March to late September) and south of the celestial equator for the other half of the year (late September to late March).

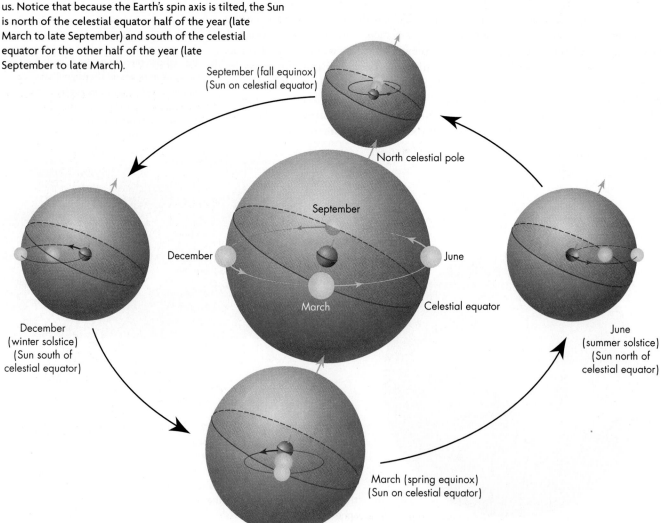

sunset, you can identify the stars near the Sun on any given date. By watching over several weeks, you will be able to see that the Sun changes its position with respect to the stars.*

Another complication we will discover when trying to track the Sun relative to the stars is that the Sun's path does not neatly align with the path of the stars. Instead, the path is tipped with respect to the celestial equator, as you can see in figure 1.7. The tip is small ($23.5°$) but it makes the Sun drift gradually north and south at the same time that it circles the sky against the background stars.

Why is the Sun's path tipped in this way? To understand, we must look more closely at the Earth's motion around the Sun. At the same time that the Earth moves along its orbit, our planet spins. The axis about which it turns is tipped by $23.5°$ with respect to the Earth's orbit. It turns out that this tilt remains the same as the Earth moves around the Sun, as shown in figure 1.7. Moreover, the tilt is such that in June the northern hemisphere points toward the Sun. As a result, the Sun lies in the northern part of the sky. That is, the Sun is north of the celestial equator, as you can see in figure 1.7. On the other hand, in December, when the Earth has shifted to the other side of the Sun, the Sun lies south of the celestial equator. It is this change in the Sun's position north and south in the sky that makes the ecliptic tilted, as is shown by the sequence of sketches in figure 1.8. Finally, the tilt angle is $23.5°$ with respect to the celestial equator because our planet's rotation axis is tilted by that much with respect to its orbit.

Solstices and equinoxes. The tilt of the ecliptic with respect to the celestial equator means that during the year the points on the horizon where we see the Sun rise and set will not in general be due east and west. Rather, for half the year the Sun will rise in a regular and predictable fashion first to the north of east and then to the south, as illustrated in figure 1.9A. For example, during the northern hemisphere summer, the Sun rises in the northeast and sets in the northwest. In the winter, the Sun rises in the southeast and sets in the southwest.

Another consequence of the Sun's annual motion around the celestial sphere is that at some times of the year the noonday Sun will appear high in the Northern Hemisphere sky

FIGURE 1.9
(A) The direction of the rising and setting Sun changes throughout the year. At the equinoxes the rising and setting points are due east and west. The sunrise direction shifts slowly northeast from March to the summer solstice, whereupon it shifts back, reaching due east at the autumn equinox. The sunrise direction continues moving southeast until the winter solstice. The sunset point similarly shifts north and south. (B) Sunrise on the summer solstice at Stonehenge. (Courtesy English Heritage.)

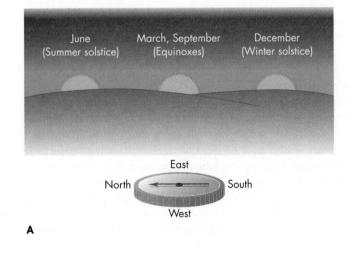

---

*Plotting the Sun's path through the stars is easier if you measure its position at noon and compare that position to the midnight sky half a year later.

The temple of Amen-Ra at Karnak, Egypt. The temple was aligned so that its main axis points to the position of sunrise on the winter solstice. (Courtesy E. C. Krupp.)

and low in the Southern Hemisphere sky, whereas at other times the exact opposite will be true. In fact, as we will see in chapter 4, this variation is the cause of the seasons.

Astronomers give special names to the Sun's position at the times when the seasons change. For example, spring and autumn begin when the Sun crosses the celestial equator. At these times of year the days and nights are of equal length (approximately), and so they are called the **equinoxes**. The spring or vernal equinox occurs near March 21, but the fall or autumnal equinox occurs near September 23. Likewise the beginning of summer and winter are given special names. At these times of year the Sun pauses in its steady north-south motion and changes direction. Accordingly, these times are called the **solstices**, meaning the Sun (sol) has stopped its motion north or south and is *static* and about to reverse direction. The dates of the solstices (summer and winter) also change slightly from one year to the next, but they are always close to June 21 and December 21.

Just as the changing position of the Sun against the constellations could be used as an indicator of the seasons, so too could the position of the rising and setting Sun. One well known example is Stonehenge, the ancient stone circle in England, illustrated in figure 1.1. Although we cannot know for certain how this ancient monument was used, it seems likely that it was laid out so that such seasonal changes in the Sun's position could be observed by noting through which stone arches the Sun rose or set. For example, figure 1.9B (a photograph taken on the summer solstice) shows that on that day, an observer standing at the center of this immense circle of standing stones sees the rising Sun directly over a marker stone and framed by an arch. Similarly, some ancient Egyptian temples and pyramids have astronomical alignments, such as the Temple of Amen-Ra at Karnak, whose main chamber points toward the position of sunrise at the winter solstice.

Astronomical alignment of buildings occurs in many other places too. For example, the Mayans, native peoples of Central America, and their neighbors built pyramids from the summits of which they could get a clear view of the sky over the surrounding rain forest. The Mayans devised a precise and complex calendar based on the planet Venus, which they believed determined the course of their history. Through such interest in the sky, the Mayans and other ancient cultures learned to recognize another type of celestial motion: that of the planets.

The Planets and the Zodiac

All cultures whose astronomy has been studied have noted that some bright objects that looked like stars did not follow the same cyclic motion associated with the rest of the stars in the sky. We do not know when humans first recognized the special motion of these objects, but the Greeks are credited with giving them the name *planētai* meaning "wanderers," from which our word *planet* comes.

Planets move across the background stars because of a combination of the Earth's and their own orbital motion around the Sun. One of the more striking features of this motion is that the planets *always* move within a very narrow band on the celestial sphere. We call this band the **zodiac**. The motion of the planets lies in this narrow zone because their orbits, including that of the Earth, all lie in nearly the same plane.

The zodiac circles the heavens following the ecliptic, which lies exactly along its centerline, and both are tilted by 23.5° with respect to the celestial equator, as shown in figure 1.10. This tilt arises because the Earth's rotation axis is tilted with respect to our orbit around the Sun.

The word *zodiac* is from Greek *zoidion,* "little animal," leading to "animal sign." That is, zodiac refers to a circle of animals, which for the most part, its constellations represent. The names of these constellations are Aries, Taurus, Gemini, Cancer, Leo, Virgo, Libra, Scorpius, Sagittarius, Capricornus, Aquarius, and Pisces.

The motion of the planets along the zodiac can be seen easily by marking off the position of a planet on the celestial sphere over a period of a week or so. Figure 1.11 illustrates such a plot and shows that planets normally move eastward through the stars as a result of their orbital motion around the Sun.

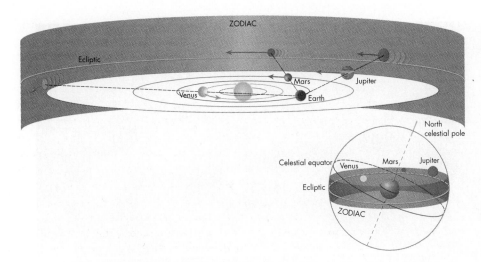

FIGURE 1.10
The zodiac is tilted with respect to the celestial equator. Notice that a planet can *never* appear outside the zodiac.

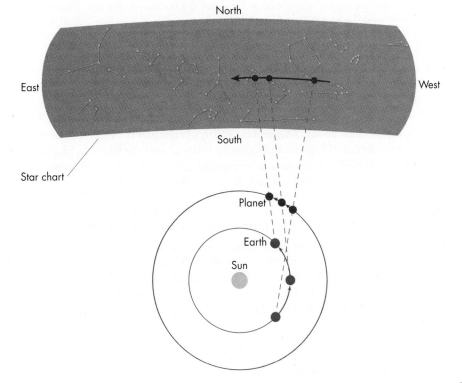

FIGURE 1.11
A planet's eastward drift against the background stars plotted on the celestial sphere. Note: Star maps usually have east on the left and west on the right, so that they depict the sky when looking south.

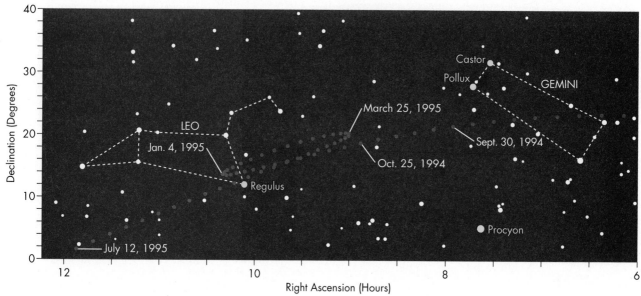

FIGURE 1.12

The position of Mars marked out on the background stars and showing its retrograde motion. In what constellation is Mars in October 1994? (Use the star charts on the inside covers of the book to identify the constellations.)

Although the apparent motion is usually from west to east through the stars, this does *not* mean that the planet rises in the west and sets in the east. As seen from Earth, planets *always* rise in the east and set in the west because they are carried across the sky—just as the stars are—by the Earth's rotation. However, the motion of the planets is usually slower than that of the stars because their orbital motion partly offsets the rotation of the Earth that causes the apparent motion of the stars. Thus, if a star and planet rise side by side, at some later time the planet will not be as far above the horizon as the star. Therefore *with respect to the stars* the planet has moved to the east because of its orbital motion around the Sun.

This simple pattern of movement is sometimes interrupted. Occasionally, a planet will move west with respect to the stars, a condition known as **retrograde motion,** and shown in figure 1.12. The word *retrograde* means "backward," and when a planet is in retrograde motion, its path through the stars may bend backward for a few days or it may even form a loop (refer to fig. 1.12). All planets undergo retrograde motion for at least some portion of their paths around the sky. This motion greatly complicates the otherwise straightforward idea that the celestial sphere and its bodies rotate around the Earth. In fact, the need for a simple, plausible explanation of retrograde motion was what ultimately led astronomers to reject models of the Solar System with the Earth at the center.

The Moon

Like all celestial objects, the Moon rises in the east and sets in the west. Also, like the planets and the Sun, the Moon shifts its position across the background stars from west to east. You can verify this motion by observing the Moon at the same time each evening and noticing its change in position with respect to nearby stars. In fact, if the Moon happens to lie close to a bright star, its motion may be seen in only a few minutes, because in 1 hour the Moon moves a distance, on the sky, that is approximately equal to its own apparent diameter.

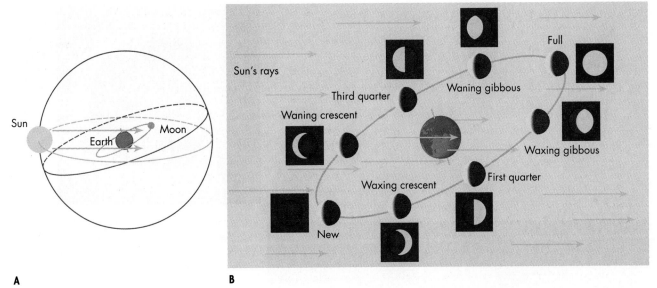

A

B

FIGURE 1.13
(A) The cycle of the phases of the Moon from new to full and back again. (B) The Moon's phases are caused by our seeing different amounts of its illuminated surface. The pictures in the dark squares show how the Moon looks to us on Earth.

Perhaps the most striking feature of the Moon is that, unlike the Sun, its shape seems to change throughout the month in what is called the "cycle of lunar **phases**." That is, the Moon appears alternately as a thin crescent, a fully illuminated disk, and then as a crescent again. During a period that is approximately 30 days the Moon goes through a complete set of phases from invisibility (new) to fully lit (full) and back to invisibility. This is the origin of the month as a time period and also the source of the name "month," which was derived from the word "Moon."

The cycle of the phases and the Moon's changing position against the stars is caused by the Moon's orbital motion around the Earth. A new moon occurs when the Moon lies approximately between us and the Sun. A full moon occurs when the Moon is on the other side of the Earth from the Sun, opposite it in the sky. Many people believe these changes in shape are caused by the Earth's shadow falling on the Moon. However, that is *not* the explanation, as you can deduce from the fact that crescent phases occur when the Moon and Sun lie approximately in the same direction in the sky and the Earth's shadow must therefore point *away* from the Moon. In fact, we see the Moon's shape change because, as it moves around us, we see different amounts of its illuminated half. For example, when the Moon lies approximately opposite the Sun in the sky, the side of the Moon toward the Earth is fully lit. On the other hand, when the Moon lies approximately between us and the Sun, its fully lit side is turned nearly completely away from us, and therefore we glimpse only a sliver of its illuminated side, as illustrated in figure 1.13. This figure also shows the cycle of the lunar phases and the names used to describe the Moon's appearance.

The Moon's motion around the Earth has other effects as well. For example, because the Moon shifts eastward through the stars, the Earth itself must rotate eastward a little extra each evening to bring the Moon above the horizon. This extra rotation takes about an hour (actually 50 minutes, on the average) each day, and so moonrise is later each night by about that amount. The most beautiful and dramatic effect of the Moon's motion, however, is that it leads to eclipses.

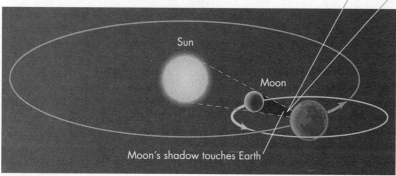

FIGURE 1.14
A solar eclipse occurs when the Moon passes between the Sun and the Earth so that the Moon's shadow strikes the Earth. The photo inset shows what the eclipse looks like from Earth. (Photo courtesy of Dennis di Cicco.)

Eclipses

An eclipse occurs when the Moon happens to lie exactly between the Earth and the Sun, or when the Earth lies exactly between the Sun and the Moon so that all three bodies lie on a straight line.* Thus there are two type of eclipses: solar and lunar.

A **solar eclipse** occurs whenever the Moon passes directly between the Sun and the Earth and blocks our view of the Sun, as depicted in figure 1.14. A **lunar eclipse** occurs when the Earth passes between the Sun and the Moon and casts its shadow on the Moon, as shown in figure 1.15. Further details about eclipses can be found in chapter 6, The Moon.

Eclipses can be amazing sights. During a solar eclipse the midday sky may become dark as night for a few minutes. During a lunar eclipse, the Moon may become a dull red or disappear altogether for over an hour. Therefore it is not surprising that early people recorded such events and sought (successfully) to predict them.

The astronomical phenomena that we have discussed so far (rising and setting of Sun, Moon, and stars; the constellations; annual motion of Sun; motion of planets through the zodiac; phases of the Moon and eclipses) were the basis of ancient knowledge of the heavens. With these observations we can now describe the attempts of early people to explain the heavens. We will see that many of their conclusions were incorrect, just as we today are

---

*When new, the Moon is on the side of the Earth toward the Sun but not necessarily exactly in a straight line between them. Likewise when full, the Moon lies on the opposite side of the Earth from the Sun but not necessarily on precisely the same line.

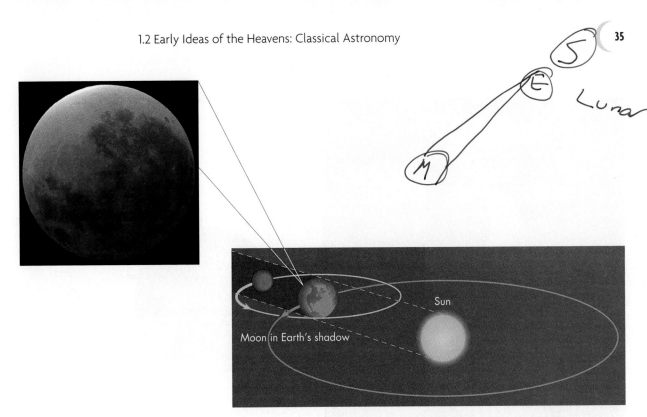

FIGURE 1.15
A lunar eclipse occurs when the Earth passes between the Sun and Moon, causing the Earth's shadow to fall on the Moon. Some sunlight leaks through the Earth's atmosphere casting a deep reddish light on the Moon. The photo inset shows what the eclipse looks like from Earth. (Photo courtesy of Dennis di Cicco.)

probably in error about some aspects of astronomy. We study ancient ideas of the heavens not so much for what they tell us about the heavens, but to learn how observation and reasoning can lead us to an understanding of the Universe.

1.2 EARLY IDEAS OF THE HEAVENS: CLASSICAL ASTRONOMY

As far as we know, the ancient Greek astronomers of classical times were the first people to try to explain the workings of the heavens in a careful, systematic manner, using models and observations. Given the limitations of naked-eye observation, these astronomers were extraordinarily successful, and their use of logic and mathematics (geometry) as tools of inquiry created a method for studying the world around us that we continue to use even today. This method is in many ways as important as the discoveries themselves.

The Shape of the Earth

The ancient Greeks knew that the Earth was round. As long ago as about 500 B.C., the mathematician Pythagoras (about 560–480 B.C.) was teaching that the Earth was spherical, but the reason for his belief was as much mystical as rational. He, like many of the ancient philosophers, believed that the sphere was the perfect shape and that the gods would therefore have utilized that perfect form in the creation of the Earth.

By 300 B.C., however, Aristotle (384–322 B.C.) was presenting arguments for the Earth's spherical shape that were based on simple naked-eye observations that anyone could make. Such reliance on careful, first-hand observation was the first step toward acquiring scientifically valid knowledge of the contents and workings of the Universe. For instance, Aristotle noted that if you look at an eclipse of the Moon when the Earth's

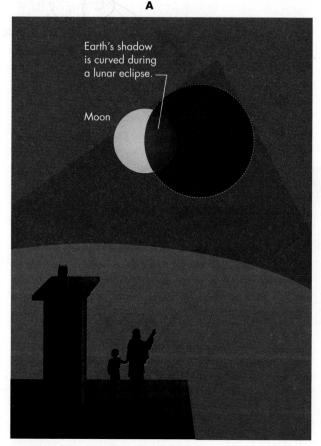

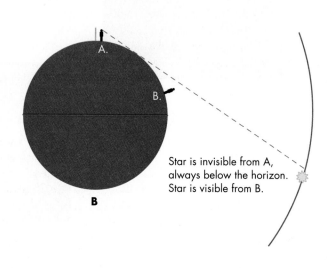

FIGURE 1.16
(A) During a lunar eclipse, we see that the Earth's shadow on the Moon is curved. Thus the Earth must be round. (B) As a traveler moves from north to south on the Earth, the stars that are visible change. Some disappear below the northern horizon, whereas others, previously hidden, become visible above the southern horizon. This variation would not occur on a flat Earth.

shadow falls upon the Moon, the shadow can be clearly seen as curved, as figure 1.16 shows. As he wrote in his treatise "On the Heavens":

> The shapes that the Moon itself each month shows are of every kind—straight, gibbous, and concave—but in eclipses the outline is always curved: and, since it is the interposition of the Earth that makes the eclipse, the form of this line will be caused by the form of the Earth's surface, which is therefore spherical.

Another of Aristotle's arguments that the Earth is spherical was based on the observation that a traveler who moves south will see stars that were previously hidden below the southern horizon, as illustrated in figure 1.16. For example, the bright star Canopus is easily seen in Miami but is invisible in Boston. This could not happen on a flat Earth.

The Size of the Earth

Knowing that the Earth was spherical, the Greeks desired to know its size. Eratosthenes (276–195 B.C.), head of the famous Library at Alexandria in Egypt, made the first measurement of the Earth's size. He obtained a value for its circumference of about 25,000 miles, which is roughly the presently known value. Eratosthenes's demonstration is one of the most beautiful ever performed. Because it so superbly illustrates how science links observation and logic, the demonstration is worth describing in some detail.

You must realize that at this point in time, astronomers were very well acquainted with the yearly movement of the Sun and could predict accurately the times of the solstices and equinoxes. The summer solstice marked the day in the year when the Sun would reach its highest point in the noonday sky. However, despite its height, the Sun still cast a shadow at noon. Eratosthenes, a geographer as well as an astronomer, heard that lying to the south, in the Egyptian town of Syene (the present city of Aswân), the Sun would be directly overhead at noon and cast no shadow. Proof of this was the fact that at that time the Sun shone exactly down a well near there. Knowing the distance between Alexandria and Syene and appreciating the power of geometry, Eratosthenes realized he could deduce the size (circumference) of the Earth. He analyzed the problem as follows: Because the Sun is far away from the Earth, its light travels in parallel rays toward the Earth. Thus, two rays of sunlight, one hitting Alexandria and the other shining down the well, are parallel lines, as depicted in figure 1.17.

Now imagine drawing a straight line from the center of the Earth outward so that it passes vertically through the Earth's surface in Alexandria. The angle, call it A, between that line and the Sun's rays in Alexandria is the same as the angle between that line and the line from the center of the Earth up through the well in southern Egypt (see fig. 1.17). The reason is that the angles formed where a single line crosses two parallel lines are equal (a geometric theorem).

The angle A can be measured with sticks and a protractor (or its ancient equivalent) and is the angle between the direction to the Sun and the vertical to the ground (see fig. 1.17). Eratosthenes found this angle to be about 7°. Therefore the angle formed by a line from Alexandria to the Earth's center and a line from the well to the Earth's center must also be 7°, as deduced from the parallel line—equal angle theorem.

To find the circumference of the Earth, all that is needed is to recall that there are 360° around a whole circle. The 7° between Alexandria and the well thus represent 7/360, or about 1/50, of the distance around the Earth. Because Eratosthenes knew the distance between Alexandria and the well to be 5000 stadia (where a stadium is about 0.1 mile), the distance around the entire Earth is 5000 stadia times 50, or 250,000, stadia. When expressed in miles, this is roughly 25,000 miles, which is approximately the circumference of the Earth as we know it today.

By modern standards there is absolutely nothing wrong with this technique. You can use it yourself to measure the size of the Earth as described in the projects at the end of this chapter. Eratosthenes's success was a triumph of logic and the scientific technique, but it suffered from one weakness. The method required that he assume the Sun was so far away that its light reached Earth along parallel lines. That assumption, however, was supported by another set of measurements made by the ancient Greeks, namely, a rough measurement of the relative diameters and distances of the Sun and Moon.

Distance and Size of the Sun and Moon

About 75 years before Eratosthenes measured the Earth's size, Aristarchus of Samos (an island in the Mediterranean) had estimated the relative size of the Earth, Moon, and Sun, and the relative distances to the Moon and Sun. His values for these numbers gave at least the correct sense of their proportionate sizes and their relative distances from Earth. For example, by comparing the size of the Earth's shadow on the Moon during a lunar eclipse to the size of the Moon's disk, illustrated in figure 1.18, Aristarchus calculated that the Moon's diameter was about one-third of the Earth's.

Aristarchus also calculated the Sun to be about 20 times farther away from the Earth than the Moon is. He did this calculation by measuring the angle between the Sun and the Moon when the Moon was exactly half lit, as shown in figure 1.19. From the apparent size in the sky of the Sun and Moon and their relative distances, and his calculation of the Moon's diameter relative to the Earth's, Aristarchus was able to deduce that the Sun's diameter was about 7 times that of the Earth's, far too small as we know today (it is actually about 100 times greater), but, nevertheless, evidence that the Sun is bigger than the Earth.

Although Aristarchus's determination of the Moon's relative size was quite accurate (he found the Moon's diameter to be about 0.35 that of the Earth, whereas the correct ratio is about 0.27), his distance to the Sun and consequently its size were too small by a factor of

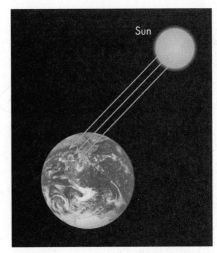

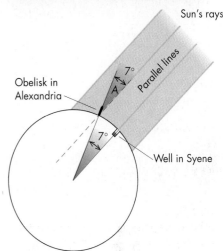

FIGURE 1.17
Eratosthenes's calculation of the circumference of the Earth. The Sun is directly overhead on the summer solstice at Syene, in southern Egypt. On that same day, Eratosthenes found the Sun to be 7° from the vertical in Alexandria, in northern Egypt. Eratosthenes deduced that the angle between two verticals placed in northern and southern Egypt must be 7°.

FIGURE 1.18
Aristarchus used the size of the Earth's shadow on the Moon during a lunar eclipse to estimate the relative size of the Earth and Moon.

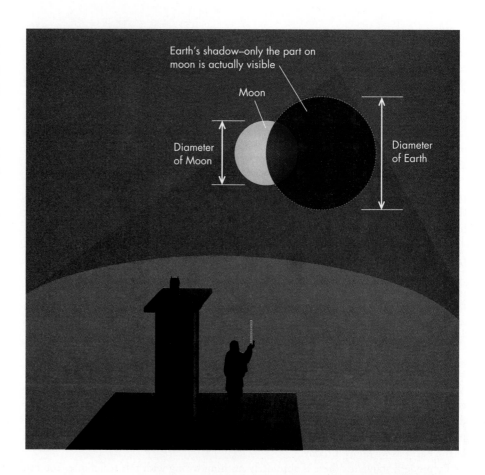

FIGURE 1.19
Aristarchus estimated the relative distance of the Sun and Moon by observing the angle *A* between the Sun and the Moon when the the Moon is exactly half lit. Angle *B* must be 90° for the Moon to be half lit. Knowing the Angle *A*, he could then set the scale of the triangle and thus the relative lengths of the sides.

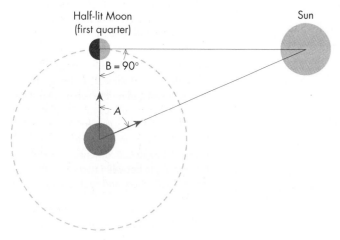

about 20. Nevertheless, he was the first person to establish that the Sun was not only very far away, but also that it was much larger than the Earth.

It was perhaps his recognition of the vast size of the Sun that led Aristarchus to the revolutionary idea that the Sun and not the Earth was the center of the heavens. Aristarchus was of course correct, but his idea was too revolutionary, and another 2000 years passed before scientists became convinced of its correctness.

However, this was not mere pigheadedness. There was a good reason for not believing that the Earth moves around the Sun. If it did, the positions of stars should change during the course of the year. Looking at figure 1.20, you can see that star A should appear to

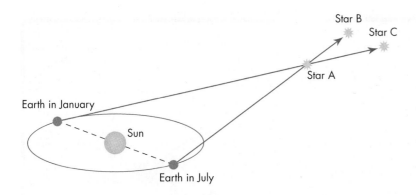

FIGURE 1.20
Motion of the Earth around the Sun causes stellar parallax. Because the stars are so remote, this is too small to be seen by the naked eye. Thus the ancient Greeks incorrectly deduced that the Sun could not be the center of the Solar System.

lie at a different distance from and in a different position with respect to stars B and C in January and in July.

This shift in a star's apparent position is called **parallax,** and Aristarchus's critics were absolutely right in supposing that it should occur. His critics argued that his Sun-centered system must be wrong because they could see no parallax as the Earth moved around the Sun. However, they erred in their estimate of the amount of parallax. Because of the immense distance to the stars, the parallax (which *does* occur just as predicted) is so tiny that it requires a powerful telescope and very precise equipment to measure it. In fact, parallax of stars was not successfully measured until about 1838. Thus an idea may be rejected for reasons that are logically correct but are based on data of limited accuracy.

EXTENDING OUR REACH

MEASURING THE DIAMETER OF ASTRONOMICAL OBJECTS

One of the most basic properties of astronomical bodies is their diameter. But there is no astronomical body for which this fundamental observation can be made directly. Since we cannot stretch a tape measure across the disk of the Sun or Moon, how do we know the size of such heavenly bodies?

The basic method for measuring the size of a distant object was worked out long ago and is still used today. This method involves first measuring how big an object looks, a quantity called its **angular size.** Astronomers measure an object's angular size by drawing imaginary lines to each side of it, as shown in figure 1.21, and then measuring the angle between the lines using a protractor. For example, we can measure the angular diameter of the Moon with two straight sticks connected at one end with a nut and bolt. We sight along the edge of one stick to one edge of the Moon and then sight along the other stick to the Moon's opposite edge. Measuring the angle between the sticks, we will find that the Moon's angular diameter is about one half degree.

FIGURE 1.21
Definition of angular size.

FIGURE 1.22
How angular size varies with distance.

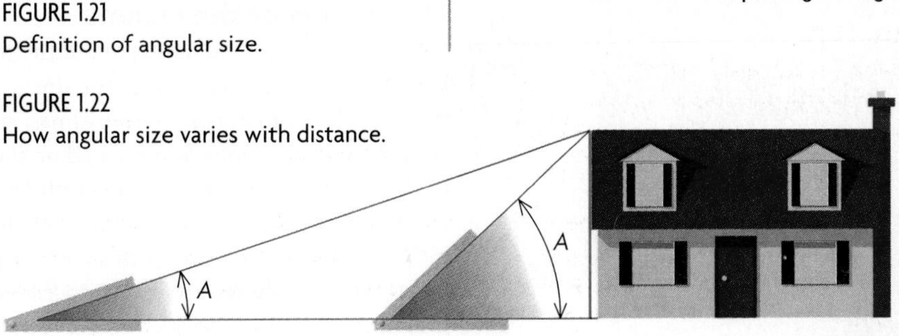

We can find an astronomical body's true diameter from its angular diameter if we know its distance. We need the distance because a body's angular size changes with distance. For example, a building looks big when we are near it and small when we are far away, as shown in figure 1.22. Furthermore it is easy to verify that the angular size changes inversely with the distance. That is, if we double the distance to an object, its angular size is halved. These basic features of angular size were known in antiquity and were in fact used by Aristarchus to determine the size of the Moon and Sun compared to the Earth.

To find an object's true diameter from its angular diameter and distance, imagine we are at the center of a circle passing through

Continued

EXTENDING OUR REACH

MEASURING THE DIAMETER OF ASTRONOMICAL OBJECTS *Continued*

the object, as illustrated in figure 1.23. Let L be the diameter of the body, and D the distance to the body, the radius of the circle. Next draw lines from the center to each end of L, letting the angle between the lines be A, the object's angular diameter.

We now determine the object's true size, L, by forming the following proportion: L is to the circumference of the circle as A is to the total number of degrees around the circle, which we know is 360.

Thus,

$$\frac{\text{Object's diameter}}{\text{Circumference}} = \frac{\text{Angle between lines}}{360}$$

$$\frac{L}{\text{Circumference}} = \frac{A}{360}$$

However, we know from geometry that the circle's circumference is $2\pi D$. Thus,

$$\frac{L}{2\pi D} = \frac{A}{360}$$

We can now solve for L and find that

$$L = \frac{2\pi DA}{360}$$

Thus, given a body's angular diameter and distance, we can calculate its true diameter. For example, suppose we apply this method to measure the Moon's diameter. We stated above that the Moon's angular diameter is

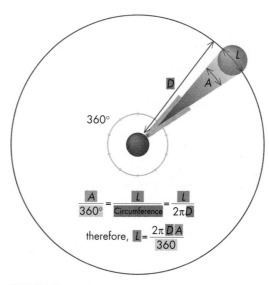

FIGURE 1.23
How to determine linear size from angular size.

about $1/2°$. Its distance is about 384,000 kilometers. Thus, its real diameter is

$$L = \frac{2\pi(384,000)(0.5)}{360}$$

$$= \text{about } 3350 \text{ km}$$

Suppose we try another example. The angular diameter of the Sun also turns out to be about $1/2°$, an interesting coincidence. The Sun's distance is 150 million kilometers.

The Sun's diameter must therefore be 2π (150,000,000) (0.5)/360, or about 1,309,000 kilometers.

An easy way to get at this same answer is to recognize that because the Sun and Moon have the same angular size, their true sizes must be in direct proportion to their distances. The Sun is about (150,000,000/384,000) = 391 times further away. Therefore it must be about 391 times larger than the Moon.

The Motion of the Planets

Following the basic discoveries about the size and distance of the Sun and Moon, the main thread of astronomical research for almost the next 2000 years centered on the motion of the planets. From earliest Greek times, one simple model formed the basis for understanding those motions. It was based on the observation that as one looks at the sky, everything seems to move around the Earth from east to west. This led naturally to the theory that the Earth was at the center of the Universe and the planets and stars moved around it. Models of the Universe of this type are called **geocentric theories**.

Figure 1.24 shows a typical geocentric model based on the work of the Greek astronomer Eudoxus, who lived about 400–347 B.C. The Sun, Moon, and planets all revolve around the Earth. The bodies that move fastest across the sky are those that are nearest to

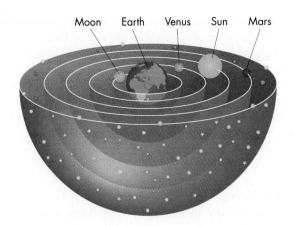

Moon   Earth   Venus   Sun   Mars

FIGURE 1.24
Cutaway view of the geocentric model of the Solar System according to Eudoxus. (Some spheres omitted for clarity.)

Epicycle

Planet

Earth

FIGURE 1.25
Epicycles are a bit like a bicycle wheel on which a Frisbee is bolted.

the Earth. Thus the Moon, whose path through the stars takes only about 28 days, is nearest to the Earth, whereas Saturn, whose path through the stars takes roughly 29 years, is located the farthest out. By assuming that each body was mounted on its own revolving transparent (crystalline) sphere and by tipping the spheres slightly with respect to one another, Eudoxus was able to give a quite satisfactory explanation of the motions of the heavenly bodies at any given time.

Unfortunately, such models do not work well if they are used to explain retrograde motion unless one believes that the planets sometimes stop moving, reverse direction, pause, and then resume their original motion. This idea is clumsy and unappealing. Eudoxus was able to explain retrograde motion only by requiring that each planet moved on *two* spheres, one inside the other. By adjusting their rotation rates and sizes, he was able to get reasonable agreement with the observed positions of the planets as they shifted across the sky. However, about A.D. 150, far more elaborate models, which could be used to predict with much better accuracy, were developed by Ptolemy, the great astronomer of Greco-Roman times.

Ptolemy

Ptolemy lived in Alexandria, Egypt, which at that time was one of the intellectual centers of the world, in part because of its magnificent library. Ptolemy fashioned a model of planetary motions in which each planet moved on one small circle, which in turn moved on a larger one (fig. 1.25). The small circle, called an **epicycle,** was supposed to be carried along on the large circle like a Frisbee spinning on the rim of a bicycle wheel.*

According to Ptolemy's model, the motion of the planet from east to west across the night sky is caused by the rotation of the large circle (the bicycle wheel in our analogy). Retrograde motion occurs when the epicycle carries the planet in a reverse direction (caused by the rotation of the Frisbee, in our model). Thus, with epicycles, it is possible to account for retrograde motion, and Ptolemy's model was able to predict planetary motions with fair precision.

Unfortunately, discrepancies remained between the predicted and true positions of the planets. This led to further modifications of the model, each of which led to slightly better agreement, but at the cost of adding much greater complexity. Nevertheless, Ptolemy's theory survived until the 1500s and ultimately collapsed from a loss of faith. It had

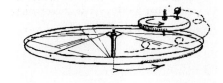

---

*Ptolemy seems to have got the idea of epicycles from the writings of Hipparchus, who lived about 150 B.C. Hipparchus is best known to astronomers for his invention of the magnitude system (see chapter 12) for measuring stellar brightness and for his discovery of precession (see chapter 4). The latter was made possible by his meticulous observations of star positions and the care with which he compared his data to that of his predecessors.

become far too complex to be plausible, and simplicity is an important element of scientific theory. As the medieval British philospher William of Occam wrote in the 1300s, "Entities must not be unnecessarily multiplied," a principle known as "Occam's razor."

Ptolemy's era was one of decay and general political instability for the Greco-Roman civilization, which accounts for our uncertainty of the year of his birth or death. We know of him mainly through his great book, *Almagest*, a compendium of the astronomical knowledge of the ancient Greeks. The book includes tables of star positions and brightnesses. In fact, much of what we know of Ptolemy (and of Greek and Roman civilization in general) we owe to the Islamic civilization that flourished around the southern edge of the Mediterranean from about 700 to 1200.

Islamic Contributions

Islamic civilization, like so many others, relied on celestial phenomena to set its religious calendar. For example, Islam's influence is very evident in astronomy through Arabic words such as *zenith* and the names of nearly all the bright stars—Betelgeuse, Aldebaran, and so on. In addition, Arabic scholars revolutionized mathematical techniques through innovations such as algebra (another Arabic word), Arabic numerals, and the concept of zero (yet another Arabic word).

Asian Contributions

The early people of Asia, like their contemporaries to the west, studied the heavens to devise constellations based on the Asian mythologies and to make maps of the sky. Although the ancient astronomers of the East did not devise elaborate geometric models of the heavens, their careful observations of celestial events nevertheless prove useful to astronomers today. For example, Chinese, Japanese, and Korean astronomers kept detailed records of unusual celestial events, such as eclipses, comets, and exploding stars, and they even noted dark spots on the Sun (sunspots) that they could occasionally see with the naked eye when the Sun was low in the sky and its glare was dimmed by dust or haze. Such records, especially those of exploding stars, allow today's astronomers to date the remnants created by these celestial outbursts. In addition, ancient Chinese astronomers could predict eclipses.

1.3 ASTRONOMY IN THE RENAISSANCE

Copernicus

The man who began the demolition of the geocentric theory and the revolution in astronomical ideas that continues to this day was a Polish physician and lawyer by the name of Copernicus (1473–1543). Copernicus made many attempts to reconcile the centuries of data on planetary positions that had been collected since Ptolemy's geocentric model. All of the theories were failures. Thus he was led to reconsider Aristarchus's ancient idea that the Earth moves around the Sun.

You will remember that **heliocentric models** in which the Sun (*hēlios,* in Greek) was the center of the system had been proposed nearly 2000 years earlier by Aristarchus but had been rejected partly because the observational tools available at that time were inadequate to detect stellar parallax. Nevertheless, such models offer an enormously simpler explanation of retrograde motion. In fact, if the planets orbit the Sun, retrograde motion becomes a simple consequence of one planet on a smaller orbit overtaking and passing another on a larger orbit, as Copernicus was able to show.

To see why retrograde motion occurs, look at figure 1.26. Here we see the Earth and Mars moving around the Sun. The Earth completes its orbit, circling the Sun in 1 year, whereas Mars takes 1.88 years to complete an orbit, with the Earth overtaking and passing Mars every 780 days.

If we draw lines from the Earth to Mars, we see that Mars will appear to change its direction of motion against the background stars each time the Earth overtakes it. A very

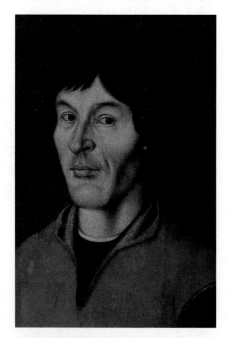

Copernicus. (Courtesy Erich Lessing, Art Resource.)

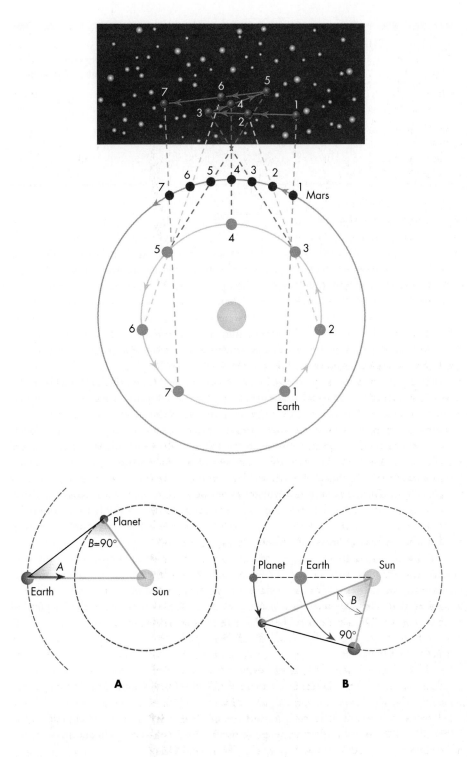

FIGURE 1.26
Why we see retrograde motion. (Object sizes and distances are exaggerated for clarity.)

FIGURE 1.27
How Copernicus calculated the distance to the planets. (A) When an inner planet appears in the sky at it's farthest point from the Sun, the planet's angle on the sky away from the Sun, A, can be measured. You can see from the figure that at the same time angle B is 90°. The planet's distance from the Sun can then be calculated with geometry, if one knows the measured value of angle A and the fact that the Earth-Sun distance is 1 AU. (B) Finding the distance to an outer planet requires determining how long it takes the planet to move from being opposite the Sun in the sky (the planet rises at sunset) to when the Sun-Earth-planet angle is 90° (the planet rises at noon or midnight). Knowing that time interval, one then calculates what fraction of their orbits the Earth and planet moved in that time. Multiplying those fractions by 360° gives the angles the planet and Earth moved. The difference between those angles gives angle B. Finally, using geometry and the value of angle B as just determined, the planet's distance from the Sun can be calculated.

similar phenomenon happens when you drive on a highway and pass a slower car. Both cars are, of course, moving in the same direction. However, as you pass the slower car, it *looks* as if it is shifting backward against the stationary objects behind it.

With his heliocentric theory, Copernicus not only could give a simple explanation of retrograde motion but also could determine each planet's distance from the Sun by a geometric construction, as shown in figure 1.27. The distances found in this manner must be expressed in terms of the Earth's distance from the Sun (whose value was not known until

TABLE 1.1	Planetary Distances According to Copernicus	
Planet	**Distance in AU According to Copernicus**	**Actual Distance**
Mercury	0.38	0.39
Venus	0.72	0.72
Earth	1.00	1.00
Mars	1.52	1.52
Jupiter	5.22	5.20
Saturn	9.17	9.54

several hundred years later), but table 1.1 illustrates that they agree well with modern values. These distances became an important starting place for those who came later.

Copernicus described his theory of a Sun-centered universe in one of the most influential scientific books of all time, *De revolutionibus orbium coelestium (On the Revolutions of the Celestial Orbs)*. Because his ideas were counter to the teaching of the Roman Catholic church, they were met with hostility and skepticism. The book itself was not published until shortly before his death (which was perhaps just as well), and he saw the first copy while on his deathbed.

Ironically, some of the criticism of Copernicus's work was justified. Although the idea was basically correct, his model did not account for the observed positions of the planets much better than did the more complicated but incorrect theory of Ptolemy . This lack of complete agreement between theory and observation arose at least in part because Copernicus insisted that the planetary orbits were circles. Furthermore, his model again raised the question of why no stellar parallax could be seen. Finally, his views of planetary motion ran counter to the teachings of Aristotle, views supported both by "common sense" and by the Catholic church at that time. After all, when we observe the sky it *looks* as if it moves around us. Moreover, we do not detect any sensations caused by the Earth's motion—it feels at rest. This mixture of rational and irrational objections made even scientists slow to accept the Copernican view.

However, by this time, there was a growing recognition of the immensity of the Universe. Astronomers, such as the Englishman Thomas Digges and the Italian Giordano Bruno, went so far as to claim that the stars were other suns, perhaps with other worlds around them. This new scientific open-mindedness coupled with the aesthetic appeal of the simpler system led to a growing belief in the Copernican system.

Tycho and Kepler

Copernicus's theory, although not the only stimulus, marked the opening of a new era in the history of astronomy. Conditions were favorable for new ideas: the cultural renaissance in Europe was at its height; the Protestant reformation had just begun; the new world was being settled. In such an intellectually stimulating environment, new ideas flourished and found, at least among scientists, a more receptive climate than in earlier times.

One scientist whose ideas flourished in this more intellectually open environment was the sixteenth-century Danish astronomer, Tycho Brahe (1546–1601). Born into the Danish nobility, Tycho utilized his position and wealth to indulge his passion for study of the heavens, a passion based in part on his professed belief that God placed the planets in the heavens to be used as signs to mankind of events on Earth. Driven by this interest in the skies, Tycho designed and had built instruments of far greater accuracy than any yet devised in Europe. Tycho then used these devices to make precise measurements of planetary positions. His meticulous observations turned out to be crucial not only for distinguishing the superiority of the heliocentric over the geocentric system, but also for revealing the true shape of planetary orbits.

Tycho was more than just a recorder of planetary positions; he recognized opportunity when he saw it. In 1572 when an exploding star (what we would now call a supernova) became visible, Tycho demonstrated from its lack of motion with respect to the other stars that it was far beyond the supposed spheres on which the planets move. Likewise, when a bright comet appeared in 1577, he showed that it would have had to pass through several of the alleged perfect crystalline spheres (see Eudoxus's model of the Solar System, fig. 1.24). These observations suggested that the heavens were both changeable and more complex than was previously believed.

Although Tycho could see the virtues of the simplicity offered by the Copernican model, he was also unconvinced of its validity because he could find no evidence for stellar parallax. Therefore he offered a compromise model in which all of the planets except the Earth went around the Sun, while the Sun, as in earlier models, circled the Earth. Tycho was the last of the great astronomers to hold that the Earth was at the center of the Universe.

Upon Tycho's death, his observational data were passed to his young assistant, Johannes Kepler. Kepler (1571–1630) was able to derive from this huge set of precise information a detailed picture of the path of the planet Mars. Whereas all previous investigators had struggled to fit the planetary paths to circles, Kepler, by using Tycho's superb data, was able to show that Mars did not move in a circular path but rather in an elliptical one.

The shape of an **ellipse** is described by its long and short dimensions, called its major and minor axes respectively. An ellipse can be drawn with a pencil inserted in a loop of string that is hooked around two thumbtacks. If you move the pencil while keeping it tight against the string, as shown in figure 1.28A, you will draw an ellipse. Each point marked by a tack is called a **focus** of the ellipse.

Having measured that the orbit of Mars was elliptical, with the Sun located at a spot that was *not* the center of the ellipse but off center at a focus, Kepler was able to obtain excellent agreement between the calculated and the observed position not only of Mars but also of the other planets.

Along with discovering the shape of planetary orbits, Kepler also measured the relative sizes of the orbits. Because the orbits are elliptical, he could not describe them by a single radius. He therefore used the orbit's **semimajor axis**—half the major axis, as shown in figure 1.28B—as a measure of its size. When he compared these orbital sizes with how long it takes the planets to orbit the Sun, their periods, Kepler noticed the relation illustrated in table 1.2: the square of the orbital period was proportional to the cube of the orbital size, as measured by the semimajor axis.

Stars much more massive than our Sun explode as supernovas at the end of their lives when they have used up all their fuel. This process is discussed in chapter 13.

Tycho Brahe. (Courtesy Bettmann Archive.)

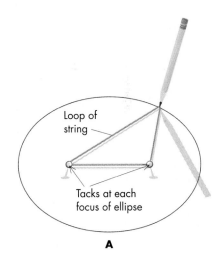

Loop of string

Tacks at each focus of ellipse

A

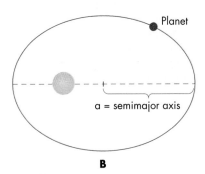

Planet

a = semimajor axis

B

FIGURE 1.28
(A) Drawing an ellipse. (B) The Sun lies at one focus of the ellipse.

Table 1.2   Table Illustrating Kepler's Third Law*

Planet	Distance from Sun (a) (in astronomical units)	Orbital period (P) (in years)	a^3	P^2
Mercury	0.387	0.241	0.058	0.058
Venus	0.723	0.615	0.378	0.378
Earth	1.0	1.0	1.0	1.0
Mars	1.52	1.88	3.51	3.53
Jupiter	5.2	11.86	141.0	141.0
Saturn	9.54	29.46	868.0	868.0

*Because the numbers have been rounded off, the agreement is not exact in all cases.

Johannes Kepler. (Art Resource/Erich Lessing.)

Kepler's discoveries of the nature of planetary motions are expressed in what are known today as **Kepler's three laws:**

I. Planets move in *elliptical* orbits with the Sun at one focus of the ellipse (see fig. 1.29A).
II. The orbital speed of a planet varies so that a line joining the Sun and the planet will sweep over equal areas in equal time intervals (see fig. 1.29B).
III. The amount of time a planet takes to orbit the Sun is related to its orbit's size, such that the period, P, squared is proportional to the semimajor axis, a, cubed (fig. 1.29C). Mathematically,

$$P^2 = a^3$$

where P is measured in years, and a is measured in astronomical units.

These three laws describe the essential features of planetary motion around our Sun. They describe not only the shape of a planet's path, but also its speed and distance from the Sun. For example, the second law—in its statement that a line from the planet to the Sun sweeps out equal areas in equal times—implies that when a planet is near the Sun it moves more rapidly than when it is farther away. We can see this by considering the shaded areas in figure 1.29B. For the areas to be equal, the distance traveled along the orbit in a given time must be larger when the planet is near the Sun. Thus, according to Kepler's second law, as a planet moves along its elliptical orbit, its speed changes, increasing as it nears the Sun and decreasing as it moves away from the Sun.

The third law also has implications for planetary speeds, but it deals with the relative speeds of planets whose orbits are at different distances from the Sun, not the speed of a given planet. Because the third law states that $P^2 = a^3$, a planet far from the Sun (larger a) has a longer orbital period (P) than one near the Sun. For example, the Earth takes 1 year to complete its orbit, but Jupiter, whose distance from the Sun is slightly more than 5 times Earth's distance, takes about 12 years. Thus a planet orbiting near the Sun overtakes and passes a planet orbiting farther out, leading to the phenomenon of retrograde motion, as discussed earlier.

Kepler's third law has other implications. For example, as we shall see later, the law gives information about the nature of the force holding the planets in orbit. Finally, the third law allows us to calculate the distance from the Sun of any body orbiting it if we measure the body's orbital period. The distance we obtain will be relative only to the Earth's distance, but the law thereby gives us at least the relative scale of the Solar System.

For example, suppose we wish to determine how far Pluto is from the Sun compared with the Earth's distance from the Sun. We measure Pluto's motion against the background stars to determine how long Pluto takes to circle the Sun, finding from such observations that its orbital period, P, is 248 years. Kepler's third law tells us that $P^2 = a^3$. If $P = 248$ yrs, then $(248)^2 = a^3$. Squaring 248 and then taking the cube root gives us $a = 39.5$ astronomical units. That is, Pluto is about 40 times farther from the Sun than the Earth is.

FIGURE 1.29
Kepler's three laws. (A) A planet moves in an elliptical orbit with the Sun at one focus. (B) A planet moves so that a line from it to the Sun sweeps out equal areas in equal times. Thus the planet moves fastest when nearest the Sun. (C) The square of a planet's orbital period (in years) equals the cube of the semimajor axis of its orbit (in AU), the planet's distance from the Sun if the orbit is a circle.

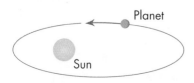

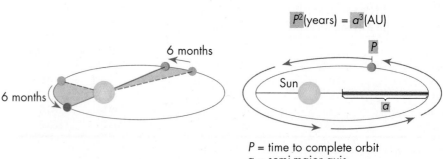

A   B   C

Apart from such astronomical applications, Kepler's laws have an additional significance. Kepler's laws are the first mathematical formulas to describe the heavens correctly, and as such they revolutionized our way of thinking about the Universe. Without such mathematical formulations of physical laws, much of our technological society would be impossible. These laws are therefore a major breakthrough in our quest to understand the world around us.

It is perhaps ironic that such mathematical laws should come from Kepler because so much of his work is tinged with mysticism. For example, as a young man, he sought to explain the spacing of the planets as described in Copernicus's work in terms of nested geometrical figures, the sphere, the cube, and so on. In fact, it was Tycho's notice of this work that led to his association with Kepler. Moreover, Kepler's third law evolved from his attempts to link planetary motion to music, using the mathematical relations known to exist between different notes of the musical scale. Kepler even attempted to compose "music of the spheres" based upon such a supposed link. Nevertheless, despite such excursions into these nonastronomical matters, Kepler's discoveries remain the foundation for our understanding of how planets move. As a final remark about his contributions, we owe credit to Kepler for the word *satellite*. When he saw the moons of Jupiter with a small telescope, their motion around the planet made him think of attendants or bodyguards—*satellēs*, in Latin.

Galileo

At about the same time that Tycho Brahe and Kepler were striving to understand the motion of heavenly bodies, the Italian scientist Galileo Galilei (1564–1642) was likewise trying to understand the heavens. However, his approach was dramatically different.

Galileo demonstrating his discoveries with his telescope. (Courtesy Jean-Leon Huens © National Geographic Society.)

Galileo was interested not just in celestial motion, but in all aspects of motion. He studied falling bodies, swinging weights hung on strings, and so on. In addition, he used the newly invented telescope for astronomical problems. Galileo did not invent the telescope himself. That invention seems to have been the work of the Dutch spectacle maker, Johannes (Hans) Lippershey. However, Galileo was the first person we know of who used the telescope to study the heavens. He met with astonishing results.

In looking at the Moon, Galileo saw that its surface had mountains and was in that sense similar to the surface of the Earth. Therefore he concluded that the Moon was not some mysterious ethereal body but a ball of rock. He looked (without taking adequate precaution) at the Sun, and saw dark spots (now known as sunspots) on its surface. He noticed that the position of the spots changed from day to day showing that the Sun not only had blemishes and was not a perfect celestial orb, but that it also changed. Both these observations were in disagreement with previously held conceptions of the heavens as perfect and immutable. In fact, by observing the changing position of the spots from day to day, Galileo deduced that the Sun rotated.

Galileo looked at Jupiter and saw four smaller objects orbiting it, which he concluded were moons of the planet. These bodies, known today in his honor as the Galilean satellites, proved unambiguously that there were at least *some* bodies in the heavens that did not orbit the Earth. However, perhaps as importantly, the motion of these bodies raised the crucial problem of what held them in orbit. He looked at Saturn and discovered that it did not appear as a perfectly round disk but that it had blobs off the edge. However, his telescope was too small and too crudely made (inferior to inexpensive binoculars of today) to show these as rings, a discovery that had to wait until 1656 when they were first seen by the Dutch scientist Christiaan Huygens as features that were detached from the planet.

Galileo also examined the Milky Way and saw that it was populated with an uncountable number of stars. This single observation, by demonstrating that there were far more stars than previously thought, shook the complacency of those who believed in the simple Earth-centered universe.

Galileo observed that Venus went through a cycle of phases, like the Moon, as illustrated in figure 1.30. The relation between the phase of the planet and its position with respect to the Sun left absolutely no doubt that Venus must be in orbit around the Sun exactly as the Moon is in orbit around the Earth. Perhaps more than any other observation, this one dealt the death blow to the old geocentric model of planetary motion.

FIGURE 1.30
As Venus orbits the Sun, it undergoes a cycle of phases. The phase and its position with respect to Sun show conclusively that Venus cannot be orbiting the Earth. The gibbous phases Galileo observed occur for the heliocentric model but cannot happen in the Earth-centered Ptolemaic model.

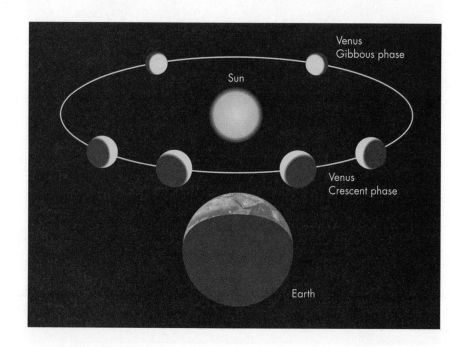

Galileo's contributions to science would be honored even had he not made all these important observational discoveries, for he is often credited with originating the experimental method for studying scientific problems. From his experiments on the manner in which bodies move and fall, Galileo deduced the first correct "laws of motion," laws that ultimately led Newton to his explanation of why the planets obey the laws of planetary motion that Kepler discovered.

Galileo's probings into the laws of nature led him into trouble with religious "law." He was a vocal supporter of the Copernican view of a Sun-centered universe and wrote and circulated his views widely and somewhat tactlessly. His exposition followed the style of Plato, presenting his arguments as a dialog between a wise teacher (patterned after himself) and an unbeliever in the Copernican system whom he called Simplicio and who, according to his detractors, was patterned after the Pope. Although the Pope was actually a friend of Galileo, more conservative churchmen urged that Galileo be brought before the Inquisition because his views that the Earth moved were counter to the teachings of the Catholic Church. Considering that his trial took place at a time when the Papacy was attempting to stamp out heresy, Galileo escaped lightly. He was made to recant his "heresy" and was put under house arrest for the remainder of his life. Only in 1992 did the Catholic Church admit it had erred in condemning Galileo for his ideas.

1.4 ISAAC NEWTON AND THE BIRTH OF ASTROPHYSICS

Isaac Newton (1642–1727), who was born the year that Galileo died, is arguably the greatest scientist of all time. Newton's contributions span mathematics, physics, and astronomy. Moreover, Newton pioneered the modern studies of motion, optics, and gravity. In his attempts to understand the motion of the Moon, Newton not only deduced the law of gravity but also discovered that he needed mathematical methods for calculating the gravitational force of a spherical body and that no such methods were then available. This realization led him to invent what we now know as calculus.

What is especially remarkable about Newton's work is that the discoveries he made in the seventeenth century still form the core for most of our understanding of gravity and the motion of bodies, discoveries we will discuss in more detail in chapter 2. In chapter 3 we will discuss some of Newton's ideas and discoveries about light, ideas that are still in use.

Newton was a fascinating individual. He came from very modest origins and rose to high positions not only in academia but also in the government. He was Warden of the Mint and is alleged to have invented milling, the process whereby grooves are cut in the edge of coins to detect metal being pared off them, thereby debasing their value. He was also a deeply religious man and wrote prolifically on theological matters as well as science.

Newton's laws of motion, when combined with his law of gravity, successfully applied for the next 200 years to essentially all problems of the motion of astronomical bodies. Only at the beginning of the twentieth century, with the discovery of a tiny discrepancy in the motion of Mercury as calculated using Newton's work, did scientists recognize that Newton's laws may not have been the last word on planetary motion. For example, his descriptions of motion require modification if we are to correctly describe motion at velocities near the speed of light or where gravitational fields are very intense. These modifications are incorporated in Einstein's theories of relativity.

Isaac Newton. (Courtesy Bausch and Lomb Optical Co.)

1.5 THE GROWTH OF ASTROPHYSICS

Newton's enormous contributions tend to overshadow other advances in astronomy during the eighteenth and nineteenth centuries. That period began with observational discoveries that increased astronomers' faith in using physical laws to understand the structure and workings of astronomical bodies. However, by the end of the period, newly found physical laws gave astronomers totally new tools for studying the heavens. In fact, the increasing use of the word *astrophysics* describes that shift well.

New Discoveries

Unexpected discoveries play a major role today in expanding our knowledge of the heavens, no less so than in the time immediately after Newton's death. For example, in 1781 the English astronomer Sir William Herschel discovered the planet Uranus. He also discovered that some stars have companions in orbit around them. The motion of such double stars offered additional tests of Newton's laws, but the most striking triumph of these laws of motion was their explanation of irregularities in the orbital motion of Uranus. Such irregularities hinted that another body was exerting a gravitational force on Uranus and, from Newton's laws, astronomers could calculate the position of the unseen body. As we will discuss further in chapter 9, a search of the sky near the calculated position revealed the planet Neptune.

New Technologies

New technologies also played an important role during this period. For example, improvements in optics allowed astronomers to build bigger telescopes and thereby observe much fainter objects. Among these objects, astronomers found nebulas, some of which were gas clouds within the Milky Way. However, others turned out to be external star systems similar to the Milky Way.

Another important technological advance was the application of photography to astronomy, starting in the middle of the nineteenth century. Photographic film gave astronomers permanent records of what they saw, and because film could store light during long exposures, astronomers were now able to detect objects much fainter than the eye could see in a single moment.

The scientific and technical advances described above have a direct bearing on astronomy, but scientific discoveries often influence totally unconnected areas. For example, during the eighteenth and nineteenth centuries many scientists were studying the nature of matter and heat. The study of heat was prompted, at least in part, by a desire to improve the newly invented steam engine.

The Nature of Matter and Heat

Like so many of our ideas about the nature of the Universe, our ideas of matter date back to the ancient Greeks. For example, Leucippus, who lived about the fifth century B.C. in Greece, and his student Democritus taught that matter was composed of tiny indivisible particles. They called these particles *atoms,* which means "uncuttable" in Greek. Our current model for the nature of atoms dates back to the early 1900s, with the work of the British physicist Ernest Rutherford. Rutherford showed with a series of experiments that atoms have a tiny core, the nucleus, around which yet smaller particles, called electrons, orbit. We will discuss this model of the atom more fully as we describe the production of light (chapter 3) and as we discuss the source of energy for stars (chapter 11).

The recognition of a link between energy and atoms began in the eighteenth and nineteenth centuries as scientists attempted to understand the behavior of gases. For example, why does heating a gas raise its pressure? The answer comes from the recognition (discussed more fully in chapter 11) that heating increases the motion of a gas's molecules and that pressure arises from that motion. One of the most important contributors to the understanding of heat and molecular motion was the English physicist Lord Kelvin.

The Kelvin Temperature Scale

Kelvin studied numerous problems in physics and astronomy, ranging from the motion of fluids to the properties of gases. Much of this latter work was motivated by his attempts to improve the energy efficiency of steam engines. In the course of studying the energy content of gases, Kelvin devised a temperature scale that is used today in virtually all the physical sciences. The reason for this wide usage is that on the Kelvin scale, a body's temperature is directly related to its energy content and to the speed of its molecular motion. That is, the greater a body's Kelvin temperature, the more rapidly its atoms move and the more energy it possesses. Similarly, if the body is cooled toward a temperature of zero on the Kelvin scale, molecular motion within it slows to a virtual halt and its energy approaches zero. Partly as a result of this, the Kelvin scale has no negative temperatures.

Temperatures on the Kelvin scale are not given in degrees, but are simply called "Kelvin." For example, the freezing and boiling points of water are very nearly 273 and 373 Kelvin, respectively. Room temperature is about 300 Kelvin. Relatively simple formulas allow conversion between Kelvin and the more familiar Fahrenheit and Celsius scales. Celsius temperatures are simply Kelvin temperatures minus 273. Fahrenheit temperatures, F, can be calculated by using the formula $F = 9/5\ K - 459.4$, where K is the Kelvin temperature. Because of its direct relation to so many physical processes, we will use the Kelvin scale in most of the remainder of this book.

SUMMARY

We create a mental model of the night sky overhead as a giant dome, forming part of the celestial sphere. Star patterns on the celestial sphere are called *constellations*. Using this model, stars rise in the east and set in the west as the celestial sphere rotates overhead. This apparent motion is actually caused by the Earth's spin. The points about which the celestial sphere appears to pivot are called the *celestial poles*.

At any given time of year, the Sun's glare hides those stars lying near it on the celestial sphere. However, as the Earth moves around the Sun it looks to us as if the Sun changes its position with respect to the stars. Thus stars previously visible are lost in the Sun's glare, whereas stars previously hidden become visible. Therefore different constellations are visible at different times of year. We call the path that the Sun follows around the celestial sphere the *ecliptic*.

Planets also move with respect to the constellations but always within the narrow band of the zodiac. The zodiac follows the Sun's path around the sky, but it is slightly tilted with respect to the celestial equator, an imaginary line on the celestial sphere that lies directly above the Earth's equator. The usual direction of planetary motion is from west to east with respect to the stars. However, during part of the year, each planet shifts in the other direction, undergoing apparent retrograde motion.

The zodiac and the Sun's path are tipped at an angle of 23.5° to the celestial equator because the Earth's rotation axis is tipped by that amount with respect to its orbit. The solstices and equinoxes mark when the Sun reaches its maximum distance from the celestial equator and when it crosses the equator, respectively. These points define the onsets of the seasons.

Ancient peoples noted the basic patterns of the night sky discussed above, but the Greeks appear to have been the first to give explanations of planetary motion, based on a combination of observational and geometric analysis. The Greeks pictured the planets, Sun, and Moon all orbiting the Earth on crystalline spheres.

Through the work of Aristotle and Eratosthenes respectively, the Greeks knew the shape and size of the Earth. Aristarchus measured the relative size and distance of the Moon and Sun and proposed about 300 B.C. that the Earth orbited the Sun. However, his model was rejected because the expected shift in star positions (parallax) was unobservable at that time.

Ptolemy (about A.D.150) developed a complex model of planetary motion with the Earth at the center (geocentric) and with retrograde motion explained by planets moving on epicycles.

The geocentric model began to crumble in the 1500s with the revival of the heliocentric model by Copernicus. Better observations by Tycho Brahe and detailed mathematical models by Kepler based on those observations placed the heliocentric model on a firmer basis. Galileo's observations with the recently invented telescope helped prove the heliocentric model.

Newton's discovery in the 1700s of the law of gravity and the laws of motion allowed him to explain why Kepler's laws worked, thereby completing the understanding of planetary motions.

QUESTIONS FOR REVIEW

1. What is the celestial sphere?
2. What is a constellation?
3. What is the zodiac?
4. What is the ecliptic?
5. Where on the celestial sphere would you look for the planets?
6. How long does it take the Moon to go through a cycle of phases?
7. Sketch the path on the sky that a planet makes when undergoing retrograde motion.
8. Will a planet in retrograde motion rise in the east or west?
9. If you see a bright "star" in the sky, how could you tell whether it is a star and not, for example, Venus?
10. Describe the major astronomical contribution(s) of the following in a sentence or two for each: Aristotle, Aristarchus, Eratosthenes, Ptolemy, Copernicus, Tycho, Kepler, Galileo, and Newton.
11. What is meant by the phrase *angular diameter?*
12. If you triple your distance from an object, what happens to its angular size?

THOUGHT QUESTIONS

1. If you were standing on the Earth's equator, where would you look to see the north celestial pole? Could you see this pole from Australia?
2. Draw a sketch of the Earth and persuade yourself that your latitude is the angle of the north celestial pole above the northern horizon.
3. Can you think of an astronomical reason why the zodiac may have been divided into 12 signs rather than 8 or 16?
4. Some people still believe the Earth is flat. What "proof" would you offer them that it is round? Could you persuade them?
5. Why does the position of sunrise along the eastern horizon change during the year?
6. Suppose the stars were very much closer than they really are. How might that have made it easier for Aristarchus to persuade people that the Earth moves around the Sun?
7. Tycho argued that the Sun orbited the Earth but that the other planets orbited the Sun. Could Tycho's model explain the phases of Venus as observed by Galileo? Why?

PROBLEMS

1. Suppose you were an alien living on the fictitious warlike planet Myrmidon and you wanted to measure its size. The Sun is shining directly down a missile silo 1000 miles to your south, while at your location, the Sun is 36° from straight overhead. What is the circumference of Myrmidon? What is its radius?
2. Make a table listing the astronomers named in the review questions above and then add the approximate dates of their births and deaths. Then add a few historic events of each period, as well as names of famous artists, writers, musicians, or politicians, who lived about the same time.
3. Suppose a planet were found with an orbital period of 64 years. How might you estimate its distance from the Sun? If its orbit is circular, what is its radius?
4. Suppose you received a message from aliens living on a planet orbiting a star identical to our Sun. They say they live 4 times further from their star than the Earth is from the Sun. What is the length of their year compared to ours?
5. The great galaxy in Andromeda has an angular diameter along its long axis of about 5°. Its distance is about 2.2 million light years. What is its linear diameter?
6. A shell of gas blown out of a star has an angular diameter of 0.1° and a linear diameter of 1 light-years. How far away is it?

TEST YOURSELF

1. If you are standing at the Earth's north pole, which of the following will be directly overhead?
   (a) The celestial equator.
   (b) The ecliptic.
   (c) The zodiac.
   (d) The north celestial pole.
   (e) The Sun.
2. A planet in retrograde motion
   (a) rises in the west and sets in the east.
   (b) shifts westward with respect to the stars.
   (c) shifts eastward with respect to the stars.
   (d) will be at the north celestial pole.
   (e) will be exactly overhead no matter where you are on Earth.
3. The circular shape of the Earth's shadow on the Moon led early astronomers to conclude that
   (a) the Earth is a sphere.
   (b) the Earth is at the center of the Solar System.
   (c) the Earth must be at rest.
   (d) the Moon must orbit the Sun.
   (e) the Moon is a sphere.
4. Galileo used his observations of the changing phases of Venus to demonstrate that
   (a) the Sun moves around the Earth.
   (b) the Universe is infinite in size.
   (c) the Earth is a sphere.
   (d) the Moon orbits the Earth.
   (e) Venus follows an orbit around the Sun rather than around the Earth.
5. Kepler's third law
   (a) relates a planet's orbital period to the size of its orbit around the Sun.
   (b) relates a body's mass to its gravitational attraction.
   (c) allowed him to predict when eclipses occur.
   (d) allowed him to measure the distance to nearby stars.
   (e) showed that the Sun is much farther away than the Moon.

FURTHER EXPLORATIONS

Ahmad, Imad A. "The Science of Knowing God: Astronomy in the Golden Era of Islam." *Mercury* 24 (March/April 1995):28.

Aveni, Anthony F. "Emissaries to the Stars: the Astronomers of the Ancient Maya." *Mercury* 24 (January/February 1995): 15.

Aveni, Anthony F. "Native American Astronomy." *Physics Today* 37:24, 1984.

Farrell, Charlotte, "The Ninth-Century Renaissance in Astronomy." *The Physics Teacher* 34 (May 1996): 268.

Gingerich, Owen. "Astronomy in the Age of Columbus." *Scientific American* 267 (November 1992): 100.

Gingerich, Owen. "Islamic Astronomy." *Scientific American* 254 (April 1986): 74.

Gingerich, Owen. "Copernicus and Tycho." *Scientific American* 229 (December 1973): 86.

Gingerich, Owen. *The Great Copernicus Chase.* Cambridge, New York: Cambridge University Press, 1992.

Gould, Stephen J. "The Persistently Flat Earth." *Natural History* 103 (March 1994): 12.

Hoskin, Michael, ed. *The Cambridge Illustrated History of Astronomy.* Cambridge, England; New York: Cambridge University Press, 1997.

Krupp, E. C. *In Search of Ancient Astronomies.* New York: McGraw Hill, 1978.

Krupp, E. C. *Echoes of the Ancient Skies: The Astronomy of Lost Civilizations.* New York: Harper and Row, 1983.

McPeak, W. J. "Tycho Brahe Lights up the Universe." *Astronomy* 18 (December 1990): 28.

Reston, James. *Galileo: a Life.* New York: HarperCollins Publishers, 1994.

Web Sites

"The Brief History of Astronomy,'" by Marek Dudka.
   www.bios.niu.edu/orion/history.html
A detailed chronology and list of astronomers and discoveries.

"History of Astronomy and Space Science" by Alan Cairns
   www-hpcc.astro.washingron.edu/scied/astro/ astrohistory.html
A general site for astronomy history with many links.

KEY TERMS

angular size, 39
celestial equator, 25
celestial poles, 25
celestial sphere, 23
constellations, 24
ecliptic, 27
ellipse, 45
epicycle, 41
equinoxes, 30
focus, 45
geocentric theories, 40

heliocentric models, 42
Kepler's three laws, 46
lunar eclipse, 34
model, 23
parallax, 39
phases, 33
retrograde motion, 32
semimajor axis, 45
solar eclipse, 34
solstices, 30
zodiac, 30

 PROJECTS

1. Motion of the Moon and planets. Obtain a star map (Sky Publishing Company's SC1 available from Sky Publishing Co., P.O. Box 9111, Belmont, MA 02178-9118, is excellent and inexpensive). Determine from the calendar when the Moon is a few days past new so that it will be visible in the early evening. Go outside shortly after sunset and look for the Moon in the west near where the Sun went down. Next check the dates along the bottom of the star map to find what region of the sky is visible at that date and time. As the sky darkens, locate the brighter stars near the Moon and mark on the chart where the Moon is with respect to those stars. Finally, make a sketch of the Moon's shape.

Repeat this process for the next four or five nights. The Moon will set a little later each evening, and so you need to adjust your observing time accordingly. After watching a few nights, mark out the Moon's path on the star map. Ideally, you might want to follow the Moon's track for about 2 weeks, although as the Moon reaches third quarter (3 weeks after new moon) you'll have to stay up pretty late because the third-quarter moon does not rise until midnight. If you are really ambitious, get up before dawn and watch the crescent moon shrink as it approaches the new phase.

You can also use this method to study the motion of the planets. Venus is a good choice because it is bright and it moves rapidly across the sky. However, Venus may lie too close to the Sun to be easily visible.

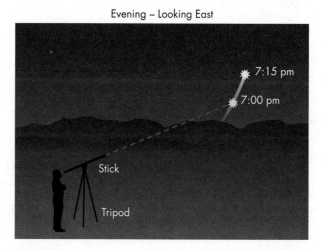

Evening – Looking East

7:15 pm
7:00 pm

Stick

Tripod

FIGURE 1.31
Sketch illustrating how to observe the motion of the stars across the sky by sighting along a stick.

To locate the planets you can look in an almanac, which will tell you what constellation the planet is in on the day you are observing. You may also find this information in the local newspaper, or you can ask your instructor to mark off the planet's approximate position on the star chart to get you started. Because the outer planets move relatively slowly across the sky, you should space out your observations, perhaps marking positions once a week rather than every night.

As you progress in your project you might ask whether the planets really follow the ecliptic? Can you see retrograde motion?

2. Motion of the stars across the sky. Many people are startled when they are told that the stars move across the sky the same way that the Sun does. However, it is easy to show that they do.

Get a stick that you can poke into the ground so it will stand upright. Get a second smaller stick that you can tape or affix to the upright in some manner. A ruler taped to a camera tripod would be ideal, as sketched in figure 1.31.

Find a bright star and sight along the smaller stick toward the star. If you now wait 5 or so minutes and again sight along the stick, you will see that it no longer

Continued

PROJECTS—Continued

points to the star. That is, the star has moved so that it now lies west of where the stick is pointing. You could do this experiment indoors if you have a window on which you can put a small mark. Set up a chair by the window so that you can watch a star through the glass. While you remain seated, have a friend place a mark on the glass with a grease pencil or piece of tape where the star appears to be. Again, wait a few minutes. The star's motion will be clearly visible.

3. Observe the changing location of sunset. Find a spot, perhaps a window facing west, where you can see the western horizon in the evening. Make a sketch of the horizon, noting hills, buildings, or trees that might serve as reference marks. From your chosen viewing spot, watch the sunset and mark on your sketch where the Sun goes down. Label the date and time. Make an observation each week over a period of a month or so. Does the Sun set at the same point each night? If not, which direction along the horizon is the Sun moving? Does the Sun go down at the same time each night? Does it set earlier or later? If you enjoy photography, you might try taking a photograph or videotaping the sunset or sunrise.

4. Measure the diameter of the Earth. This project duplicates the method used by Er-atosthenes; he used the length of a shadow at two different locations to determine the Earth's size. Thus you need either to collaborate with someone relatively far away (at least several hundred miles north or south of you) or be able to travel such a large distance yourself. The larger the distance, the better.

Begin by setting up a vertical stick where you live. Tape down a piece of paper beside the stick and mark off on it the length of the stick's shadow over a few hours around noon time. Record the length of the stick and the smallest shadow.

On the same day as you do your measurement, have a friend do the same. If you are going to travel yourself—for example, during a school break—make your measurements at the two different locations as close together in time as possible.

From a road atlas, measure the straight-line, north-south distance between the two places where you made the measurements. On a piece of paper, make a careful *scale* drawing of the stick and shadow at each location. A sample pair of drawings is shown in figure 1.32A. From these drawings, measure the angles *A* and *B* at the top of the triangles.

Now look at figure 1.32B. This shows a beam of sunlight striking the Earth and the sticks you used in your experiment at the two different locations. Lines 1, 2, and 3 rep-resent rays of light from the Sun, with lines 1 and 2 just cutting the top of the stick at the two locations, and line 3 passing through the center of the Earth. Notice that the difference in latitude between the two locations is just the angle *AB*. To see why this is so, notice that a line from the center of Earth to *A* makes equal angles with lines 1 and 3 because these lines are parallel. Similarly a line from the center to *B* makes equal angles with lines 2 and 3. If we let *l* be the difference in latitude between the two locations, then *l* = *A* − *B*. You could of course just look up that latitude difference in an atlas, but there were no atlases when Eratosthenes lived.

Once you know the angle of the difference in latitude, *l*, you can now find the circumference of the Earth. A little geometry will show you that 360/*l* = the Earth's circumference/*D*, where *D* is the distance in miles or kilometers between the two places at which you made the observations. Thus you can solve for the Earth's circumference. With the Earth's circumference known, you can then find its radius from the formula Circumference = 2π*R*. How does your answer compare with the value you find in the Appendix? What might explain the difference? How might you conduct the experiment more accurately?

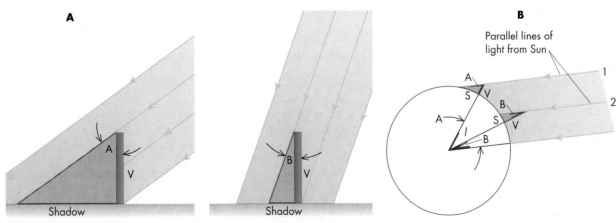

FIGURE 1.32
Sketch illustrating a method for finding the radius of the Earth.

Backyard Astronomy

You can learn many of the same things classical astronomers did by simply watching the night sky. But there is a bonus as well. Backyard astronomy is just plain fun, as evidenced by the many thousands of amateur astronomers who in their spare time pursue activities ranging from simply stargazing to searching for new comets.

This section is intended to give you some hints on how to become a backyard astronomer, beginning with learning the constellations and some of the stories associated with them. We will then briefly discuss small telescopes and star charts, introducing some of the terms used to describe the location of the planets. We will conclude with a description of some of the physical changes in your eye when you are observing in very dim light.

Learning the Constellations

One of the best ways to get started as a backyard astronomer is to learn the constellations. All it takes is a star chart, a dim flashlight, and a place that is dark and has an unobstructed view of the night sky. The star chart will generally tell you how to hold it so that it matches the sky for the date and time that you are observing.

Start by determining which way is north, using a compass if necessary. Then try to locate a few of the brighter stars, matching them up with the chart. This will give you some sense of how big a piece of the sky the chart corresponds to. Next, try to identify a few of the constellations. Focus at first on just a few of the brighter ones. For example, if you live at midlatitude in the Northern Hemisphere, the Big Dipper—part of the constellation Ursa Major—is a good group to start with because it is always in the northern part of the sky.

As you attempt to find and identify stars, your spread hand held at arm's length makes a useful scale. For most people, a fully spread hand at arm's length covers about 20° of sky, or about the length of the Big Dipper from tip of handle to bowl, as shown in figure E1.1. For smaller distances, you can use finger widths (a few degrees).

This scaling of sky distances with your hand makes it easy to point out stars to other people. For example, you can say that a star is two hands away from the Moon and at the 4 o'clock position, as illustrated in figure E1.2.

Once you have come to recognize a few of the constellations, you may be interested in the stories associated with them, that is, star lore.

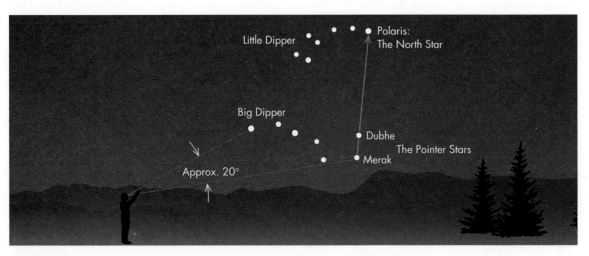

FIGURE E1.1
The Big Dipper, part of the constellation Ursa Major, the Great Bear. A line through the two pointer stars points toward Polaris. The Big Dipper spans about 20° on the sky, about a hand width at arm's length for most people.

FIGURE E1.2
Describing the location of stars by clock position. Star A is two hands from the Moon and at 4 o'clock position.

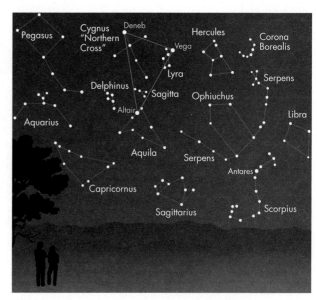

FIGURE E1.3
Dominating the night sky in July, August, and September are the three bright stars Vega, Altair, and Deneb, which form the summer triangle.

Star Lore

Star lore is part of virtually all cultures. The ancient Greeks, the Pawnee tribes of the American Midwest, the Australian aborigines, all created stories about the star groupings they saw in the sky.

Because the star groupings do not change except on time scales of tens of thousands of years, the night sky we see is essentially the same night sky that ancient peoples saw. Star lore can therefore link us to the remote past. Furthermore, star lore can help us to remember constellations. In fact, it has been suggested that many of the stories were created as aids to memory, especially important in a time when familiarity with the stars could be literally a matter of life or death to a farmer or a navigator. Scientists have shown that baby birds learn to recognize star patterns and movements and use them to navigate safely—unguided by their parents—across thousands of miles of ocean to their winter homes. Perhaps we too have such instinctive faculties that help us learn the stars.

Probably the most familiar star grouping is the Big Dipper. It is not a constellation but rather is called an **asterism.** An asterism is an easily recognized grouping of stars that may be part of one constellation or may incorporate pieces of several. For example, the Big Dipper is part of the constellation Ursa Major, the Great Bear. The asterism of the Summer Triangle, on the other hand, spans three constellations. It consists of the three bright stars conspicuous in the summer evening: Deneb (in Cygnus, the Swan), Altair (in Aquila, the Eagle), and Vega (in Lyra, the Harp), shown in figure E1.3 .

The Big Dipper is not only easy to spot, but it is also an excellent signpost to other asterisms and stars. For example, two of the stars in its "bowl" (see fig. E1.1) are called the "pointers" because they point, roughly, to the North Star, Polaris.

Polaris lies almost exactly above the Earth's north pole, and because of its position there, it is the only star in the northern sky that shows, to the naked eye, no obvious motion during the night. Its relatively fixed position is illustrated by the time exposure in figure E1.4, showing the other stars rotating around it. Because Polaris always lies nearly true north, it is useful in orienting yourself to compass directions.

Polaris marks the end of the handle of the Little Dipper, another asterism and part of the constellation Ursa Minor, the Little Bear. The Little Dipper is not easy to see because its stars are dim, but it curves back toward the Big Dipper so that these two star groupings fill most of the sky directly around the north celestial pole.

The native inhabitants of North America had a story about the Big Dipper. Its bowl represented a huge bear, and the handle represented three warriors in pursuit of the bear. They had wounded it, and it was bleeding. The red color of the leaves in autumn was said to be caused by the bear's blood dripping on them when the constellation lies low in the sky during the evening hours of the autumn months.

If you follow the pointer stars in the Big Dipper past the Little Dipper and Polaris, you will come to a set of constellations tied together by an ancient Greek myth, the story of Perseus and Andromeda. The cast includes a king (Cepheus), a queen (Cassiopeia), the hero (Perseus), the princess (Andromeda), a sea monster

Polar

FIGURE E1.4

A time exposure showing how Polaris remains essentially fixed while the sky pivots around it. (Hermann Eisenbeiss/Photo Researchers, Inc.)

(Cetus), and the winged horse (Pegasus). The constellations are shown in figure E1.5, and their story goes as follows:

In ancient days there lived a queen of Ethiopia, Cassiopeia, who was very beautiful but also very vain. She and king Cepheus, her husband, and their daughter, Andromeda, lived happily until one day the queen boasted that she was more beautiful than the daughters of Nereus, a sea god. In punishment for such pride, the sea-god Neptune sent a sea monster, Cetus, to ravage the kingdom. To save his people and appease the gods, Cepheus was instructed to tie his daughter, Andromeda, to a rock for the monster to devour. Meanwhile, Perseus was returning home from a quest in which he slew the snake-haired Gorgon, Medusa. On her death, her blood dripped into the sea and turned into the flying winged horse, Pegasus. Medieval versions of the story have Perseus riding Pegasus, but the classical myth has him borne on winged sandals. Regardless of his means of travel, Perseus saw the maiden's peril and landed, slaying the monster and delivering the kingdom. They all lived as happily ever after as most mythological families, and their astronomical tableau is most easily seen in the late autumn, when the brighter of its constellations are high in the sky.

Stories are also told about stars in other parts of the sky. For example, in the winter sky you can see the sad legend of the Hunter, Orion, and the maiden who refused to fall in love with him. The story also involves Orion's hunting dogs (Canis Major and Canis Minor), a bull (Taurus), a rabbit (Lepus), the maiden's sisters (the Pleiades—a cluster of stars in the constellation Taurus), and a scorpion (Scorpius).

The king of the island Chios had a lovely daughter, Merope. His island was filled with savage beasts, and to rid his kingdom of these dangerous animals, the king called upon Orion to kill the beasts and make his kingdom safe. When the task was done, Orion met Merope and made unwelcome advances. In punishment, he was blinded by the king, but after doing penance he had his sight restored. After reaching an old age, however, Orion one day stepped on a scorpion, which stung and killed him. On his death, the gods placed him in the sky with his faithful dogs (one of whom chases Lepus, the rabbit), forever attacking the wild bull, Taurus. Beyond the bull, Merope and her sisters (the Pleiades) run from the hunter, who pursues them each night across the sky. The scorpion was also placed in the sky, but on the other side of the heavens so that Orion would never again be threatened by it (Orion is visible in the evening only in the winter, whereas Scorpius is visible in the evening only in the summer).

The Orion myth has several versions, but the one described here fits together many of the astronomical

FIGURE E1.5
Perseus, Andromeda, Cassiopeia, Cepheus, Cetus, and Pegasus.

☀ Can you spot the Big Dipper?

references. Another myth with several versions involves the late summer constellations Boötes and Virgo. They are linked by the gloomy story of Icarius, the first cultivator of grapes for wine. According to legend, his recompense for sharing this knowledge was to be killed by drunken peasants, who buried his body under a tree. His dog, Maera, led Erigone, his daughter, to the spot, where, on discovering her father's body, she killed herself out of grief. Icarius was placed in the sky as Boötes, and Erigone as Virgo. The dog Maera became the star Procyon.

There are many other stories about constellations, but the above may give you some sense of the ones that have been handed down for thousands of years of written and oral history. Pass them on.

Amateur Astronomy

Anyone with access to even very modest equipment such as binoculars or low-powered telescopes has better equipment than Galileo ever had. With such equipment and access to a dark sky that person can become an amateur astronomer. The pleasures of the hobby can range from the aesthetic satisfaction of taking a lovely photograph (fig. E1.6) to the thrill of discovering a new comet or an exploding star.

You can take surprisingly fine photographs with a simple 35-millimeter camera mounted on a tripod. Set the focus for infinity, open the diaphragm (f stop) all the way, and expose for 10 or so seconds with one of the high-speed color films (technically, ISO [ASA] 400 or higher). For many cameras, you can easily do this by putting the camera on the B (for "flashbulb") setting and holding the shutter release down for the desired time. A cable release will help you expose without shaking the camera but is not essential.

If you expose for more than about 15 seconds, the Earth's rotation will smear the star image into a streak. Deliberately allowing the smearing to occur can produce dramatic pictures of what are called "star trails" (see fig. E1.4). To make star-trail pictures, leave the shutter open for 20 minutes or so.

To take untrailed long exposures or to use a telephoto lens, you will need a way to compensate for the Earth's rotation. You can make a simple device to do this

FIGURE E1.6
Picture of the Orion Nebula taken with a small backyard telescope. (Courtesy Carol B. Ivers and Gary Oleski.)

from hinged boards and a carriage bolt through them. Mount the camera on one board and set the other on a firm surface. Slowly turning the bolt will shift the camera, allowing you to take exposures of half an hour or so. For best results, however, you need a motor drive. Check your local library or the references at the end of the chapter for books or articles on drives and astrophotography. You may also find in such sources suggestions for scientifically useful projects, such as variable star observing or comet hunting. Projects like these are ideal for amateur astronomers because the results are not only valuable, but also require basically only patience and care without the need for a big telescope.

Small Telescopes

A small telescope will greatly increase the number and interest of objects you can observe. Such telescopes come in a wide range of styles and prices, but selecting the best one for your needs can be confusing. Moreover, it is a sad truth that you generally get what you pay for. Many amateur astronomers begin with a 3- to 4-inch reflecting telescope. Such a telescope uses a mirror to collect and focus the light; thus its name, "reflecting." The numbers refer to the diameter of the mirror, important because a larger mirror collects more light, al-

lowing you to see fainter objects. A larger mirror also permits seeing finer details with other things being equal (see chapter 5 for how a telescope works). With such a telescope you can easily see the moons of Jupiter, the rings of Saturn, and many lovely star clusters and galaxies. However, these later objects will not look like the pictures in books because your eyes—unlike a photograph or electronic detector—cannot store up light.

Notice in the above we have said nothing of magnifying power. For most amateur telescopes, the maximum useful power is limited by distortions to the light as it passes through our atmosphere. These distortions make a magnification of about 100 to 200 the useful limit. Beyond that, the distortions dominate and higher power gives no increase in clarity.

Whatever type of telescope you choose, be sure to get a sturdy mounting for it. Even at 100 power, tiny vibrations of the telescope caused by wind or the touch of your hand will make the image jiggle, hopelessly blurring it.

Before you actually buy a telescope, you might want to talk with your instructor or a local amateur astronomer. Such people often belong to an astronomy club, some of whose members may have second-hand telescopes they are willing to sell at reduced prices. For

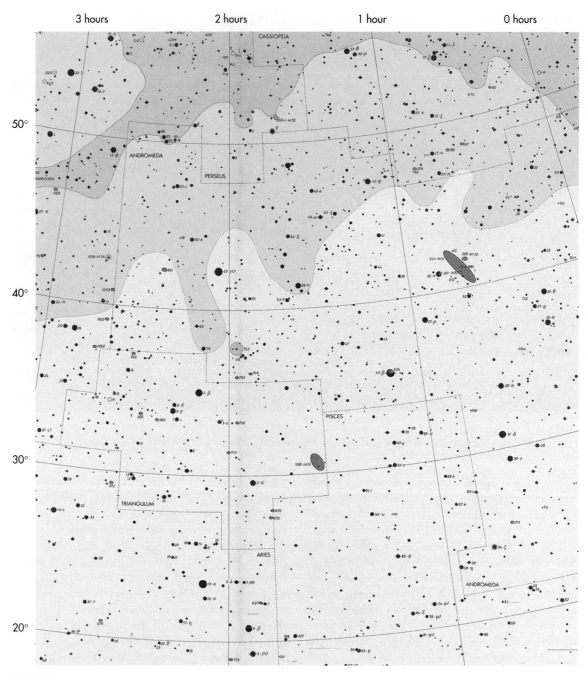

FIGURE E1.7
A modern star chart showing stars, galaxies, and coordinates. (from Tirion, Wil. *Sky Atlas 2000.0*, Cambridge, Mass.: Sky Publishing Corp., 1981.)

information about new telescopes, browse through the magazines *Astronomy* or *Sky and Telescope*—publications widely read by both amateur and professional astronomers—which contain many advertisements for small telescopes.

Star Charts

One of the pleasures of using a telescope is to look at objects too faint to be seen by the naked eye, such as most galaxies, faint star clusters, and remnants of dying stars. To find many of these dim objects you will need a good star chart.

Astronomers use star charts to find objects in the sky much as navigators use charts to find places on Earth. A typical star chart (fig. E1.7) shows the location of the constellations, the stars and other objects. It also gives some indication of the relative brightness of the stars. Finally, many charts also have information about the season and time of night at which the stars are visible.

Star charts are designed much like maps of the Earth. For example, both represent on a flat surface a map of something curved—in one case the celestial sphere; in the other the surface of the Earth. Also, both use a coordinate grid.

Celestial Coordinates

The coordinate grid used by astronomers is similar to that used by navigators. The grid consists of one set of lines running east-west on the celestial sphere, parallel to the celestial equator, and another set running north-south, connecting one celestial pole to the other. The east-west lines play the same role as latitude on the Earth, but to avoid confusion with terrestrial coordinates, they are called lines of **declination,** or "dec" for short. The north-south lines play the same role as longitude on the Earth and are called lines of **right ascension,** or "RA" for short. Declination values run from +90° to −90° (the north and south celestial poles), with 0° being the celestial equator. Right-ascension lines divide the celestial sphere into 24 equal zones that are labeled not in degrees but in units of time. Thus the right ascension of an object is given in hours (h), minutes (m), and seconds (s). Because the 360° around the sky is divided into 24 segments, each hour of RA equals 15°; that is, 360° divided by 24 = 15°. The point $0^h\,0^m\,0^s$ of RA is arbitrarily chosen to be where the Sun's path, the ecliptic, crosses the celestial equator as the Sun moves north (fig. E1.8A).

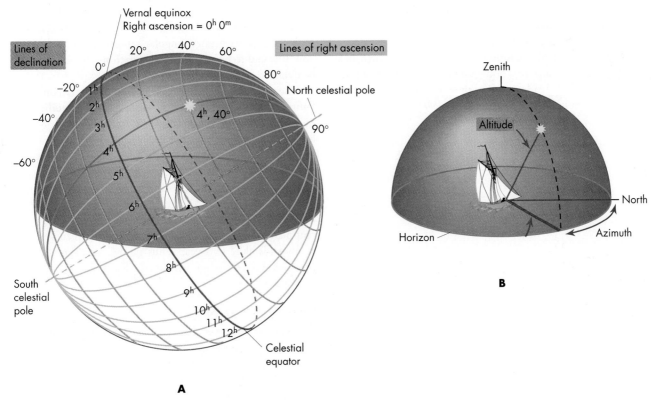

FIGURE E1.8

(A) Locating a star according to right ascension and declination. (B) Locating a star according to altitude and azimuth.

Right ascension and declination are not the only way to locate objects on the sky. For example, a celestial object may be located by its altitude and azimuth, as illustrated in figure E1.8B. In this so-called horizon system, **altitude** is an object's angle above the horizon. **Azimuth** is generally defined as the angle measured eastward along the horizon from North to the point directly below the object, although sometimes it is measured from South. Whichever convention is used for azimuth, these coordinates are useful for pointing out objects or tracking their motion. However, they have a serious drawback compared to the otherwise seemingly cumbersome right ascension and declination: an astronomical body's altitude and azimuth constantly change as it moves across the sky whereas, because they are defined with respect to the rotating celestial sphere, a body's right ascension and declination remain the same as a body rises and sets.

With a set of coordinate lines established, we can now locate astronomical objects on the sky the same way we can locate places on the Earth. For example, M31, a galaxy in the Local Group, is at right ascension 0 hours 42.7 minutes (0^h 42.7^m) declination $+41°$ $16'$, as you can see in figure E1.7.

Planetary Configurations

Because planets move across the stellar background, astronomers have invented some terms to help describe where they are located at any given time. These terms describe a planet's position with respect to the Earth and the Sun—planetary configurations—and are shown in figure E1.9. Understanding these terms when they are used can help you find planets.

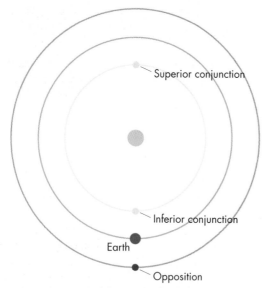

FIGURE E1.9

Planetary configurations: opposition, superior conjunction, and inferior conjunction.

If a planet lies in the sky in the same direction as the Sun, it is said to be at **conjunction.** If it lies approximately between us and the Sun, it is at **inferior conjunction.** If it is on the other side of the Sun, it is at **superior conjunction.**

Planets are very hard to see at either conjunction because they are hidden in the Sun's glare. On some very rare occasions a planet may pass directly between us and the Sun. We may then see it silhouetted against the Sun's bright disk, as shown in figure E1.10. Such an event is called a **transit.** Only Mercury and Venus can transit the Sun as seen from Earth, but we can imagine talking with an astronaut on Mars who has just witnessed the Earth transiting the Sun. This would occur when Mars is directly opposite the Sun in the sky, or at what seen from Earth is called **opposition.**

When an outer planet is at opposition, it is nearest Earth, as well as being brightest and easiest to see, rising at sunset. Inner planets, too, are easiest to see when they appear farthest away from the Sun. However, as shown in figure E1.11, they can never get very far from the Sun. For example, Mercury can never be more than 28° from the Sun, and Venus never more than 47° as seen on the sky. It is for this reason that Mercury and Venus are usually visible only in the morning or evening sky when the Sun is just below the horizon.

A planet seen close to the Sun at dawn or dusk is sometimes called the **Morning** or **Evening star.** When a planet is at its greatest angular separation from the Sun, it is said to be at **greatest elongation**—which can be either western or eastern.

The time interval between successive planetary configurations of the same type is called the **synodic period.** The synodic period differs from the planet's orbital period because both the Earth and the other planets move around the Sun. Thus the interval between oppositions is neither an Earth year nor another planet's orbital period. For example, the Earth takes about 2 years to catch up to and overtake Mars after an opposition. The Earth overtakes the slower moving more distant planets more quickly, and the interval between oppositions is close to a year. Thus the Martian synodic period is about 780 days, whereas the Saturnian synodic period is 378 days.

Your Eyes at Night

You will soon discover that the longer you stay outside in dim light, the more sensitive your eyes will become and the fainter the stars you will be able to see. This is the result of physiological changes in your eye referred to as **dark adaption.**

The simplest change in your eye occurring in dim light is that the pupil opens wider. This is easy to verify by looking at yourself in a mirror in a dimly lit room. In full sunlight your pupil normally has a diameter of

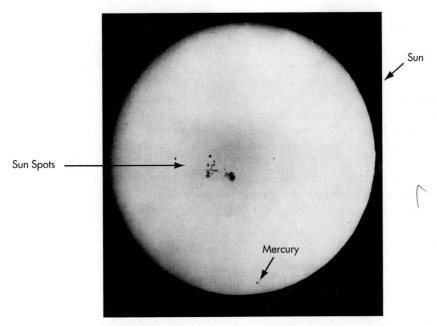

FIGURE E1.10
Transit of Mercury, November 14, 1907, photographed at Yerkes Observatory of the University of Chicago. Upcoming transits will occur on the following dates: Nov. 15, 1999; May 7, 2003; Nov. 8, 2006; May 9, 2016. (Courtesy Yerkes Observatory.)

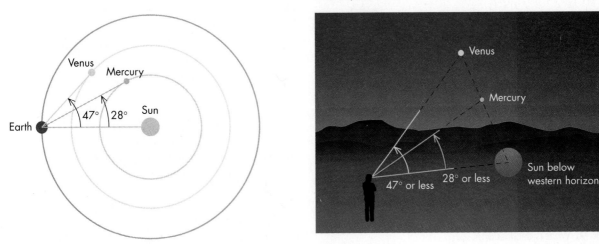

FIGURE E1.11
Elongations of Mercury and Venus and the Evening star phenomenon.

about 2 millimeters, but in total darkness its diameter may expand to about 7 or 8 millimeters, thereby allowing more light to enter your eye. The same principle is used in cameras when you adjust the aperture to match the available light.

Your eyes undergo another change in the dark. Chemical changes make the dark-adapted retina about one million times more sensitive to light than that under full daylight conditions. The process takes about 20 minutes to get well established but is undone by even

a few seconds' exposure to bright light. Thus once you are dark adapted you should stay away from bright lights for as long as you intend to observe.

In addition to becoming more sensitive to light, your eye also changes its sensitivity to color slightly, a phenomenon known as the "Purkinje effect." In full daylight, the eye responds best to greenish colors. At low light levels, it responds best to slightly bluer colors. This is probably the result of natural selection because starlight is much bluer than sunlight and eyes responsive to blue will therefore aid survival. It is certainly the case that night flying insects see blue light better than yellow. That is the reason bug-zappers use a blue light to attract insects and why a yellow light bulb is often used for outdoor night lighting to be less attractive to insects.

You may also notice that it is easier to see very faint objects if you don't look directly at them but instead look a little to one side. The greater sensitivity you enjoy from this so called **averted vision** arises because the center of your field of view is densely packed with receptors designed to allow you to see fine details. These receptors are better at showing fine structure than at seeing faint light. Thus looking slightly to one side of a faint object makes it easier to see.

Summary

Looking at the night sky is not only fun, it will help you understand some of the phenomena described in chapter 1. Star maps will help you identify constellations and bright stars, and by learning the mythology of the stars, you will be able to find your way around the night sky more easily. In addition, you will forge a link to distant and ancient cultures. For many people, backyard astronomy—even with simple equipment—is an enjoyable and exciting hobby. Perhaps you will discover a comet and have it named for you!

Questions for Review

1. Approximately where would you look for Mercury in the sky at 7 p.m.?
2. What is meant by "Morning star"?
3. If a planet is at opposition and you see it straight overhead, what time of night must it be?
4. Can you see Mercury in the western sky at dawn?
5. What is meant by "dark adaption"?
6. Why does the pupil of your eye get bigger in dim light?
7. What is averted vision?
8. What is right ascension?
9. What is declination?
10. What is azimuth?

Test Yourself

1. As a star rises and moves across the sky, which of the following change?
   (a) Its right ascension.
   (b) Its declination.
   (c) Its azimuth.
   (d) Both (a) and (b).
   (e) None of the above.
2. A planet is at inferior conjunction. It therefore rises at approximately
   (a) sunset.
   (b) sunrise.
   (c) midnight.
   (d) 2 hours before the Sun.
   (e) You can't tell from the available information.
3. If Mercury is at greatest western elongation it will be easiest to see
   (a) just before dawn.
   (b) just after sunset.
   (c) about midnight.
   (d) just before sunset.
   (e) None of the above.
4. Which of the following planets can be at inferior conjunction?
   (a) Jupiter.
   (b) Mars.
   (c) Uranus.
   (d) Venus.
   (e) All of them.
5. When your eye is dark-adapted
   (a) your pupils are smallest.
   (b) your pupils are biggest.
   (c) your color vision is at its most sensitive.
   (d) your eye is most sensitive to light.
   (e) Both (b) and (d).

Further Explorations

Allen, Richard Hinckley. *Star Names: Their Lore and Meaning*. New York: Dover, 1963.

Aveni, Anthony. *Conversing with the Planets*. New York: Times Books, 1992.

Berman, Bob. *Secrets of the Night Sky: The Most Amazing Things in the Universe You Can See with the Naked Eye*. New York: W. Morrow, 1995.

Covington, Michael. *Astrophotography for the Amateur*. Cambridge, New York: Cambridge University Press, 1985.

Harrington, P. "South for the Winter." *Sky and Telescope* 83 (February 1992): 142.

Hamilton, Edith. *Mythology*. New York: (Mentor) NAL-Dutton. 1953.

Levy, David H. *The Quest for Comets: an Explosive Trail of Beauty and Danger*. New York: Plenum Press, 1994.

Mechler, Gary. *Galaxies and Other Deep-Sky Objects*. New York: Knopf, 1995. (One in a series of National Audubon Society Pocket Guides.)

Muirden, James, ed. *Sky Watcher's Handbook: The Expert Reference Source for the Amateur Astronomer*. New York: W.H. Freeman, 1993.

Nagler, A. "Choosing Your Telescope's Magnification." *Sky and Telescope* 81 (May 1991): 553.

Newton, Jack. *Guide to Amateur Astronomy*. Cambridge: Cambridge University Press, 1995.

Observer's Handbook. Toronto: Royal Astronomical Society of Canada.

Pasachoff, Jay, and Donald H. Menzel. *Field Guide to the Stars and Planets.* 2nd. ed. Boston: Houghton Mifflin, 1992.

Ridpath, Ian. "The Origin of Our Constellations." *Mercury* 19 (November/December 1990): 163.

Schaaf, Fred. *Wonders of the Sky: Observing Rainbows, Comets, Eclipses, the Stars, and Other Phenomena.* New York: Dover, 1983.

Sesti, Giuseppe M. *The Glorious Constellations: History and Mythology.* New York: Harry N. Abrams, 1991. *This beautiful book contains extensive descriptions of the constellations and their mythology from many ancient cultures.*

Staal, Julius D. W. *The New Patterns in the Sky: Myths and Legends of the Stars.* Blacksburg, Va.: McDonald and Woodward, 1988.

Strom, Karen D. "Photographing the Stars." *The Physics Teacher* 34 (September 1996): 340.

Tirion, Wil. The Cambridge Star Atlas, 2nd. ed. Cambridge; New York: Cambridge University Press, 1996.

Two excellent magazines for the amateur astronomer are the following:

Astronomy, published by Kalmbach Publishing Co., 21027 Crossroads Circle, P.O. Box 1612, Waukesha, WI 53187.

Sky and Telescope, published by Sky Publishing Corporation, P.O. Box 9111, Belmont, MA 02178-9111.

Web Sites

"Backyard Astronomy: Tips on observing the Universe," by *Sky and Telescope.* **www.skypub.com/backyard/backyard.shtml.** An excellent site with many use links and good advice about choosing telescopes.

Key Terms

altitude, 64
asterism, 58
averted vision, 66
azimuth, 64
conjunction, 64
dark adaption, 64
declination, 63
greatest elongation, 64
inferior conjunction, 64
Morning/Evening Star, 64
opposition, 64
right ascension, 63
superior conjunction, 64
synodic period, 64
transit, 64

Atoms, Forces, Light, and How We Learn about the Universe

Planets and stars are so far away that we cannot directly explore them. Instead, we learn about their properties from their motion and the light we receive from them. In the next three chapters we will explore some basic laws of light and motion and learn what they can tell us about distant planets, stars, and galaxies.

Matter

All matter in the Universe is made up of submicroscopic particles called atoms. Each atom is in turn composed of yet more fundamental particles. At the atom's core lies a dense clump of subatomic particles called protons and neutrons that form its nucleus. Around the nucleus orbit yet other subatomic particles called electrons (fig. OV2.1).

About ninety kinds of atoms exist in nature. Each kind differs from the others in the number of protons in its nucleus. For example, hydrogen atoms (which make up most of the Sun's bulk) contain one proton. On the other hand, carbon atoms (which make up much of our bodies) contain 6 protons (fig. OV2.2).

Forces

The matter that composes our Universe is given structure by three forces: gravity, electromagnetism, and the strong force (fig. OV2.3). You see gravity in action when you drop something. On the astronomical scale, you see it at work because gravity holds the Earth, Sun, and all large astronomical bodies together and in their places: the Moon in orbit around the Earth, the Earth in orbit around the Sun, and the Sun in orbit within our galaxy.

FIGURE OV2.1
Schematic view of an atom, showing its nucleus and orbiting electrons.

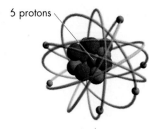

Hydrogen

Carbon

FIGURE OV2.2
Different kinds of atoms, such as hydrogen and carbon, differ because their nuclei contain different numbers of protons.

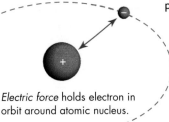

Gravitational force holds Earth and Sun together and Earth in orbit around Sun.

Electric force holds electron in orbit around atomic nucleus.

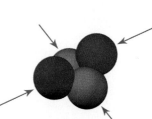

Strong force holds protons and neutrons together in atomic nucleus.

FIGURE OV2.3
The fundamental forces: gravity; electro-magnetism; and the strong or nuclear force.

You can see the electromagnetic force in action when you take clothes from a drier and notice that some stick to-gether or when you stick notes to a refrigerator with a magnet. You see indirectly the electromagnetic force at work in everything around you because it holds an atom's electrons to its nucleus and binds atoms and molecules together to give most things their structure.

The strong force is important only at submicroscopic distances, but it is critical because it holds the protons and neutrons together in an atom's nucleus.

What creates these forces? An object's gravitational force comes from its mass. Because all objects have mass, every object exerts a gravitational tug on every other object (fig. OV2.4). The collective force of all the atoms within the Earth pulls on the atoms in your body to give you weight. Similarly, the collective force of all the atoms in the Sun pulls on all the atoms in the Earth and holds our planet in its orbit around the Sun.

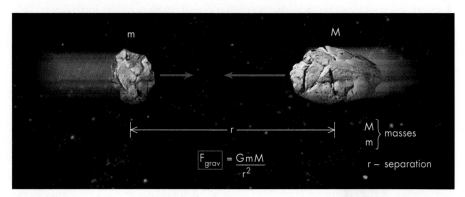

$$F_{grav} = \frac{GmM}{r^2}$$

$\left.\begin{matrix} M \\ m \end{matrix}\right\}$ masses

r – separation

FIGURE OV2.4
The gravitational force.

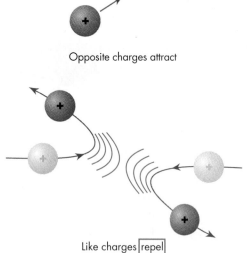

Opposite charges attract

Like charges repel

Although gravity gives the large objects in the Universe their structure, its effects are insignificant within atoms. For these tiny structures, the electric part of the electromagnetic force dominates. That force originates from electric charges—positive and nega-tive—and it can either attract or repel (fig. OV2.5). If the charges are the same (both positive, for example), the force repels. If the charges are opposite, the force attracts.

The attractive force within atoms arises because protons carry a positive charge and electrons carry a negative charge. Thus, the electric force between these oppositely charged particles holds the electrons in their orbits. Similarly, electric forces create the attrac-tion between articles of clothing taken from a dryer. As the clothes tumble, charges rub off some items and collect on others. That im-balance creates an electric force that leads to "static cling."

FIGURE OV2.5
The electromagnetic force.

Motion

Most matter is in motion. The Earth moves around the Sun. Air and water currents stir the atmosphere and oceans of our planet. By experiment and observation, scientists have found that *all* motion obeys fundamental laws. For example, one law states that an object at rest stays at rest unless forces act to set it moving. Similarly, once set in motion, an object always moves in a straight line at a constant speed unless a force acts on it (fig. OV2.6). Thus, once the Earth was set in motion at its origin, it has continued to move. It tries to follow a straight line, but the Sun's gravitational force deflects the Earth into its orbit around the Sun. Because the amount of that deflection depends on the Sun's mass, the path followed by a planet orbiting the Sun gives a clue to the Sun's mass (fig. OV2.7). Similarly, by analyzing the orbital motion of stars and galaxies, astronomers can "weigh" even these immensely distant objects.

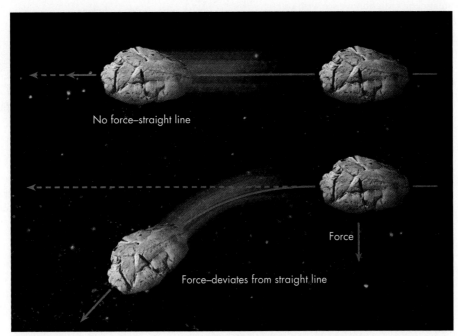

No force–straight line

Force–deviates from straight line

Force

FIGURE OV2.6
A moving object travels in a straight line at constant speed if no net forces act on it. If its path is not straight, a force must be acting on it.

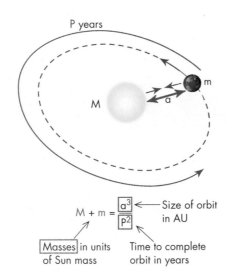

P years

M

m

a

FIGURE OV2.7
Measuring the mass ("weighing") of a star using a modified form of Kepler's third law of planetary motion.

$$M + m = \frac{a^3}{P^2}$$

Size of orbit in AU

Masses in units of Sun mass

Time to complete orbit in years

Light

Our entire knowledge of stars and galaxies comes from the light we receive from them. You can think of that light in two ways: either as a tiny moving wave of electric and magnetic energy or as a stream of submicroscopic particles called photons (fig. OV2.8). Because light may be thought of as a wave of electric and magnetic energy, scientists often refer to it as electromagnetic radiation.

The electromagnetic radiation we see with our eyes is referred to as visible light. Many other forms of "light" exist, however, such as ultraviolet, infrared (heat), and even radio waves. These other forms of light are also electromagnetic radiation; they differ from visible light only in the spacing between the electric and magnetic ripples (their wavelengths) that spread them across space (fig. OV2.9).

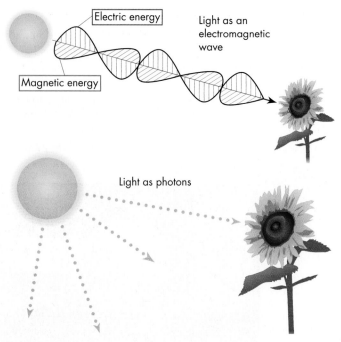

FIGURE OV2.8
Light can be thought of in two ways: as a wave of electric and magnetic energy or as a steam of particles called photons.

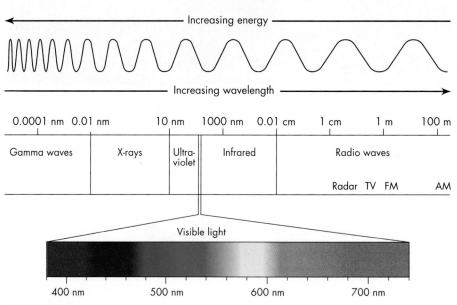

FIGURE OV2.9
The wavelength of electromagnetic radiation (light) sets its color. Our eyes see only a narrow range of wavelengths. Wavelengths much shorter than visible light correspond to x-rays and ultraviolet radiation. Wavelengths longer than visible light correspond to infrared radiation and radio waves.

Until the middle of the twentieth century, astronomers depended almost entirely on visible light to "see" astronomical bodies. Today, however, they routinely observe stars, planets, and galaxies with radio, infrared, ultraviolet, and x-ray wavelengths (fig. OV2.10). For example, with radio waves astronomers can "see" cold, dark clouds of gas in space. With x-rays, they can detect gas swirling around black holes.

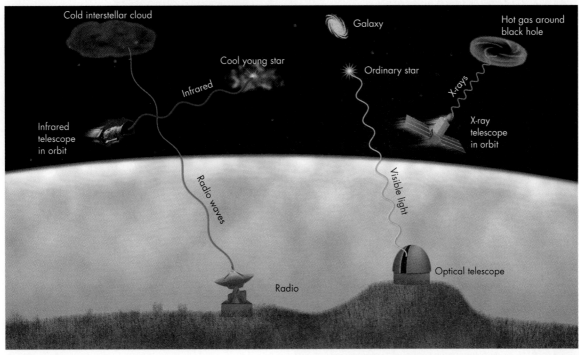

FIGURE OV2.10
Astronomers use nearly all the many wavelengths of electromagnetic radiation to study the Universe.

Telescopes and Instruments

Astronomers observe astronomical objects with tools that are as diverse as the objects themselves. They may collect the faint radio signals from a distant galaxy with a single antenna 1000 feet across or with a set of linked antennas thousands of miles apart (fig. OV2.11). They may collect visible light from a dying star with a telescope whose polished mirror is 30 feet across (fig. OV2.12). They may collect the faint glow of a forming planetary system around some other star with an infrared telescope in orbit above the Earth (fig. OV2.13).

FIGURE OV2.11 A
Radio telescope.

A

FIGURE OV2.11 B
Radio telescope.

B

FIGURE OV2.12
The 10-meter Keck telescope.

FIGURE OV2.13
The Infrared Satellite Observatory, operating in
space because our atmosphere blocks most
infrared energy from reaching the ground.

Astronomers use telescopes for more than just getting pictures of distant stars and galaxies. They can also often learn how hot and how big such objects are. For example, the color of a star gives clues to its temperature, just as the color of a glowing stove burner gives a clue to how hot it is (fig. OV2.14).

The hotter burner glows more yellow than the cooler burner.

FIGURE OV2.14
The color that a hot object glows gives clues to its temperature. Bluer is hotter. Redder is cooler.

Perhaps the most powerful method astronomers use is to spread the light from a planet, star, or galaxy into a spectrum (fig. OV2.15). We see the Sun's spectrum in a rainbow with its colors spread into the familiar red, orange, yellow, green, blue, and violet. Any kind of light can be spread into a spectrum, and because the light is given off by the atoms making up an object, the pattern of the spectrum can reveal the kind of atoms it contains. This is how we know that the Sun and most other astronomical objects are made mainly of hydrogen.

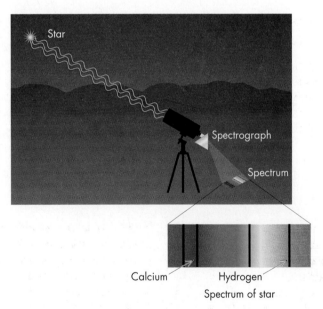

Star

Spectrograph

Spectrum

Calcium   Hydrogen

Spectrum of star

FIGURE OV2.15
The prism in a spectrograph spreads light into its component colors to reveal what substances are emitting the light.

CHAPTER TWO
GRAVITY AND MOTION

The Space Shuttle blasts off (Courtesy of NASA)

Gravity gives the Universe its structure. It is a universal force that acts on all the objects in the Universe so that every particle is drawn toward every other particle by its pull. Gravity holds together astronomical bodies of all sizes, from the Earth to the Universe itself. But the role of gravity extends beyond giving structure to astronomical bodies. Gravity also controls their motions, holding the Earth in orbit around the Sun, the Sun in orbit around the Milky Way, and the Milky Way within the Local Group. Thus gravity and motion are tightly connected in the Universe. This connection is the theme of this chapter.

2.1 SOLVING THE PROBLEM OF ASTRONOMICAL MOTION

Astronomers of antiquity did not make the connection between gravity and astronomical motion that we recognize today. The astronomers were puzzled as to why, if the Earth moved, they and it did not simply fly off into space or fall into the Sun, and they were also mystified about what kept the planets moving.

   The solutions to these mysteries began with a series of careful experiments conducted by Galileo in the 1600s. Apart from his famous—but probably fictitious—demonstration of weights dropped from the Leaning Tower of Pisa, Galileo experimented with projectiles and with balls rolling down planks. Such experiments led him to propose several laws of motion. More importantly, perhaps, these experiments demonstrated the power and importance of the experimental method for verifying scientific conjectures.

2.2 INERTIA

Central to Galileo's laws of motion is the concept of **inertia.** Inertia is the tendency of a body at rest to remain at rest and a body in motion to keep moving. Aristotle noted that bodies at rest resist being moved, but he failed to link this property to the tendency of objects to keep moving once they are set in motion. Kepler also recognized inertia's importance and in fact was first to use that term. However, Galileo not only proposed this property of matter, but also demonstrated it by real experiment.

   In one such experiment, Galileo rolled a ball down a sloping board over and over again and noticed that it always sped up as it rolled down the slope (fig. 2.1). He next rolled the

FIGURE 2.1
A ball rolling down a slope speeds up. A ball rolling up a slope slows down. A ball rolling on a flat surface rolls at a constant speed if no forces act on it.

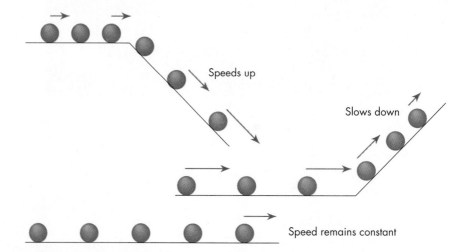

Speeds up

Slows down

Speed remains constant

ball up a sloping board and noticed that it always slowed down as it approached the top. He hypothesized that if a ball rolled on a flat surface and no forces—such as friction—acted on it, its speed would neither increase nor decrease but remain constant. That is, in the absence of forces, inertia keeps an object already in motion moving at a fixed speed. Inertia is familiar to us all even in everyday life. Apply the brakes of your car suddenly, and the inertia of the bag of groceries beside you keeps the bag moving forward at its previous speed until it hits the dashboard or spills onto the floor.

Newton recognized the special importance of inertia. He described it in what is now called **Newton's first law of motion** (sometimes referred to simply as the law of inertia). The law can be stated as follows:

A body continues in a state of rest or uniform motion in a straight line unless made to change that state by forces acting on it.

In applying Newton's first law, we should note two important points. First, we have not defined force yet but have relied on our intuitive feeling that a force is anything that pushes or pulls. Second, we need to note that when we use the term *force* we are talking about *net* force; that is, the total of all forces acting on a body. For example, if a brick at rest is pushed equally by two opposing forces, the forces are balanced. Therefore the brick experiences no net force and accordingly does not move (fig. 2.2).

Newton's first law may not sound impressive at first, but it carries the idea that is crucial in astronomy, that if a body is *not* moving in a straight line at constant speed, some net force *must* be acting on it.

Actually, Newton was preceded in stating this law by the seventeenth-century Dutch scientist Christiaan Huygens. However, Newton went on to develop additional physical laws and—more important for astronomy—showed how to apply them to the Universe. For example, let us look at what happens if we swing a mass tied to a string in a circle. Newton's law tells us that the mass's inertia will carry it in a straight line if no forces act. What force, then, is acting on the circling mass? The force is the one exerted by the string, preventing the mass from moving in a straight line and keeping it in a circle. We can feel that force as a tug on the string, and we can see its importance if we suddenly let go of the string. With the force no longer acting on it, the mass flies off in a straight line, demonstrating the first law, as illustrated in figure 2.3.

Balanced forces = no acceleration

FIGURE 2.2
Balanced forces lead to no acceleration.

Side view

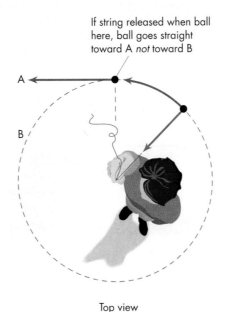

If string released when ball here, ball goes straight toward A *not* toward B

A

B

Top view

FIGURE 2.3
For a mass on a string to travel in a circle, a force must act along the string to overcome inertia. Without that force, inertia makes the mass move in a straight line.

We can translate this example to an astronomical setting and apply it to the orbit of the Moon around the Earth, or of the Earth around the Sun, or to the Sun in orbit around the Milky Way. Each of these bodies follows a curved path. Therefore each must have a force acting on it, the origin and nature of which we will now describe.

2.3 ORBITAL MOTION AND GRAVITY

Newton was not the first person to attempt to discover and define the force that holds planets in orbit around the Sun. Nearly 100 years before Newton, Kepler recognized that some force must hold the planets in their orbits and proposed that something similar to magnetism might be responsible. Newton was not even the first person to suggest that gravity was responsible. That honor belongs to Robert Hooke, another Englishman, who noted gravity's role in celestial motions several years before Newton published his law of gravity in 1687. Newton's contribution is nevertheless special because he demonstrated the *properties* that gravity must have if it is to control planetary motion. Moreover, Newton went on to derive equations that describe not only gravity but also its effects on motion. The solution of these equations allowed astronomers to predict the position and motion of the planets and other astronomical bodies.

According to legend, Newton realized gravity's role when he saw an apple falling from a tree. The falling apple drawn downward to the Earth's surface presumably made him speculate whether Earth's gravity might extend to the Moon. Influenced by an apple or not, Newton correctly deduced that Earth's gravity, if weakened by distance, could explain the Moon's motion.

Most of Newton's work is highly mathematical, but as part of his discussion of orbital motion he described a thought experiment to demonstrate how a body can move in orbit. Thought experiments are not actually performed; rather, they serve as a way to think about problems. In Newton's thought experiment, we imagine a cannon on a mountain peak firing a projectile (fig. 2.4A). From our everyday experience, we know that whenever a body is thrown horizontally, gravity pulls it downward so that its path is an arc. Moreover, the faster we throw, the farther the body travels before striking the ground.

We now imagine increasing the projectile speed more and more, allowing it to travel ever farther. However, as the distance traveled becomes very large, we see that the Earth's surface curves away below the projectile (fig. 2.4B). Therefore if the projectile moves sufficiently fast, the Earth's surface may curve away from it in such a way that the projectile will never hit the ground. Such is the nature of orbital motion and how the Moon orbits the Earth. The balance between inertia and the force of gravity maintains the orbit.

We can analyze this thought experiment more specifically with Newton's first law of motion. According to that law, in the absence of forces, the projectile will travel in a

FIGURE 2.4
(A) A cannon on a mountain peak fires a projectile. If the projectile is fired faster, it travels further before hitting the ground. (B) At a sufficiently high speed, the projectile travels so far that the Earth's surface curves out from under it, and the projectile is in orbit.

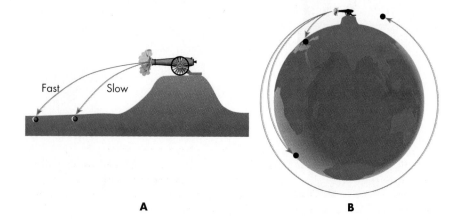

A   B

straight line at constant speed. But because a force, gravity, is acting on the projectile, its path is not straight but curved. Moreover, the law helps us understand that the projectile does not stop because its inertia carries it forward.

Notice that in the above discussion we used no formulas. All we needed was Newton's first law and the idea that gravity supplies the deflecting force. However, if we are to understand the particulars of orbital motion, we require additional laws. For example, to determine how rapidly the projectile must move to be in orbit, we need laws that have a mathematical formulation.

2.4 NEWTON'S SECOND LAW OF MOTION

We showed that an object's inertia means that it will move uniformly in a straight line in the absence of forces. However, suppose forces do act on the object. How much deviation from straight-line motion will such forces produce? To answer that question, we need to define more carefully what we mean by motion.

Motion of an object is a change in its position, which we can characterize in two ways: by the direction of the object and by its speed. For example, a car is moving east at 40 miles per hour. If the car's speed and direction remain constant, we say it is in uniform motion. If the car changes either its speed or direction, it is no longer moving uniformly, as depicted in figure 2.5. Such nonuniform motion is defined as **acceleration.**

Acceleration

We are all familiar with acceleration as a change in speed. For example, when we step on the accelerator in a car and it speeds up from 30 to 40 mph, we say the car is accelerating. Its speed has changed, and its motion is therefore nonuniform. Although in everyday usage acceleration implies an increase in speed, scientifically it simply means a *change* in speed. Thus, technically, a car also "accelerates" when we apply the brakes, and it slows down.

In the above example we produced an acceleration by changing a body's speed.* We can also produce an acceleration by changing a body's direction of motion. For example, suppose we drive a car around a circular track at a steady speed of 30 miles per hour. At

FIGURE 2.5
Views of a car in uniform motion and accelerating. (A) Uniform motion implies no change in speed or direction. The car moves in a straight line at a constant speed. If either an object's (B) speed or (C) direction changes, the object undergoes an acceleration.

Uniform motion
(Same speed, same direction)

A

Acceleration
(A change in speed)

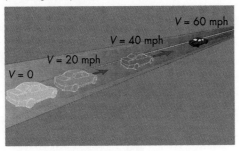

B

Acceleration
(A change in direction)

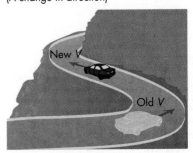

C

*In our discussion, we use the word *speed* to denote the rate of motion, irrespective of its direction. Were we to be more technically correct, we would use the term *velocity,* which means the speed in a *given direction.* Thus a body's velocity changes if *either* its speed or direction changes. With velocity so defined, acceleration is simply a change in velocity.

each moment, the car's direction of travel is changing, and therefore it is not in uniform motion. Similarly, a mass swung on a string or a planet orbiting the Sun is experiencing nonuniform motion and is therefore accelerating. In fact, a body moving in a circular orbit constantly accelerates even though its speed is not changing.

How do we produce an acceleration? Newton realized that for a body to accelerate, a force must act on it. For example, to accelerate—change the direction of—the mass whirling on a string, we must constantly exert a pull on the string. Similarly, to accelerate a shopping cart, we must exert a force on it. In addition, experiments show that the acceleration we get is proportional to the force we apply. That is, a larger force produces a larger acceleration. For example, if we push a shopping cart gently, its acceleration is slight. If we push harder, its acceleration is greater. But experience shows us that more than just force is at work here. For a given push, the amount of acceleration also depends on how full the cart is. A lightly loaded cart may scoot away under a slight push, but a heavily loaded cart hardly budges, as illustrated in figure 2.6. Thus the acceleration produced by a given force also depends on the amount of matter being accelerated.

Mass

The amount of matter an object contains is determined by a quantity which scientists call mass. Technically, **mass** measures an object's inertia. The more inertia, the more mass, and vice versa.

Scientists generally measure the mass of ordinary objects in grams, or kilograms (1 kilogram—abbreviated kg—equals 1000 grams). For example, under normal conditions, a liter of water (roughly a quart) has a mass of 1 kilogram, but it is important to remember that mass is not the same as weight. Because an object's measured mass describes the amount of matter in it, its mass in kilograms is a fixed quantity. An object's weight, however, measures the force of gravity on it, a point we will explore more later. Thus, although a body's mass is fixed, its weight changes if the local gravity changes. For example, on Earth we have one weight but on the Moon, where gravity is less, we have a lesser weight, but no matter where we are we have the same mass.

Mass is the final quantity we need to define before we can understand **Newton's second law of motion** in its full form. Mathematically, the law states that

F = Any force
m = Mass of the body being accelerated
a = Amount of acceleration

$$F = ma$$

or, in words:

The amount of acceleration (a) that a force (F) can produce, depends on the mass (m) of the object being accelerated.

FIGURE 2.6
A loaded cart will not accelerate as easily as an empty cart

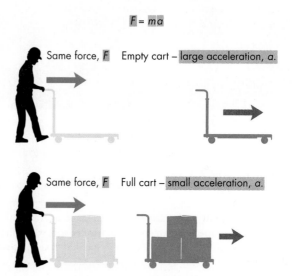

$F = ma$

Same force, F Empty cart – large acceleration, a.

Same force, F Full cart – small acceleration, a.

This astonishingly simple equation allows scientists to predict virtually all features of a body's motion. With $F = ma$ and with knowledge of the masses and the forces in action, engineers and scientists can, for example, drop a spacecraft safely between Saturn and its rings or use a computer to design an airplane that will fly successfully without being tested.

2.5 THE LAW OF GRAVITY

Using Newton's second law, we can now determine an object's motion if we know its initial state of motion and the forces acting on it. For astronomical bodies, that force is often limited to gravity, and so to predict their motion we need to know how to calculate gravity's force. Once again we encounter Newton's work, for it was he who first worked out the **law of gravity**. On the basis of his study of the Moon's motion, Newton concluded the following:

Every mass exerts a force of attraction on every other mass. The strength of the force is directly proportional to the product of the masses divided by the square of their separation.

We can write this extremely important result in a shorthand mathematical manner as follows:

Let m and M be the masses of the two bodies (fig. 2.7) and let the separation between their centers be r. Then the strength of the force between them, F, is

$$F = \frac{GMm}{r^2}$$

The factor G is a constant whose value is found by measuring the force between two bodies of known mass and separation. The resulting number for G depends on the units chosen to measure M, m, r, and F, but once determined, G is the same as long as the same units are used. For example, if M and m are measured in kilograms, r in meters, and F in SI units,* then $G = 6.67 \times 10^{-11}$ meters^3/(kilogram-second^2) [m^3-kg^{-1}-s^{-2}].

Writing the law of gravity as an equation helps us see several important points. If either M or m increases, and the other factors remain the same, the force increases. If r (the distance between two objects) increases, the force gets weaker. Moreover, the force weakens as the square of the distance. That is, if the distance between two masses is doubled, the gravitational force between them decreases by a factor of four, not two. Finally, although one body's gravitational force on another weakens with increasing distance, the gravitational force never completely disappears. Thus the gravitational attraction of a body reaches across the entire Universe, so the Earth's gravity not only holds you on to its surface, but also extends to the Moon and exerts the force that holds the Moon in orbit around the Earth.

2.6 NEWTON'S THIRD LAW

Newton's studies of motion and gravity led him to yet another critical law, which relates the forces that bodies exert on each other. This additional relation, **Newton's third law of motion**, is sometimes called the law of action-reaction. This law states:

When two bodies interact, they create equal and opposite forces on each other.

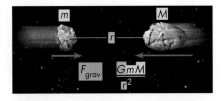

FIGURE 2.7
Gravity produces a force of attraction between bodies. The strength of the force depends on the product of their masses, m and M, and the square of their separation, r. G is the universal gravitational constant.

F = Strength of the force between two bodies
M = Mass of one body
m = Mass of second body
r = Separation between bodies' centers
G = Constant

---

* SI units (System International) use the kilogram, meter, and second as the fundamental units.

FIGURE 2.8
Skateboarders illustrate Newton's third
law of motion. When A pushes on B an
equal push is given to A by B.

FIGURE 2.8
Skateboarders illustrate Newton's third
law of motion. When A pushes on B an
equal push is given to A by B.

Two skateboarders side by side may serve as a simple example of the third law (fig. 2.8). If A pushes on B, *both* move. According to Newton's law, when A exerts a force on B, B exerts a force on A so that both are accelerated.

The gravitational force between the Earth and the Sun affords an astronomical example of Newton's third law and at the same time leads us a step closer to understanding orbital motion. The gravitational force of the Earth on the Sun is exactly equal to the gravitational force of the Sun on the Earth. We can see this perhaps surprising result from Newton's law of gravity where the gravitational force between two bodies depends on the product of their masses. We thus get the same force regardless of whether we let the Earth act on the Sun or vice versa. Why then does the Earth orbit the Sun and not the other way around?

The second law supplies the answer. If we translate $F = ma$ into $a = F/m$, we see that the acceleration a body feels is inversely proportional to its mass; that is, the more massive it is, the more force is required to accelerate the mass. Thus, even though the forces acting on the Earth and Sun are precisely equal, the Sun accelerates 300,000 times less because it is 300,000 times more massive than the Earth. Thus, because the Earth's acceleration is so much larger than the Sun's, the Earth does most of the moving. In fact, however, the Sun does move a little bit as the Earth orbits it, much as you must move if you swing a child around you in play.

2.7 MEASURING A BODY'S MASS USING ORBITAL MOTION

One goal of studying orbits is to relate their properties (for example, orbital radius and velocity) to the properties of the bodies in orbit (such as their masses). To pursue that goal we will consider only the following very simple case of orbital motion: motion in a perfect circle with the orbiting body having a mass so small it can be ignored compared with that of the central body. These restrictions are met to high precision in many astronomical systems, such as the Earth's motion around the Sun and the Sun's motion around the Milky Way. By assuming that the mass of the central body is large compared with the orbiting body, we can ignore the acceleration of the central body (as we just discussed above) and assume it is at rest. These assumptions simplify the problem but are not essential for its solution.

To work out the orbital properties of a body moving around another we use Newton's laws of motion and his law of gravity. From the first law we know that if a body moves along a circular path, there is a net force (an unbalanced force) acting on it because balanced forces give straight-line motion. This force* must be applied to any body moving in a circle, whether it is a car rounding a curve, a mass swung on a string, or the Earth orbiting the Sun.

Using Newton's second law of motion and some algebra gives us an equation that shows that if a mass (m) moves with a velocity (v) around a circle of radius (r), the force needed to hold it in a circular orbit is

$$F = \frac{mv^2}{r}$$

F = Force needed to hold a body in orbit
m = Mass of the body
r = Radius of the orbit
v = Velocity of the body

Using the above equation, we can find the orbital velocity of a planet around the Sun. Let the Sun's mass be M and the planet's mass be m, the latter of which is assumed to be much smaller than M. Assume the planet moves in a circular orbit of radius (r) at a velocity (v) so that the force required to hold the planet in orbit is mv^2/r. That force—the force that deflects the planet from its tendency to move in a straight line—is supplied by the gravitational force between the Sun and the planet. We have already defined that force as GMm/r^2.

Because mv^2/r is the force required to hold the planet in orbit, it must equal the force of gravity.

$$\frac{mv^2}{r} = \frac{GMm}{r^2}$$

We can cancel m out, and one of the r's to obtain

$$v^2 = \frac{GM}{r}$$

Finally, we take the square root of both sides to obtain the orbital velocity.

$$v = \sqrt{\frac{GM}{r}}$$

The above equation, giving the orbital velocity in terms of M and r, can be used to determine the mass of the central body if the orbiting object's velocity and distance from it are known. As an example, we will find the mass of the Sun.

To determine M, we go back to our equation for v^2, cross-multiply by r, and divide by G to obtain

$$M = \frac{v^2 r}{G}$$

Therefore, to evaluate M we need v, the Earth's orbital velocity (the speed with which it is moving in its orbit). We find v by dividing the circumference of its orbit, C, by the time it takes the Earth to complete one orbit, a length of time we call the orbital period, P. Thus

$$v = \frac{C}{P}$$

The formula for the circumference of a circle is $C = 2\pi r$. Thus

$$v = \frac{2\pi r}{P}$$

---

*Sometimes this force is called a "centripetal force," which is the force applied to draw an object *inward* toward the center of the orbit.

To evaluate v, we need to know the Earth's orbital period, which is one year. But for use in the equation, we need P in seconds, a number we can find by multiplying the number of seconds in a minute (60), times the number of minutes in an hour (60), times the number of hours in a day (24), times the number of days in a year (365.25). The result of that calculation, rounded off to three significant figures, is

$$P = 3.16 \times 10^7 \text{ seconds}$$

Now setting $r = 1.5 \times 10^{11}$ meters (the Earth-Sun distance), multiplying by 2π and dividing by P, we find that

$$v = 3 \times 10^4 \text{ m/s} = 30 \text{ km/s}$$
$$(\text{roughly } 70{,}000 \text{ miles/hr})$$

With these values for v and r, and recalling that $G = 6.67 \times 10^{-11} \text{ m}^3\text{-kg}^{-1}\text{-s}^{-2}$, we find that the Sun's mass, M is

$$M = \frac{v^2 r}{G}$$

$$M = \frac{(3 \times 10^4 \text{ m/s})^2 \times 1.5 \times 10^{11} \text{ m}}{6.67 \times 10^{-11} \text{ m}^3\text{-kg}^{-1}\text{-s}^{-2}}$$

$$M = \frac{9 \times 1.5 \times 10^{19} \text{ kg}}{6.67 \times 10^{-11}}$$

$$M = 2 \times 10^{30} \text{ kg}$$

The above method can be used to find the mass of any body around which another object orbits. Thus gravity becomes a tool for determining the mass of astronomical bodies, and we shall use this method many times throughout our study of the Universe.

There are other ways we can use the law of gravity. For example, we can use it to find out how much we would weigh on another planet and how fast a spacecraft must move to escape from a planet's surface.

2.8 SURFACE GRAVITY

Surface gravity measures the gravitational attraction at a planet's or star's surface. It is the *acceleration* on a mass created by the local gravitational force, not the force itself. This acceleration determines how fast objects fall (fig. 2.9). To understand the importance of surface gravity, recall that mass measures the amount of material an object contains and is therefore constant. On the other hand, a body's weight depends on its mass and the acceleration of gravity. Thus surface gravity determines what a mass weighs. In addition to determining weight on a planet, surface gravity also influences a planet's shape and whether it has an atmosphere. For example, small bodies such as asteroids are not spherical because their surface gravity is too weak to compress them into round shapes. Likewise, a small surface gravity makes it hard for a planet to keep an atmosphere.

We determine the strength of a planet's surface gravity as follows.

The law of gravity states that a planet of mass M exerts a gravitational force F on a body of mass m at a distance r from its *center* given by

$$F = \frac{GMm}{r^2}$$

At the planet's surface, $r = R$, the planet's radius, so

$$F = \frac{GMm}{R^2}$$

FIGURE 2.9
The speed of a falling object is set by the local gravity and does not depend on the object's mass. Thus, in a tube pumped free of air to remove the effect of air resistance, a feather and a ball both fall at the same rate, as the photograph shows. (Courtesy of James Sugar/Black Star.)

Newton's second law ($F = ma$) tells us that for any force, F, the acceleration it produces on a body of mass m is $a = F/m$. Therefore,

$$ma = \frac{GMm}{R^2}$$

Canceling out the m's then gives

$$a = \frac{GM}{R^2}$$

Thus a planet's surface gravity depends on its mass and radius. That dependence is such that two planets with the same radius but different masses will have different surface gravities. In particular, if two planets have the same radius but different masses, the planet with the larger mass has the larger surface gravity. Similarly, if two planets have the same mass but different radii, the planet with the larger radius has less surface gravity.

Surface gravity is usually denoted by the letter g (the origin of the phrase "Pulling g's," which is used by pilots). We can therefore write that

$$g = \frac{GM}{R^2}$$

g = Surface gravity
G = Constant
M = Mass of the attracting body
R = Radius of the attracting body

Because the surface gravity depends on the mass and radius of the attracting body, the strength of the surface gravity is different from body to body. For example, we can compare the surface gravity of the Moon with that of the Earth to show why you would weigh less on the Moon than on the Earth.

To make the calculation we need to know the mass and radius of the Earth and the Moon. Those numbers are given in table 2.1.

Because the Earth has greater mass, we might guess that its surface gravity, g, is greater than the Moon's. However, the Moon's radius is smaller, so its surface gravity might be greater. Thus, to determine which body has the larger surface gravity, we need to evaluate g mathematically.

Table 2.1	Mass and Radius of the Earth and Moon*	
	Mass	**Radius**
Earth	6.0×10^{24} kg	6.4×10^{6} meters
Moon	7.3×10^{22} kg	1.7×10^{6} meters
Ratio Earth/Moon	81	3.8

*Numbers have been rounded slightly.

If we put numbers into the equation for surface gravity, we find that

$$g_{\text{Earth}} = \frac{GM}{R^2}$$

$$g_{\text{Earth}} = \frac{6.7 \times 10^{-11} \text{ m}^3\text{-s}^{-2}\text{-kg}^{-1} \times 6.0 \times 10^{24} \text{ kg}}{(6.4 \times 10^6 \text{ m})^2} = 9.8 \text{ m-s}^{-2}$$

and

$$g_{\text{Moon}} = \frac{6.7 \times 10^{-11} \text{ m}^3\text{-s}^{-2}\text{-kg}^{-1} \times 7.4 \times 10^{22} \text{ kg}}{(1.7 \times 10^6 \text{ m})^2} = 1.7 \text{ m-s}^{-2}$$

Thus the ratio of g on the Earth to g on the Moon is about 6 : 1 so that you weigh about 6 times more on the Earth than you would on the Moon. That fact allowed the astronauts to make such huge leaps on the Moon.

You should notice that the above calculation could have been made more quickly by taking the ratio of the g's directly. In that case we would have

$$\frac{g_{\text{Earth}}}{g_{\text{Moon}}} = \frac{GM_{\text{Earth}}/(R_{\text{Earth}})^2}{GM_{\text{Moon}}/(R_{\text{Moon}})^2}$$

$$= \frac{M_{\text{Earth}}/M_{\text{Moon}}}{(R_{\text{Earth}}/R_{\text{Moon}})^2}$$

$$= \frac{(81)}{(3.8)^2} = 5.6 \sim 6$$

To illustrate this point once more, let's compare our weight on the Earth to our weight on Jupiter (assuming we could stand on Jupiter). Jupiter's mass is about 300 times that of the Earth. Its radius is about 10 times larger than the Earth's. The mass ratio contributes a factor of 300 toward a greater g on Jupiter, but its larger radius when squared reduces g by a factor of 100, yielding a net change in g of $300/100 = 3$. Thus the surface gravity on Jupiter is 3 times that of the Earth and a 150-pound person on Earth would weigh about 450 pounds on Jupiter. By the same reasoning, a 10-ton rocket on Earth would weigh 30 tons on Jupiter. Thus, if we ever hope to explore such massive bodies, we must understand how we can overcome greater gravity.

An astronaut can make huge leaps in the Moon's low gravity. (Courtesy NASA.)

2.9 ESCAPE VELOCITY

To overcome a planet's gravitational force and escape into space, a rocket must achieve a critical speed known as the **escape velocity**. Escape velocity is the speed an object needs to move away from a body and not be drawn back by its gravity. We can understand how such a speed might exist if we think about throwing an object into the air. The faster the object is tossed upward, the higher it goes and the longer it takes to fall back. Escape velocity is the speed an object needs so that it will never fall back, as depicted in figure 2.10. Thus, escape velocity is of great importance in space travel if craft are to move away from

FIGURE 2.10
Escape velocity is the speed an object must have to overcome the gravitational force of a planet or star and not fall back.

one body and not be drawn back to it. However, escape velocity is also important in many astronomical phenomena, such as whether a planet has an atmosphere and the nature of black holes.*

The escape velocity, V_{esc}, for a spherical body such as a planet or star, can be found from the law of gravity and Newton's laws of motion and is given by the following formula:

$$V_{esc} = \sqrt{2GM/R}$$

V_{esc} = Escape velocity
G　= Constant
M　= Mass of the body to be escaped from
R　= Radius of the body to be escaped from

Here, M stands for the mass of the body from which we are attempting to escape, and R is its radius, as shown in figure 2.11.

Notice in the equation for V_{esc} that if two bodies of the same radius are compared, the larger mass will have the larger escape velocity. Likewise, if two bodies of the same mass are compared, the one with the smaller radius will have the greater escape velocity. We will see in chapter 14 that the huge escape velocity of a black hole arises from its abnormally small radius.

To illustrate the use of the formula, we calculate the escape velocity from the Moon. From the data in the table 2.1, we find the Moon's radius and mass. We insert these values in the formula for escape velocity and find

$$V_{esc}\,(Moon) = \sqrt{2GM/R}$$
$$V_{esc}\,(Moon) = \sqrt{2 \times 6.7 \times 10^{-11}\ m^3\text{-}s^{-2}\text{-}kg^{-1} \times 7.4 \times 10^{22}\ kg/1.7 \times 10^6 m}$$
$$= 2.4 \times 10^3\ m/s = 2.4\ km/s$$

A similar calculation shows that the escape velocity from the Earth is 11 kilometers per second. Thus, it is much easier to blast a rocket off the Moon than the Earth.** In chapter 6 we will see that this low escape velocity is partly responsible for the Moon's lack of an atmosphere.

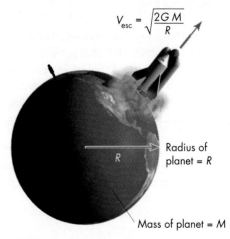

$$V_{esc} = \sqrt{\frac{2GM}{R}}$$

Radius of planet = R

Mass of planet = M

FIGURE 2.11
Calculating the escape velocity from a body.

---

*In chapter 14 we will see that a black hole is an object whose escape velocity equals the speed of light. Thus, light cannot escape from it, thereby making it black.
**If an object's mass and radius are given in units of the Earth's, then $V_{esc} \approx 11\sqrt{M/R}$ km/s.

SUMMARY

A gravitational force exists between any two objects in the Universe. The strength of this force depends on the masses of the bodies and their separation. Gravitational forces hold astronomical bodies together and in orbit about one another.

A body's inertia makes it remain at rest or move in a straight line at a constant speed unless the body is acted on by a net force. Thus, for a planet to orbit the Sun, the Sun's force of gravity must act on it. The law of inertia and Newton's other laws of motion, when combined with the law of gravity, allow us to relate the size and speed of orbital motion to the mass of the central body.

The gravitational force exerted by a planet determines its surface gravity and escape velocity. The former determines your weight on a planet. The latter is the speed necessary to leave the surface and escape without falling back.

QUESTIONS FOR REVIEW

1. What is meant by inertia?
2. What does Newton's first law of motion tell you about the difference between motion in a straight line and motion along a curve?
3. What is Newton's law of gravity?
4. How does mass differ from weight?
5. If your mass is 70 kilograms on Earth, what is it on the Moon?
6. If you weigh 110 pounds on the Earth, do you weigh 110 pounds on the Moon? Why?
7. What does surface gravity measure?
8. What is meant by escape velocity?

THOUGHT QUESTIONS

1. Which do you think has more inertia, a small, inflated balloon or a cinder block? If each were moving toward you at 1 meter/second, which would be easier to catch?
2. A cinder block can be weightless in space. Would you want to kick it with your bare foot? Even if it is weightless, does it have mass?
3. In some amusement park rides, you are spun in a cylinder and are pressed against the wall as a result of the spin. People sometimes describe that effect as being due to "centrifugal force." What is really holding you against the wall of the spinning cylinder? Drawing a sketch and using Newton's first law may help you answer the question.

PROBLEMS

1. How many times greater is the Earth's gravitational force on the Moon than the Moon's gravitational force on the Earth? Think about Newton's third law of motion before answering this.
2. Calculate the escape velocity from the Earth, given that the mass of the Earth is 6×10^{24} kilograms and its radius 6×10^{6} meters. In this problem, round off G to 7×10^{-11} meters^3/(kg-s^2).
3. Calculate the escape velocity from the Sun, given that its mass is 2×10^{30} kg and its radius is 7×10^{8} meters.
4. Which body has a larger escape velocity, Mars or Saturn?

$$M_{\text{Mars}} = 0.1\, M_{\text{Earth}}$$
$$M_{\text{Saturn}} = 95\, M_{\text{Earth}}$$
$$R_{\text{Mars}} = 0.5\, R_{\text{Earth}}$$
$$R_{\text{Saturn}} = 9.4\, R_{\text{Earth}}$$

5. Calculate the ratio of the escape velocities from the Moon and Earth.
6. Calculate your weight on the Moon.
7. Given that Jupiter is about 5 times farther from the Sun than the Earth, calculate its orbital velocity. How many years does it take Jupiter to complete an orbit around the Sun?
8. Given that the mass of the Milky Way galaxy is 10^{11} times that of the Sun and that the Sun is 2.6×10^{20} meters from its center, what is the Sun's orbital speed around the center of the galaxy? How long does it take the Sun to orbit the Milky Way? (In this problem we assume that the galaxy can be treated as a single spherical blob of matter. Strictly speaking this isn't correct, but the far more elaborate math needed to calculate the problem properly ends up giving almost the same answer.)

TEST YOURSELF

1. Which of the following demonstrate the property of inertia?
   (a) A car skidding on a slippery road.
   (b) The oil tanker *Exxon Valdez* running aground.
   (c) A brick sitting on a table top.
   (d) Whipping a table cloth out from under the dishes set on a table.
   (e) All of the above.

2. If an object moves along a curved path at a constant speed, you can infer that
   (a) a force is acting on it.
   (b) it is accelerating.
   (c) it is in uniform motion.
   (d) both (a) and (b) are true.
   (e) neither (a) or (b) is true.

3. If the distance between two bodies is quadrupled, the gravitational force between them is
   (a) increased by a factor of 4.
   (b) decreased by a factor of 4.
   (c) decreased by a factor of 8.
   (d) decreased by a factor of 16.
   (e) decreased by a factor of 64.

4. The gravitational force exerted by the Sun on the Earth is the same as the gravitational force exerted by the Earth on the Sun.
   (a) True.
   (b) False.

5. Two planets have identical diameters but differ in mass by a factor of 25. The more massive planet therefore has an escape velocity
   (a) 25 times larger than the other.
   (b) 25 times smaller than the other.
   (c) 5 times larger than the other.
   (d) 625 times larger than the other.
   (e) 50 times larger than the other.

FURTHER EXPLORATIONS

Duzen, Carl, Jane Nelson, and Jim Nelson. "Classifying Motion." *Physics Teacher* 30 (October 1992): 414.

The following books discuss the laws of motion in greater detail:

Casper, Barry M., and Richard J. Noer. *Revolutions in Physics.* New York: W. W. Norton, 1972.

Hecht, Eugene. *Physics in Perspective.* Reading, Mass.: Addison-Wesley, 1980.

Hewitt, Paul G. *Conceptual Physics.* New York: Harper Collins, 1992.

Kirkpatrick, Larry D., and Gerald F. Wheeler. *Physics: A World View.* Philadelphia: Saunders, 1992.

KEY TERMS

CHAPTER THREE
LIGHT AND ATOMS

Newton studies the Sun's spectrum
(Courtesy of Bausch & Lomb Optical Co.)

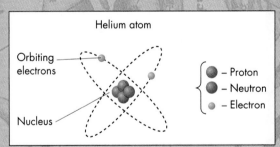

Helium atom

Orbiting electrons

Nucleus

– Proton
– Neutron
– Electron

Sketch of an atom

Our home planet is separated from other astronomical bodies by such vast distances that, with few exceptions, we cannot learn about them by direct measurements of their properties. For example, if we want to know how hot the Sun is, we cannot stick a thermometer into it. Similarly, we cannot directly sample the composition of the atmosphere of Saturn or a distant star. However, we can sample such remote bodies indirectly by analyzing their light. Light from a distant star or planet can tell us what the body is made of, its temperature, and many of its other properties. Light, therefore, is our key to studying the Universe. To use the key, however, we need to understand some of its properties.

In this chapter we will discover that light is a form of energy that can be thought of either as a wave or as a stream of particles. Furthermore, we will discover that the light we see is just part of the radiation emitted by astronomical objects. We will also learn that light can be produced within an atom by changes in its electrons' energies. These changes imprint on the light the atom's "signature." However, the light may also bear unwanted messages. For example, when light reaches our atmosphere, gases there alter its properties, blocking some rays, and bending and blurring others. These distortions place severe limits on what astronomers can learn from the ground.

The goal of this chapter is to explain the nature of light, how it is produced, and how it interacts with our atmosphere. Our first step toward this goal is to better understand what light is.

3.1 PROPERTIES OF LIGHT

Light is radiant energy; that is, it is energy that can travel through space from one point to another without the need of a direct physical link. Therefore, light is very different in its basic nature from, for example, sound. Sound can reach us only if it is carried by a medium such as air or water, whereas light can reach us even across empty space. In empty space we can see the burst of light of an explosion, but we will hear no sound from it at all.

Light's capacity to travel through the vacuum of space is paralleled by another very special property: its high speed. In fact, the speed of light is an upper limit to all motion. In empty space, light travels at the incredible speed of 299,792.5 kilometers per second. An object traveling that fast could circle the Earth in a mere seventh of a second.

The speed of light in empty space is a constant and is denoted by "c." However, in transparent materials, such as glass, water, and gases, the speed of light is reduced. Furthermore, different colors of light are slowed differently. For example, in nearly all materials, blue light travels slightly more slowly than red light. As we will see in chapter 5, lenses and prisms work because they slow the light as it travels through them.

The Nature of Light—Waves or Particles?

Observation and experimentation on light throughout the last few centuries have produced two very different models of what it is and how it works. According to one model, light is an **electromagnetic wave** that is an alternation of electric and magnetic energy, changing in synchrony, as depicted in figure 3.1. The ability of such a wave to travel through empty space comes from the interrelatedness of electricity and magnetism.

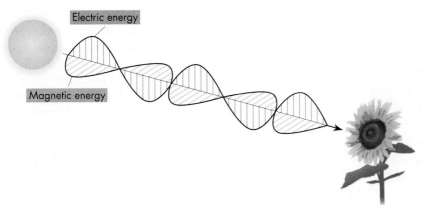

FIGURE 3.1
A wave of electromagnetic energy moves through empty space at the speed of light, 299,792.5 kilometers per second. The wave carries itself along by continually changing its electric energy into magnetic energy and vice versa.

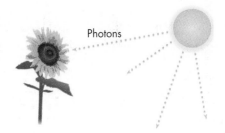

FIGURE 3.2
Photons—particles of energy—stream away from a light source at the speed of light.

You can see this relationship between electricity and magnetism in everyday life. For example, when you start your car, turning the ignition key sends an electric current from the battery to the starter. There, the current generates a magnetic force that turns over the engine. Similarly, when you pull the cord on a lawn mower, you spin a magnet that generates an electric current that creates the spark to start its engine.

This interrelatedness between electricity and magnetism is what allows light to travel through empty space. A small disturbance of an electric field creates a magnetic disturbance in the adjacent space, which in turn creates a new electric disturbance in the space adjacent to it, and so on. Thus a fluctuation of electric and magnetic field spreads out from its source carried by the fields. In this fashion light can move through empty space "carrying itself by its own bootstraps."

As the electromagnetic wave travels through matter, it may disturb the atoms, causing them to vibrate the way a water wave makes a boat rock. It is from such disturbances in our eyes, a piece of film, or an electronic sensor that we detect the light.

The model of light as a wave works well to explain many phenomena, but it fails to explain some of light's properties. In those circumstances it is necessary (and easier!) to use a different model.

In this model, a light source emits a stream of particles called **photons** (fig. 3.2). The photons are packets of energy, and they move in a straight line through space at the speed of light. When they enter our eyes, the photons produce the sensation of light.

The wave model and the photon model are equally useful explanations of the physical properties of light such as brightness, color, and speed. For that reason, scientists often speak of light as having a **wave-particle duality**, and they accordingly use whichever model best describes a particular phenomenon. For example, reflection of light off a mirror is easily understood if you imagine photons striking the mirror and bouncing back just the way a ball rebounds when thrown at a wall. On the other hand, the focusing of light by a lens is best explained by the wave model.

Brightness or intensity of light can be described conveniently by either model. Both brightness and intensity measure the amount of energy carried by the wave. If we imagine light as photons, intensity is proportional to the number of photons traveling in a given direction. If we think of light as a wave, intensity is related to the strength of the wave's vibrating electric and magnetic energy.

In most of the rest of this chapter we will explain light using the wave model, so that we do not have to constantly refer first to photons and then to waves.

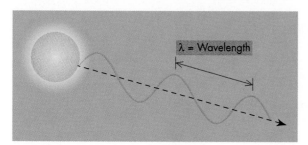

FIGURE 3.3
The distance between crests defines the wavelength, λ , for any kind of wave, be it water or electromagnetic.

Light and Color

Regardless of whether we consider light to be a wave or a stream of photons, our eyes perceive one of its most fundamental properties as color. Human beings can see colors ranging from deep red through orange and yellow into green, blue, and violet. The colors to which the human eye is sensitive define what is called the **visible spectrum.** But what property of photons or electromagnetic waves corresponds to light's different colors?

According to the wave theory, the color of light is determined by the light's **wavelength,** which is the spacing between wave crests (fig. 3.3). That is, instead of describing a quality, light's color, we can specify a quantity, its wavelength, usually denoted by the Greek letter lambda, λ. For example, the wavelength of deep red light is about 7×10^{-7} meters. The wavelength of violet light is about 4×10^{-7} meters. Intermediate colors have intermediate wavelengths. Size also determines the color of light in the photon model: photons of red light are bigger than photons of blue light.

Be sure to remember that shorter wavelengths of visible light correspond to bluer colors. We will use that information *many* times in later chapters.

The wavelengths of visible light are very small (roughly the size of a bacterium). They are therefore usually measured not in meters but in billionths of a meter, a unit called the **nanometer,** abbreviated nm. Thus the wavelength of red light is about 700 nanometers, but that of violet light is about 400 nanometers.* Table 3.1 lists the wavelengths of the primary colors in nm and some other units sometimes used to measure wavelengths.

Characterizing Electromagnetic Waves by their Frequency

Sometimes it is useful to describe electromagnetic waves by their frequency rather than their wavelength. You can see an everyday example of this on a radio dial, where you tune in a station by its frequency rather than its wavelength. **Frequency** is the number of wave crests that pass a given point in 1 second. It is measured in hertz (abbreviated as Hz) and is usually denoted by the Greek letter nu, ν . Frequency and wavelength are related to the wave speed because in one vibration a wave must travel a distance equal to one wavelength. This implies that for light, the product of the wavelength and the wave frequency is the speed of light, that is, $\lambda \nu = c$. Because all light travels at the same speed (in empty space), we can treat c as a constant. Thus, specifying λ determines ν and vice versa. We will generally use λ to characterize electromagnetic waves, but ν is just as good.

TABLE 3.1	Colors and Wavelengths*		
Red	700 nm	0.7 micrometers	7000 angstroms
Yellow	580	0.58	5800
Blue	480	0.48	4800
Violet	400	0.40	4000

*These color equivalences are only approximate.

*Scientists sometimes use other units of length to measure wavelengths. For example, astronomers have traditionally used angstrom units and micrometers. One angstrom unit is 10^{-10} meters (1 ten-billionth of a meter) and is thus the same as 1/10 of a nanometer. The wavelength of red light is thus about 7000 angstroms. The micrometer (also called a micron, and abbreviated μm) is used especially at infrared wavelengths (wavelengths longer than visible light which we perceive as heat). One micrometer is 10^{-6} meters. The wavelength of red light is about 0.7 micrometers.

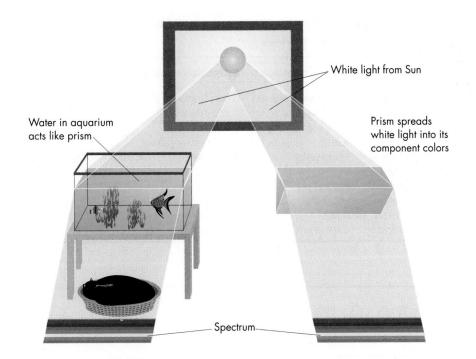

FIGURE 3.4
White light is spread into a spectrum by a prism.

White light from Sun

Water in aquarium acts like prism

Prism spreads white light into its component colors

Spectrum

White Light

Although wavelength is an excellent way to specify most colors of light, some light seems to have no color. For example, the Sun when it is seen high in the sky and an ordinary lightbulb appear to have no dominant color. Light from such sources is called **white light.**

White light is not a special color of light; rather it is a mixture of all colors. That is, the sunlight we see is made-up of all the wavelengths of visible light, a blend of red, yellow, green, blue, and so on, and our eyes perceive these as white. Newton demonstrated this property of sunlight by a very simple but elegant experiment. He passed sunlight through a prism (fig. 3.4) so that the light was spread out into the visible spectrum (or rainbow of colors). He then recombined the separated colors with a lens and reformed the beam of white light.

Why do we see sunlight as colorless? Presumably because our senses have evolved to make us aware of *changes* in our surrounding. Thus we ignore the ambient "color" of sunlight just as we come in time to ignore a steady background sound or smell.

But there is more to light than what meets the eye. Just as red is but one part of the visible spectrum, so too the visible spectrum itself is but one part of a much wider spectrum of electromagnetic radiation.

You can see how colors of light mix if you look at a color television screen close up. You will notice that the screen is covered with tiny red, green, and blue dots. In a red object, only the red spots are lit. In a blue one, only the blue spots. In a white object all three are lit, and the brain mixes these three colors to form white. Other colors are made by appropriate blending of red, green, and blue. Notice this is very different from the way that pigments of paint mix. Red, green, and blue paint when mixed give a brownish color.

3.2 THE ELECTROMAGNETIC SPECTRUM: BEYOND VISIBLE LIGHT

You are already familiar with many other forms of electromagnetic radiation from many different sources. Radio waves, x-rays, and ultraviolet light are all electromagnetic waves that differ from visible light only in their wavelengths. To indicate the fundamental unity of all these kinds of radiation, they are referred to as parts of the **electromagnetic spectrum.** The electromagnetic spectrum is the assemblage of all types of electromagnetic

TABLE 3.2   Electromagnetic Spectrum

Wavelength	Kind of radiation	Astronomical sources
100–500 meters	Radio (AM broadcast)	Pulsars (remnants of exploded stars)
10–100 meters	Short-wave radio	Active galaxies
1–10 meters	TV, FM radio	
1–100 centimeters	Radar	Planets, active galaxies
1–10 millimeters	Microwaves	Interstellar clouds, cosmic background radiation
1000–10^6 nanometers	Infrared (heat)	Young stars, planets, interstellar dust
400–700 nanometers	Visible light	Stars, Sun
1–300 nanometers	Ultraviolet	Stars
0.01–1 nanometers	X-rays	Collapsed stars, hot gas in galaxy clusters
10^{-7}–0.01 nanometers	Gamma rays	Active galaxies and gamma-ray bursters

FIGURE 3.5
The electromagnetic spectrum.

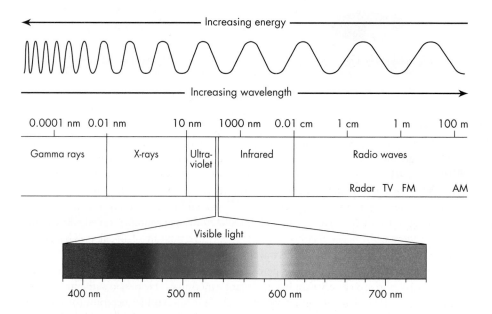

waves arranged according to their wavelength. The longest electromagnetic waves yet detected have wavelengths thousands of kilometers long.* The shortest have wavelengths of 10^{-18} meters or less. Ordinary visible light falls in a very narrow section in about the middle of the known spectral range (see table 3.2 and fig. 3.5).

As you can see in figure 3.5, there is a vast range of wavelengths in the electromagnetic spectrum that we cannot see with our eyes. Nevertheless, the development of various instruments, allows these wavelength regions to be explored. In fact, new instruments allow astronomers to "see" such astronomically important events as the formation of stars, the remnants left behind when stars die, and, indirectly, black holes.

Infrared Radiation

The exploration of the electromagnetic spectrum began in 1800, when Sir William Herschel (discoverer of the planet Uranus) showed that heat radiation, such as you feel from the Sun or from a warm radiator, though invisible, was related to visible light.

---

*Such long waves have not been detected from astronomical sources, however, and cannot pass through our atmosphere.

Herschel was trying to measure heat radiated by astronomical sources. He projected a spectrum of sunlight onto a table top and placed a thermometer in each color to measure its energy. He was surprised that when he put a thermometer just off the red end of the visible spectrum, the thermometer registered an elevated temperature there just as it did in the red part of the spectrum. He concluded that some form of invisible energy perceptible as heat existed beyond the red end of the spectrum and therefore called it **infrared.** Even though our eyes cannot see infrared light, nerves in our skin can feel it as heat.

Ultraviolet Light

Another important part of the electromagnetic spectrum, **ultraviolet** radiation, was discovered in 1801 by J. Ritter while he was experimenting with chemicals that might be sensitive to light. Ritter noted that when he shined a spectrum of sunlight on a layer of silver chloride, the chemical blackened most strongly in the region just beyond the violet end of the spectrum.

Infrared and ultraviolet radiation differ in no physical way from visible light except in their wavelength. Infrared has longer wavelengths and ultraviolet shorter wavelengths than visible light (see table 3.2). Exploration of those parts of the electromagnetic spectrum with wavelengths much larger and much smaller than visible light had to await the growth of new technology, as the development of radio astronomy demonstrates.

Radio Waves

James Clerk Maxwell, a Scottish physicist, predicted the existence of radio waves in the mid-1800s. It was some 20 years later, however, before Heinrich Hertz produced them experimentally in 1888, and another 50 years had to pass before Karl Jansky discovered naturally occurring radio waves coming from cosmic sources. Jansky's discovery in the 1930s that the center of the Milky Way was a strong source of radio emission was the birth of radio astronomy.

Radio waves range in length from millimeters to hundreds of meters, making them much longer than visible and infrared waves. Today we can generate radio waves and use them in many ways, ranging from communication to radar to microwave ovens. Astronomers too use radio waves, detecting them with radio telescopes. Such signals, generated by natural processes, allow astronomers to obtain radio "views" of forming stars, exploding stars, active galaxies, and interstellar gas clouds. Radio wavelengths are even being searched for signals that might hint at the existence of extraterrestrial civilizations (see essay 3).

Other Wavelength Regions

Many decades also passed between the discovery of x-rays by Wilhelm Roentgen in 1895 and detection in the late 1940s of x-rays coming from the Sun. X-ray wavelengths turn out to be far shorter than those of visible light, typically between 0.01 and 10 nanometers, but they too are important. Doctors and dentists use x-rays to probe our bones and organs. Astronomers use x-ray telescopes to detect x-rays emitted by the hot gas surrounding black holes and the tenuous gas in distant groups of galaxies.

Two parts of the electromagnetic spectrum remain relatively unexplored: the region between infrared and radio waves and the region of gamma rays, which are the shortest wavelengths known. Both these wavelength regions are difficult to study from the ground because, as we will see in section 3.7, they fall in wavelength bands that are strongly blocked by the Earth's atmosphere.

Despite the enormous variety of electromagnetic waves, they are all the same physical phenomenon: the vibration of electric and magnetic energy traveling at the speed of light. The essential difference between kinds of electromagnetic radiation is merely their wavelength (or frequency). This difference alters not only how we perceive them, but also how much energy they can carry.

Although humans cannot see infrared radiation, several kinds of snakes, including the rattlesnake, have special infrared sensors located just below their eyes. These allow the snake to "see" in total darkness, helping it to find warm-blooded prey such as rats.

E = Energy carried by an electromagnetic wave

h = Constant

c = Speed of light (constant)

λ = Wavelength

If we describe light by the photon model rather than the wave model, the light's energy is carried by the photons. Each photon is a bundle of energy, or "energy-packet." Similarly, we can talk about ultraviolet photons, radio photons, and so forth.

Energy Carried by Electromagnetic Radiation

The warmth we feel on our face from a beam of sunlight demonstrates that light carries energy, but not all wavelengths carry the same amount of energy. It turns out that the amount of energy, E, carried by an electromagnetic wave depends on its wavelength, λ, such that

$$E = \frac{hc}{\lambda}$$

The speed of light c, and the constant, h, are unchanging,* so if the wavelength of the light decreases, the energy it carries increases. Thus

Short wavelength radiation carries proportionally more energy than long wavelength radiation.

Ultraviolet light with its short wavelength therefore carries proportionally more energy than infrared light with its longer wavelength. In fact, ultraviolet light of sufficiently short wavelengths can carry so much energy that it can break apart atomic and molecular bonds. As we will see in chapter 13, this may cause intense heating of gas near stars. Nearer to home, it is the reason ultraviolet light gives you a sunburn while an infrared heat lamp does not.

Wien's Law: A Wavelength-Temperature Relation

Heated bodies generally radiate across a range of wavelengths, but there is usually a particular wavelength at which their radiation is most intense. That wavelength, which for visible wavelengths affects the color of the body's light, is given by a relation called **Wien's law.**

Wien's law states that the wavelength at which a body radiates most strongly is inversely proportional to the body's temperature. That is,

Hotter bodies radiate more strongly at shorter wavelengths.

As a body is heated, the color of the visible light it emits shifts gradually from red to orange to yellow. At sufficiently high temperatures, its light would look bluish white. This does *not* mean that a very hot body emits only blue light and no red light. Rather, it means that it emits *more* blue than red.

You have almost certainly used Wien's law instinctively and subconsciously when you cook on an electric stove. When the burner of an electric stove first is turned on, it glows dull red, but as it heats up, the *glow becomes a bright orange and eventually yellow.* This effect is the consequence of Wien's law and shows that as a body gets hotter, it emits more light at shorter and hence yellower colors.

We can see this effect illustrated in figure 3.6. In the diagram we see the amount of energy radiated at each wavelength (color) by three bodies of different temperatures. Notice that the hotter body has its most intense emission (highest point) at a shorter wavelength than the cooler body. This is what gives it a different color.

You might note that the wavelength at which the Sun radiates most strongly corresponds to a blue-green color, yet the Sun looks yellow-white to us. The reason we see it as whitish is related to how our eyes perceive color. Physiologists have found that the human eye interprets sunlight (and light from all extremely hot bodies) as whitish, with only tints of color. The light from such hot bodies obeys Wien's law but not with pure colors. Thus cool stars look white tinged with red, while very hot stars look white tinged with blue.

---

*If E is measured in joules and λ in meters, h = 6.63×10^{-34} joule-second (known as Planck's constant) and hc = 1.99×10^{-25} joule-meters.

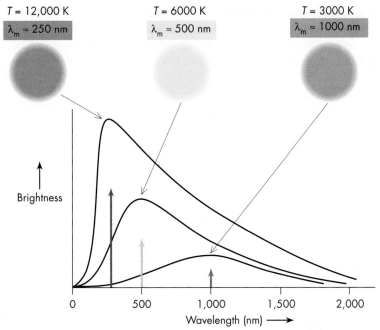

FIGURE 3.6

As a body is heated, the wavelength at which it radiates most strongly, λ_m, shifts to shorter wavelengths, a relation known as "Wien's Law." Thus the color of an electric stove burner changes from red to yellow as it heats up.

EXTENDING OUR REACH

TAKING THE TEMPERATURE OF THE SUN

Wien's law, named for the German physicist who discovered it near the turn of the century, is extremely important because we can use it to measure how hot something is simply from the color of light it radiates most strongly. The law has a few very important exceptions, which we will discuss later, but it works accurately for most stars and planets.

To measure a distant body's temperature using Wien's law, we proceed as follows. First we measure the body's brightness at many different wavelengths to find at which particular wavelength it is brightest (that is, its wavelength of maximum emission). Then we use the law to calculate the body's temperature. To see how this is done, however, we need a mathematical expression for the law.

If we let T be the body's temperature, measured in Kelvin* and λ_m be the wavelength in nanometers at which it radiates most strongly (fig. 3.6).

Wien's law can be written in the form

$$T = \frac{3 \times 10^{6\dagger}}{\lambda_m}$$

The subscript m on λ is to remind us that it is the wavelength of maximum emission.

As an example, let's measure the Sun's temperature. The Sun turns out to radiate most strongly at a wavelength of about 500 nanometers. Then, substituting $\lambda_m = 500$ nanometers, we find

$$T = \frac{3 \times 10^6}{500} = \frac{3 \times 10^6}{5 \times 10^2}$$

$$= 0.6 \times 10^4 = 6000 \text{ Kelvin}$$

This is within a few hundred degrees of the actual value.

---

*See p. 50–51 for discussion of the Kelvin temperature scale.

†The constant 3×10^6 K-nm is more accurately 2.898×10^6. We round it off here to make calculations easier. The error this creates is small.

We must also be careful in applying Wien's law when we look at the light that objects *reflect,* as opposed to light that they emit. For example, the red color of an apple and the green color of a lime come from the light they reflect and have nothing to do with temperature. The apple does emit some radiation, but if it is at normal room temperature, its radiation will be mostly in the infrared, as we will now show.

By cross multiplying, we can change the form of Wien's law to

$$\lambda_{m} = \frac{3 \times 10^{6}}{T}$$

Room temperature is approximately 300 on the Kelvin temperature scale. If we therefore set $T = 300$, we find that the apple, or any room-temperature object, radiates most strongly at

$$\lambda_{m} = \frac{3 \times 10^{6}}{300} = 10{,}000 \text{ nm}$$

an infrared wavelength.

Wien's law makes good sense if you think about the relation between energy and temperature. Hotter things carry more energy (other quantities being equal) than cooler things. Also, bluer light carries more energy than red. Thus it is reasonable to expect that hotter bodies emit bluer light.

Our discussion above has been qualified several times by terms like "usually" and "most." The reason for these qualifications is that Wien's law applies only to a class of objects known as blackbodies.

Blackbodies and Wien's law

A **blackbody** is an object that absorbs all the radiation falling upon it. Because such a body reflects no light, it looks black to us when it is cold; hence its name. Experiments show that when blackbodies are heated, they radiate more efficiently than other kinds of objects. Thus they are both excellent absorbers and excellent emitters. Moreover, the intensity of their radiation changes smoothly from one wavelength to the next with no gaps or narrow peaks of brightness. Very few objects are perfect blackbodies, but many of the objects we will study are near enough to being blackbodies that we can use Wien's law with little fear of its being in error. For example, the electric stove burner, the Sun, and the Earth all obey Wien's law quite satisfactorily.

On the other hand, gases (unless compressed to a very high density) are generally not blackbodies and do not obey Wien's law. Interstellar clouds, for example, can radiate strongly only in narrow wavelength ranges, such as the red part of the visible spectrum or the millimeter wavelength part of the radio spectrum. The clouds' color is determined by composition more than temperature. You can easily demonstrate the importance of composition in determining color with a gas flame on a stove or Bunsen burner. Normally the flame has a blue part and a yellow part. The yellow part is blackbody radiation from very hot specks of carbon soot. However, the blue part is caused by nonblackbody emission from carbon atoms. If you add chemicals to the flame, the flame's color may change dramatically. For example, if you hold some copper sulfate crystals with a pair of pliers in the flame, the flame will take on a greenish-blue color caused by the emission wavelengths of copper.

So far we have described some of the general properties of electromagnetic radiation, such as its wave nature and the existence of different wavelength regions, but we have not explained the origin of the radiation. Radiation originates in matter, and we must thus look more closely at the structure of matter if we are to understand radiation's origin. That is our next goal, and, as we will see, it will lead us not only to an understanding of the nature of atoms, but also to an explanation of how we can detect those atoms in remote astronomical bodies.

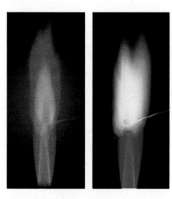

(Courtesy Stephen Frisch.)

Bunsen invented the Bunsen burner for just this purpose: the study of the colors created by chemicals in a flame. His work was instrumental in the development of the basic ideas of spectra as probes of chemical composition.
The photographs above show the effect of strontium (red) and copper (green) on the burner's flame.

3.3 ATOMS

The structure of atoms determines both their chemical properties and their light-emitting and light-absorbing properties. For example, iron and hydrogen not only have very different atomic structures, but also emit very different wavelengths of light. From those differences astronomers can deduce whether an astronomical body—a star or a planet—contains iron, hydrogen, or whatever chemicals happen to be present. Therefore an understanding of the structure of atoms ultimately leads us to an understanding of the nature of stars.

Structure of Atoms

We described in the introduction to this text that an atom has a dense core called a *nucleus* around which smaller particles called *electrons* orbit (fig. 3.7). The nucleus is in turn composed of particles called *protons* and *neutrons;* the protons have a positive charge, the neutrons have no charge, and the electrons have a negative charge. Moreover, it is the positive electrical charge of the protons that attracts and hold the negatively charged electrons in their orbits.

Those orbits are generally extremely small. For example, the diameter of the smallest electron orbit in a hydrogen atom is only about 10^{-10} (one ten-billionth) meter. This infinitesimal size leads to effects that operate at an atomic level that have no counterpart in larger systems. The most important of these effects is that the electron orbits may have only certain prescribed sizes. Although a planet may orbit the Sun at any distance, an electron may orbit an atomic nucleus at only certain distances, as when you climb a set of stairs, you can be only at certain discrete heights. For example, in a hydrogen atom the electron can have an orbital radius of 0.0529 n^2 nanometers where n is 1, 2, 3, etc., but *it cannot have intermediate values;* that is, the orbits are said to be **quantized.**

The above restriction on orbital sizes results from the electron's acting not just as a particle, but also as a wave. That is, just as light itself has a wave-particle duality, so too does an electron. The electron's wave nature forces the electron to move only in orbits whose circumference is a whole number of wavelengths. If it were to move in other orbits, the electron's wave nature would "cancel" it out.

Electrons in orbit have another property totally unlike those of planets in orbit: they routinely shift from one orbit to another. This shifting changes their energy, as can be understood by a simple analogy.

The electrical attraction between the nucleus and the electron creates a force between them like a spring. If the electron increases its distance from the nucleus, the spring must stretch. This requires giving energy to the atom. Likewise, if the electron moves closer to the nucleus, the spring contracts and the atom must give up, or emit, energy. We perceive that emitted energy as light or, more generally, electromagnetic radiation. The wavelength of that radiation is not the same for all atoms. But to understand why, we need to say a bit more about atomic structure. In particular, we need to describe what makes one kind of atom different from another. For example, what makes iron different from hydrogen.

The Chemical Elements

Iron and hydrogen are examples of what are called chemical **elements.** A chemical element is a substance composed only of atoms that all have the same number of protons. For example, hydrogen consists exclusively of atoms that contain 1 proton; helium, of atoms that contain 2 protons; carbon 6; oxygen 8; and so forth. Although the identity of an element is determined by the number of protons in its nucleus, the chemical properties of each element are determined by the number of electrons orbiting its nucleus. However, the number of electrons must equal the number of protons. This means that each atom has an equal number of positive and negative electrical charges and is therefore electrically neutral.

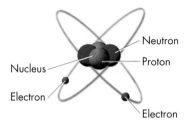

FIGURE 3.7
Sketch of an atom's structure, showing electrons orbiting the nucleus. The electrons are held in orbit by the electrical attraction between their negative charge and the positive charge of the protons in the nucleus. "Orbits" are in reality more like clouds.

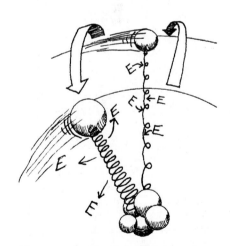

You can see the effect of electrical charges in action when you take clothes out of a dryer. As the clothes tumble in the machine, electrical charges rub off some articles and accumulate on others. These charges remain for a while on the clothes even after the machine is turned off, causing, for example, socks to stick to shirts and lingerie. When you pull them apart, you can feel the attraction of the electricity that makes them stick to each other.

TABLE 3.3	Astronomically Important Elements	
Element	**Number of protons**	**Number of neutrons***
Hydrogen	1	0
Helium	2	2
Carbon	6	6
Nitrogen	7	7
Oxygen	8	8
Silicon	14	14
Iron	26	30

*The number of neutrons listed here is the number found in the most abundant form of the element. Different neutron numbers can occur and lead to what are called *isotopes* of the element.

Table 3.3 lists some of the more important elements we will discuss during our exploration of the Universe and the number of protons each contains.

In summary then, atoms consist of a nucleus containing protons and neutrons around which electrons orbit. The electrons are bound to the nucleus by the electric attraction between the protons and electrons. Electrons may shift from one orbit to another accompanied by a change in the atom's energy. The identity of the atom—the element—is determined by the number of protons in its nucleus. With this picture of the atom in mind, we can now turn to how light is generated within atoms.

3.4 THE ORIGIN OF LIGHT

We saw above that when an electron moves from one orbit to another the energy of the atom changes. If the atom's energy is increased, the electron moves outward from an inner orbit. Such an atom is said to be **excited.** On the other hand, if the electron moves inward toward the nucleus, the atom's energy is decreased.

Although the energy of an atom may change, the energy cannot just disappear. One of the fundamental laws of nature is the **conservation of energy.** This law states that energy can never be created or destroyed, it can only be changed in form. According to this principle, if an atom loses energy, that energy *must reappear in some other form*. One important form in which the energy reappears is light, or, more generally, electromagnetic radiation.

How is the electromagnetic radiation created? When the electron drops from one orbit to another, it alters the electric energy of the atom. As we described earlier, such an electrical disturbance generates a magnetic disturbance, which in turn generates a new electrical disturbance. Thus, the energy released when an electron drops from a higher to a lower orbit becomes an electromagnetic wave, a process called **emission** (fig. 3.8).

Emission plays an important role in many astronomical phenomena. The aurora (northern lights) is an example of emission by atoms in the Earth's upper atmosphere, and sunlight and starlight are examples of emission in those bodies.

The reverse process, in which light is stored in an atom as energy, is called **absorption** (fig. 3.9). Absorption lifts an electron from a lower to a higher orbit and excites the atom by increasing the electron's energy. Absorption is important in understanding such diverse phenomena as the temperature of a planet and the identification of star types, as we will discover in later chapters.

Emission and absorption are particularly easy to understand if we use the photon model of light. According to this model, an atom emits a photon when one of its electrons drops from an upper to a lower orbit. Similarly, an atom absorbs light when a photon collides with it and "knocks" one of its electrons into an upper level.

You may find it helpful in understanding emission and absorption if you think of an analogy. Absorption is a bit like drawing an arrow back preparatory to shooting it from a bow. Emission is like the arrow being shot. In one case, energy of your muscles is transferred to and stored in the flexed bow. In the other, it is released as the arrow takes flight.

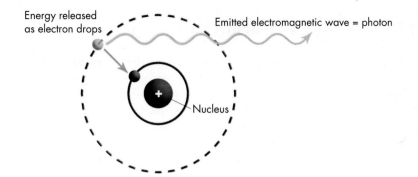

Energy released as electron drops

Emitted electromagnetic wave = photon

Nucleus

FIGURE 3.8
Energy is released when an electron drops from an upper to a lower orbit, causing the atom to emit electromagnetic radiation.

FIGURE 3.9
An atom can absorb light, using the light's energy to lift an electron from a lower to a higher orbit. To be absorbed, the energy of the light's photons must equal the energy difference between the atom's electron orbits. In this example, the green light's energy matches the energy difference, but the red and blue light's energy does not. Therefore only the green is absorbed.

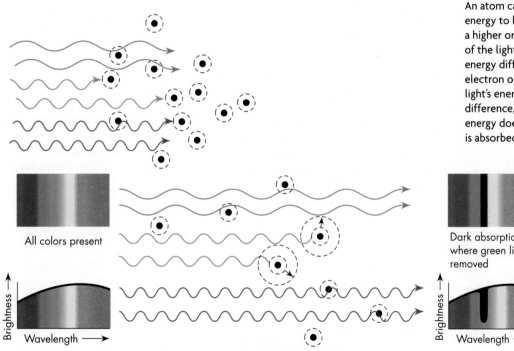

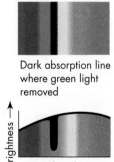

All colors present

Brightness

Wavelength

Dark absorption line where green light removed

Brightness

Wavelength

3.5 FORMATION OF A SPECTRUM

The key to determining the composition and conditions of an astronomical body is its spectrum. The technique used to capture and analyze such a spectrum is called **spectroscopy**. In spectroscopy, the light (or more generally the electromagnetic radiation) emitted or reflected by the object being studied is collected with a telescope and spread into its component colors to form a spectrum by passing it through a prism or a grating consisting of numerous, tiny, parallel lines (fig. 3.10). Since light is emitted from atoms as electrons shift between orbits, we might expect that the light will bear some imprint of the kind of atom that creates it. That is usually the case, and astronomers can search for the atom's "signature" by measuring how much light is present at each wavelength.

Spectroscopy is such an important tool for astronomers that we should look in greater detail at how it works. Specifically, why does an atom produce a unique spectral signature? To understand that, we need to recall how light is produced.

How a Spectrum Is Formed

We saw earlier that each kind of atom has a different number of electrons. This means that each kind of atom has a different set of electron orbits. Figure 3.11 shows schematically some of the possible orbits of the electron of a hydrogen atom and a few of the orbits of the two electrons in a helium atom. Because the atom's energy determines what orbit its electrons move in, orbits are sometimes referred to as **energy levels.**

When an electron moves from one energy level (orbit) to another, the atom's energy changes by an amount equal to the difference in the energy between the two levels. As an example, suppose we look at light from heated hydrogen. Heating speeds up the atoms, causing more forceful and frequent collisions, knocking each excited atom's electron to outer orbits. However, the electrical attraction between the nucleus and the electron draws the electron back almost at once. Suppose we look at an electron shifting from orbit 3 to orbit 2, as shown in figure 3.12. As the electron shifts downward, the atom's energy decreases, and the energy lost appears as light.

FIGURE 3.10
Sketch of a spectroscope and how it forms a spectrum. Either a prism or grating may be used to spread the light into its component colors.

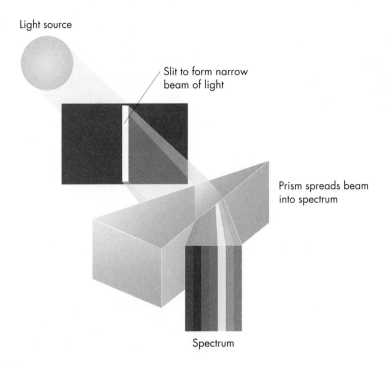

Light source

Slit to form narrow beam of light

Prism spreads beam into spectrum

Spectrum

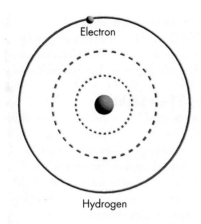

FIGURE 3.11
Sketch of electron orbits in hydrogen and helium.

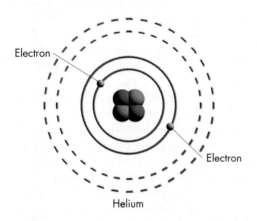

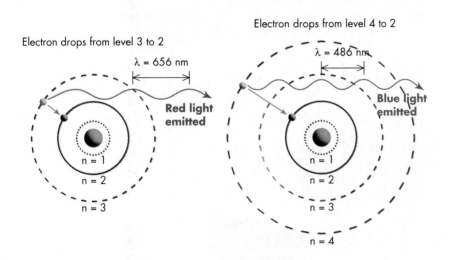

FIGURE 3.12
Emission of light from a hydrogen atom. The energy of an electron dropping from an upper to lower orbit is converted to light.

The wavelength of the emitted light can be calculated from the energy difference of the levels and the relation we mentioned earlier between energy and wavelength ($E = hc/\lambda$). If we evaluate the wavelength of this light, we find that it is 656 nanometers, a bright red color. An electron dropping from orbit 3 to orbit 2 in a hydrogen atom will always produce light of this wavelength.

If instead the electron moves between orbit 4 and orbit 2, there will be a different change in energy because orbit 4 has a different energy from that of orbit 3. That different energy will have a wavelength different from 656 nanometers. A calculation of its energy change leads in this case to a wavelength of 486 nanometers, a turquoise blue color.

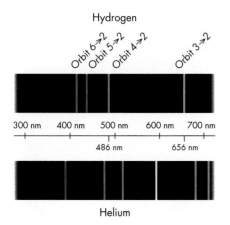

Hydrogen

Orbit 6→2
Orbit 5→2
Orbit 4→2
Orbit 3→2

300 nm   400 nm   500 nm   600 nm   700 nm
                    486 nm   656 nm

Helium

FIGURE 3.13
The emission spectra of hydrogen and of helium.

FIGURE 3.14
Cool gas between an observer and a hotter source of light creates an absorption-line spectrum. Atoms in the cool gas absorb only those wavelengths whose energy equals the energy difference between their electron orbits. The absorbed energy lifts the electrons to upper orbits. The lost light makes the spectrum darker at the wavelengths where it is absorbed.

If we made a similar calculation for a different kind of atom, we would discover that its wavelengths in general differ from those of hydrogen (see fig. 3.13). Thus a signature of hydrogen is its red 656-nanometer and blue 486-nanometer lines, and that signature offers astronomers a way to determine what astronomical objects are made of.

Identifying Atoms by Their Light

If we spread the light from a hot gas into a spectrum, we will see that in general the spectrum contains light at only certain wavelengths. For example, if the gas is hydrogen, we will see in the spectrum the red and turquoise blue colors described above as well as violet light corresponding to electrons dropping from orbit 5 to 2 (fig. 3.13). We will see no light at most other colors because hydrogen has no electron orbits corresponding to those energies. Therefore, the spectrum shows a set of brightly colored lines separated by wide, dark gaps. This is how an emission-line spectrum is formed.

Other kinds of atoms have other electron orbits and therefore other energy levels. This makes them emit light at different wavelengths. For example, if we were to look at a gas of heated helium atoms, we would again see a bright line spectrum, but because the orbits in helium atoms differ from those in hydrogen atoms, the helium spectrum is different from the hydrogen spectrum, as you can see in figure 3.13. The spectrum thus becomes a means of identifying what atoms are present in a gas.

It is also possible to identify atoms in a gas from the way they absorb light. Light is absorbed if the energy of its wavelength corresponds to an energy that matches the difference between two energy levels in the atom. If the wavelength *does not* match, the light will not be absorbed, and it will simply move past the atom, leaving itself and the atom unaffected.

For example, suppose we shine a beam of light that initially contains all the colors of the visible spectrum through a box full of hydrogen atoms. If we examine the spectrum of the light after it has passed through the box we will find that certain wavelengths of the light have been removed and are missing from the spectrum (fig. 3.14). In particular, the spectrum will contain gaps that appear as dark lines at 656 nanometers and 486 nanometers, precisely the wavelengths at which the hydrogen atoms emit. The absorption spectrum is, in effect, the opposite of the emission spectrum.

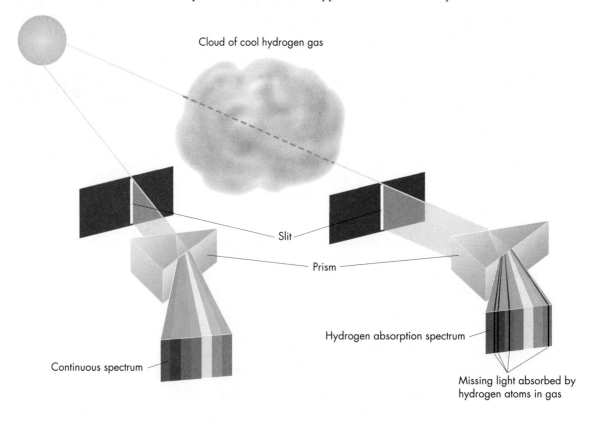

Cloud of cool hydrogen gas

Slit

Prism

Hydrogen absorption spectrum

Continuous spectrum

Missing light absorbed by hydrogen atoms in gas

These gaps are created by the light at 656 nanometers and 486 nanometers interacting with the hydrogen and lifting electrons from orbit 2 to 3 and orbit 2 to 4, respectively. Light at other wavelengths has no effect on the atom.

Thus we can tell that hydrogen is present from either its emission or its absorption spectral lines. In our discussion above we have considered light emitted and absorbed by individual atoms in a gas. If the atoms are linked to one another to form molecules, such as water or carbon dioxide, they too produce emission and absorption lines. In fact even solid objects may imprint spectral lines on light that reflects off them.

In the above examples we have considered light emitted from or absorbed by the body we wish to identify. But even reflected light generally bears some imprint of the surface from which it came. For example, when light from the Sun reflects from an asteroid, spectral features appear that were not present in the original sunlight. This gives astronomers information about the surface composition of bodies too cool to emit significant light of their own.

We conclude that in general we can identify the kind of atoms or molecules that are present by examining either the bright or dark spectrum lines. Gaps in the spectrum at 656 nanometers and 486 nanometers imply that hydrogen is present. Similar gaps at other wavelengths would show that other elements are present. By matching the observed gaps to a directory of absorption lines we can identify the atoms that are present. *This is the fundamental way astronomers determine the chemical composition of astronomical bodies.*

Types of Spectra

Although a spectrum may bear the imprint of the matter that emitted the light, it may also have certain general properties. For example, the spectrum of a hot, tenuous gas is always different from that of a hot dense solid, regardless of the composition of either the gas or the solid. Therefore it will be helpful to learn to recognize such gross properties of spectra as well.

Spectra have the three basic forms listed below:

1. For some sources, the light is emitted in such a way that the intensity changes smoothly with wavelength and all colors are present. We say such a light has a **continuous spectrum** (fig. 3.15A). For a source to emit a continuous spectrum, its atoms must in general be packed so closely that the electron orbits of one atom are distorted by the presence of neighboring atoms. Such conditions are typical of solid or dense objects such as the heated filament of an incandescent lightbulb or a nail heated by a blowtorch.
2. Some heated objects have a spectrum in which light is emitted at only a few particular wavelengths while most of the other wavelengths remain dark (fig. 3.15B). This type of spectrum is called an **emission-line spectrum.** Emission-line spectra are usually produced by hot, tenuous gas, such as that in a fluorescent tube, the aurora, and many interstellar gas clouds.
3. A still different type of spectrum arises when light from a hot dense body passes through cooler gas between it and the observer. In this case nearly all the colors are present, but light is either missing or much dimmer at some wavelengths (fig. 3.15C). This causes the bright background to be crossed with narrow dark lines where the light of some colors is fainter or absent altogether. The resulting spectrum is therefore called a **dark-line** or **absorption-line spectrum.**

Absorption lines were first detected astronomically in 1802 when the English scientist William H. Wollaston viewed sunlight through a prism and a narrow slit. Interested in learning whether each color blended smoothly into its neighbor (red into orange, orange into yellow, and so forth), he noticed dark lines between some of the colors but paid little attention to them. These dark lines in the Sun's spectrum were independently discovered a few years later by the German scientist Joseph Fraunhofer, who catalogued them and discovered similar lines in other stars. In fact, because nearly all stars have absorption-line

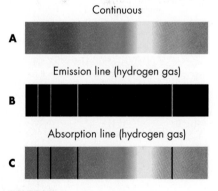

FIGURE 3.15
Types of spectra: (A) continuous, (B) emission-line, and (C) absorption-line.

A

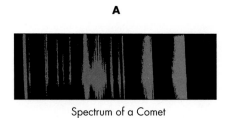

Spectrum of a Comet

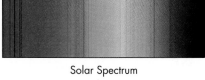

Solar Spectrum

B

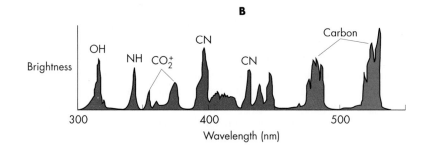

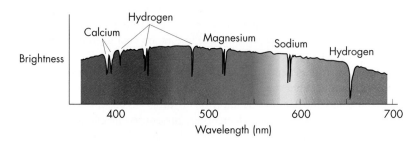

FIGURE 3.16
(A) The spectrum of a comet (Courtesy
Stephen M. Larson, University of Arizona)
and the Sun. (Courtesy Mees Solar Obs.,
U. of Hawaii.) (B) Graphical representation
of their spectra.

spectra, this spectrum type is especially important in astronomy. However, if we consider the physical process of spectrum formation, such absorption-line spectra are really just a special case of continuous spectra with light missing at some wavelengths.

Depicting Spectra

Spectra can be displayed in several ways. One method is simply to take a photograph of the light after it has been spread out by a prism or grating. Figure 3.16A shows two spectra, one an emission-line spectrum from a comet and the other an absorption-line spectrum from the Sun. Notice that in the former there are bright regions separated by large regions where there is no light, but in the latter there are certain very narrow regions (the dark absorption lines) where there is little or no light.

Another extremely useful way to depict a spectrum is to plot the brightness of the light at each wavelength. Figure 3.16B shows the spectra of the comet and the Sun illustrated above depicted in this latter fashion.

Analyzing the Spectrum

Regardless of how the spectrum is depicted, the first step facing an astronomer is to identify the spectral features. This is done by measurement of the wavelengths and then consultation of a directory of spectral lines. By matching the wavelength of the line of interest to a line in the table, astronomers can determine what kind of atom or molecule created the line. A look at a typical spectrum will show you that some lines are hard to see, being faint and weak. On the other hand, some lines may be very obvious and strong. The strength or weakness of a given line turns out to depend on the number of atoms or molecules absorbing (or emitting, if we are looking at an emission line) at that wavelength. That is, a faint line implies that the object contains few atoms capable of absorbing at that wavelength, while a strong line implies that the object contains many such atoms. This makes it possible for astronomers to calculate the chemical abundance (the number of atoms of a given kind) present in the object. From the abundance of each kind of atom or molecule astronomers can then tabulate the percentage of each different substance in the light source and thereby determine its composition. Table 3.4 shows typical results.

TABLE 3.4	Composition of a Typical Star, our Sun*	
Element	**Relative number of atoms**	**Percent by mass**
Hydrogen	10^{12}	70.6%
Helium	9.77×10^{10}	27.4%
Carbon	3.63×10^8	0.31%
Nitrogen	1.12×10^8	0.11%
Oxygen	8.51×10^8	0.96%
Neon	1.23×10^8	0.18%
Silicon	3.55×10^7	0.07%
Iron	4.68×10^7	0.18%
Gold	10.0	1.4×10^{-7}%
Uranium	less than 0.3	less than 5.7×10^{-9}%

*Only the most common elements are listed. Gold and uranium are included only to illustrate how extremely rare they are. Notice that they are a million or more times rarer than iron, which is itself thousands of times rarer than hydrogen. (From Anders and Grevesse, *Geochim Cosmochim Acta* 53 [1989]:197.)

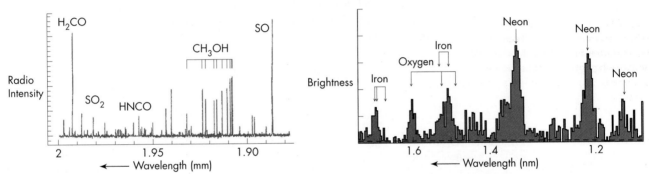

FIGURE 3.17
(A) A radio spectrum of a cold interstellar cloud. (Courtesy Doug McGonagle, FCRAO.) (B) An X-ray spectrum of hot gas from an exploding star. (Courtesy P. F. Winkler, Middlebury College.)

Astronomical Spectra

Let us now apply what we know about spectra to astronomical bodies. We begin by using a telescope to obtain a spectrum of the object of interest. Next we measure the wavelengths and identify the lines. As an example, consider the spectrum of the Sun in figure 3.16A.

We can see from the spectral lines that the Sun contains hydrogen. In fact, when a detailed calculation is made of the strength of the lines, it turns out that the Sun is about 71% hydrogen.* Similar observations show that the spectrum of a comet consists mainly of emission lines from such substances as carbon dioxide and CN. Thus, we know that comets contain these substances. Moreover, recalling our earlier discussion of types of spectra (continuous, emission-line, or absorption-line), we can tell that the CN and carbon dioxide must be gaseous because the spectrum consists of emission lines. There may be other gases present too, but without seeing their spectral features, we cannot tell for sure.

Although the examples we have used above involve spectra of visible light, one of the most useful features of spectroscopy is that it may be used in *any* wavelength region. For example, figure 3.17 shows a radio spectrum and an x-ray spectrum.

Regardless of the wavelength region we use, the spectrum allows us to determine what kind of material is present. In addition, it can sometimes reveal the speed of motion of that material.

---

*Expressed in a slightly different way, this means that roughly 9 out of 10 atoms in the Sun are hydrogen. Hydrogen contributes only about 71% of the mass because it is the lightest element.

3.6 THE DOPPLER SHIFT

If a source of light is set in motion, its spectral lines shift to new wavelengths in a phenomenon known as the **Doppler shift.** If the light source moves away from us, we observe its wavelengths as longer, while if the source approaches, we observe that the wavelengths decrease, as illustrated in figure 3.18A. The amount of shift we observe depends on the source's speed, so by measuring the shift we can determine how fast the source is moving toward or away from us. Astronomers refer to the speed of such motion toward or away from an observer as radial velocity.

The Doppler shift occurs for any relative motion between source and observer and for *any* kind of wave. For example, you have probably heard the Doppler shift of sound waves as the horn of a passing car changes pitch (fig. 3.18B). Likewise, you may have been caught in a speed trap when a police officer bounced radar waves off your car and from their observed shift determined that you were speeding (fig. 3.18C). But why does motion create a shift in wavelength?

A simple way to think of how the Doppler shift occurs is to imagine that when a light source moves away from you it "stretches" the waves—think of a toy Slinky being stretched—and that when the source moves toward you it "compresses" them, as shown in figure 3.18D. Notice that the stretching or compression occurs if either you or the source moves.

FIGURE 3.18
(A) The Doppler shift: waves appear to shorten as a source approaches and lengthen as it recedes. (B) Doppler shift of sound waves from a passing car. (C) Doppler shift of radar waves in a speed trap. (D) A Slinky illustrates the shortening of the space between its coils as its ends move toward each other and a lengthening of the space as the ends move apart.

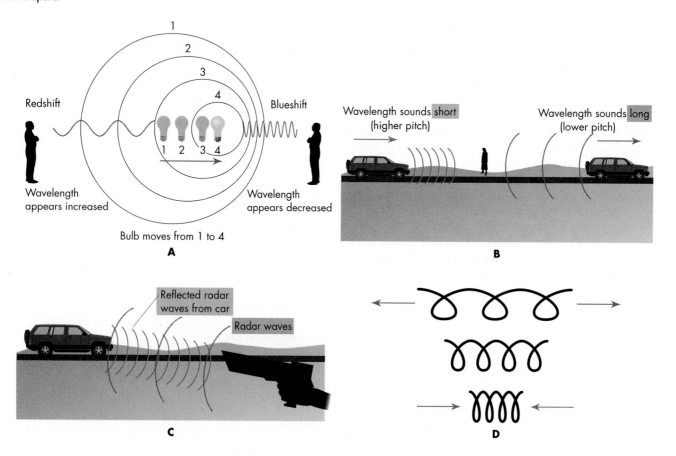

Mathematically, the wavelength shift occurs because the wavelength we observe, λ, is a combination of the true wavelength, λ_0, plus the distance the source moves during the time a single wave is emitted. With this interpretation and a little algebra we can derive a formula for the radial velocity of the source, v, given its wavelength shift, $\lambda - \lambda_0$. The result of such a calculation is that the wavelength shift, often written $\Delta\lambda$,* is

$$\Delta\lambda = (\lambda - \lambda_0) = \frac{\lambda_0 v}{c}$$

$\Delta\lambda$ = Wavelength shift
λ = Measured wavelength (what we observe)
λ_0 = Emitted wavelength
c = Speed of light
v = Velocity of source along the line of sight (radial velocity)

Therefore, if we can measure the wavelength shift, we can solve for the source's radial velocity (that is, its velocity along the line of sight to the object).

Astronomers often refer to Doppler shifts that increase the measured wavelength as redshifts and those that decrease the measured wavelength as blueshifts, regardless of whether the waves are of visible light. Thus, even though we may be describing radio waves, we will say that an approaching source is blueshifted and a receding one is redshifted. We will save for later chapters how to measure the shift and calculate v.

The previous sections show how important a source of information electromagnetic radiation is. It provides information about the temperature, composition, and motion of astronomical objects. However, we often observe that radiation after it has passed through our atmosphere. Thus we must ask how the atmosphere affects the radiation passing through it.

3.7 ABSORPTION IN THE ATMOSPHERE

The gases in the Earth's atmosphere absorb electromagnetic radiation, affecting the flow of heat and light through it. Very little visible light is absorbed, and so visible light can pass through the atmosphere and reach the ground with little difficulty. However, infrared and ultraviolet radiation are strongly absorbed by carbon dioxide and water vapor, whereas x-rays and gamma rays are absorbed by oxygen and nitrogen. Molecules in general can absorb and emit radiation not only because of their changing electron levels but also because the entire molecule spins or vibrates. This ability to store energy by spinning or vibrating as well as by the orbital motion of electrons gives molecules a very strong absorbing power. Thus, even a relatively small number of molecules can absorb infrared wavelengths so well that the atmosphere lets almost no infrared radiation through. Similarly, ordinary oxygen (O_2) and ozone molecules (O_3) absorb ultraviolet radiation. These gases absorb so strongly that virtually no electromagnetic radiation with a wavelength shorter than about 300 nanometers gets through the Earth's atmosphere.

The transparency of the atmosphere to visible light compared to its opacity (nontransparency) to infrared and ultraviolet radiation creates what is called an **atmospheric window.** An atmospheric window is a wavelength region in which energy comes through easily compared to other wavelengths (fig. 3.19).

Without atmospheric windows, it would be impossible for us to study astronomical objects from the ground. As it is, the visible window allows us to study stars and galaxies (which radiate lots of visible energy), but the lack of ultraviolet and the rarity of infrared windows makes it very difficult to observe objects that radiate strongly in those spectral regions. This is one of several reasons why astronomers so badly need telescopes in space where there is no absorption by our atmosphere.

---

*The Greek letter Δ, or delta, is widely used to stand for "the change in quantity."

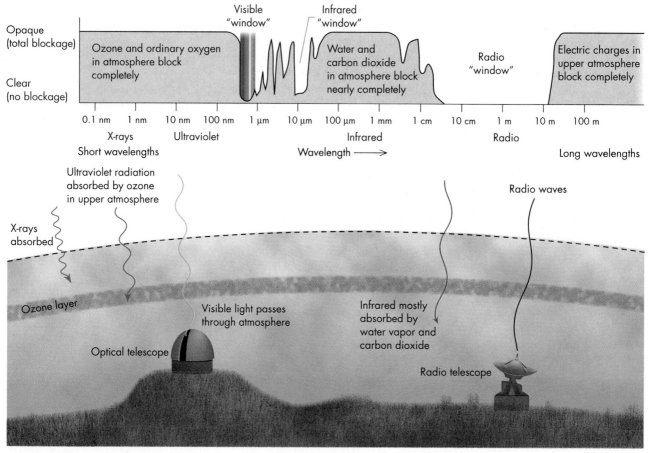

FIGURE 3.19
Atmospheric absorption. Wavelength regions where the atmosphere is essentially transparent, such as the visible spectrum, are called "atmospheric windows." Wavelengths and atmosphere not drawn to scale.

SUMMARY

Light can be described in two complementary ways: as a stream of particles called photons, or as electromagnetic waves. In the wave picture, the energy increases as the wavelength decreases. The wavelength of light determines its color. Red light has a longer wavelength than blue light. In addition to the electromagnetic radiation that we see as visible light, many electromagnetic waves are invisible to the eye, such as infrared, ultraviolet, radio, x-rays, and gamma rays. The entire assemblage of electromagnetic waves is called the electromagnetic spectrum.

Energy can be absorbed by or released from an atom when an electron moves to a higher or lower orbit, respectively. If an electron drops from an upper to a lower orbit, the energy appears as light. If light of the appropriate energy (wavelength) hits an atom, it may lift an electron in the atom from a lower to an upper orbit. The generation of light is called emission. The removal of light from a beam of radiation is called absorption.

Each atom has a unique set of wavelengths at which it emits and absorbs. These create the atom's spectrum and allow it to be identified. Motion of the emitting material alters the wavelengths, creating a Doppler shift from which the material's speed and direction of motion can be deduced.

Atmospheric gases also absorb light although there is little absorption in the wavelengths of the visible atmospheric window. Carbon dioxide and water absorb infrared radiation, and oxygen and ozone in our upper atmosphere absorb ultraviolet radiation. These gases therefore strongly hinder observations of astronomical objects at infrared and ultraviolet wavelengths from the ground.

QUESTIONS FOR REVIEW

1. Why is light called electromagnetic radiation?
2. How are color and wavelength related?
3. What is a photon?
4. Name the regions of the electromagnetic spectrum from short to long wavelengths.
5. What is the difference between emission and absorption in terms of what happens to an electron in an atom?
6. What is the Doppler shift?
7. Which gases in the atmosphere absorb infrared radiation? Which gases absorb ultraviolet?
8. What is meant by the *electromagnetic spectrum*?
9. What are some of the things astronomers can learn about astronomical objects from their spectra?

THOUGHT QUESTIONS

1. Why don't atoms emit a continuous spectrum?
2. How can you tell what sort of gas is emitting light?
3. How would a spectrum help you learn what the atmosphere of Venus is made of ?
4. If you added more water or carbon dioxide to our atmosphere, how would it alter the loss of heat from our planet? Would you expect the Earth to get warmer or colder? Why?
5. Given that water absorbs microwaves very strongly, can you explain why a Pop-Tart gets very hot inside while its crust stays cool if you heat it in a microwave oven?
6. Your body temperature is about 300 K. At what wavelength do you radiate most strongly? What region of the electromagnetic spectrum is this? Do you understand now how a rattlesnake can bite you in the dark?
7. Why do night-vision cameras use infrared detectors?
8. If you were to look at the spectrum of the gas flame of a stove or the blue part of a Bunsen burner flame, what sort of spectrum would you expect to see—Absorption, emission, or continuous? Why?
9. Can you explain why the atmospheric layer containing ozone is much warmer than the levels above and below it?

PROBLEMS

1. Given that the Sun is 150 million kilometers from the Earth, how long does it take light to travel from the Earth to the Sun?
2. Suppose you are operating a remote-controlled spacecraft on Mars from a station here on Earth. How long will it take the craft to respond to your command if Mars is at its nearest point to Earth? You will need to look up a few numbers here.
3. Sketch an atom emitting light. Does the electron end up in a higher or lower orbit? Repeat for an atom absorbing light.
4. A lightbulb radiates most strongly at a wavelength of about 3000 nanometers. How hot is its filament?

TEST YOURSELF

1. Which type of electromagnetic radiation has the longest wavelength?
   (a) Ultraviolet.
   (b) Visible.
   (c) X-ray.
   (d) Infrared.
   (e) Radio.

2. Which kind of photon has the highest energy?
   (a) Ultraviolet.
   (b) Visible.
   (c) X-ray.
   (d) Infrared.
   (e) Radio.

3. An astronomer finds that the spectrum of a mysterious object shows bright emission lines. What can she conclude about the source?
   (a) It contains cold gas.
   (b) It is an incandescent solid body.
   (c) It is rotating very fast.
   (d) It contains hot, relatively tenuous gas.
   (e) It is moving toward Earth at high speed.

4. A star's radiation is brightest at a wavelength of 400 nanometers. Its temperature is about
   (a) 4000 K.
   b) 12000 K.
   (c) 1500 K.
   (d) 750 K.
   (e) 7500 K.

5. If an object's spectral lines are shifted to longer wavelengths, the object is
   (a) moving away from us.
   (b) moving toward us.
   (c) very hot.
   (d) very cold.
   (e) emitting x-rays.

FURTHER EXPLORATIONS

Steffy, Philip C. "The Truth about Star Colors." *Sky and Telescope* 84 (September 1992): 266.

Zajonc, Arthur. *Catching the Light: The Entwined History of Light and Mind.* New York: Bantam Books, 1993.

KEY TERMS

The Earth and Moon

Of all astronomical objects, we know our home planet Earth the best. That knowledge, gained at first hand, is a tremendous help when we study other worlds.

The Earth's shape and size, two of its most important properties, were known to the ancient Greeks thousands of years ago. From the Earth's shadow on the Moon at a lunar eclipse, they deduced that our planet is a sphere. From the length of shadows in sunlight, they measured its diameter to be about 13,000 km (approximately 8,000 miles), see figure OV3.1.

The Earth is composed mainly of rock and iron. The rock in turn is made of the elements silicon, aluminum, and oxygen linked into compounds called silicates. Although scientists can directly determine that silicates make up the outer part of the Earth, they rely on indirect means to study the core. The most important such indirect probe is earthquake waves, which travel through even the innermost part of the Earth. From the speed of these waves, scientists deduce that the Earth's interior is mostly iron and nickel and that much of the core is molten (fig. OV3.2).

12,800 kilometers
(about 8,000 miles)

FIGURE OV3.1
The Earth. (Courtesy NASA)

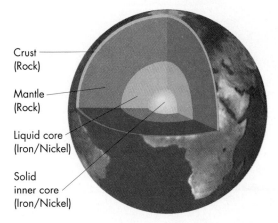

Crust
(Rock)

Mantle
(Rock)

Liquid core
(Iron/Nickel)

Solid
inner core
(Iron/Nickel)

FIGURE OV3.2
Structure of the Earth's interior as deduced from earthquake (seismic) waves.

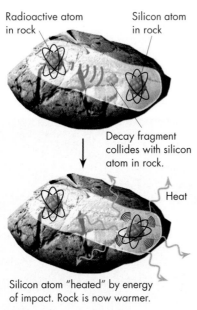

Radioactive atom in rock

Silicon atom in rock

Decay fragment collides with silicon atom in rock.

Heat

Silicon atom "heated" by energy of impact. Rock is now warmer.

FIGURE OV3.3
Radioactive atoms in rock decay and release energy that heats the rock.

Two processes appear to have played a role in heating and melting the Earth's interior. The first is the tiny amounts of natural radioactivity found in most rock. When that radioactive material decays, it generates heat (fig. OV3.3), just as in an atomic powerplant . The heat from any given small piece of rock is insignificant, but when the heat from the vast bulk of rock in the Earth's interior is accumulated over billions of years, it is sufficient to melt our planet's interior.

The heat supplied by radioactivity has been enhanced by a second source: heat left over from the birth of our planet. Astronomers think that the Earth and other planets formed along with the Sun about 4.5 billion years ago from a huge, slowly spinning cloud of gas and dust (fig. OV3.4A). Gravity made this cloud shrink, but its rotation made it flatten. The result was a flat, slowly turning disk surrounding a central core (fig. OV3.4B).

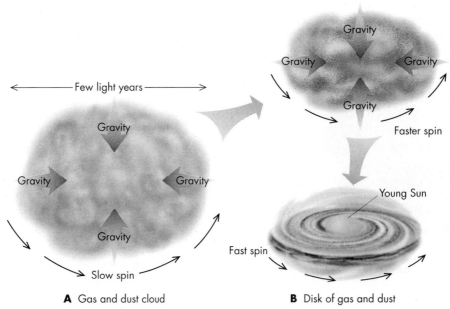

Few light years

Gravity

Gravity   Gravity

Gravity

Slow spin

A Gas and dust cloud

Gravity

Gravity   Gravity

Gravity

Faster spin

Young Sun

Fast spin

B Disk of gas and dust

FIGURE OV3.4
The formation of the Solar System from an interstellar gas and dust cloud.

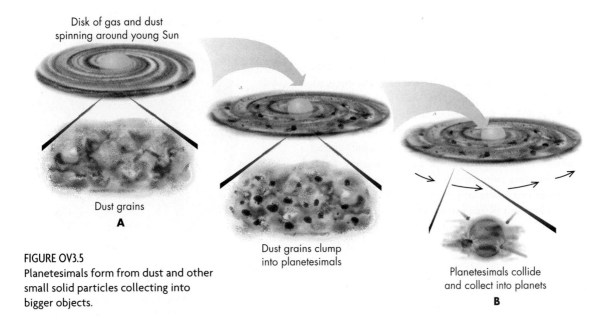

Disk of gas and dust
spinning around young Sun

Dust grains
A

Dust grains clump
into planetesimals

Planetesimals collide
and collect into planets
B

FIGURE OV3.5
Planetesimals form from dust and other
small solid particles collecting into
bigger objects.

Within the disk, dust particles of rock and flecks of ice began to stick together and gradually grew into first pebble-size and then boulder-size chunks. Over hundreds of thousands of years, these larger pieces of rock and ice collided and stuck together and gradually accumulated into objects ten or so miles across (fig. OV3.5). Astronomers call these small, primitive objects that once orbited the Sun planetesimals. Some planetesimals were big enough that their gravity attracted smaller ones. The impact of these smaller pieces generated heat, which eventually melted the larger bodies.

As the Earth and other planets grew in size, they swept up most of the smaller planetesimals near them. With fewer objects striking them, the planets cooled and formed a crust. The last impacts blasted holes in the crust and made craters, which remain to this day on many of the Solar System's moons and planets. But these impacts did more than scar a planet's crust. The heat of impact released gases trapped within the body and formed the first atmosphere of planets like the Earth (fig. OV3.6A). Later, comets struck these planets and vaporized, adding additional gases.

Heat within the Earth triggered volcanic eruptions that liberated yet more gas (fig. OV3.6B), so that gradually our planet accumulated an atmosphere rich in nitrogen, carbon dioxide, and water

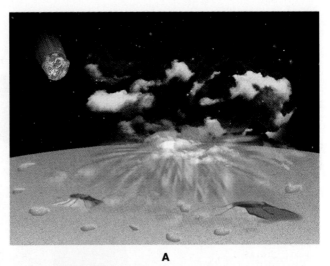

A

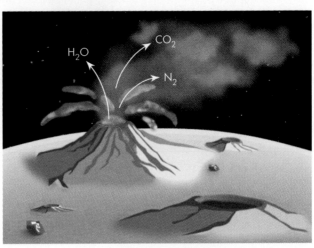

H_2O CO_2 N_2

B

FIGURE OV3.6
The Earth's atmosphere probably formed from the release of gases by volcanos and by the impact and vaporization of planetesimals and comets.

vapor. Continued cooling of the ancient Earth allowed water to condense to make oceans. Eventually life formed, and plants added oxygen.

Deep inside, the Earth remains hot and the material there circulates much as soup does in a heated pan. The circulation in the outer parts of our planet slowly shifts the crust in large pieces (fig. OV3.7). Where these pieces collide, they buckle upward to form mountains. The circulation in the liquid core, when combined with our planet's rotation, creates a natural dynamo that generates the Earth's magnetic field. A compass therefore points north because of swirling motions in the molten iron core of our planet, 3000 miles beneath our feet. Thus, our planet's magnetism, mountains, and atmosphere all result in part from its core being hot.

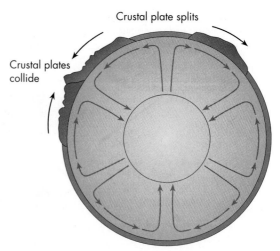

FIGURE OV3.7
Hot material in the Earth's core rises toward the surface, cools, and sinks in a process called convection. This motion shifts the crust of our planet, crumpling it into mountain ranges.

Impacts not only heated our planet, they may also have created the Moon. According to current ideas, at some point late in its formation process, the Earth collided with an exceptionally large planetesimal, an object roughly the size of Mars (fig. OV3.8). That impact "splashed" quintillions of tons of rocky debris into space, where it cooled and its gravity drew it together into a new object orbiting the Earth—our Moon.

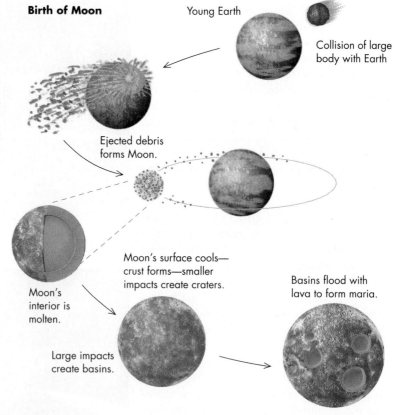

FIGURE OV3.8
Debris blasted out of the young Earth by a giant impact gave birth to our Moon.

The Moon too was probably molten at its birth, but because of its small size, it cooled much faster than the Earth, and rapidly formed a thick and solid crust. This ancient crust—like the Earth's—was pelted by planetesimals that made the craters we see today (fig. OV3.9). Earth too once had craters, but crustal motion and erosion by wind and rain erased most of them. Because the Moon is small and cool, it has no hot circulating rock in its core to alter its surface, and so it still bears the scars of its birth.

Although by earthly standards the Moon is dead, it creates several dramatic phenomena on Earth. As the Moon orbits the Earth, it occasionally passes directly between us and the Sun. When this happens, the Moon's shadow falls on the Earth and causes a solar eclipse (fig. OV3.10A). At those times, conditions are such that either two weeks earlier or two weeks later, when the Moon is on the other side of its orbit, the Earth lies exactly between the Sun and the Moon. Now it is the Earth's turn to cast its shadow on the Moon, creating a lunar eclipse (fig. OV3.10B).

FIGURE OV3.9
The Moon. Note the many craters caused by impacts that scar its surface. (Courtesy NASA.)

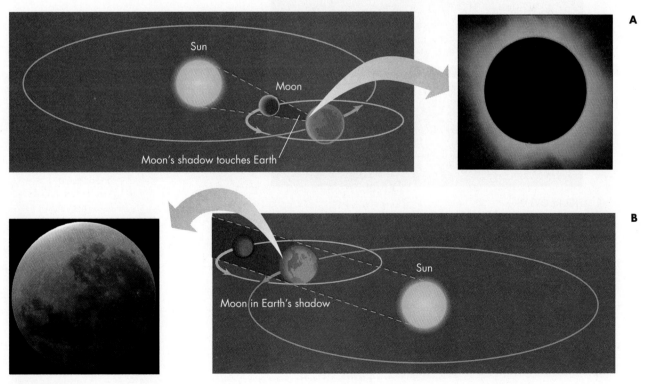

FIGURE OV3.10
(A) Solar and (B) Lunar eclipses. (Photos courtesy Dennis diCicco.)

The Moon creates tides (fig. OV3.11) by its gravitational force on the Earth and its oceans. The Moon's gravity pulls the oceans into two bulges (fig. OV3.12). As our planet spins, the watery bulges remain approximately fixed, pointing toward the Moon, but the solid body of the Earth turns under them. The Earth's rotation thus carries us alternately into and out of these high water locations creating the daily cycle of the tides.

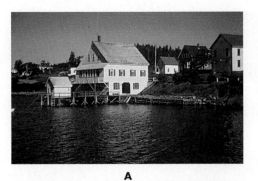

A

B

FIGURE OV3.11
High and low tides.

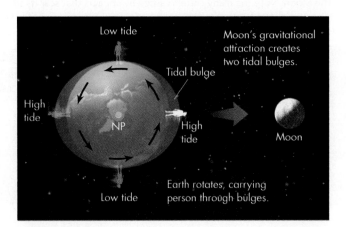

FIGURE OV3.12
Tides are caused by the Moon's gravity pulling on our oceans and distorting them into bulges. As the Earth spins under these bulges, the water level along the coast rises and falls.

CHAPTER FOUR

THE EARTH

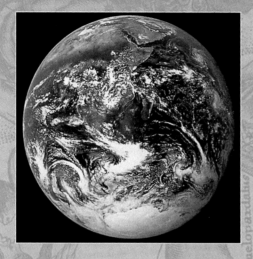

The Earth (Courtesy of NASA)

Volcanic activity shapes much of our planet
(Soames Summerhays/Photo Researchers)

Earth is a beautiful planet. Even from space we can see its beauty—blue seas, green jungles, red deserts, and white clouds. Much of our appreciation of the Earth comes from knowing that it is home for us and the billions of other living things that share this special and precious corner of the Universe. But why study Earth in an astronomy course? The reason is simple: we know Earth better than any other astronomical body; we can learn from it about many of the properties that shape other worlds.

Earth's special characteristics result in large measure from its dynamic nature. The Earth is *not* a dead ball of rock; both its surface and its atmosphere have changed greatly during its vast lifetime. Below our feet the ground sometimes trembles and wrenches in response to dynamic forces even today, crumpling our planet's crust into mountains, stretching it, and tearing it open to form new ocean basins.

These slow but violent motions within the Earth arise from heat generated deep within. That heat also drives volcanic eruptions, which vent gases and molten rock. Over billions of years such gases accumulated, in part creating our atmosphere—an atmosphere that has itself changed by the presence of abundant water and life.

4.1 THE EARTH AS A PLANET

In the simplest terms, Earth is a huge, rocky sphere spinning in space and hurtling around the Sun. In the time it took you to read that sentence, the Earth carried you about 100 miles through the black hostile space around the Sun. But you were protected by a blanket of air, a screen of magnetism, and filters of molecules that blocked most of the hazards of interplanetary space. Other planets share many of these properties but not in the right mix to make it possible for life as we know it to live on them.

Shape and Size of the Earth

The ancient Greeks knew that the Earth is a sphere with a radius of about 6400 kilometers (4000 miles). But what gave it that shape? The Greeks believed that the Earth was round because a sphere was the perfect shape. Today we know that gravity is the cause.

Gravity is the great leveler. Over millions of years, the force of gravity crushes and deforms rock, pulling high points down and rounding large bodies off. However, this shaping process is effective only if the body exceeds a critical size. For bodies made of rock, the critical radius is about 350 kilometers. An object with a radius larger than about 350 kilometers has a strong enough gravity that it can pull itself into a sphere even if it was initially irregular. Smaller objects retain their irregular shape, as you can see by comparing the Earth with the asteroid Gaspra (fig. 4.1), which has a longest dimension of about 19 km (about 12 miles).

Although the Earth is approximately a sphere, it is not a perfect one. It bulges out at the equator, as illustrated in figure 4.2A. The existence of this bulge was first demonstrated in the eighteenth century, when detailed mappings of the Earth showed that its equatorial radius is about 21 kilometers (about 13 miles) greater than its polar radius. This discovery was not a complete surprise because astronomers such as Newton and Hooke had already suggested that the Earth's spinning motion might make its equator bulge into a shape technically known as an oblate spheroid. Moreover, telescopic views of other planets showed that they bulged (fig. 4.2B), making it likely the Earth did too.

You can demonstrate rotationally caused bulges with a water balloon. If you toss a water balloon (gingerly!) into the air, it will take on an almost spherical shape. If you set it spinning as you toss it up, it will become noticeably bulged. This is a great demonstration, but choose an appropriate place.

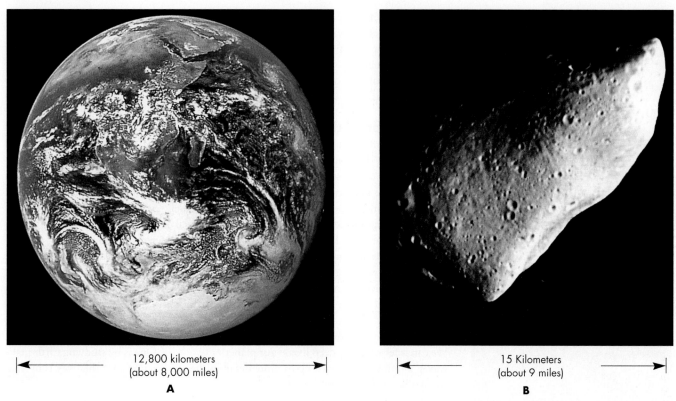

12,800 kilometers
(about 8,000 miles)

A

15 Kilometers
(about 9 miles)

B

FIGURE 4.1
Photographs show that the Earth is round but the asteroid Gaspra is not. Gaspra is too small for its gravity to make it spherical. (Courtesy NASA.)

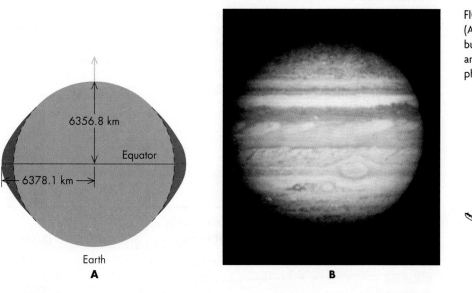

6356.8 km

Equator

6378.1 km

Earth

A

B

FIGURE 4.2
(A) Rotation makes the Earth's equator bulge. (B) Jupiter's rapid rotation creates an equatorial bulge visible in this photograph. (Courtesy NASA.)

To understand why the Earth's equator bulges, think of what happens when you lift a spinning electric beater from a bowl of cake-mix. Particles on such a spinning object fly outward as a result of their inertia, the tendency of all moving objects to keep moving in a straight line, as described by Newton's laws of motion. This same tendency moves matter away from the Earth's rotation axis and is strongest at the equator because the Earth rotates fastest there.

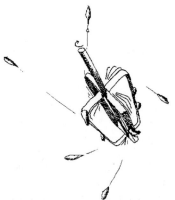

Inertia makes material move away from a spinning object.

All points on the Earth take the same time (24 hours) to rotate once around its axis, but because points near the equator travel farther than points near the pole, they must travel faster. At the equator, a point on the Earth's surface moves at about 1000 miles per hour, while at middle latitudes such a point moves at about 700 miles per hour. The greater speed of the equator is harder for the Earth's gravity to overcome so the equator bulges outward.

Composition of the Earth

Although we may call the Earth a ball of rock, the statement is not very informative because so many different kinds of rock exist. Rocks are composed of minerals, and minerals in turn are composed of chemical elements. Analysis of the surface rocks of the Earth shows that the most common elements in them are oxygen, silicon, aluminum, magnesium, and iron. Furthermore, silicon and oxygen usually occur together as **silicates.** For example, ordinary sand (particles of the silicate mineral quartz) is nearly pure silicon dioxide (SiO_2). Table 4.1 lists a few of the most abundant elements in the Earth.

Other kinds of minerals are more complicated, with atoms of calcium, magnesium, or iron included with the silicates. For example, much of the Earth's interior is composed of the mineral olivine, which is an iron-magnesium silicate. It gets its name from its color: in some solidified lavas it forms tiny green crystals that look like little pieces of chopped olive (fig. 4.3).

At this point you might ask how we can tell what the interior of the Earth is made of. We can infer what lies deep inside our planet in two ways. One is by studying earthquake waves, a point we will take up in more detail in the next section. Another way is by analyzing the Earth's density.

Density of the Earth

Density is a measure of how much material is packed into a given volume. It is defined as an object's mass divided by its volume and is usually measured in terms of the mass in grams of one cubic centimeter of the substance. In these units, for example, water has a density of 1 gram per cubic centimeter, ordinary rocks have a density of about 3 grams per cubic centimeter, and iron has a density of almost 8 grams per cubic centimeter. In other words, a volume of iron has about 8 times as much mass as a similar volume of water; iron is much more dense than water.

FIGURE 4.3
Olivine (the greenish crystals) in a rock sample.

TABLE 4.1 Composition of Earth		
Chemical element (symbol)	% of element in crust by mass	Density (g/cm³)
Oxygen (O)	45.5%	
Silicon (Si)	27.2%	2.42
Aluminum (Al)	8.3%	2.70
Iron (Fe)	6.2%	7.9
Calcium (Ca)	4.66%	1.55
Magnesium (Mg)	2.76%	1.74
Sodium (Na)	2.27%	0.97
Potassium (K)	1.84%	0.87
Titanium (Ti)	0.63%	4.5
Others	1%	

●●◖◖○

EXTENDING OUR REACH

MEASURING THE EARTH'S MASS

We can measure the Earth's mass by comparing its gravitational attraction to that of a huge lead ball of known mass. In such an experiment, a sensitive laboratory weighing scale is carefully balanced with identical standard masses in its opposite pans. Then a huge lead ball is moved under one of the pans. The lead ball exerts a tiny extra gravi-

tational force on that pan, making it dip slightly lower. A tiny extra mass is then added to the opposite pan to bring the scale back into balance. Using the mass required to balance the pans with the lead ball present and the equation for Newton's law of gravity, we can then calculate the relative mass of the Earth and the lead ball.

We see from this that density gives some clue to an object's composition. For example, it would be easy to tell if a closed box contained a block of iron or a block of wood. This is the basis of the famous story of the ancient scientist Archimedes leaping from his bath and shouting "Eureka!" after realizing that he could test whether the king's crown was pure gold by measuring its density. Likewise, we can use the density of the Earth to estimate *its* composition.

We find the Earth's density by dividing its mass by its volume. Its mass is 6×10^{21} tons (metric), or 6×10^{27} grams, and its radius is about 6400 kilometers, or 6.4×10^8 centimeters. To actually make the calculation, we divide the mass of the Earth by its volume, $(\frac{4}{3}) \pi R^3$, assuming it is a sphere. Thus the density is

$$\frac{M}{(\frac{4}{3})\pi R^3} = \frac{6 \times 10^{27} \text{g}}{(\frac{4}{3})\pi (6.4 \times 10^8)^3 \text{ cm}^3}$$

$$= 5.5 \text{ g/cm}^3$$

M = Mass of the Earth
R = Radius of the Earth
$\frac{4}{3}\pi R^3$ = Volume of a sphere

That is about twice the density of ordinary rock.

The density as defined above is really an *average density* over the whole planet. Because we can measure directly that the average density of surface rocks is much less than 5.5 grams per centimeter^3, we can therefore infer that other parts of the Earth must have a much greater density than 5.5 grams per centimeter^3. That by itself does not tell us what lies inside the Earth, but if we ask ourselves what substances are both dense and abundant in nature, we find that iron is a likely choice, as table 4.1 shows. We therefore deduce that the Earth has an iron core. But we can do better than merely deduce. We can test that hypothesis by taking advantage of one of nature's most violent phenomena: earthquakes.

4.2 THE EARTH'S INTERIOR

If we ask how the Earth's core can be studied, your first reaction might be to say, "Why not drill a very deep hole and take a look, or pull out samples?" Unfortunately, the deepest hole yet drilled in the Earth penetrates only 12 kilometers, a mere scratch when compared to the Earth's 6400-kilometer radius. If the Earth were an apple, the deepest holes yet drilled would not have broken the apple's skin. Thus, to study the Earth's core, we rely on indirect means such as earthquake waves.

Probing the Interior with Earthquake Waves

When earthquakes shake and shatter rock within the Earth, they generate **seismic waves** which travel outward from the location of the quake through the body of the Earth (fig. 4.4). Seismic waves are disturbances inside the Earth that slightly compress rock or cause it to vibrate up and down. The speed of the waves depends on the properties of the material through which they move. The wave's speed can be determined by carefully timing its arrival at remote points of the world. From that speed scientists can deduce a picture of the Earth's interior along the path of that wave. Thus seismic waves allow us to "see" inside the Earth much as doctors use sound waves to "see" inside our bodies.

To make a picture of your internal organs, sound waves are sent through your body and are then picked up with a sensitive microphone. Because the sound travels at different speeds in bone, tissue, cartilage, and so forth, a medical technician can analyze the signals with a computer to make a picture of your anatomy or of an unborn child (fig. 4.5). You can use a similar, though obviously much cruder, technique to locate wall studs by thumping areas of a plaster wall with your knuckle.

Seismic waves in the Earth are of two main types: S waves and P waves. P waves form as matter in one place pushes against adjacent matter—whether solid or liquid—compressing it. They travel easily through both solids and liquids. By contrast, S waves form as matter "jerks" adjacent material up and down or from side to side, like a wriggle in a shaken rope.

FIGURE 4.4
Seismic waves spread out through Earth from an earthquake.

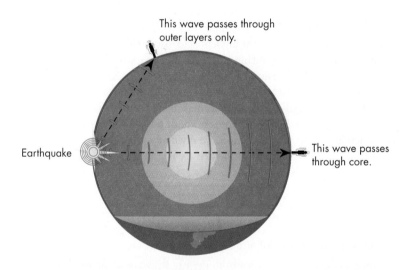

FIGURE 4.5
A sonogram allows a doctor to "see" inside a patient. (Photo Researchers/ P. Saada/Eurelios.)

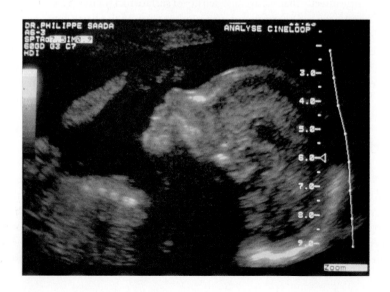

In a liquid, material easily slips past adjacent matter, preventing S waves from spreading. Thus S waves can travel only through solids. Therefore, if a laboratory on the far side of the Earth from an earthquake detects P waves but no S waves, the seismic waves must have encountered a region of liquid on their way from the earthquake to the detecting station (fig. 4.6), an indication that the Earth has a liquid interior. Observations show precisely this effect, from which we infer that the Earth has a liquid core. More complicated analyses can then reveal the density of the material and give clues to its composition.

Seismic studies show that the Earth's interior has four distinct regions. The surface layer is a solid, low-density **crust** about 20 to 70 kilometers (12 to 43 miles) thick and composed of rocks that are mainly silicates. Beneath the crust is a region of hot but not quite molten rock called the **mantle.** This region is also composed of silicates, the most common of which is the mineral olivine. The mantle extends roughly halfway to the Earth's center and, despite not being liquid, is capable of slow flow when stressed, much the way a wax candle can be bent by a steady pressure.

Beneath the mantle is a region of dense liquid material, probably a mixture of iron, nickel, and perhaps sulfur, called the **liquid, or outer, core.** At the very center is a **solid, or inner, core,** probably also composed of iron and nickel. Figure 4.7 illustrates these different layers and their relative sizes.

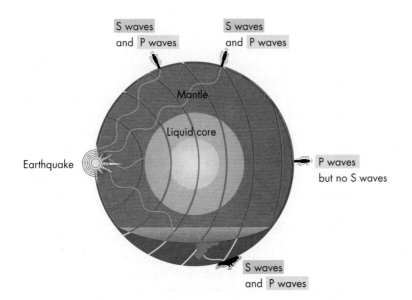

FIGURE 4.6
P and S waves move through the Earth, but the S waves cannot travel through the liquid core.

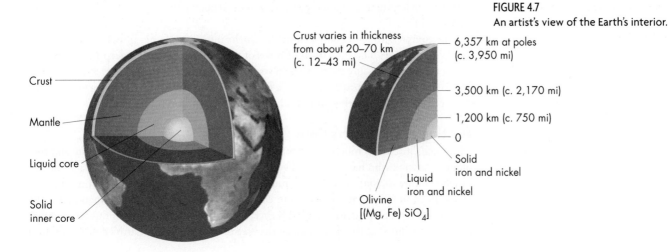

FIGURE 4.7
An artist's view of the Earth's interior.

FIGURE 4.8
Melting ice cream "differentiates" as the dense chocolate chips sink to the bottom of the box. So too, melting has made much of the Earth's iron sink to its core.

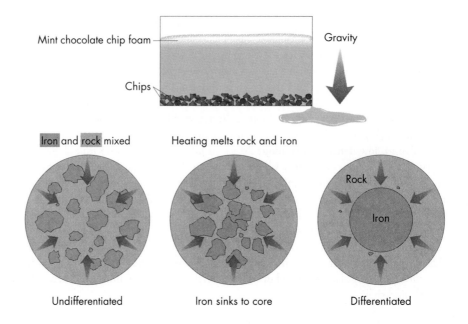

You can see from this discussion that the Earth's interior structure is layered so that the dense heavier material (the iron and nickel) is at the center and the lower-density lighter materials (silicates) are near the surface in the crust and mantle. Scientists call this separation of materials by density **differentiation.**

Differentiation can occur if a mixture of heavy and light material melts, allowing the heavy substances to sink and the lighter ones to rise. You have seen differentiation at work if you have ever had the misfortune to have a half gallon of mint chocolate chip ice cream melt. When you open the box, you find all the chips have sunk to the bottom and the air in the ice cream has risen to the top as foam. Because the Earth is differentiated, we can infer that it must have been almost entirely melted at some time in the past (fig. 4.8).

You may be puzzled as to why there is a solid core inside a liquid core at the Earth's center. If it is hot enough to melt part of the interior of the Earth, why is the very center not liquid as well? The solid core is not cooler, but rather it has a higher melting point because it is under greater pressure. At very high pressures, a previously melted material may re-solidify. You can understand why this happens in the following way. For a solid to form, the atoms composing it must be able to link up to their neighbors to form rigid bonds. Heating makes the atoms move faster, breaking their bonds, and so the material becomes liquid. However, if the material is highly compressed, the atoms may be forced so close together that, despite the high temperature, bonds to neighbors may hold and keep the substance solid.

The compression needed to solidify the Earth's inner core comes from the weight of the overlying material. The thousands of miles of rock above the Earth's deep interior generate an enormous pressure there. To help visualize that pressure, imagine what it would feel like to have a pile of cinderblocks a mile high put on your stomach. In the Earth's core, the huge pressure squeezes what would otherwise be molten iron into a solid.

Heating of the Earth's Core

The Earth's interior is much hotter than its surface (a fact that figures in folklore and theology!). Anyone who has ever seen a volcano erupt can hardly doubt this. In fact, the rise in temperature as you move deeper into the Earth can be measured easily in deep mines where air conditioning must be used to create a tolerable working environment. Near the surface, the temperature rises about 2 K every 100 meters you descend. If this temperature increase continued at this same rate all the way to the center, the Earth's core would be a

FIGURE 4.9
Heat readily escapes from small rocks but is retained in larger bodies. (Courtesy NASA.)

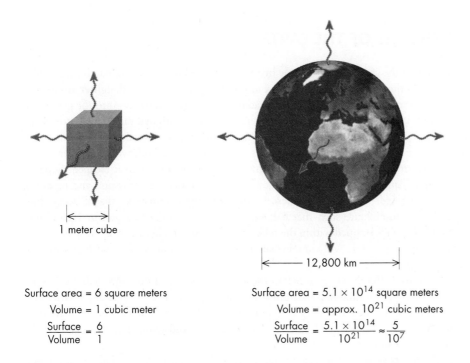

1 meter cube

Surface area = 6 square meters

Volume = 1 cubic meter

$$\frac{\text{Surface}}{\text{Volume}} = \frac{6}{1}$$

12,800 km

Surface area = 5.1×10^{14} square meters

Volume = approx. 10^{21} cubic meters

$$\frac{\text{Surface}}{\text{Volume}} = \frac{5.1 \times 10^{14}}{10^{21}} \approx \frac{5}{10^7}$$

torrid 120,000 K. However, by measuring the amount of heat escaping from the deep interior and from laboratory studies of the properties of heated rock, geologists estimate that the temperature in the core of the Earth is "only" about 6500 K (hotter than the Sun's surface). What makes our planet's heart so hot?

Some scientists think that the Earth began hot. According to this theory the Earth formed from many smaller bodies drawn together by their mutual gravity. As each body hit the accumulating young Earth, the impact generated heat.* When the bombardment stopped, the Earth's surface cooled, but its interior has remained hot. That is, the Earth has behaved much like a baked potato taken from the oven, cooling on the outside but remaining hot inside because heat leaks only slowly from its interior to its surface.

According to another theory, even if the Earth were cold at its birth, it has heated up since its formation because the heat has been supplied by small amounts of natural radioactivity in the Earth's interior. That is, the Earth generates heat much as a nuclear reactor does. All rock contains trace amounts of naturally occurring **radioactive elements** such as uranium. A radioactive element is one that breaks down into another element by ejecting a subatomic particle from its nucleus, a process called **radioactive decay**. Radioactive decay releases energy, generating heat. The heat so created in a small piece of rock at the Earth's surface simply escapes into the surroundings, and so the rock's temperature does not increase. However, in the Earth's deep interior, the heat is trapped by the outer rocky layers, slowing its escape. The amount of heat lost depends on the surface area, but the amount of heat contained depends on the volume. Because a smaller body has proportionately a much larger surface area compared to its volume than a larger body, the smaller body cools quickly, but the larger one remains hot (fig. 4.9). Thus, with the heat trapped, the temperature of the Earth's interior gradually rises and the rock eventually melts.

Before we look at the consequences of heating in the Earth, however, we should examine another aspect of radioactive decay: it can be used to measure the age of the Earth.

---

*You can demonstrate that impact generates heat by hitting a small piece of metal repeatedly with a hammer and then feeling the metal. It will be warmer than it was before you began hitting it.

4.3 THE AGE OF THE EARTH

One can find the age of a rock sample by measuring the amount of radioactive material it contains. As the rock ages, its radioactive atoms decay into so-called **daughter atoms.** For example, uranium decays into lead, and radioactive potassium decays into the gas argon. The more daughter atoms a rock contains relative to the original radioactive atoms, the older the rock is. For instance, suppose we have a rock sample containing both potassium and argon trapped within the rock crystals. To apply the method, we assume that when the rock formed from molten material any argon—being a gas—in it escaped. Suppose the sample happened to contain a million atoms of radioactive potassium and no argon when it solidified. Laboratory studies of radioactive potassium show that it decays into argon at a constant rate such that half the potassium changes to argon in 1.28 billion years. The argon is trapped within the rock unless the rock melts, and so at the end of 1.28 billion years the sample will contain half a million potassium atoms and half a million argon atoms.

In another 1.28 billion years, half of those surviving potassium atoms will decay, leaving only one-quarter million potassium atoms and creating an additional one-quarter million argon atoms. The ratio of argon to potassium in the rock therefore changes with time from 0, when it solidified, to 1, after 1.28 billion years, and so on. This ratio gives us the age of the rock sample, as table 4.2 shows.

Such studies show that the oldest rocks on the Earth have an age of about 3.6 billion years. These primordial rocks are found in such diverse places as northern Canada, southern Africa, and Australia. Thus the Earth must be at least 3.6 billion years old.

However, geologists believe that the Earth is even older because rock samples from other Solar System bodies, such as the Moon, have ages of as much as 4.5 billion years. Although we have no samples from it, the Sun also appears to be this old, as we will see in chapter 11. Such evidence suggests a common age for the Solar System of 4.5 billion years, an age that we will also take to be the Earth's. Why then are there no rocks this old on Earth? They were probably destroyed by processes we will discuss in the next section.

Four and one half billion years is an immense age. To illustrate, if those billions of years were compressed into a single year, all of human existence would be a mere 3 hours, and all human recorded history would have happened during the last minute of the year.

The brevity of human life compared to the vast age of the Earth prevents us from seeing how dynamic our planet is. Mountains and seas appear to us permanent and unchanging, but even they change over the vast epochs Earth has existed. Such changes have their ultimate cause in the heat liberated in the Earth's interior by radioactive decay, heat that creates motions in the Earth's interior and crust.

TABLE 4.2   Radioactive Decay and Age of Rock

Number of radioactive atoms (Potassium)	Number decayed (Argon)	Time
1,000,000	0	0
500,000	500,000	1.28 billion years
250,000	750,000	2.56 billion years
125,000	875,000	3.84 billion years
62,500	937,500	5.12 billion years

4.4 MOTIONS IN THE EARTH'S INTERIOR

The heat generated in the Earth's core by radioactive decay creates movement of the rock in the Earth's interior. Heating often causes motion, as you can see by watching a pan of soup on a hot stove. If you look into the pan as it heats, you will see some of the soup, usually right over the burner, slowly rising from the bottom to the top, while some will be sinking again (fig. 4.10B). Such circulating movement of a heated liquid or gas is called **convection.**

Convection in the Earth's Interior

Convection occurs because heated matter expands and becomes slightly less dense than the cooler material around it. Being less dense, it rises, the principle on which a hot-air balloon operates. As the hotter material flows upward, it carries heat along with it. Thus convection not only causes motion, but also carries heat.

Convective motions in a pot of soup are easy to see—here a lima bean rises; there a noodle sinks. Such motions are less obvious in the Earth. Our planet's crust and mantle are not bubbling and heaving like the soup, rather they are solid rock. Nevertheless, when rock is heated, it too may develop convective motions though they are very, very slow and therefore difficult to observe. Despite its slowness, the results of convective motion are evident around us. They create such diverse phenomena as earthquakes, volcanoes, the Earth's magnetic field, and perhaps even the atmosphere itself.

Plate Tectonics

Deep in the Earth's interior, hot molten material rises in great, slow plumes. When such a plume nears the crust, it spreads out and flows parallel to the surface below the crust (fig. 4.10). There the hot material drags the surface layers with it, stretching and spreading the crust and breaking it apart in a phenomenon called **rifting** (fig. 4.11A).* Gradually the hot material loses its heat to the outside and then cools and sinks downward again to be rewarmed and rise once more.

Where cool material sinks, it may drag pieces of crustal material together, buckling them upward where they collide to form mountain ranges. Sometimes one piece of crust may slip beneath another and be pushed under it in a process called **subduction** (fig. 4.11B). Rifting and subduction are the dominant forces that sculpt the landscape of our planet and contribute to Earth's uniqueness in the Solar System. These motions also trigger earthquakes and cause volcanoes. For example, friction may temporarily stop two plates from sliding past each other, causing them to stick. Pressure may then build until the rock breaks, freeing the stuck plates and generating a sudden lurch in the crust, an earthquake.

FIGURE 4.10

Examples of convection: (A) In our atmosphere, puffy cumulus clouds form when the Sun heats the ground and warms the air so that it rises. (B) You can see rising and sinking motions in a pan of heated soup. (C) An artist's view of convection in the Earth's interior.

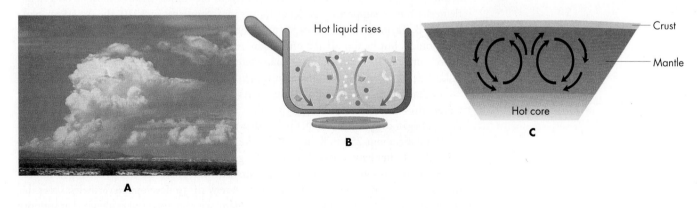

*The mid-Atlantic Ridge (fig. 4.14) is an excellent example of rifting.

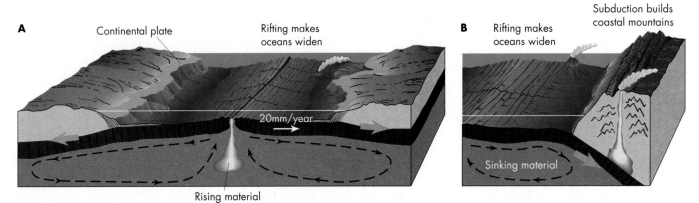

A Continental plate   Rifting makes oceans widen

20mm/year →

Rising material

B Rifting makes oceans widen   Subduction builds coastal mountains

Sinking material

FIGURE 4.11
(A) Rifting may occur where rising material reaches a planet's surface. (B) Subduction builds mountains where material sinks back toward the interior of the Earth.

The shifting of large blocks of the Earth's surface used to be called *continental drift,* but contemporary geologists prefer the term **plate tectonics.** The term *plate* is used because the pieces of the Earth's crust that move are very thin (perhaps only 50 kilometers) but many thousands of kilometers across. Some of the more important plates are those that form the continents, such as the North American and African plates, as seen in figure 4.12. The ocean bottoms also are composed of plates, such as the Pacific and Caribbean plates.

Plate motion is a little like the sliding of the crust on chocolate pudding. You know if you play with your food that pudding often forms a stiff crust over the sticky pudding beneath it. Poking it causes the crust to slip over the pudding below. Similarly the plates slide over the underlying elastic mantle of the Earth.

The development of plate tectonic theory is an interesting example of how science works. As early as 1596, Abraham Ortelius, a geographer from Antwerp, in what is now Belgium, noticed on the new maps of the period that the coast lines of South America and Africa approximately matched like two pieces of a giant jigsaw puzzle. In 1858, a French scientist, Antonio Snider-Pellegrini, also remarked how similar the coast lines were and noted that fossils found at matching locales on both sides of the Atlantic were also very alike. He conjectured that the continents had broken apart, creating the Atlantic ocean in the opening rift between, but apart from the similarities in fossils, he offered little in the way of supporting evidence for his idea. Similarly, in 1910, the American geologist F. B. Taylor published a paper proposing that South America and Africa had once been joined, but he too offered only slight evidence supporting his hypothesis. That evidence came in 1912, when the German meteorologist Alfred Wegener published a paper called *The Origin of the Continents,* in which he developed the modern theory.

Wegener proposed that all the continents were originally assembled in a single supercontinent, which he called Pangea (literally "all-Earth"). For reasons that are still obscure, Pangea began to split into smaller plates that became the familiar continents of today (fig. 4.13), taking about 250 million years for the plates to move into their present locations.

Although he had amassed fossil and geological evidence to support his theory, Wegener's ideas were not well received at first. In fact, there are still problems in understanding what drives the motions and causes a new rift to develop. Nevertheless, over the last several decades, the cumulative effect of many separate pieces of evidence that support the model of plate tectonics have made the theory almost universally accepted.

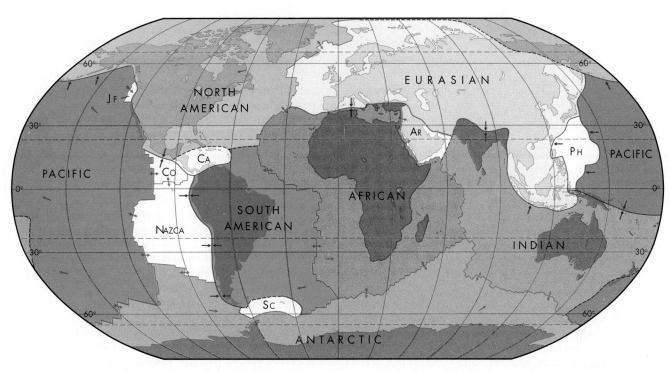

FIGURE 4.12
Map of the Earth, showing its plates. Smaller plates include the Cocos (Co), Caribbean (Ca), Juan de Fuca (Jf), Arabia (Ar), Philippines (Ph), and Scotia (Sc).

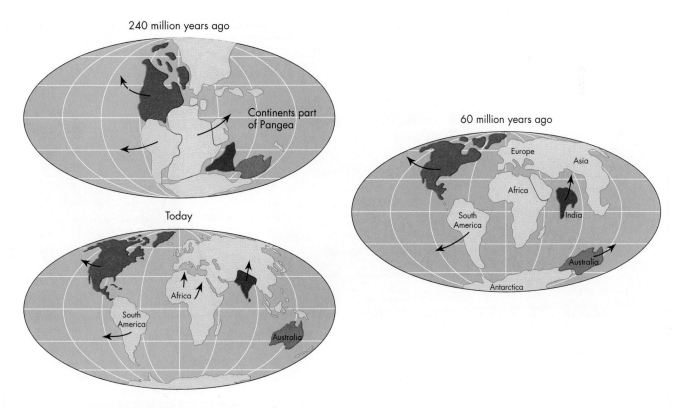

FIGURE 4.13
Breakup of Pangea and the Earth today. Notice the close match of the African and South American Coastlines.

Thus our world is literally changing beneath our feet, growing new crust at midoceanic ridges and devouring it at subduction zones. But this devoured rock is not lost. As it is carried downward, it is heated, causing it to rise again toward the surface. Such molten, rising rock creates volcanic eruptions. Volcanic eruptions do more than just spew out lava, however. They also eject enormous quantities of gases, which over billions of years in the past may have given birth to some part of our atmosphere.

EXTENDING OUR REACH

MEASURING THE MOTION OF PLATES ACROSS TIME

Geologists have at least two ways to deduce that the Earth's plates move. Probably the most persuasive evidence comes from measurements of the age of rocks across the bottom of the Atlantic ocean. Running north-south in a zigzag right down the middle of the Atlantic ocean is a huge underwater mountain range called the **mid-Atlantic ridge.** The ridge originates from hot material rising from the Earth's interior to its surface (fig. 4.14).

The age of rock on the ocean floor increases with distance from the ridge on both sides. New crust is created at the ridge and spreads away from it like paper peeling off a roll. One can calculate the speed of the spreading motion by dividing a rock's distance from the ridge by its age. For example, if rocks 100 kilometers from the ridge are 10 million years old, the plates must have shifted at an average speed of 100 kilometers per 10 million years or a speed of 1 centimeter per year, a fairly typical plate velocity.

Geologists can also measure directly plate motion by using radio signals from the orbiting satellites of the Global Positioning System (GPS). With specialized intruments similar to those used by airline pilots, hikers, and boaters—but far more precise—they can measure the distance between continents to an accuracy of millimeters and can "see" the separation grow from one year to the next. Moreover, from the change in the separation, geologists can calculate how rapidly the continents are moving. Such calculations agree closely with the speed determined when we divide the width of the Atlantic by its age.

FIGURE 4.14
Artist's view of the mid-Atlantic ridge (from *World Ocean Floor* by Bruce C. Heezen and Marie Tharp, 1977) and the increasing age of rocks away from it.

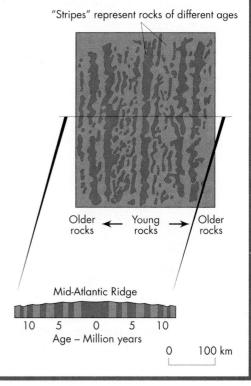

"Stripes" represent rocks of different ages

Older rocks ← Young rocks → Older rocks

Mid-Atlantic Ridge

10   5   0   5   10
Age – Million years

0   100 km

4.5 THE EARTH'S ATMOSPHERE

Surrounding the solid body of the Earth is a veil of gases that constitutes our atmosphere. Most planets in the Solar System have an atmosphere, but the Earth's has many unique features. The Earth's atmosphere is of interest not just for its unique properties, but also for what it can tell us about atmospheres in general.

Composition of the Atmosphere

One of the most striking differences between the atmosphere of the Earth and that of other planets is its composition. For example, the atmospheres of Mars and Venus are nearly completely carbon dioxide, while the atmospheres of Jupiter and Saturn are mostly hydrogen and hydrogen compounds. On the other hand, the atmosphere of the Earth is primarily a mixture of the two gases nitrogen and oxygen. Nitrogen molecules make up about 78% of the gas and oxygen about 21%. The remaining 1% includes carbon dioxide, ozone, and water, gases crucial for protecting us and making life possible. Table 4.3 lists the most important gases in our atmosphere.

Origin of the Atmosphere

Astronomers have proposed several theories to explain how our atmosphere formed. According to one theory, the gases of our atmosphere were originally trapped inside the solid material that eventually became the Earth. When that material was heated—either by volcanic activity (fig. 4.15A) or by the violent impact of asteroids hitting the surface of the young Earth (fig. 4.15B)—the gases escaped and formed our atmosphere.

Recently, some astronomers have proposed a very different explanation to account for our atmosphere. According to this theory, the gases were not originally part of the Earth but were brought here by comets.

As we will see in chapter 10, comets are made mostly of a mixture of frozen water and gases. When a comet strikes the Earth, the impact melts the ices and vaporizes the frozen gas. Given a large enough number of impacts, comets could have delivered enough gas to form the atmosphere (fig. 4.15C). We know from the collision of Comet Shoemaker-Levy with Jupiter in July 1994 that comets collide with planets even today and such collisions were almost certainly far more common billions of years ago when the Earth was young.

In both these theories the early atmosphere had a very different composition than the air we breathe today. For example, our planet's ancient atmosphere probably contained far more methane (CH_4) and ammonia (NH_3) than it does now. Although these gases are still abundant in the giant planets such as Jupiter and Saturn, they have all but disappeared from our atmosphere, which is fortunate, because both methane and ammonia are poisonous.

Volcanoes, such as this one viewed from space, add gases to our atmosphere. (Courtesy NASA.)

TABLE 4.3	Atmospheric Gases
Gas	**Percentage of molecules by number in dry air**
Nitrogen (N_2)	78.08
Oxygen (O_2)	20.95
Argon (Ar)	0.93
Water (H_2O)	Variable—typically between about 0.1 and 3.0
Carbon dioxide (CO_2)	0.03
Trace gases (less than 100 parts per million)	
Neon, helium, ozone, krypton, hydrogen, methane, carbon monoxide, and many pollutants both natural and man-made.	

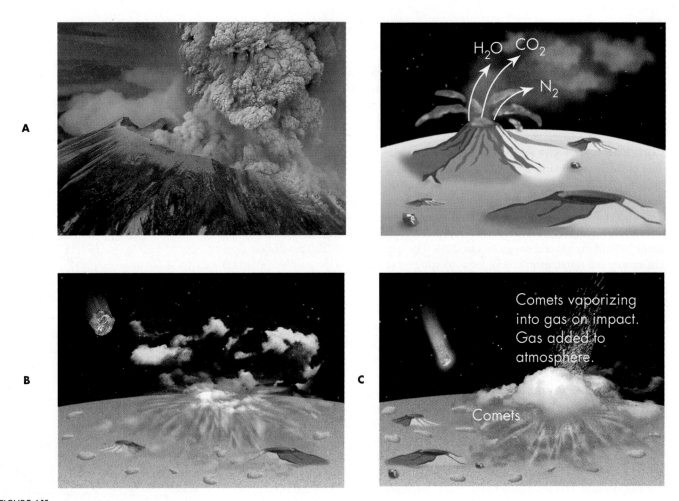

FIGURE 4.15
(A) Volcanic gas vent today. Gas from ancient eruptions built our atmosphere. (USGS.) (B) Planetesimals collide with young Earth and release gas—another source of our atmosphere. (C) Comets striking young Earth and vaporizing. The released gases contributed to our atmosphere.

What has rid Earth of these of these noxious gases? Astronomers think that sunlight is responsible. Solar ultraviolet radiation is intense enough at Earth's distance from the Sun to break the hydrogen atoms out of both methane and ammonia, leaving carbon and nitrogen atoms respectively. The nitrogen and carbon remains behind, supplying at least some of the nitrogen in our atmosphere. The hydrogen, however, gradually escapes into space because Earth's gravity is too weak to hold it. Only huge planets like Jupiter and Saturn have strong enough gravities to retain their hydrogen and thus preserve the large amounts of methane and ammonia we see there today.

Which of these theories for the origin of our atmosphere—delivered by comet or liberated from Earth's own material—is correct? Scientists have tried hard to test these very different theories. For example, they have studied whether volcanoes erupt enough gas to have supplied our atmosphere and whether the composition of these gases can explain the mix of molecules in the air around us.

It requires a certain nonchalance to walk up to the lip of a bubbling volcano and hold a collecting tube over the edge to sample the foul-smelling exhalation, but when geologists make such dangerous tests, they find that nitrogen, water, and carbon dioxide are added to the atmosphere even now by volcanic eruptions. Thus the theory can account for several of the gases in our atmosphere. The theory requires, however, a second important test: Can it account for the amount of these gases we see today? Again the answer is yes, if we assume that eruptions have been about as frequent in the past as they are now and that they ejected comparable amounts of gas.

Thus, these tests confirm that the gases of our atmosphere might have been released by heating the material from which our planet formed.

Testing the comet delivery theory is difficult because of our lack of precise knowledge about what frozen gases and ices comets contain. The evidence currently available,* however, confirms that comet impacts could account for at least part of our planet's atmospheric gases. Thus, scientists remain divided about which theory is correct. Perhaps, as with so many differences of opinion, each side is partly right and some gases came from comets while others came from volcanoes and the heating of rocks during our planet's birth.

Neither of the above hypotheses for the origin of our atmosphere—volcanic exhalations or comet impacts—can account for the large amount of oxygen in our atmosphere. Where, therefore, did that vital ingredient originate? Chemical analysis of ancient rocks, particularly those rich in iron compounds that react with oxygen, shows that our atmosphere once contained much less oxygen than it does today. In fact, over the past 3 billion years, the amount of oxygen in our atmosphere has steadily increased, a rise paralleled by the spread of plant life across our planet. Most scientists therefore agree that the bulk of the free oxygen, which we breathe, was created from H_2O by photosynthesis of ancient plants. This intimate connection between life and the environment of our planet is a fact that we ignore at our peril.

Plants have created most of our oxygen by photosynthesis, but not all. Some has come from water molecules split by solar ultraviolet radiation into hydrogen and oxygen. The lighter hydrogen slowly drifts to the top of the atmosphere and escapes, leaving oxygen behind. This mechanism for adding oxygen to the atmosphere was probably the dominant source of that gas in the early history of the Earth.

The Ozone Layer

The oxygen in our atmosphere not only is important to us for breathing, but it also forms a vitally protective blanket, shielding us from harsh solar ultraviolet radiation. Some of that shielding is provided by O_2 (the normal form of oxygen), but most of it comes from another molecular form of oxygen, O_3, or **ozone.**

Most of the ozone in our atmosphere is located in the ozone layer at an altitude of about 25 kilometers (80,000 feet). Ozone is formed because at these upper levels solar ultraviolet radiation is intense enough to split O_2 into individual oxygen atoms.** The splitting occurs because the ultraviolet radiation makes the molecule vibrate so energetically that it flies apart. These individual oxygen atoms then combine with other O_2 molecules to form O_3.

Ozone is important because it is a strong absorber of ultraviolet radiation; without the ozone layer, solar ultraviolet radiation would pour into the lower atmosphere. The short wavelength (and therefore the high energy) of the radiation would damage living organisms. Without the protective ozone layer, you would get a severe sunburn on exposed skin simply by stepping outside. In fact, it is doubtful that life could exist on the Earth's surface without the ozone layer to shield us. For this reason world governments are cooperating to limit the use of chemicals that might mix into the upper atmosphere and chemically combine with the ozone layer and destroy it.

Ozone is not the only gas that absorbs radiation in our atmosphere. Carbon dioxide and water are also strong absorbers, but they absorb infrared rather than ultraviolet radiation. Therefore, they play a crucial role in regulating our planet's temperature through the greenhouse effect.

---

*Astronomers in several countries are planning spacecraft that will land on a comet to sample its composition. Such evidence will eventually help settle how our atmosphere formed.
**Recall that ultraviolet radiation, having shorter wavelensyths, is more energetic than visible light.

The Greenhouse Effect

The transparency of the Earth's atmosphere to visible radiation allows sunlight to enter the atmosphere and reach the surface, where it is converted to heat. The surface radiates infrared energy, but the opacity of the atmosphere (its blocking power) at infrared wavelengths reduces the heat loss and makes the surface warmer than it would be if the infrared energy could escape freely, a phenomenon shown in figure 4.16 and known as the **greenhouse effect.**

You can get some indication of how effectively water vapor traps heat by noticing how the temperature drops dramatically at night in desert regions or on very clear nights. All gardeners know that it is clear nights (with no clouds and little water vapor) that are most likely to have frost. On cloudy nights heat is retained.

It is important to recognize that the greenhouse effect does not generate heat; rather, it limits the heat loss to space. The greenhouse effect therefore warms the Earth the same way a blanket warms you. The blanket doesn't make you generate more heat; it simply slows down the loss of heat already there. Likewise, the water and carbon dioxide do not create heat of their own; they simply slow down the loss of heat from the ground by absorbing the infrared energy. Eventually they re-emit it, but more is re-emitted toward the ground than is lost into space. That extra infrared energy reradiated to the ground helps keep it warm at night.

We can see how important our atmosphere is as a heat blanket by comparing Earth to Mars and Venus. Both of these planets have atmospheres that are primarily carbon dioxide. However, the atmosphere of Mars is about 100 times less dense than the Earth's, whereas Venus's is about 90 times more dense. On Mars there is consequently very little heat trapping, while on Venus the heat trapping is enormous. As a result, Mars is much colder than the Earth and Venus is much, much hotter (about 750 Kelvin, or 900 degrees Fahrenheit!). It is true that Mars is farther from the Sun than the Earth is, and Venus is closer, but for the Earth and Venus, distance affects the temperature less than the greenhouse effect does.

Ironically, the heat trapping of an atmosphere by the blockage of infrared energy is not what makes an actual greenhouse warm. Air inside a greenhouse is warmer than air outside because the greenhouse confines the air within it, preventing it from rising and cooling the way outside air can. Accordingly, some scientists prefer the term *atmosphere effect* rather than greenhouse effect.

Structure of the Atmosphere

Our atmosphere extends from the ground to an elevation of hundreds of kilometers, but at the highest altitudes the air is extremely tenuous. In fact, the density of the atmospheric gases decreases steadily with height. Gases near the ground are compressed by the weight of gases above them. Thus the atmosphere is a little like a tremendous pile of pillows. The pillow at the bottom is squashed by the weight of all those above. Likewise, a block of

FIGURE 4.16
The greenhouse effect. Radiation at visible wavelengths passes freely through the atmosphere and is absorbed at the ground. The ground heats up and emits infrared radiation. Atmospheric gases absorb the infrared radiation and warm the atmosphere, which in turn warms the ground.

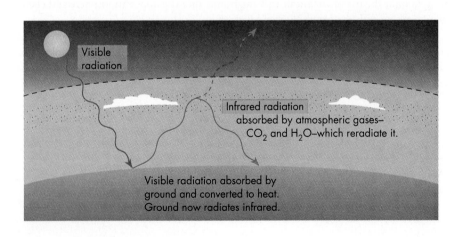

Visible radiation

Infrared radiation absorbed by atmospheric gases— CO_2 and H_2O—which reradiate it.

Visible radiation absorbed by ground and converted to heat. Ground now radiates infrared.

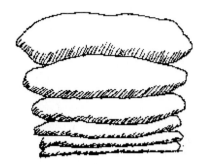

air near sea level is more compressed and therefore has a greater density than a block of air near the top of the atmosphere. That is why it is so difficult to breathe at 30,000 feet where the air is far less dense.* Because of compression by the overlying layers, roughly half the mass of the atmosphere is within 4 kilometers of the Earth's surface. Another way you can picture this is to notice that a jet airplane cruises above three fourths of the atmosphere's bulk.

Although traces of atmosphere can be detected by satellites above 100 km, the air there is extremely thin and gradually merges with the near vacuum of interplanetary space. Nevertheless, this very rarified gas has an important cumulative effect on spacecraft in low orbits, slowing them down and gradually causing them to spiral into the denser lower regions where they burn up. But, although these tenuous gases may influence the motion of spacecraft, they may themselves be influenced by magnetic forces of the planet.

4.6 THE EARTH'S MAGNETIC FIELD

Many astronomical bodies have magnetic properties, and the Earth is no exception. The English natural philosopher William Gilbert (1540–1603) was the first to appreciate that the Earth acted like a magnet, though the ancient Chinese had used the Earth's magnetism in their invention of the compass many hundreds of years earlier.

Magnetic forces are communicated by what is called a **magnetic field**. Although some forces are transmitted directly from one body to another (for example, when two billiard balls collide), other forces, such as gravity, need no such direct physical link.

Magnetic fields are often depicted by a diagram showing **magnetic lines of force**. Each line represents the direction that a tiny compass would point in response to the field. The concentration of lines indicates the field's strength, with more lines implying a stronger field. For example, the field lines of an ordinary toy magnet emanate from one end of the magnet, loop out into the space around it, and return to the other end. The Earth's magnetic field has a similar shape, as represented in figure 4.17.

FIGURE 4.17
Schematic view of Earth's magnetic field lines and photograph of iron filings sprinkled on a toy magnet, revealing its magnetic field lines.

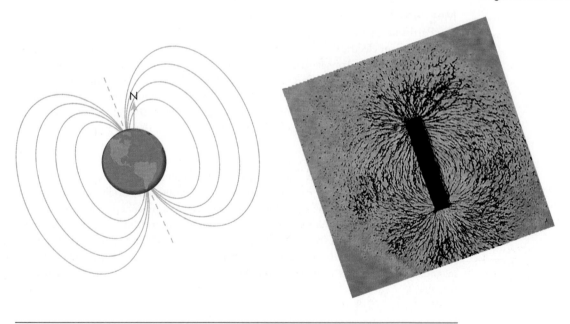

*One cubic centimeter (a volume roughly the size of the end of your little finger) of the air around you contains about 10^{19} molecules. When you take even a tiny sniff, you inhale a number of molecules roughly comparable to the number of grains of sand in a pile the size of the Astrodome.

Magnetic fields have an important property called **polarity**. Polarity gives field lines a direction, and so they always start at a north magnetic pole and end at a south magnetic pole. Thus all magnets must have both a north pole and a south pole. The existence of north and south poles allows magnets to either attract or repel. Two north poles or two south poles repel each other, but a north and a south pole attract. A compass works on this principle. Its needle is a magnet, and its north pole is attracted to the Earth's magnetic south pole, and its south pole is attracted to the Earth's magnetic north pole.

An important question about any field is, How is it generated? Gravitational fields are generated by masses. Magnetic fields are generated by electric currents, either large-scale currents or currents on the scale of atoms. You can easily demonstrate the former by wrapping a few coils of insulated wire around an iron nail and attaching the wire ends to a battery. The nail will now act like a magnet and be able to pick up small pieces of iron and will deflect a compass needle. Microscopic currents on the atomic scale create the magnetism of toy magnets.

Origin of Earth's Magnetic Field

The magnetic field of the Earth is generated by currents flowing in its molten iron core. Scientists are still unsure about how the currents originate but hypothesize that they originate from a combination of rotational motion and convection.* Studies of the magnetic fields of other Solar System bodies support this view. For example, bodies with weak or no magnetic fields, such as the Moon and Venus, are either too small to have a large convecting core or rotate very slowly. On the other hand, bodies with large magnetic fields, such as Jupiter and Saturn, rotate very rapidly and probably have very active cores.

The Earth's magnetic poles do not coincide with the poles determined by its rotation axis (true north and south). Therefore a compass needle does not in general point true north but, instead, several degrees away to what is called "magnetic north." Both the position and strength of the Earth's magnetic poles change slightly but measurably, from year to year, even reversing their polarity about every 10,000 years. Thus, at some time in the distant future a compass that now points north will point south. These changes in our planet's magnetic field probably come from irregular motions of the molten iron swirling in the Earth's core.

Magnetic Effects in the Upper Atmosphere

The Earth's magnetic field does more than make a compass work. For example, it partially screens us from electrically charged particles emitted by the Sun. Such particles are often energetic enough to damage living cells and are therefore potentially harmful to us. The magnetic field protects us because when charged particles encounter it, they are deflected by its magnetic force into a spiraling motion around the field lines (fig. 4.18). This slows the particles streaming from the Sun, causing them to flow around the Earth much as water in a stream is diverted by a rock. Thus we are spared the full impact of the Sun's charged particles.

As the solar particles stream past the Earth, they generate electric currents in our upper atmosphere. These currents, circulating around the magnetic poles, force electrons down the magnetic field lines. The moving electrons spiral around the field lines, colliding with molecules of nitrogen and oxygen. Such collisions excite atmospheric gases, lifting their electrons to upper orbits. As the electrons drop back to lower orbits, they emit the lovely light we see as the **aurora** (fig. 4.19). The exact process by which the aurora forms is still controversial, but there is no doubt that its beautiful streamers are shaped by the Earth's magnetic field, much as sprinkled iron filings outline the field of an ordinary toy magnet (fig. 4.17).

---

*Astronomers call this process a magnetic dynamo.

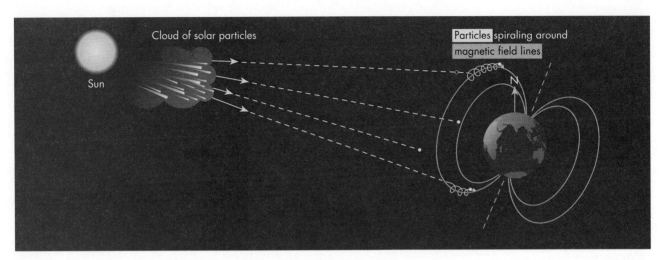

FIGURE 4.18
Electrically charged particles from the Sun spiral in the Earth's magnetic field.

A

B

FIGURE 4.19
Photographs of an aurora from (A) the ground (courtesy Eugene Lauria) and (B) from space. (Courtesy NASA.)

The magnetic field exerts an even stronger influence on particles at the top of the atmosphere in what is called the **magnetosphere.** Here, roughly two or three Earth radii out in space (fig. 4.20), the magnetic field controls the structure of the atmosphere, trapping charged particles in two doughnut-shaped rings called the **Van Allen radiation belts.**

The particles trapped in the Van Allen belts are energetic enough to penetrate spacecraft and be a hazard to space travelers, damaging their genetic material or other tissue. Astronauts therefore try to avoid passing through the belts or to go through them as quickly as possible.

We have described above how the combined motion of the molten material in the Earth's core and our planet's spin create its magnetic field. The Earth undergoes many other motions as well.

FIGURE 4.20
Artist's view of the Van Allen radiation belts (side view).

You may have seen a demonstration that proves the Earth rotates. Many science museums and college campuses have a Foucault Pendulum— a huge ball swinging at the end of a long cord. As time passes, the ball swings in a slowly changing direction. However, if you could watch from space, you would see that it is not the ball's path that changes—the ball's inertia keeps its path fixed in space. Rather, the Earth (and therefore you and the building) turns beneath the swinging ball.

Putting spin on a thrown football helps keep it oriented in the same direction for the same reason that the spinning Earth keeps its orientation.

4.7 MOTIONS OF THE EARTH

Considering what happens to many of us on an amusement park ride, it is just as well that we are unaware of the Earth's many motions. Our planet spins on its axis, orbits the Sun, is dragged along by the Sun around the galaxy, and in ways as yet not fully understood, moves with the Milky Way galaxy (fig. 4.21). We have already discussed how the Earth's rotational and orbital motion define the day and year, but they also cause seasons and influence winds and ocean currents.

The Seasons

Many people believe that we have seasons because the Earth's orbit is elliptical. They argue that summer occurs when we are closest to the Sun and winter when we are farthest away. It turns out, however, that the Earth is nearest the Sun in early January , when the Northern Hemisphere is coldest. Thus seasons must have some other cause.

In thinking about alternative explanations for the seasons, we might recall that the Sun is higher in the sky during the summer than in the winter. For example, at midlatitudes (40°) the Sun's elevation above the horizon at noon is about 74° in summer but only about 27° in winter, an indication that seasons may be linked to the Sun's elevation. To understand how, we need to think about the Earth's orientation in space.

In chapter 1 we stated that the Earth is tilted with respect to its orbit around the Sun. We describe that tilt in terms of the Earth's **rotation axis**, the imaginary line about which it spins, running through the Earth from its north pole to its south pole. The rotation axis is not perpendicular to the Earth's orbit around the Sun but is tilted by about 23.5° (fig. 4.22A), and in this epoch the axis happens to point nearly toward the North Star, Polaris.

The Earth's rotation axis maintains nearly exactly the same tilt and direction as we orbit the Sun (fig. 4.22B). That is, the Earth behaves much like a giant spinning top. This tendency to preserve tilt is shared by all spinning objects. For example, it is what keeps a rolling coin upright. Moreover, you can easily feel this tendency of a spinning object to resist changes in its orientation by lifting a spinning bicycle wheel off the ground and trying to twist it from side to side.

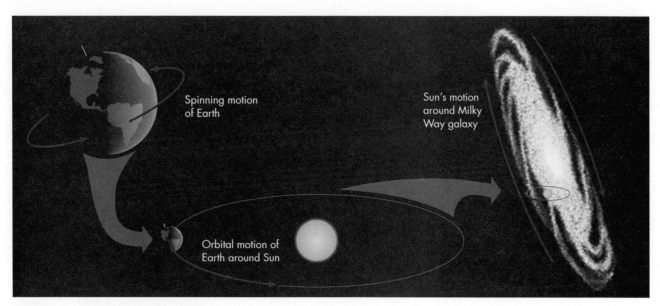

FIGURE 4.21
The Earth's many motions in space.

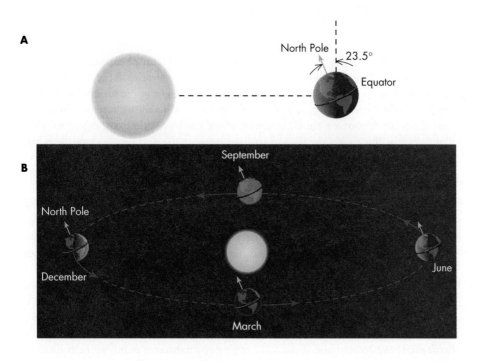

FIGURE 4.22
(A) The Earth's rotation axis is tilted 23.5°
to its orbit around the Sun. (B) The Earth's
rotation axis keeps nearly the same tilt
and direction as it moves around the Sun.

Because the Earth's tilt remains nearly constant as we move around the Sun, sunlight
falls more directly on the Northern Hemisphere for part of the year and more directly on
the Southern Hemisphere for another part of the year, as illustrated in figure 4.23A. This
changes the amount of heat we receive from the Sun.

A flat surface area presented directly to the Sun intercepts a larger portion of the Sun's
radiation, and hence its heat, than the same surface area when tilted, as figure 4.23B illustrates. You take advantage of this effect instinctively when you warm your hands at a fire
by holding your palms flat toward the fire, not edgewise. Similarly, the Northern Hemisphere receives its greatest heating at the time of year when the Sun shines most directly

FIGURE 4.23
(A) Because the Earth's rotation axis keeps the same tilt as we orbit the Sun, sunlight falls more directly on the Northern Hemisphere during part of the year and on the Southern Hemisphere during the other part of the year. (B) A surface directly facing the Sun receives more light (and thus heat) than a tilted surface.

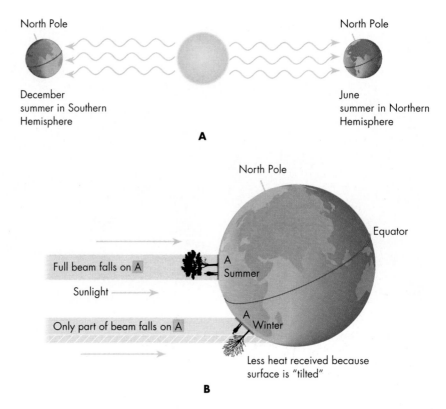

North Pole

North Pole

December
summer in Southern
Hemisphere

June
summer in Northern
Hemisphere

A

North Pole

Equator

Full beam falls on A

A
Summer

Sunlight

Only part of beam falls on A

A Winter

Less heat received because
surface is "tilted"

B

Although the seasons begin on the solstices and equinoxes, the hottest and coldest times of year occur roughly 6 weeks after the solstices. The delay, known as the lag of the seasons, results from the oceans and land being slow to warm up in summer and slow to cool down in winter.

on it, making it summer. Six months later, the Northern Hemisphere receives its sunlight least directly, and so it is colder and therefore winter. This heating difference is enhanced because the Earth's tilt leads to many more hours of daylight in the summer than in winter. As a result, we not only receive the Sun's light more directly, we receive it for a longer time. Thus the seasons are caused by the tilt of the Earth's rotation axis. Figure 4.23A also illustrates that this makes the seasons reversed between the northern and southern hemispheres; when it is summer in one, it is winter in the other.

Air and Ocean Circulation: The Coriolis Effect

If you sit with a friend on a rotating schoolyard merry-go-round and toss a ball back and forth, you will discover that the ball does not travel in the direction you aim your throw but instead curves off to the side. Similarly, ocean and air currents sweeping across our planet are deflected from their original direction of motion. This phenomenon is called the **Coriolis effect.**

The Coriolis effect, named for the French engineer who first studied it, alters the path of objects moving over a rotating body, such as the Earth, other planets, or stars. To understand why the Coriolis effect occurs, imagine standing at the North Pole and throwing a rock as far as you can toward the equator (fig. 4.24). As the rock arcs through the air, the Earth rotates under it. Thus, if you were aiming at a particular point on the equator, you will miss because the surface has turned beneath the rock's path, making the rock appear to have been pushed to the right.* Air, water, rockets, or anything moving across the rotating Earth in any direction are affected similarly.

---

*We must not be guilty of "hemispherism." The Coriolis effect deflects objects to the *left* in the Southern Hemisphere. There is no truth to the story that water spirals down the drain clockwise in the Northern Hemisphere and counterclockwise in the Southern Hemisphere . . . at least, unless you ensure that the water is initially absolutely motionless or use an ocean-sized basin! The Coriolis effect is totally overwhelmed by the persistence of the motion from filling the basin unless you allow the water to come to a complete rest, a state that can take weeks to achieve.

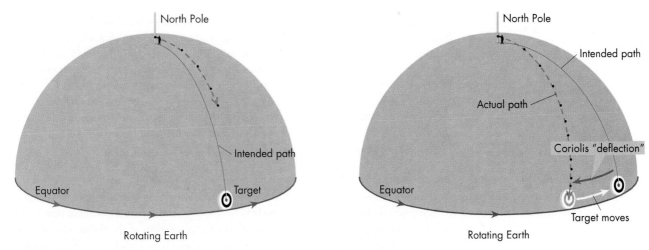

FIGURE 4.24
Coriolis effect on a rock thrown toward the equator from the North Pole.

FIGURE 4.25
Weather satellite pictures show clearly the spiral pattern of spinning air around a storm that results from the Coriolis effect. (Courtesy NOAA.)

You can clearly see the results of the Coriolis effect when you see weather satellite pictures of a large storm. The Coriolis effect spins the air flowing into the storm, creating the spiral pattern seen in figure 4.25. Likewise, the Coriolis effect deflects ocean currents, creating flows like the Gulf Stream. It also creates atmospheric currents such as the trade winds.

The trade winds blow relatively steadily around our planet from east to west in two broad bands, one north and the other south of the equator. They form in response to the Sun's heating of equatorial air, which rises and expands, flowing away from the equator toward the poles. However, before this air can reach the poles, it cools and sinks toward the surface, where it flows back toward the equator. As that air approaches the equator, the Coriolis force deflects it into the pattern we observe in the trade winds.

The Coriolis force also establishes the direction of the **jet streams,** which are narrow bands of rapid, high-altitude winds. Jet streams are an important feature of the Earth's weather and are found on other planets as well. On rapidly rotating Jupiter, Saturn, and Neptune, the Coriolis effect is much stronger than that on the Earth, creating extremely fast jet streams. The striking cloud bands we see on these planets are partly caused by this effect (fig. 4.26).

At one time the Coriolis effect was of special interest because it was an indirect proof that the Earth rotates. Now, we can see the spin of our planet from space pictures and need no longer rely on such indirect proofs.

Precession

As the Earth moves around the Sun over long periods of time the direction in which its rotation axis points changes very slowly. This motion, similar to the wobble that occurs when a spinning coin begins to slow down, is called **precession.** If the Earth were perfectly spherical, precession would not occur. The Earth's equator is slightly bulged, however, so the Sun and Moon exert an unbalanced gravitational attraction on our planet, twisting it slightly. That twisting makes the Earth's rotation axis slowly change direction, completing one swing in about 26,000 years (fig. 4.27). Currently, the North Pole points almost at the star Polaris. In about A.D. 14,000, the North Pole will point instead nearly at the bright star Vega. Thirteen thousand years later the North Pole will again point nearly at Polaris.

FIGURE 4.26
Cloud bands on Jupiter, Saturn, and Neptune, created in part by the Coriolis effect. (Courtesy NASA/JPL.)

A   B   C

FIGURE 4.27
Precession makes the Earth's rotation axis swing slowly in a circle.

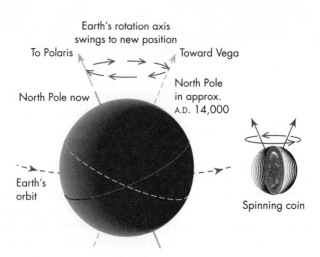

Precession is of minor importance in day-to-day life, but over long periods of time it may cause climate changes. As noted earlier, we are currently nearest the Sun when it is winter in the Northern Hemisphere. In 13,000 years the Earth will be farthest from the Sun in Northern Hemisphere winter. That will make winters slightly colder in the Northern Hemisphere and perhaps trigger a new ice age.

SUMMARY

The Earth is roughly spherical, and its radius is about 6400 kilometers (4000 miles). Trace amounts of radioactivity in surface rocks reveal that the Earth's age is about 4.5 billion years. Radioactive material also heats its interior. The heat stirs slow convective motions, which shift the Earth's crust (plate tectonics) creating mountains, volcanoes, ocean basins, and earthquakes.

The waves generated by earthquakes (seismic waves) allow us to study the Earth's interior. They show it is stratified into four distinct regions: a very thin crust of ordinary rock, a mantle of hot but essentially solid silicates, an outer core of liquid iron and nickel, and an inner core of solid iron and nickel.

Currents created by motions in the Earth's core generate the Earth's magnetic field. That field in turn affects the motion of charged particles in the upper atmosphere. Such particles may create auroral displays when they collide with oxygen and nitrogen in the upper atmosphere.

The nitrogen, carbon dioxide, and water of our atmosphere may have come from volcanic gases vented over the Earth's history. Alternatively, these atmospheric gases may be the evaporated remains of comets that hit the Earth in its infancy. Plant life has created the atmosphere's oxygen by photosynthesis.

Some atmospheric gases absorb radiation. High-altitude ozone absorbs ultraviolet radiation, thereby protecting us from its biologically harmful effects. Carbon dioxide and water vapor absorb infrared radiation, trapping heat radiated from the Earth's surface. By slowing heat loss from our planet into space, these gases create the greenhouse effect and make Earth slightly warmer than it would be if the infrared radiation could escape freely.

The Earth's spinning motion creates a Coriolis effect that deflects objects moving over its surface. The Coriolis effect makes large storm systems rotate and is essential for driving the circulation of the atmosphere and the oceans. The Earth's rotation also keeps its rotation axis pointing in approximately the same direction as we move around the Sun. Thus the northern and southern hemispheres are heated alternately, causing the seasons.

QUESTIONS FOR REVIEW

1. Why is the Earth not perfectly round?
2. What are some of the more common elements composing the Earth?
3. What causes the interior of the Earth to be hot? How hot is it?
4. What is convection? What are some other examples of convection besides hot soup?
5. What is the relation between rising and sinking material in the Earth's interior and subduction and rifting?
6. When it is winter in Australia, what season is it in New York? Is it April in Paris?
7. What is precession? What are some of its possible consequences?
8. What is the mid-Atlantic ridge?
9. On what plate of the crust are you located? Which way is it taking you?
10. What is happening where your plate is smashing into another?
11. How do we know the Earth has a liquid core?
12. Why is the inner core solid even though it is hotter than the outer liquid core?
13. What factors are thought to be responsible for the Earth's magnetic field?
14. How is the aurora related to the Earth's magnetic field?
15. If the Earth rotated more slowly, would you expect it to have as strong a magnetic field?

THOUGHT QUESTIONS

1. Thinking about a stone thrown from the pole toward the equator, convince yourself that if the Earth rotated faster the Coriolis effect would be larger.
2. As seen from space, the Earth rotates counterclockwise as seen from above the North Pole. Using the thrown-stone argument of Question 1, explain why the Coriolis effect deflects objects to the right of their motion in the Northern Hemisphere.
3. When you choose fruit at a supermarket, you may heft the fruit in your hand to test its weight. How does this tell you whether the fruit is dried out inside? How is that similar to using mean density as an indicator of the composition of the Earth's interior?
4. Flicking your finger against your cheek makes a different sound from flicking it against your forehead. How is that similar to studying the interior of the Earth with seismic waves.

5. How does the eventual acceptance of the plate tectonic theory illustrate some aspects of the scientific method?
6. Astronomers are still uncertain about how the Earth's atmosphere formed. How does this illustrate the workings of the scientific method?
7. According to the *Guiness Book of Mountains and Mountaineering* the summit of the volcano Chimborazo in Ecuador is the point on the Earth's surface farthest from the center. However, the book also states that the summit of Mt. Everest is the highest point above sea level. Is something wrong here? Why?

PROBLEM

Given that Pluto's mass is 1.3×10^{25} grams and its radius is 1.1×10^{8} centimeters, what is its average density? Does this indicate that Pluto has a large iron core like the Earth? Why?

TEST YOURSELF

1. Scientists think the Earth's core is composed mainly of
   (a) silicate rocks.
   (b) uranium.
   (c) lead.
   (d) sulfur.
   (e) iron.
2. Our knowledge of the composition of the Earth's core comes from
   (a) analysis of earthquake waves.
   (b) x-ray pictures taken with a powerful device in Russia.
   (c) samples obtained by drilling deep holes.
   (d) analysis of the material erupted from volcanoes.
   e) Both (c) and (d).
3. The slow shifts of our planet's crust are believed to arise from
   (a) the gravitational force of the Moon pulling on the crust.
   (b) the gravitational force of the Sun pulling on our planet's crust.
   (c) the Earth's magnetic field drawing iron in crustal rocks toward the poles.
   (d) heat from the interior causing convective motion, which pushes on the crust.
   (e) the great weight of mountain ranges forcing the crust down and outward from their bases.

4. If the shape of the Earth's orbit was unaltered but its rotation axis was shifted so that it had no tilt with respect to the orbit, seasons would be
   (a) essentially as they are now.
   (b) more extreme, with colder winters and hotter summers.
   (c) essentially nonexistent, with little temperature change at a given place throughout the year.
   (d) Not enough information is given to make a reasonable guess.
   (e) Either (a) or (b) is most likely.
5. What evidence indicates that part of the Earth's interior is liquid?
   (a) With sensitive microphones, sloshing sounds can be heard.
   (b) We know the core is lead, and we know the core's temperature is far above lead's melting point.
   (c) Deep bore holes have brought up liquid from a depth of about 4000 kilometers.
   (d) No S-type seismic waves are detectable at some locations after an earthquake.
   (e) S-type waves are especially pronounced at all locations around the Earth after an earthquake.

FURTHER EXPLORATIONS

Akasofu, Syun-Ichi. "The Dynamic Aurora." *Scientific American* 260 (May 1989): 90.

Allègre, Claude J., and Stephen H. Schneider. "The Evolution of the Earth." *Scientific American* 271 (October 1994): 66.

Davis, Neil. *The Aurora Watcher's Handbook.* Fairbanks, Alaska: University of Alaska Press, 1992.

Fernie, J. D. "The Shape of the Earth (Parts I–III)." *American Scientist* 79 (March/April 1991): 108; ibid, 79 (September/October 1991): 393; ibid, 80 (March/April 1992): 125.

Hartmann, W. K. "Piecing Together Earth's Early History." *Astronomy* 17 (June 1989): 24.

Jeanloz, Raymond, and Thorne Lay. "The Core-Mantle Boundary." *Scientific American* 268 (May 1993): 48.

Murphy, J. Brenden, and R. Damian Nance. "Mountain Belts and the Supercontinent Cycle." *Scientific American* 266 (April 1992): 84.

Powell, Corey S. "Peeping Inward." *Scientific American* 264 (June 1991): 100.

Web Sites

"The Nine Planets: A Multimedia Tour of the Solar System," by Bill Arnett.
www.seds.org/billa/tnp
Excellent pictures, glossary, and other info. One of the best astromony web sites.

"Views of the Solar System," by Calvin Hamilton.
bang.lanl.gov/solarsys/earth.htm
Nice pictures, lots of text, fact sheets, and glossary. Essentially an on-line text.

"Welcome to the Planets."
pds.jpl.nasa.gov/planets
Nice pictures, fact sheets, and other info.

National Space Science Data Center.
nssdc.gsfc.nasa.gov/photo_gallery/photo_gallery-earth.html
A NASA site with lots of pictures, fact sheets, glossary.

All the above have many additional links to other sites.

VIDEO

Planet Earth, produced by WQED in association with the National Academy of Sciences, funded by the Annenberg/CPB Project, 1986.

KEY TERMS

aurora, 142
convection, 133
Coriolis effect, 146
crust, 129
daughter atoms, 132
density, 126
differentiation, 130
greenhouse effect, 140
jet streams, 148
liquid or outer core, 129

magnetic field, 141
magnetic lines of force, 141
magnetosphere, 143
mantle, 129
mid-Atlantic ridge, 136
ozone, 139
plate tectonics, 134
polarity, 142
precession, 148
radioactive decay, 131

radioactive elements, 131
rifting, 133
rotation axis, 144
seismic waves, 128
silicates, 126
solid or inner core, 129
subduction, 133
Van Allen radiation belts, 143

 PROJECTS

Make a Xerox copy or tracing of a map of the world (fig. E2.5 showing world time zones will do). Cut out North and South America and place them against the western edge of Europe and Africa. Do they fit? Does this help you see what led early scientists to the idea that the continents shift?

CHAPTER FIVE

TELESCOPES

Orion in the nightime sky (Roger Ressmeyer/Corbis)

Astronomers, like all scientists, rely heavily on observations to guide them in theorizing and in testing theories already developed. Unlike most scientists, however, astronomers cannot directly probe the objects they study. Rather, they must perform their observations from vast distances and can only passively collect radiation emitted by the bodies they seek to study. Collecting enough radiation to be useful in studying astronomical objects is difficult because most objects are so remote that their radiation is extremely faint by the time it reaches Earth. Moreover, special instruments must be used to extract the information desired from the radiation: instruments that can measure the brightness, the spectrum, and the position of objects to high precision. For example, to collect enough light to detect remote galaxies, astronomers use telescopes with mirrors the size of a large truck. To avoid the blurring and blocking effects of our atmosphere, they use orbiting observatories. To analyze and display the observations, they use a computer. This section describes some of the more important devices and how they work. We will see that modern telescopes bear little resemblance to the long tubes depicted in cartoons. Moreover, modern astronomers rarely sit at the eyepiece of a telescope. They are more likely to be sitting at a computer terminal operating a telescope remotely, examining the data collected, or solving equations that describe, for example, the paths of stars when two galaxies collide.

5.1 TELESCOPES

A telescope enables the astronomer to observe things not visible to the naked eye. Although our eyes are superb detectors, they cannot see extremely faint objects or fine details on distant sources. For example, we are unable to read standard newspaper print on the opposite wall in a dimly lit room. A telescope overcomes these difficulties, first, by collecting more light than the eye can and, second, by increasing the detail discernible. The first of these properties is called "collecting power"; the second is called "resolving power."

Collecting Power

For our eyes to see an object, photons (light) from it must strike the retina. How bright that object appears to us depends on the number of its photons that enter our eye per second, a number limited by the size of our eyes. Astronomers overcome that limit by "collecting" photons with a telescope, which then "funnels" the photons to our eye. The bigger the telescope's collecting area, be it a lens or mirror, the more photons it gathers, as shown in figure 5.1. The result is a brighter image, which allows us to see dim objects that are invisible in telescopes with smaller collecting areas. Because the collecting area of a circular collector of radius r is πr^2, a small increase in the radius of the collecting area gives a large increase in the number of photons caught. For example, doubling the radius of a lens or mirror increases its collecting area by a factor of 4. Because the collecting area is so important to a telescope's performance, astronomers often describe a telescope by the diameter of its lens or mirror. Thus the 10-meter Keck Telescope in Hawaii has mirrors spanning 10 meters (roughly 30 feet) in diameter.

Focusing the Light

Once light has been collected, it must be focused to form an image or to concentrate it on a detector. Telescopes in which light is collected and focused by a lens are called "refracting telescopes," or **refractors** for short. The lens of a refractor focuses the light by bending the rays, as shown in figure 5.2. This bending is called **refraction,** and it happens when light moves from one substance (such as air) into a different substance (such as glass), as discussed in the "Extending our Reach" box. It is this refraction that gives such telescopes their name.

Lenses have many serious disadvantages in large telescopes, however. First, large-diameter lenses are extremely expensive to fabricate. Moreover, a lens must be supported at its edges so as not to block light passing through it. This makes the lens "sag" in the middle (though by only tiny amounts), distorting its images. A third difficulty with lenses is that most transparent materials focus light of different colors to different spots, creating images fringed with color. Finally, many lens materials completely absorb short-wavelength light, making them, for some purposes, as useless in a telescope as a chunk of concrete. Figure 5.3 shows a photograph of the world's largest refractor, the 1-meter (40-inch) diameter Yerkes telescope of the University of Chicago.

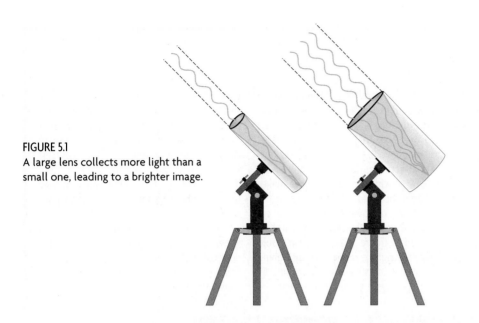

FIGURE 5.1
A large lens collects more light than a small one, leading to a brighter image.

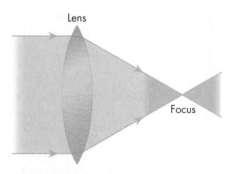

FIGURE 5.2
How a lens focuses light.

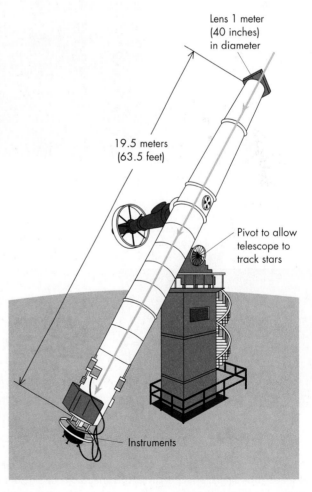

FIGURE 5.3
Photograph and a sketch of a refracting telescope. Completed in 1897 for the University of Chicago's Yerkes Observatory in Williams Bay, Wisconsin, this refractor has a lens approximately 1 meter (40 inches) in diameter, making it the world's largest refracting telescope. (Photo courtesy Yerkes Observatory.)

EXTENDING OUR REACH

REFRACTION

When light moves at an angle from one material into another (for example, from air into water) its direction of travel generally bends. This phenomenon is called refraction. Refraction is the principle by which our eyes and eyeglasses focus light. You can easily see its effects by sticking a pencil in a glass of water and noticing that the pencil appears bent, as seen in figure 5.4.

Refraction occurs because the speed of light changes as it enters matter, generally becoming slower in denser material. This decrease in the speed of light arises from its interaction with the atoms through which it moves. To understand how this reduction in light's speed makes it bend, imagine a light wave approaching a slab of material. The part of the wave that enters the material first is slowed while the part remaining outside is unaffected, as depicted in figure 5.5A. To see why slowing part of the wave makes it bend, imagine what would happen if the wheels on the right hand side of your car turned more slowly than those on the left, as sketched in figure 5.5B. Your car would swerve to the right, a result that lies behind the reason why cars have a differential. By allowing one wheel to turn faster than the other, the differential "swings" your car smoothly around corners. A similar effect occurs if you walk hand-in-hand with a friend, and your friend walks more slowly than you do. You will soon find yourself traveling in a curve (see fig. 5.5C). So, too, if one portion of a light wave moves more slowly than another, the light's path will bend.

Refraction not only bends light but also generally spreads the light into its component colors, breaking white light into a spectrum, or rainbow. This spreading occurs because different colors of light travel at

FIGURE 5.4
Refraction of light in a glass of water. Note how the pencil appears bent.

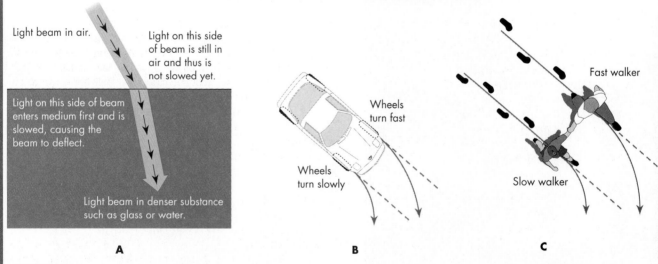

Light beam in air.

Light on this side of beam is still in air and thus is not slowed yet.

Light on this side of beam enters medium first and is slowed, causing the beam to deflect.

Light beam in denser substance such as glass or water.

A

Wheels turn fast

Wheels turn slowly

B

Fast walker

Slow walker

C

FIGURE 5.5
Cause of refraction. (A) Light entering the denser medium is slowed, while the portion still in the less dense medium proceeds at its original speed. (B) A similar effect occurs if the wheels on one side of a car turn more slowly than those on the other side, or (C) when you walk hand-in-hand with someone who walks more slowly than you do.

Continued

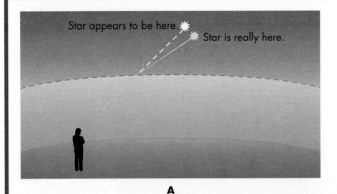

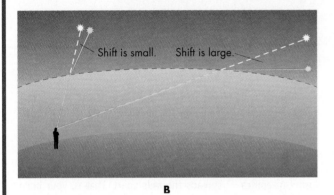

FIGURE 5.6
(A) Atmospheric refraction makes the Sun or a star look higher in the sky than it really is. (B) Refraction is stronger for objects nearer the horizon because their light passes through more atmosphere. (C) The Sun is flattened because refraction "lifts" its lower edge more than its upper edge. (Photo courtesy Patrick Watson.)

different speeds in most materials and are therefore bent by different amounts in a process called **dispersion.** Thus, if light consisting of a mix of colors enters a block of glass, each color is slowed to a different speed and is therefore deflected differently. The result is that colors initially traveling together separate into different beams. This is how a prism creates a spectrum.

Atmospheric Refraction
Refraction is easy to see in the Earth's atmosphere where it distorts the shape of the Sun when it rises and sets and makes the stars twinkle at night.

Distortion of the Sun's Shape
Refraction distorts the shape of the rising or setting Sun because when sunlight enters the Earth's atmosphere from space, it is re-

fracted and bent slightly toward the ground. Thus, if you are on the ground and look back along the light ray, the light seems to come from slightly higher than it really does (fig. 5.6A). That is, refraction makes astronomical objects look higher in the sky than they actually are. This effect is greatest when objects are near the horizon because light then passes through more of the atmosphere and is therefore refracted more, as can be seen in figure 5.6B. The result is that light from the lower edge of the Sun is refracted more than light from its upper edge, which "lifts" the lower edge more than the upper and makes the Sun look flattened (fig. 5.6C).

Refraction also slightly alters the time at which the Sun seems to rise or set. When it is at the horizon, the Sun is "lifted" by about the height of its diameter. Thus, at the mo-

ment we see the Sun touch the horizon, it has actually set. By "lifting" the Sun's image above the horizon, even though it has set, refraction slightly affects the length of the daylight hours. As a result, the day of the year the Sun is above the horizon for exactly 12 hours is not the equinox, but rather a few days before the spring and a few days after the autumnal equinox. It turns out that near latitude 40 N, St. Patrick's Day (March 17) is the day with almost exactly 12 hours between sunrise and sunset in the spring.

The Moon Illusion
Atmospheric refraction is not, however, the reason why the Moon sometimes appears to be so huge when you see it rising. In fact, if you measure the Moon's apparent diameter carefully, you will find it to be smaller when it is near the horizon than when it is
Continued

EXTENDING OUR REACH—*Continued*

FIGURE 5.7
Circles beside converging rails illustrate how your perception may be fooled. The bottom circle looks smaller than the circle on the horizon but is in fact the same size. Similarly, the circle high in the sky looks smaller than the circle on the horizon.

overhead, regardless of how huge it looks. This misperception, known as the **Moon illusion,** is still not well understood but is an optical illusion caused, at least in part, by the observer's comparing the Moon with objects seen near it on the horizon, such as distant hills and buildings. You know those objects are big even though their distance makes them appear small. Therefore you un-

consciously magnify both them and the Moon, making the Moon seem larger. You can verify this sense of illusory magnification by looking at the Moon through a narrow tube that blocks out objects near it on the sky line. Seen through such a tube, the Moon appears to be its usual size.

Figure 5.7 shows a similar effect. Because you know that the rails are really parallel,

your brain ignores the apparent convergence of the railroad tracks and mentally spreads the rails apart. That is, your brain provides the same kind of enlargement to the circle near the rails' convergence point as it does to the rails, causing you to perceive the upper circle as larger than the lower one, even though they are both the same size.

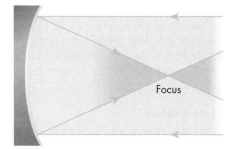

FIGURE 5.8
How a mirror focuses light.

To avoid such difficulties, most modern telescopes use mirrors rather than lenses to collect and focus light, and they are therefore called **reflectors**. The mirrors are made of glass that has been shaped to a smooth curve, polished, and then coated with a thin layer of aluminum or some other highly reflective material. As figure 5.8 shows, such a curved mirror can focus light rays reflected from it, creating an image just as well as a lens can. Moreover, because the light does not pass through the mirror, it focuses all colors equally well and does not absorb short-wavelength light. Furthermore, because the light does not have to pass through the mirror, the mirror can be supported from behind, thereby reducing the sagging problem that affects large lenses. For these and other reasons, astronomers now use reflecting telescopes almost exclusively. Figure 5.9 is a photograph of a large reflecting telescope, the 10-meter Keck reflector at Mauna Kea, Hawaii.

Figure 5.10 shows that a reflecting telescope focuses light in front of the mirror. Thus to observe the image the observer would ordinarily have to stand in front of the mirror, thereby blocking some of the light. To overcome this difficulty, a secondary mirror is often used to deflect the light either off to the side or back toward the mirror and out through a hole in its center. In one large reflector, the 4-meter (150-inch) National Optical

FIGURE 5.9
Photograph of one of the twin Keck telescopes. The 36 mirrors cover an area 10 meters (about 33 feet) in diameter, making them the world's largest optical telescopes. (Courtesy Roger Ressmeyer/Corbis.)

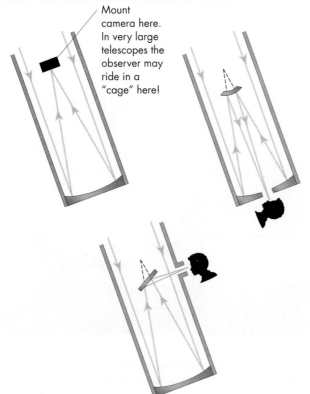

Mount camera here. In very large telescopes the observer may ride in a "cage" here!

FIGURE 5.10
Sketches of different focus arrangements for reflectors.

Astronomy Observatory telescope shown in figure 5.11, the astronomer may work in a cage directly behind the mirror.

Most telescopes are mounted on huge pivots that allow them to follow astronomical objects as they move across the sky. Swinging the many tons of metal and glass smoothly and with precision requires great care in construction and design. Moreover, as the telescope moves, its lenses or mirrors must keep their same precise shapes and relative positions if the images are to be sharp. This is one of the most technically demanding parts of building a large telescope, because the large pieces of glass used in lenses and mirrors bend slightly when their positions are shifted. One way to counteract this effect in a reflector is to collect and focus the light with several smaller mirrors and then align each one individually. That is, instead of a telescope having a single large mirror, it may have many small ones. Telescopes designed this way are called multimirror instruments. Currently, the largest such multimirror instruments are the twin 10-meter Keck Telescopes (fig. 5.12), operated by Caltech and the University of California, and located on the 14,000 foot volcanic peak Mauna Kea, in Hawaii. Each Keck telescope consists of 36 separate mirrors

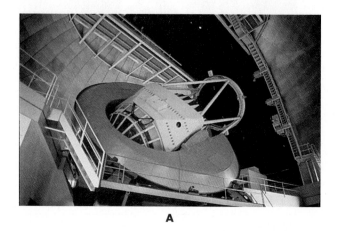

A

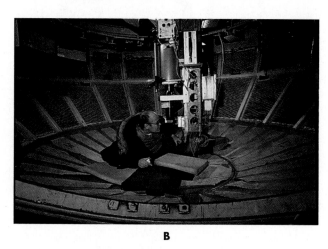

B

FIGURE 5.11
(A) Photograph of the 4-meter diameter reflecting telescope at the United States National Optical Astronomy Observatory on Kitt Peak near Tucson, Arizona. (B) An astronomer riding in the observing cage of this telescope. (Courtesy NOAO.)

FIGURE 5.12
Photograph of the twin Keck telescopes on the summit of Mauna Kea in Hawaii. As the telescopes track the stars or galaxies being observed, all 36 mirrors in each telescope can be measured with a laser and then adjusted to keep them in precise alignment. (Courtesy California Association for Research in Astronomy.)

(visible in fig. 5.9) that are kept aligned by lasers that measure precisely the tilt and position of each mirror. If any misalignment is detected, tiny motors shift the offending separate mirror segment to keep the image sharply in focus.

Ironically the problem of keeping the separate segments of a multimirror telescope aligned has led astronomers back to single mirrors, but mirrors much thinner and lighter than those in most existing telescopes. In the past, astronomers made mirrors thick to make them stiff and reduce sagging of the glass. Large pieces of glass, however, weigh more than smaller pieces and thus sag even worse unless very carefully supported. Just as a smaller gob of whipped cream will keep its shape on a tilted plate whereas a large gob will slump under its own weight, so too a thin piece of glass, if properly supported, keeps its shape better when tilted than a large piece. Thus astronomers have sought ways to make extremely thin mirrors that are then kept precisely shaped by alignment systems similar to those used in the Keck. Such a procedure is used on the New Technology Telescope of the European Southern Observatory in Chile which, as a result, can achieve several times better resolution than conventional telescopes.

Up to now, we have concentrated on the collecting power of telescopes: their ability to make dim objects bright enough to see. Telescopes, however, serve another important function—they increase our eyes' ability to see details.

Resolving Power

If you mark two tiny black dots close together on a piece of paper and look at them from the other side of the room, your eye may see them as a single dark mark, not as separate spots. Likewise, stars that lie very close together or markings on planets may not be clearly distinguishable. A telescope's ability to discern such detail depends on its **resolving power**.

Resolving power is limited by the wave nature of light. For example, suppose two stars are separated by a very tiny angle. For them to be discernible as separate images their light waves must not get mixed up. Such mixing, however, always occurs when waves pass through an opening, because as each wave passes the opening, smaller, secondary waves are produced in a phenomenon called **diffraction**. Figure 5.13A shows how water waves are diffracted as they pass through a narrow opening. Light waves are similarly diffracted as they enter a telescope. The result of diffraction is that point sources of light become surrounded by rings of light. You can observe diffraction by looking at a tiny, bright light source, such as a study lamp, through a piece of cloth, such as a T-shirt. The light will be surrounded by colored diffraction rings produced as the light waves pass through the tiny openings in the weave of the fabric. Similarly, if you look at a bright light source through a few strands of hair pulled over your eye, you may see diffraction fringes with rainbowlike colors around the hairs. An even better way to see the effects of diffraction is to fog a piece of clear glass with your breath and hold it close to your eye while you look at a small, bright light source. If you wear glasses, just breathe on them, put them back on, and look at a bright light. You will see colored diffraction rings around the light, as figure 5.13B illustrates.

FIGURE 5.13
(A) Water ripples diffracted as they pass through a narrow opening. (Courtesy E. R. Degginger.) (B) Photograph showing colored rings created by the diffraction of light waves by the tiny water droplets on a fogged piece of glass.

A

B

Diffraction seriously limits the detail visible through a telescope. In fact, diffraction theory shows that if two points of light that are separated by an angle α (measured in arc seconds*) are observed at a wavelength λ through a telescope whose diameter is D, the points cannot be seen as separate sources unless $D > 2.5 \times 10^5\, \lambda/\alpha$. If D is expressed in centimeters, λ in nanometers, and α in seconds of arc, then $D > 0.02\, \lambda/\alpha$. For example, to resolve two stars separated by 0.1 seconds of arc when observing with visible light ($\lambda = 500$ nanometers), you need a telescope whose lens has a diameter greater than 100 centimeters (about 40 inches). Notice that the diameter needed to resolve two sources increases as the sources get closer together.**

5.2 INTERFEROMETERS

Diffraction effects can never be totally eliminated, but they can be reduced by enlarging the opening through which the light passes, so that its waves do not mix as severely. Astronomers sometimes accomplish the same end with a device called an **interferometer.** With an interferometer, observations are made simultaneously through two widely spaced mirrors that direct the light to a common detector that combines the separate light beams, as shown in figure 5.14.

The interferometer is so named because when it mixes the separate beams, the light waves of one "interfere" with the waves from the other. Where the crests of two waves arrive together, they create a bright region. Where the crest of one wave arrives simultaneously with the trough of another, they cancel and create a dark patch. The result is a com-

FIGURE 5.14
Drawing of an astronomical interferometer for optical and infrared observations. (Courtesy IOTA project.)

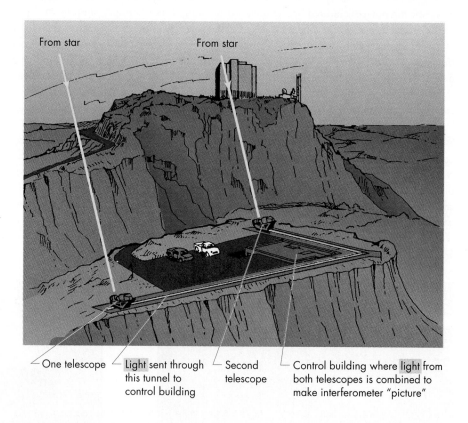

From star

From star

One telescope   Light sent through this tunnel to control building   Second telescope   Control building where light from both telescopes is combined to make interferometer "picture"

---

*An arc second is a unit of angle and is equal to 1/3600 of a degree.

**As we will discover later in this chapter, our atmosphere seriously blurs fine details in astronomical objects, degrading the resolving power of large ground-based telescopes to far below their diffraction limits.

plex pattern of alternating light and dark regions, which can be analyzed by a computer to create an image of the object observed. The result of this complicated process is an image in which the resolution is set not by the size of the individual mirrors but rather by their separation. For example, if the mirrors are 100 meters apart, the interferometer has the same resolving power as a telescope 100 meters in diameter. The resulting high resolution is far beyond what can be obtained in other ways. For example, figure 5.15A shows a view of two closely spaced stars as observed with a small telescope. Their images are severely blended as a result of diffraction. Figure 5.15B shows the same stars observed with an interferometer and after the image has been processed by a computer. The two stars can now be easily distinguished: the two separate mirrors produce the resolving power of a single mirror whose diameter equals the spacing between them.

5.3 OBSERVATORIES

The immense telescopes and the associated equipment astronomers use are extremely expensive. Therefore the largest telescopes are often national or international facilities, for example, the National Optical Astronomical Observatory of the United States and the Anglo-Australian Telescope. Despite the great expense of such facilities, many colleges and universities have their own large research telescopes (in addition to smaller ones near campus for instructional purposes). In addition, some large private groups such as the Carnegie Institution operate observatories. Altogether, several thousand observatories exist around the world on literally every continent. There is even a telescope at the South Pole in Antarctica to take advantage of the extreme dryness of the bitterly cold Antarctic air.

The three largest telescopes in the United States at this time are the 5-meter Palomar Mountain instrument operated by Caltech and located about 50 miles northeast of San Diego, California, and the two 10-meter Keck Telescopes, described earlier. Several other large, new telescopes are under construction both in the United States and in other countries. For example, the Hobby-Eberly telescope in Texas uses 91 mirrors 1-meter in diameter set into an 11-meter disk. To avoid distortions to this immense assembly of mirrors, the telescope can only by swiveled to point to a limited part of the sky. Moreover, the main mirror is only pointed approximately toward the star or galaxy of interest and then a much smaller "secondary" mirror adjusts the beam of light so that it falls on the detectors. Also, a consortium of European countries is building four telescopes, each with a single mirror 8 meters (more than 25 feet) in diameter, that will operate together on a mountain peak in Chile. Several other single-mirror, 8-meter telescopes are in the design stage in the United States, Canada, Europe, and Japan. Other large existing telescopes are the 236-inch Russian telescope located in the Caucasus area east of the Black Sea; the two 4-meter (150-inch) telescopes of the National Optical Astronomy Observatory, one located about 50 miles west of Tucson, Arizona, on Kitt Peak, and its twin, located in Chile; and the Anglo-Australian telescope operated jointly by Great Britain and Australia and located west of Sydney in the dry Australian outback. Astronomers need these large instruments in the Southern Hemisphere for complete sky coverage: a telescope located in the United States cannot see the sky near the south celestial pole.

5.4 DETECTING THE LIGHT

Once light has been collected, it must be detected and recorded. In olden days, the detector was the eye of an astronomer who sat at the telescope eyepiece and wrote down data or made sketches of the object being observed. The eye, marvel that it is, has difficulty seeing very faint light. Many astronomical bodies are too distant or too dim for their few photons to create a sensible effect on the eye. For example, if you were to look at any but the nearest galaxies through even the Palomar 5-meter telescope, they would appear

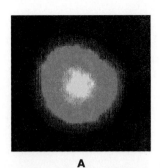

A

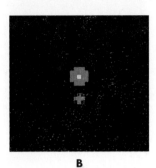

B

FIGURE 5.15
(A) A young star observed with an ordinary telescope. (B) The same star observed with an interferometer. The higher resolution of the interferometer reveals that the "star" is actually two stars in orbit around one another. (Courtesy Andrea Ghez, UCLA.)

merely as dim smudges. Only by storing up their light, sometimes for hours, can a quality picture of them be taken. Thus, to see very faint objects, astronomers use detectors that can store light in some manner. Such storage can be done chemically with film or electronically with detectors similar to those used in video camcorders.

From the later part of the last century until the 1980s, astronomers generally used photographic film to record the light from the bodies they were studying. Film forms an image by absorbing photons that cause a chemical change, making the film dark where light has fallen on it. This process, however, is very inefficient: fewer than 4% of the photons striking the film produce a useful image. The result of such low efficiency is that many hours are needed to accumulate enough light to create an image of faint objects. Moreover, the film must be developed, thereby delaying the observing process even further.

Astronomers today, however, almost always use electronic detectors similar to those in camcorders. In these devices, the incoming light strikes a semiconductor surface, allowing electrons to move within the material. The surface is divided into thousands of little squares, or pixels, in which the electrons are temporarily stored. The number of electrons in each pixel is proportional to the number of photons hitting it (that is, proportional to the intensity of the light). An electronic device coupled to a computer then scans the detector, counting the number of electrons in each pixel and generating a picture, much the way a dot-matrix printer creates a picture made up of many separate tiny dots.

Such electronic devices are extremely efficient, recording about 75% (or more) of the photons striking them, allowing astronomers to record images much faster than with film. Electronic detectors have other advantages as well. For example, they record the signal digitally, essentially counting every photon that falls on each part of the detector. Such digital images can be processed by computers to sharpen them, remove extraneous light, and enhance contrast.

Astronomers use many kinds of electronic detectors, but one of the best types is the **CCD**, which stands for "charge-coupled device." Modern CCDs can make pictures virtually indistinguishable from photographs in their detail and with a sensitivity to faint light approximately 20 times greater.

5.5 OBSERVING AT NONVISIBLE WAVELENGTHS

Visible light, which we can see because its wavelengths are detectable by our eyes, is just one of many wave bands of electromagnetic radiation. Many astronomical objects, however, radiate at wavelengths that our eyes cannot see, and so astronomers have devised ways to observe such objects. For example, cold clouds of gas in interstellar space emit little visible light but large amounts of radio energy. Thus, to observe them, astronomers use radio telescopes. Likewise, dust clouds in space are too cold to emit visible light, but they do radiate infrared energy, which astronomers observe with infrared telescopes. Similarly, astronomers use x-ray telescopes to observe the hot gas that accumulates around black holes. Radio, infrared, and x-ray telescopes have the same fundamental purposes as optical telescopes: to collect radiation and to resolve details. Figure 5.16 shows a typical radio telescope with its large "mirror" to collect the radiation and focus it on a detector. However, because our eyes cannot see these other wavelengths, astronomers must devise ways to depict what such instruments record.

One way astronomers depict observations made at nonvisible wavelengths is with false-color pictures, as shown in figure 5.17. In a false-color picture the colors used represent the intensity of radiation that we cannot see. For example, in figure 5.17A (a radio "picture" of a galaxy and the jet of hot gas spurting from its core) astronomers color the regions emitting the most intense radio energy red; they color areas emitting somewhat less energy yellow and color the faintest areas blue. Thus, if we could "see" radio waves, the red areas would look brightest and the blue areas dimmest. Similar techniques may be used to depict astronomical objects at other wavelengths. For example, figure 5.17B shows a false-color x-ray "photograph" of the gas shell ejected by an exploding star.

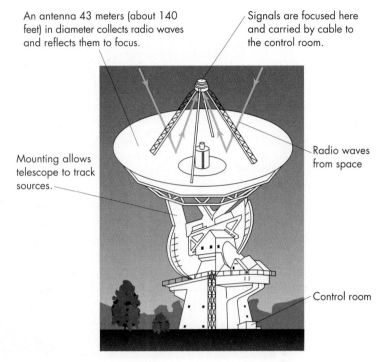

An antenna 43 meters (about 140 feet) in diameter collects radio waves and reflects them to focus.

Signals are focused here and carried by cable to the control room.

Mounting allows telescope to track sources.

Radio waves from space

Control room

FIGURE 5.16
Photograph of a radio telescope, the 43-meter (140-foot) diameter National Radio Astronomy Observatory telescope in Green Bank, West Virginia. (Courtesy NRAO/AUI.)

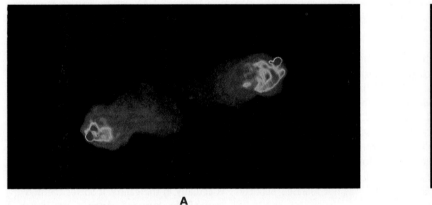

A

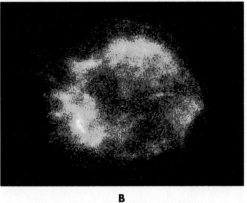

B

FIGURE 5.17
(A) False-color picture of a radio galaxy. (Courtesy NRAO/AUI.) (B) False-color X-ray picture of an exploding star. (Courtesy Stephen S. Murray, Smithsonian Astrophysical Observatory.)

Telescopes operating at infrared, ultraviolet, and x-ray wavelengths face an additional obstacle: most of the radiation they seek to measure cannot penetrate the Earth's atmosphere. Gases in our atmosphere such as ozone, carbon dioxide, and water strongly absorb infrared, ultraviolet, and shorter wavelengths. For example, infrared radiation whose wavelength is 50 micrometers is almost totally absorbed by water and carbon dioxide in our atmosphere. However, some infrared radiation penetrates the atmosphere in narrow-wavelength regions called "atmospheric windows," as shown in figure 5.18. Thus radiation whose wavelength is between 8 and 15 micrometers is only weakly absorbed. If astronomers want to view an object in a blocked wavelength, they must use a telescope in space, operated by a scientist-astronaut in space or remotely from the ground.

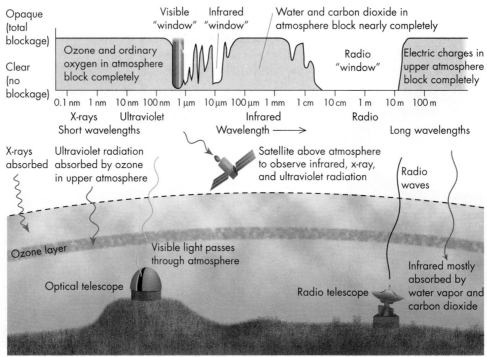

FIGURE 5.18
Diagram showing the transmission of light through the Earth's atmosphere.

EXTENDING OUR REACH

EXPLORING NEW WAVELENGTHS: GAMMA RAYS

Astronomers have made many of their most exciting discoveries when new telescopes allowed them to observe the sky at wavelengths not previously detectable. Gamma-ray wavelengths are among the last of the wavelength regions to be explored, and astronomers are still trying to interpret what they see.

Gamma-ray astronomy began in 1965 when a small and (by modern standards) primitive satellite detected cosmic gamma rays. A few years later, a slightly more advanced satellite detected gamma rays coming from the center of our galaxy, the Milky Way. By the 1970s astronomers had discovered that many familiar sources, such as the Crab Nebula and the remnants of other exploded stars, emitted gamma rays. Ironically, perhaps the most interesting gamma-ray sources, and ones that remain unexplained to this day, were discovered accidentally earlier.

In 1967 the United States placed several military surveillance satellites in orbit to watch for the gamma rays produced when a nuclear bomb explodes. The satellites were designed to monitor the United States/Soviet Union ban on nuclear bomb tests in the atmosphere. Curiously, on a number of occasions, the satellites detected gamma-ray bursts coming not from the Earth, but from space. Unfortunately for astronomers, the discovery of the bursts was top secret at the time and was not released publicly until 1973.

Astronomers' thirst for more information about these high-energy sources was unsatisfied for many years because our atmosphere absorbs gamma rays, and ordinary telescopes cannot focus gamma rays. Nevertheless, with ever more complex instruments in satellites, astronomers discovered that gamma-ray sources—apart from the bursts—coincided with known astronomical objects such as dying stars and some peculiar galaxies. The gamma-ray burst, on the other hand, would appear suddenly in otherwise blank areas of the sky, flare in intensity for a few seconds, and then fade to invisibility. Despite more than twenty years of study, astronomers still do not know what causes the bursts or from where they come. They outline no known sources like the zodiac or Milky Way. Astronomers cannot even answer the simple question, "Are they near or far?"—although not for lack of trying. Recently astronomers detected a gamma-ray burst that appears to coincide with a distant galaxy. If the source is actually in the galaxy, it solves the mystery of the bursters' distance but leaves unanswered what they are. Theories to explain the bursts abound. According to one of the most popular hypotheses, the bursts are formed when a pair of dead, collapsed objects called neutron stars* collide in a distant galaxy. According to a different theory, the bursts occur much closer to us and are triggered by a comet falling onto the surface of a neutron star at the outer fringes of the Milky Way. Mysteries like this are what make doing science exciting.

_____
*Neutron stars are the remnants of massive stars that have exploded. They are discussed in more detail in chapter 14.

EXTENDING OUR REACH

OBSERVING THE CRAB NEBULA AT MANY WAVELENGTHS

In midsummer, A.D. 1054 just after sunset, astronomers in ancient China and other far eastern countries noticed a brilliant star near the crescent moon in a part of the sky where no bright star had previously been seen. They wrote of this event:

"In the last year of the period Chih-ho, . . . a guest star appeared. . . . After more than a year it became invisible."

We know today that these astronomers of long ago witnessed a supernova explosion, the violent event that marks the death of a massive star. Their record — nearly 1000 years old—begins a story that continues today as astronomers try to understand what causes a star explosion. Although the story began with naked-eye observations, it continues with observations made with telescopes on the ground and in space.

Moreover, the story illustrates how astronomers have come to rely on observing radiation at many wavelengths, not just visible light.

Despite its initial brilliance, the dying star seen so long ago, faded and disappeared from the sky and astronomical records. Then, in 1731, John Bevis, a British physician and amateur astronomer, noticed with his telescope a faint dim patch of light in the constellation Taurus. Twenty-seven years later, Charles Messier, a French astronomer and comet hunter, rediscovered the glowing cloud and made it the first entry in his catalog of fuzzy patches of light that were not comets. In 1844, Lord Rosse, a British astronomer and telescope builder, noticed that the fuzzy patch contained filaments (fig. 5.20A) that to his eye resembled a crab. He therefore named it the Crab Nebula.

In 1921, John Duncan, an American astronomer, compared two photographs of the

FIGURE 5.19
Rock painting made by the Anasazi people, ancient pueblo dewllers of northern New Mexico. The star near the Moon may represent the supernova explosion of 1054 that gave birth to the Crab Nebula. (Courtesy Mike Zeilik.)

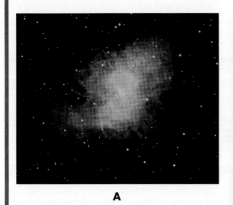

A

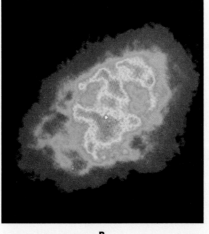

B

C

FIGURE 5.20
(A) Visible-light photograph of the Crab Nebula. (Courtesy R. Wainscoat.) (B) Radio image of the Crab Nebula. (Courtesy NRAO.) (C) X-ray image in false-color of the Crab Nebula. (Courtesy ROSAT.)

Continued

EXTENDING OUR REACH—*Continued*

nebula taken twelve years apart and noticed that it had increased slightly in diameter. He therefore deduced that the nebula was expanding. At the same time, several other astronomers came across the ancient Chinese records and noticed the coincidence in position of the nebula with the report of the exploding star. Then, seven years later, Edwin Hubble, at Mt. Wilson Observatory in California, measured the increase of size more accurately and calculated from the rate of expansion that the nebula was about 900 years old — roughly the same age as the dying star seen nearly a millennium earlier by the Chinese astronomers. Thus, astronomers realized that the Chinese astronomers had witnessed a supernova explosion, marking the death of a star.

Since then, astronomers have examined the Crab Nebula with telescopes using virtually all wavelength bands and, in doing so, have added yet more to their understanding of a star's demise. For example, in 1948, Australian astronomers using a radio telescope discovered that the Crab Nebula is a powerful source of radio waves (fig. 5.20B). In 1968, further observations at radio wavelengths revealed that a faint peculiar star near the center of the nebula is spinning about 30 times per second and that it is the core of the star whose explosion created the Crab Nebula. Likewise, with an x-ray telescope above the atmosphere, astronomers discovered that it is a source of x-ray radiation (fig. 5.20C).

What have all these observations shown? They have given astronomers their best view yet of the last moments of a star's life. From visible-wavelength observations, astronomers measure that the gas ejected when the star exploded is expanding with a speed of about 1000 km/s. From radio-wavelength observations they deduce that the nebula contains charged particles moving at nearly the speed of light and that the central star pulses on and off about 30 times per second. The x-ray observations confirm this picture. Thus, by observing the Crab Nebula and its stellar corpse at a variety of wavelengths, astronomers have shown that it is a far richer and more mysterious object than could be deduced from observations at one wavelength alone.

5.6 OBSERVATORIES IN SPACE

Figure 5.21 shows several of the many orbiting telescopes that astronomers use. Some of these have been operating for decades. For example, the International Ultraviolet Explorer (IUE) was launched in 1978 and was turned off in 1996. Other orbiting telescopes, however, have much briefer useful lives. For example, the Infrared Astronomy Satellite (IRAS)—designed to map the sky at infrared wavelengths that cannot penetrate the atmosphere—required a supply of liquid helium to keep its instruments cold for maximum sensitivity. Built and operated by the United States, Great Britain, and the Netherlands, it had a successful two-year flight, obtaining beautiful pictures of gas and dust clouds in space and discovering star formation activity in distant galaxies on a scale undreamed of before its flight. IRAS is now dormant, however, because it has run out of its liquid helium coolant. Its task of studying the sky at infrared wavelengths has been taken over now by ISO, the Infrared Space Observatory, operated by a group of European nations. In the meantime, other wavelength regions not observable from the ground are being probed, such as the x-ray and gamma-ray parts of the electromagnetic spectrum.

X-rays are absorbed in our atmosphere by ozone and oxygen. To observe cosmic x-ray sources, astronomers in Germany, Great Britain, and the United States have therefore launched the Roentgen x-ray Satellite Observatory, named for the German physicist who first detected x-rays. The Roentgen satellite, ROSAT for short, makes x-ray maps of the sky as well as detailed observations of individual x-ray-emitting objects such as exploding stars. ROSAT is actually only one of several smaller x-ray observatories launched by several countries, including Great Britain, The United States, Russia, and Japan.

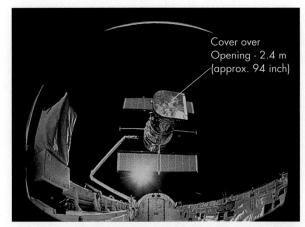

Hubble Space Telescope (13.6 m long)–HST

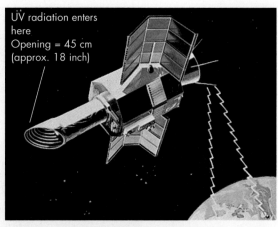

International Ultraviolet Explorer–IUE

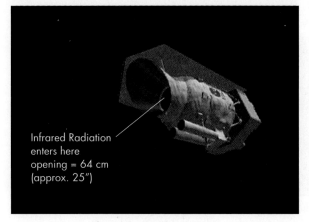

Infrared Space Observatory

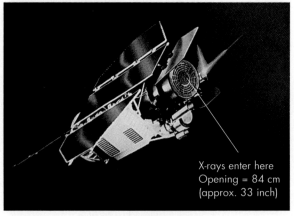

Roentgen Satellite–ROSAT

FIGURE 5.21
Photograph of the Hubble Space Telescope and drawings of some other orbiting observatiories: IUE, ISO, and ROSAT. (Courtesy NASA, ESA and Max Planck Institute for Extraterrestrial Physics, Germany.)

Of the many orbiting telescopes used by astronomers, the Hubble Space Telescope (HST) is the most ambitious to date. The HST is designed to observe at visible and ultraviolet wavelengths and has a mirror 2.4 meters (about 94 inches) in diameter that directs light to a set of instruments. Its two electronic cameras, one for wide-field views and the other for detailed, large-scale images, have already taken many striking pictures. The HST also has two spectrographs for analyzing light from stars and galaxies.

Although the HST initially had a number of problems, astronauts have repaired the major defects, and astronomers are now delighted with the clarity of its images (fig. 5.22). These images reveal details never before seen by telescopes on the ground because such telescopes must peer through the blurring effects of our atmosphere.

Atmospheric Blurring

Anyone who has ever watched the stars on a clear night has seen the blurring that our atmosphere creates. The light we see from the stars flickers, making them "twinkle." Twinkling, more properly called **scintillation,** is caused by atmospheric irregularities refracting the star's light. These irregularities result from slight variations in the air's density caused by small temperature differences. As light moves through these irregularities, its speed changes by a tiny amount, and the light is slightly refracted. As the irregularities drift on the wind between you and the star, they bend its light erratically into and out of your eye,

Jupiter

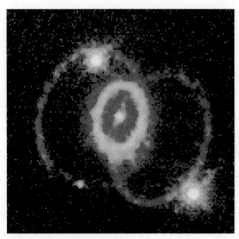

Ring of gas around Supernova 1987 A

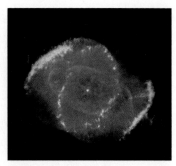

Cat's eye nebula

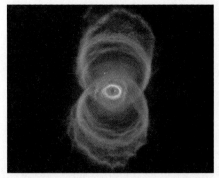

Hourglass Nebula

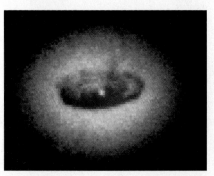

Disk of hot gas in core of the galaxy
NGC 4261; a black hole may lie in the
disk's center

FIGURE 5.22
Some of the pictures obtained with the
Hubble space telescope. (Courtesy NASA.)

making the star's brightness change (fig. 5.23). You can see a similar effect looking at something through water. If the surface of the water has even slight disturbances, a pebble or coin on the bottom seems to dance around. Atmospheric irregularities also slightly disperse the light, making the star's color dance, too. Such refractive twinkling, though very pretty to watch, seriously limits the ability of Earth-bound observers to see fine details in astronomical objects. The dancing image of a star or planet blurs its picture when recorded by a camera or other device. Outside the atmosphere (in space) this blurring, technically called **seeing**, disappears. The elimination of seeing problems caused by the turbulence of our atmosphere is one reason why astronomers find space observatories so useful.

Until recently, ground-based astronomers had to submit to the distortions of seeing, but now they can partially compensate for such seeing in several ways. One technique involves observing a star of known size simultaneously with the object of interest. By measuring carefully how the atmosphere distorts the known star's image, corrections can be made in the pictures of other objects.

Another method to improve seeing uses an artificial star created by a powerful laser beam. The laser beam is projected on the atmosphere, as shown in figure 5.24. The distortions of the artificial star image are recorded by a computer that then triggers tiny actuators on a correcting mirror placed in the telescope's light beam. The actuators create tiny distortions in the correcting mirror that cancel out those created by the atmosphere. This technique, called **adaptive optics,** has already given astronomers dramatically improved views through the turbulence of our atmosphere.

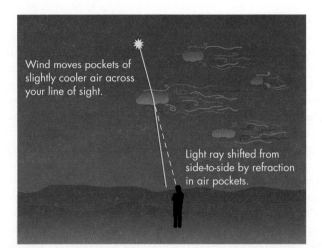

FIGURE 5.23
Twinkling of stars (seeing) is caused by moving atmospheric irregularities that refract light in random directions.

Wind moves pockets of slightly cooler air across your line of sight.

Light ray shifted from side-to-side by refraction in air pockets.

FIGURE 5.24
A laser beam creates an artificial star whose image serves as a reference to eliminate the atmosphere's distortion of real stars. This photograph was taken at the Starfire Optical Range of the Phillips Laboratory at Kirtland Air Force Base in New Mexico. (Courtesy USAF.)

Space Observatories versus Ground-Based Observatories

Despite the freedom from atmospheric blurring and absorption that space observatories enjoy, much astronomical work will be done from the ground for the foreseeable future. For many years, ground-based telescopes will be much larger than orbiting telescopes. Moreover, equipment problems can be corrected easily without the expense, delay, danger, and complexity of a space-shuttle launch.

Because huge telescopes in space or even on the Moon will remain dreams for years to come, astronomers therefore choose with care the location of ground-based observatories. Sites are picked to minimize clouds and the inevitable distortions and absorption of even clear air. Thus nearly all observatories are in dry, relatively cloud-free regions of the world such as the American southwest, the Chilean desert, Australia, and a few islands such as Hawaii and the Canaries. Moreover, astronomers try to locate observatories on mountain peaks to get them above the haze that often develops close to the ground in such dry locales.

Recently, astronomers have had to contend with another factor that affects the location of observatories: light pollution. Most inhabited areas are peppered with nighttime lighting such as street lights, advertising displays, and automobile headlights (fig. 5.25A and B). Although some such lighting may increase safety, much of it is wasted energy, illuminating unessential areas and spilling light upward into the sky, where it serves no purpose at all. Figure 5.25C shows a satellite view of North America at night and illustrates the waste of energy created by light pollution. Such stray light can seriously interfere with astronomical observations. In fact, some observatories have been essentially shut down because of light pollution. In some places, astronomers have persuaded regional planning bodies to develop lighting codes to minimize light pollution. Light pollution, however, not only wastes energy and interferes with astronomy, it also destroys a part of our heritage—the ability to see stars at night. The night sky is a beautiful sight, and it is shameful to deprive people of it.

A

B

C

FIGURE 5.25
Photographs illustrating light pollution.
(A) Los Angeles basin viewed from Mt. Wilson Observatory in 1908. (Courtesy Kitt Peak National Observatory.)
(B) Los Angeles at night in 1988.
(C) Notice the pattern of the interstate highway system visible in the satellite picture of North America at night.
(W. T. Sullivan, copyright 1993.)

5.7 GOING OBSERVING

When astronomers want to observe at a telescope, they do not just run off to the observatory. Generally they must submit an observing proposal that describes what objects they wish to look at and why such observations are important. Proposals are then screened by a committee that allocates telescope time. If the proposal is accepted, the astronomers must prepare for the observing run, being sure that all necessary equipment will be available. This may include backup equipment and computers as well as charts for finding the objects to be observed. Because many observatories are at remote locations, sometimes travel plans must also be made.

Observing sessions at a large telescope are typically assigned in blocks of several nights so that the astronomer's risk of clouds or equipment failure is minimized. If the astronomer is using an optical telescope and must therefore observe at night, he or she must also get used to switching schedules to be awake all night and sleep during the day. Most observatories have small dorm rooms with special shades to make the room dark and quiet during daylight hours for astronomers trying to catch up on sleep. They also often have small cafeterias where food (and coffee) is available at odd hours. Sometimes observing runs go smoothly, and the astronomer can return to his or her home institution laden down with data. At other times weather or equipment will not cooperate, and the run is a waste of both time and money.

5.8 COMPUTERS

In the last few decades, the computer has become one of the astronomer's most important tools (fig. 5.26). In fact, for many astronomers today, operating a computer and being able to program is more important than knowing how to use a telescope. Astronomers use computers not only to solve equations, but also to move the telescope, feed the information to detectors, and convert the data obtained to a useful form. Moreover, electronic computer networks allow astronomers to communicate and exchange their data and ideas with other astronomers around the world in a matter of minutes. For such purposes,

FIGURE 5.26
Using a computer, perhaps the astronomer's most important tool. (Courtesy Debra Elmgreen, Vassar College.)

observatories typically have a range of computers for different tasks, ranging from pocket calculators through desktop personal computers to work stations. If a huge computer is needed, for example, for the mathematical simulation of the collision of two galaxies or the explosion of a star, the astronomer may use the computer network to link his or her desktop machine to a supercomputer, perhaps thousands of miles away. The hard calculations are then done on the remote machine, but the output ends up at the desktop.

5.9 ASTRONOMERS

Who are these astronomers so often referred to in this book? They come from many backgrounds and countries and include both women and men. They have generally studied physics and astronomy in college and have then gone to graduate school for more specialized course work. Some are your professors, people with a deep passion for understanding what the Universe is like and how it works.

SUMMARY

Astronomers use telescopes to collect radiation from astronomical sources. Telescopes generally have large collecting areas so as to gather as much radiation as possible and allow faint objects to be seen. A large collecting area also increases the telescope's ability to resolve detail, giving sharper images, but such gains are seriously limited for ground-based visible light telescopes by the blurring effects of our atmosphere. Interferometers give the resolving power of a single large area by combining radiation detected with two or more small but widely separated collectors.

Although many astronomical objects radiate visible light, some of the most interesting bodies radiate in radio, infrared, x-ray or gamma-ray regions of the electromagnetic spectrum. Astronomers use special telescopes to observe in these other wavelength regions. Because many of these wavelengths do not penetrate our atmosphere, telescopes have been put in space to observe them.

QUESTIONS FOR REVIEW

1. What is meant by the term *collecting area?* How does it affect the ability to see faint objects?
2. What is resolution of a telescope? What physical process limits it?
3. How is resolution affected by the size of a telescope's mirror or lens?
4. What is the difference between a reflecting and a refracting telescope?
5. What are some reasons for using mirrors rather than lenses in telescopes?
6. What is the purpose of an interferometer?
7. Why do astronomers put x-ray observatories in space rather than just on a high mountain?
8. What is meant by adaptive optics?

THOUGHT QUESTIONS

1. How, apart from magnification, do binoculars help you see better?
2. Which binoculars should you choose to observe at night, a pair with 50-millimeter lenses or a pair with 35-millimeter lenses? Why?
3. Why does the useful resolving power of a ground-based telescope with a 2-meter diameter mirror not match its theoretical value?
4. If you look with binoculars down a beach on a hot day, you will see that distant objects appear shimmery. How is this related to astronomical "seeing"?

PROBLEMS

1. Compare the collecting power of a telescope with a 10-centimeter (about 4-inch) diameter mirror to that of the human eye. (Take the diameter of the pupil of the eye to be about 5 millimeters.)

2. Estimate your eye's resolving power by drawing two lines 1 millimeter apart on a piece of paper. Put the paper on the wall and then step back until the two lines appear as one, measuring that distance. From the distance and the separation of the lines (1 millimeter), estimate their angular separation. How does your result for the eye's resolving power compare with that calculated from the resolving-power formula, using a pupil diameter of 5 millimeters and a wavelength of 500 nanometers?

3. Can the unaided human eye resolve a crater on the Moon whose angular diameter is 1 minute of arc (=60 seconds of arc)? (Take the diameter of the pupil of the eye to be about 5 millimeters and the wavelength of the light to be 500 nanometers.)

TEST YOURSELF

1. Telescope A's mirror has three times the diameter of telescope B's. How much greater is A's light-gathering power?
   (a) 3 times.
   (b) 6 times.
   (c) 8 times.
   (d) 9 times.
   (e) 27 times.

2. A telescope's resolving power measures its ability to see
   (a) fainter sources.
   (b) more distant sources.
   (c) finer details in sources.
   (d) larger sources.
   (e) more rapidly moving sources.

3. A ground-based telescope to observe x-rays would
   (a) be a powerful tool for studying abnormally cold stars or distant planets.
   (b) give astronomers the chance to study the insides of stars and planets.
   (c) be worthless because no astronomical objects emit x-rays.
   (d) be worthless because x-rays cannot get through the Earth's atmosphere.
   (e) be worthless because astronomers have not yet devised detectors for x-rays.

4. Astronomers use interferometers to
   (a) observe extremely dim sources.
   (b) measure the speed of remote objects.
   (c) detect radiation that otherwise cannot pass through our atmosphere.
   (d) enhance the resolving power (see fine details) in sources.
   (e) measure accurately the composition of sources.

5. One way to increase the resolving power of a telescope is to
   (a) make its mirror bigger.
   (b) make its mirror smaller.
   (c) replace its mirror with a lens of the same diameter.
   (d) use a mirror made of gold.
   (e) observe objects using longer wavelengths.

FURTHER EXPLORATIONS

Chaisson, Eric J. "Early Results from the Hubble Space Telescope." *Scientific American* 266 (June 1992): 44.

Finkbeiner, A. "The Future of a Science." *Mosaic* 22 (Winter 1991): 12.

Friedman, H. "Discovering the Invisible Universe." *Mercury* 20 (January/February 1991): 2.

Fugate, Robert Q., and Walter J. Wild. "Untwinkling the Stars—Part I." *Sky & Telescope* 87 (May 1994): 24.

Martin, B., et al. "The New Ground-Based Optical Telescopes." *Physics Today* 44 (March 1991): 22.

McAlister, Harold A. "Twenty Years of Seeing Double." *Sky and Telescope* 92 (November 1996): 28.

Mitton, Jacqueline, and Stephen P. Maran. *Gems of Hubble*. New York: Cambridge University Press, 1996.

Parker, Barry R. *Stairway to the Stars: The Story of the World's Largest Observatory*. New York: Plenum Press, 1994.

Physics Today 44 (April 1991): Four articles devoted to observational astronomy.

Powell, Corey S. "Mirroring the Cosmos." *Scientific American* 265 (November 1991): 112.

Smith, Bradford A., and Roger H. Ressmeyer. "New Eyes on the Universe." *National Geographic* 185 (January 1994): 2.

Strom, Stephen. E. "New Frontiers in Ground-Based Optical Astronomy." *Sky and Telescope* 82 (July 1991): 18.

Web Sites

You can look at many of the pictures taken by the instruments described in this chapter at the following websites.

Hubble Space Telescope pictures and information about the telescope are available at

www.oposite.stsci.edu/pubinfo/Pictures.html
and
www.stsci.edu/pubinfo

Information about the National Optical Observatories (Kitt Peak, for example) can be found at

www.noao.edu/noao.html

Information about the European Southern Observatory can be found at

http.hq.eso.org/eso.homepage.html

(Note: This address may look odd, but it does begin with http. This is a fairly technical site.)

Information about the National Radio Astronomy Observatory can be found at

info.aoc.nrao.edu

KEY TERMS

adaptive optics, 170
CCD, 164
diffraction, 161
dispersion, 157
interferometer, 162
Moon illusion, 158

reflectors, 158
refraction, 154
refractors, 154
resolving power, 161
scintillation, 169
seeing, 170

 **PROJECTS**

1. Make a simple spectroscope using aluminum foil, a cardboard tube, and a grating. Plans are sketched below. Put aluminum foil over a small hole at one end of the tube. Cut a very thin slit in the foil with a razor blade or sharp knife. The slit should be straight, about an inch long, and as narrow as you can make it and still have light get through it. At the other end of tube mount a small square of grating material. Tape it over a small hole cut in the tube's end. If you use a toilet-paper tube, just bunch the foil over one end and tape it on. Fold some thin cardboard over the other end and cut a small square hole about an inch across. Hold the grating material up to light and notice to which side it creates a rainbow. Tape the grating along its edges oriented so that the rainbow runs perpendicular to the slit. Ask your instructor about details of construction and how to get the grating. To use, hold grating end of tube right up to your eye and look at the light source. Rotate tube so that you see a rainbow off to the side. Sketch the spectra you see from (a) a fluorescent light, (b) a mercury vapor street light, (c) an ordinary incandescent light bulb, (d) the blue sky.

2. Use your spectroscope to look at the spectrum of flames. Use a gas burning stove and add a pinch of table salt to see sodium emission lines (yellow) or use copper sulfate crystals or a scrap of copper wire to see green copper emission lines.

Old oatmeal box–or even toilet paper tube. A square box will work as well. Length should be at least 4 inches but may be longer.

Narrow slit–as thin as possible (approximately 1 inch long).

Grating material taped over hole. Hole should be about 1 inch square.

Keeping Time

From before recorded history, people have used events in the heavens to mark the passage of time. The day, the month, and the year were all originally defined in terms of obvious astronomical phenomena. The day is the time interval from sunrise to sunrise. The month is the interval from new moon to new moon. The year is the time it takes for the Sun to complete one circle of the zodiac. Astronomical events are not perfect time markers: even the day and year need to be defined with care if we are to have reliable clocks and calendars. Otherwise, we may end up with snow in summer and heat waves in winter.

Length of the Daylight Hours

Although the day is always 24 hours long, the number of hours of daylight, or the amount of time the Sun is above the horizon, changes drastically throughout the year unless you are very close to the equator. For example, in northern middle latitudes, including most of the United States, southern Canada, and Europe, summer has about 15 hours of daylight and only 9 hours of night. In the winter, the reverse is true.

This variation in the number of daylight hours is caused by the Earth's tilted rotation axis. Remember that as the Earth moves around the Sun, its rotation axis points in roughly a fixed direction. Thus the Sun shines more directly on the Northern Hemisphere during its summer and less directly during its winter. The result (as you can see in fig. E2.1) is that only a small part of the Northern Hemisphere is unlit in the summer, but a large part is unlit in the winter. Thus, as the Earth's rotation carries us around, only a relatively few hours of a summer day are unlit, but a relatively large number of winter hours are dark. Figure E2.1 also shows that on the first day of spring and autumn (the equinoxes) the hemispheres are equally lit, so that day and night are of equal length everywhere on Earth.

If we change our perspective and look out from the Earth, we see that during the summer the Sun's path is high in the sky, so that the Sun spends a larger portion of the day above the horizon. This gives us not only more heat but also more hours of daylight. On the other hand, in winter the Sun's path across the sky is much shorter, giving us less heat and fewer hours of light.

The Day

The length of the day (24 hours) is set by the Earth's rotation on its axis. If you were to measure the time interval from one sunrise to the next with a watch, however, you would find that in general it is *not* exactly 24 hours. Furthermore, the interval would be slightly different from one day to the next.

On the other hand, the time interval between the risings of a given star is constant. However, it is a bit less than 24 hours. That is, there is a difference in the length of the day as measured by the stars and the length of the day as measured by the Sun. The length of the day measured with respect to the stars is called the **sidereal day.** The length of the day measured with respect to the Sun is called the **solar day.**

Dalí, Salvador. *The Persistence of Memory [Persistence de la Mémoire].* 1931 Oil on canvas, 9 1/2 × 13 (24.1 × cm). The Museum of Modern Art, New York. Given anonymously.

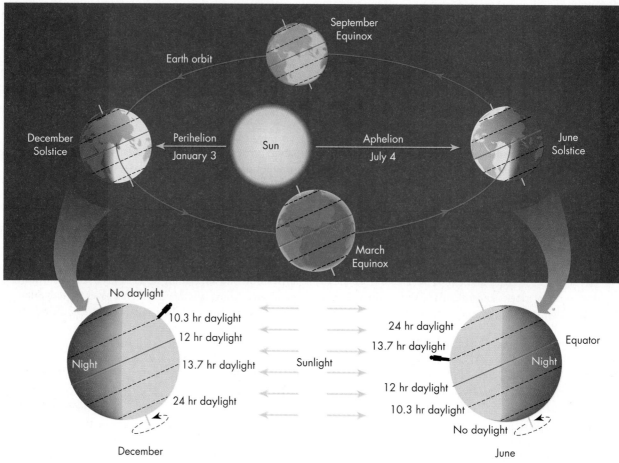

FIGURE E2.1

The tilt of the Earth affects the number of daylight hours. Locations near the equator always receive about 12 hours of daylight, but locations toward the poles have more hours of dark in winter than in summer. In fact, above latitudes 66.5°, the Sun never sets for part of the year and never rises for another part of the year (the midnight sun phenomena). At the equinoxes, all parts of the Earth receive the same number of hours of light and dark.

The difference in the length of the sidereal and solar days arises from the Earth's orbital motion around the Sun. We can see the reason by looking at figure E2.2, where instead of measuring the interval between sunrises, we measure the interval between successive apparent noons, the time when the Sun is highest in the sky. As the Earth spins on its axis, we watch the Sun pass overhead, timing its passage until it is again overhead. That time interval is a solar day. Let us imagine that at the same time we are watching the Sun, we can also watch a star and that we measure the time interval between its passages overhead, what astronomers call a sidereal day.

As we wait for the Sun and star to move back overhead, the Earth moves along its orbit. The distance the Earth moves in one day is so small compared with the star's distance that we see the star in essentially the same direction as on the previous day. However, we see the Sun in a measurably different direction, as figure E2.2 shows. The Earth must therefore rotate a bit more before the Sun is again overhead. That extra rotation, needed to compensate for the Earth's orbital motion, makes the solar day slightly longer than the sidereal day.

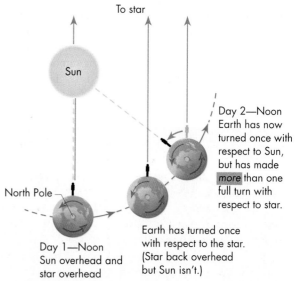

FIGURE E2.2

The length of the day measured with respect to the stars is not the same as the length measured with respect to the Sun. The Earth's orbital motion around the Sun makes it necessary for the Earth to rotate a tiny bit more before the Sun will be back overhead. (Motion is exaggerated for clarity.)

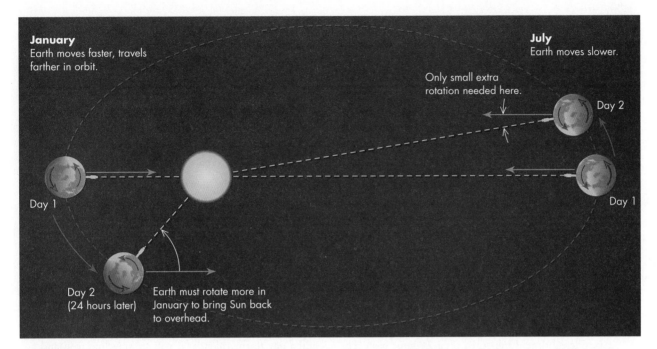

January
Earth moves faster, travels farther in orbit.

July
Earth moves slower.

Only small extra rotation needed here.

Day 2

Day 1

Day 1

Day 2
(24 hours later)

Earth must rotate more in January to bring Sun back to overhead.

FIGURE E2.3
As the Earth moves around the Sun, its orbital speed changes as a result of Kepler's second law of motion. For example, the Earth moves faster in January when it is near the Sun than in July when it is far from the Sun. Thus in 24 hours the Earth moves farther along its orbit in January than in July. As a result, the Earth must turn slightly more in January to bring the Sun back to overhead. This makes the interval between successive noons longer in January than in July and means they are not exactly 24 hours. For that reason, time is kept using a "mean Sun" that moves across the sky at the real Sun's average rate.

It is easy to figure out how much longer on the average the solar day must be. Because it takes us 365¼ days to orbit the Sun, and because there are by definition 360 degrees in a circle, the Earth moves approximately 1 degree per day in its orbit around the Sun. That means that for the Sun to be visible on the horizon, the Earth must rotate approximately one degree past its position at the previous dawn.* In 24 hours = 24 × 60 = 1440 minutes, the Earth rotates 360 degrees. Therefore to rotate 1 degree takes 1440/360 minutes, or about 4 minutes. The solar day is therefore about 4 minutes (3 minutes 56 seconds, to be precise) longer than the sidereal day.

The motion of the Earth around the Sun also alters the length of the solar day. If you measure carefully the time interval from sunrise to sunrise, in general, it is not 24 hours but may differ slightly either way. This variation arises because the Earth's orbit is not circular, and therefore our orbital velocity changes according to Kepler's second law.

The Earth moves along its orbit faster when it is near the Sun and slower when it is farther away. This means that it takes a little longer for the Earth to swing you around into the morning Sun (slightly lengthening the interval between successive sunrises) when the Earth is moving rapidly in its orbit than when it is moving slowly (fig. E2.3). Hence, the solar day is longer when we are near the Sun and shorter when we are farther away. The amount by which the length of the solar day varies is small, but it must be accounted for if our clocks are to always read about noon when the Sun is highest in the sky.

We could design clocks so that the hour is of different lengths at different times of the year. That could be done to ensure that our clocks advance to conform with the changing length of the solar day. However, it is much easier to define the length of the day differently, using not the true interval from sunrise to sunrise, but the *average* value of that interval over the year. That average daylength is called the **mean solar day,** and it has, by definition, 24 hours of clock time. We therefore use mean solar time in our daily timekeeping.

The difference in length between the mean solar day and the true solar day accumulates over time and leads to a difference of several minutes between clock time and time based on the position of the Sun. This difference is called the "equation of time" and is shown graphically in figure E2.4. The equation of time gives

---

*Another way of thinking about this is that the Sun is slowly moving eastward across the sky through the stars at the same time the Earth is rotating. Thus, in a given "day", the Earth must rotate a bit more to keep pace with the Sun than it would to keep pace with the stars.

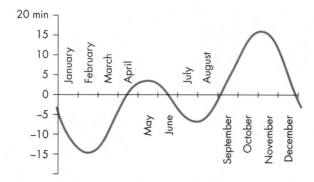

FIGURE E2.4
The equation of time is the correction that must be applied to the true Sun to determine mean solar time. It can be shown as a graph (left) or as a figure-8 shape called an "analemma."

the correction needed on a sundial if it is to give the same time as your watch.

Although we use solar time in regulating our daily activities, astronomers find sidereal time extremely useful. One reason is that a given star always rises at the same sidereal time. To avoid the nuisance of A.M. and P.M., sidereal time is measured on a 24-hour basis. For example, the bright star Betelgeuse in the constellation Orion rises at about 10 P.M. in November but at about 8 P.M. in December. However, on a clock keeping sidereal time, it always rises at the same time at a given location, about 23 hours 50 minutes.

Time Zones

Because the Sun is our basic time-keeping reference, most people like to measure time so that the Sun is highest in the sky at about noon. This is unnecessary now that we have good electronic clocks that can keep time independent of the Sun. Nevertheless, it is a tradition that is hard to break, and as a result, clocks in different parts of the world are set to read different times. Because the Earth is round, the Sun can't be "overhead" everywhere at the same time so it can't be noon everywhere at the same time.

The Earth is therefore divided into 24 major **time zones** in which the time differs by one hour from one zone to the next. There are a few exceptions, but we'll ignore them here.

Across the contiguous 48 United States the time zones are, from east to west, Eastern, Central, Mountain, and Pacific (see fig. E2.5). The time within each zone is the same and is called **standard time.** Thus, in the eastern zone, the time is denoted Eastern Standard Time (EST) whereas in the central zone it is denoted Central Standard Time (CST). As you travel across the country, it is therefore necessary to reset your watch if you cross from one time zone to another, adding 1 hour for each time zone as you move from west to east and subtracting 1 hour when you move from east to west.

If you travel a very large number of time zones, you may need to make such a large time correction that you shift your watch past midnight. For example, if you could travel fast enough and far enough, setting

your watch back each time you cross a time zone, you could end up at your starting point with your watch set to a time 24 hours before you left!

A high-speed traveler cannot actually "gain" a day, however, because when you cross longitude 180° (roughly down the middle of the Pacific Ocean), you add a day to the calendar if you are traveling west and subtract a day if you are traveling east. The precise location where the day shifts is called the **international date line** (fig. E2.5). It generally follows 180° longitude but bends around extreme eastern Siberia and some island groups to ensure they keep the same calendar time as their neighbors.

Universal Time

The nuisance of having different times at different locations can be avoided by using **Universal time,** abbreviated as UT. Universal time is the time kept in the time zone containing the longitude zero, which passes through Greenwich, England. By using UT, which is based on a 24-hour system to avoid confusion between A.M. and P.M., two people at remote locations can decide to do something at the same time without worrying about what time zone they are in.

Daylight Saving Time

In many parts of the world, people set clocks ahead of standard time during the summer months and then back again to standard time during the winter months. This has the effect of shifting sunrise and sunset to later hours during the day, thereby creating more hours of daylight during the time most people are awake. Time kept in this fashion is therefore called **daylight saving time** in the United States. In other parts of the world it is called "Summer Time."

Daylight saving time was originally established during World War I as a way to save energy. By setting clocks ahead, less artificial light was needed during work hours late in the day. Nowadays, it allows us more daylight hours for recreation after work during the summer.

Daylight saving time in the United States currently runs from the first weekend in April to the last weekend in October. However, many people advocate extending

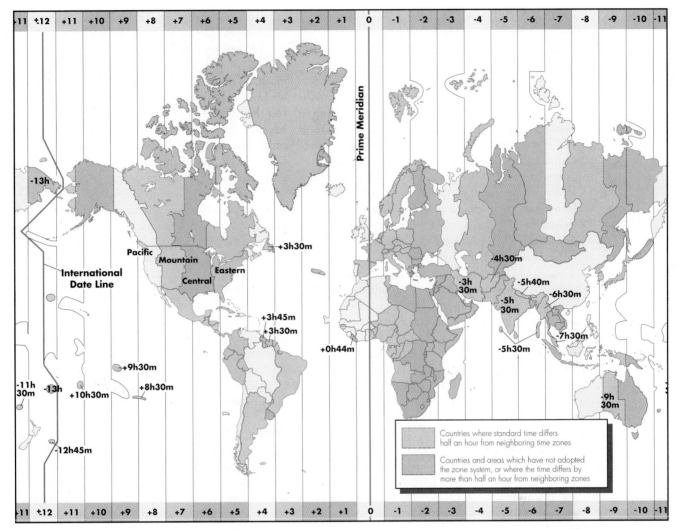

FIGURE E2.5
Time zones of the world and the international date line.

it a month earlier and later. This would presumably save energy by lessening still more the need for artificial light during waking hours.

The Month

The day with its 24 hours is excellent for keeping track of short periods of time, but for longer intervals, longer time units are helpful. Of these other units, the month is the next largest unit used by nearly all cultures.

As you know from looking at a calendar, the month is usually about 30 days long. This time interval, and its name, derives from the Moon's cycle of phases. The time interval between full moons is about 29.5 days, which for use on the calendar is rounded off to 30. Because the year has about 365 days in it, there are about 12 lunar cycles per year. That is the reason that we have 12 months in the year.

You will notice, however, that 12 lunar cycles of 30 days ends up 5 days short of a full year. For that reason,

some of the months are made 31 days long. In fact, if every other month, starting with January, were a 31-day month, the year would total 366 days. To make the days add up to 365, February is trimmed 1 day, to 29.

But, you protest, that is not the way the calendar looks. Although January, March, May, and July have 31 days, the sequence is broken in the later months. You see at work there the politics of ancient Rome.

The Calendar

Our calendar is based on one developed about 200 B.C. by the Romans. In fact, the word *calendar* is itself of Roman origin. There is some controversy about how the original Roman calendar was organized. It may have had only 10 months, and it probably began on the first day of spring (the vernal equinox) rather than in January.

Certainly the names of our months date from that calendar and its modifications. For example, if the year began in March, then September, October, November, and December were the 7th (Sept.), 8th (Oct.), 9th (Nov.), and 10th (Dec.) months, respectively.

Because it did not contain the right number of months and days to match the astronomical phenomena, this original calendar became a form of political patronage. The priests who regulated the calendar would add days and even months in order to please one group, and take days off to punish another. Such confusion resulted from these abuses that in 46 B.C., Julius Caesar asked the astronomer Sosigenes to design a calendar that would fit the astronomical events better and give less room for the priests and politicians to tinker with it. The resulting calendar, known as the **Julian calendar,** consisted of 12 months, which with the exception of February, alternated between 31 and 30 days in length.

The Julian calendar barely survived Caesar before the politicians were at it again. First, the name of the seventh month was changed to Julio to honor Julius Caesar—hence our July. Next, on the death of Julius Caesar's successor, Augustus Caesar, a very able and highly respected leader, it was decided to name the eighth month in *his* honor—hence the name August. However, because it would have been impolitic to have his month a day shorter than Julius's, August became a 31-day month, and all the following months had the number of their days changed to maintain the alternation. Unfortunately, this led to using up one more day than there were days in the year. Thus, poor February, already one day short, was trimmed a second day, leaving it with only 28 days. With only minor modifications, this is the calendar we use today. However, those modifications are important, as we will see.

Leap Year

The ancient Egyptians knew that the year is not exactly 365 days long. It turns out that it takes about 365 and ¼ days for the Earth to complete an orbit around the Sun, which is how we measure a year. Because we can't have fractions of a day in the calendar, a calendar based on a year of 365 days will come up 1 day short every 4 years.

Your first reaction might be, So what? However, the seasons are set by the orientation of the Earth's rotation axis with respect to the Sun, not by how many days have elapsed. We therefore want to make sure that we start each year with the Earth having the proper orientation. Otherwise, the seasons get out of step with the calendar. For example, because in 4 years you will lose 1 day, in 120 years you will lose a month, and in 360 years, you will lose an entire season. With a 365-day year, in a little over three centuries, April would be coming in what is now January.

This problem is corrected by leap year, a device used by the ancient Romans to keep the calendar in step with the seasons. Leap year corrects by adding a day to the calendar every fourth year. The extra day is traditionally added to February because it is the shortest month.

Unfortunately, the year is actually a little bit shorter than 365 ¼ days. Thus leap year corrects a tiny bit too much. To deal with this problem, only centuries divisible by 400 are leap years. Thus, 1900 was not a leap year, but 2000 will be. This modification of omitting leap year every 400 years was not part of the Julian calendar, but was added in 1582 at the direction of Pope Gregory XIII. The calendar we use today is thus known as the **Gregorian calendar**.

The inauguration of the Gregorian calendar in 1582 was not a peaceful affair. Because it was adopted roughly 1600 years after the Julian calendar, the error in the length of the year had grown to about 10 days. To bring the calendar back into synchrony with the seasons, Pope Gregory simply eliminated 10 days from the year 1582 so that the day after October 4 became October 15.

Although the changeover went smoothly in most places, non-Catholic countries such as Protestant England refused to abide by the Pope's edict. The calendar in England and in a few other northern European countries was not altered. This made commerce between Catholic and non-Catholic countries very difficult because the day and sometimes even the month and year were different from one country to the next.

Eventually, the Gregorian calendar was adopted essentially worldwide, but the change was not made in England until 1752, at which time rioting broke out because people feared that they would be charged a full month's rent for only 20 days. In Russia, the change was not made until the revolution in the early part of this century. Other countries (Greece and Turkey, for example) changed in the 1920s.

Religious Calendars

Although the Gregorian calendar is used nearly worldwide, many religions use traditional calendars for setting their feast and holy days. Two of the best known examples are the Jewish and Muslim calendars, which are particularly interesting to astronomers because of the role played in them by the lunar cycle.

The Muslim calendar is in fact a purely lunar one consisting of 12 months of either 29 or 30 days. The year thus comes out about 11 days shorter than those in the Gregorian calendar. This means that the Muslim calendar is totally out of synchrony with the seasons. As a result, the holy month of Ramadan can fall at any time during the year, irrespective of the season. This may seem odd to people living in climates with strong seasonal variations, but for people living in the Middle East where seasons are very unremarkable, it makes little difference.

Likewise, the Jewish calendar is based on the lunar cycle. To correct for the missing days, from time to time an extra month is added in the middle of the year to keep it in step with the seasons. The Jewish calendar is especially interesting astronomically because it begins near the autumnal equinox and the extra month is added near the vernal equinox. Also, the holy days of Yom Kippur and Passover are located near the equinoxes, a feature shared by the Gregorian calendar, wherein Easter is near the vernal equinox (actually the first Sunday after the first full moon after the equinox) and Christmas is near the winter solstice.

Other Calendars

Another calendar of note is the Chinese one which, like the Jewish calendar, combines lunar and seasonal aspects. The Chinese calendar is based on a 365.2444-day year broken up into 6 cycles of 60 days. The years are themselves grouped into 60-year cycles composed of 5 cycles repeating every 12 years. The years in each 12-year cycle are given names such as the Year of the Rat, The Year of the Dog, and so forth. Days and months are added so that it works out to match the cycle of the seasons.

Names of the Months and Days

That there are seven days in the week is probably a result of there being seven visible objects that move across the sky with respect to the stars: the Sun, the Moon, Mercury, Venus, Mars, Jupiter, and Saturn. We can see the names of some of these bodies in our English day names (Sunday, Monday, and Saturday). The influence is even clearer in the romance languages, such as Spanish (*lunes, martes, miércoles, jueves, viernes*).

Some English day names come to us through the names of Germanic gods (many of whom have a direct parallel with the Greco-Roman gods after whom the planets are named). For example, Tuesday is from *Tĩw,* god of war, like Mars (matching Spanish *martes*). Wednesday is named for *Wõden,* the chief god of Germanic peoples and identified with the Roman Mercury (matching Spanish *miércoles*). Thursday is named for *Thõr,* the thunder god (matching Spanish *jueves,* 'Jove's day'). Friday is named for *Freya,* a love goddess, like Venus (matching *viernes*).

The names of the months have less certain origins. Table E2.1 lists possible origins.

The Abbreviations A.M., P.M., B.C., and A.D.

Four abbreviations are used frequently in the measure of time and calendars. They are the familiar letters A.M., P.M., B.C., and A.D. The first two have specific astronomical meaning. The last two have cultural meaning.

A.M. and P.M. stand for "antemeridian" and "postmeridian," respectively. The meridian is the line passing from due north to due south and passing directly through the point exactly overhead (also called the zenith).

TABLE E2.1	Origin of the Names of the Months
January	Janus (gate), the two-faced god looking to the past and future; hence beginnings.
February	Februa, (expiatory offerings).
March	The god Mars.
April	Etruscan *apru* (April), probably shortened from Greek *Aphrodite,* goddess of love and earlier of the underworld.
May	Maia's month; Maia (She who is great), the eldest of the Pleiades and the mother of Hermes by Zeus.
June	Junius, an old Roman noble family (from Juno, wife and sister of Jupiter, equal to Greek Hera).
July	Julius Caesar (Julius "descended from Jupiter"; The Ju of June and July are the same: Jupiter "Sky-father").
August	Augustus Caesar (*augustus* means sacred, grand).
September-December	"Seventh-month" to "tenth-month." The *-ember* may come from the same root as "month."

As the Sun moves across the sky, it crosses the meridian at the time called apparent noon. Before noon, it lies before (ante) the meridian. After noon, it lies past (post) the meridian. Hence, A.M. and P.M.

B.C. stands for "before Christ," referring to the year of His birth. Oddly, there is no convention as to whether 1 B.C. refers to the year before or the year of his birth, although most historians make it the year of his birth to avoid having a year "0."

A.D. stands, not for "after death," but for *anno Domini,* meaning "the year of our lord." The term A.D. was introduced by the sixth-century monk Dionysius Exiguus, about A.D. 528, in his attempts to trace the chronology of the Bible.

Summary

Our system for keeping time is based on the motion of the Earth, Moon, and Sun. The day is determined by the Earth's spin, the month by the Moon's orbital motion around the Earth, and the year by the Earth's orbital motion around the Sun.

The solar day is based on the time interval between one sunrise and the next. The sidereal day, or the interval between the time of star-rise for a given star and the time of its next rising, is about 4 minutes shorter than the solar day. This difference arises because, as the Earth moves along its orbit, the direction to the Sun shifts slightly. We must therefore wait a little longer to allow the Earth's rotation to carry us into the same position with respect to the Sun.

Time zones divide the Earth into regions such that the time differs by 1 hour (in general) from zone to zone. The resulting time difference allows the Sun to be approximately at its highest point above the horizon at noon in each zone.

The Earth makes approximately 365.25 rotations in the time it takes it to complete one orbit around the Sun. Thus, every 4 years, an extra day accumulates, which in leap years we add to the calendar as February 29.

Questions for Review

1. How is the solar day defined? How is the sidereal day defined?

2. Why do the sidereal and solar days differ in length?

3. Why do we need a leap year?

4. Why isn't the solar day always exactly 24 hours long?

5. What do A.M., P.M., A.D., and B.C. stand for?

6. Suppose you were asked to revise the calendar. What changes would you make?

Test Yourself

1. Suppose that the length of the year was 365.2 days instead of 365.25 days. How often would we have leap year? Every
   (a) 2 years.
   (b) 5 years.
   (c) 10 years.
   (d) 20 years.
   (e) 50 years.

2. Suppose the Earth's rotation axis was not tilted with respect to its orbit. How would the number of daylight hours change throughout the year?
   (a) The number would be no different.
   (b) Days would be longer and nights shorter at all times of the year.
   (c) Days and nights would be of equal length throughout the year.
   (d) Days would be shorter and nights longer throughout the year.
   (e) None of the above.

3. Why is February the shortest month?
   (a) The Earth is moving most slowly in its orbit then.
   (b) The Earth is moving fastest in its orbit then.
   (c) The Earth spins faster in February than at other times of the year.
   (d) The Earth spins slower in February than at other times of the year.

   (e) When the calendar was revised, days were taken from February to make other months longer.

4. If at a given time of year the night is 24 hours long at the North Pole, how many hours long is the night at the South Pole?
   (a) 24 hours.
   (b) 12 hours.
   (c) 36 hours.
   (d) 48 hours.
   (e) There is no night then.

5. On what day(s) of the year are nights longest at the equator?
   (a) They are the same length throughout the year there.
   (b) The solstices.
   (c) The equinoxes.
   (d) Approximately June 21.
   (e) Approximately December 21.

Further Explorations

Bartky, Ian R., and Elizabeth Harrison. "Standard and Daylight-Saving Time." *Scientific American* 240 (May 1979): 46.

Cleere, G. S. "Eleven Lost Days." *Natural History* 100 (September 1991): 78.

Daniel, Glyn. "Megalithic Monuments." *Scientific American* 263 (July 1990): 78.

Jespersen, James, and Jane Fitz-Randolph. *From Sundials to Atomic Clocks: Understanding Time and Frequency.* National Bureau of Standards Monograph 155. Washington, D.C.: United States Government Printing Office, 1977.

Monson, B. "A Simple Method of Measuring the Length of the Sidereal Day." *Physics Teacher* 30 (December 1992): 558.

Moyer, Gordon. "The Gregorian Calendar." *Scientific American* 246 (May 1982): 144.

Key Terms

daylight saving time, 180
Gregorian calendar, 182
international date line, 180
Julian calendar, 182
mean solar day, 179
sidereal day, 177
solar day, 177
standard time, 180
time zones, 180
Universal time, 180

CHAPTER SIX

THE MOON

The Moon (Courtesy of © UCO/Lick Obs image)

Lunar crater (Courtesy of NASA)

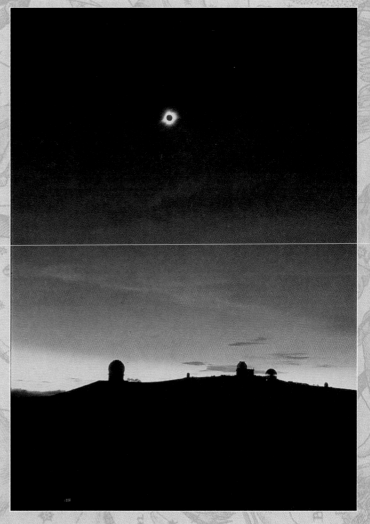

A total eclipse of the Sun (Courtesy of Richard Wainscoat, University of Hawaii)

The Moon is our nearest neighbor in space, a natural satellite orbiting the Earth. It is the frontier of direct human exploration, an outpost that we reached more than a quarter century ago but from which we have since drawn back. But despite our retreat from its surface, the Moon remains of great interest to astronomers. Although originally the Moon was molten, its small mass and radius have made it difficult for it to either generate or retain any appreciable internal heat. It is therefore a dead world, with neither plate tectonic nor volcanic activity. That inactivity when coupled with the Moon's lack of atmosphere means that its surface features are essentially unaltered since its youth. But the Moon has not always been quiescent. Shortly after its formation, it was pelted with a hail of rocky fragments whose size ranged up to 200 kilometers (about 100 miles) in diameter. The small fragments made craters, and the big fragments made huge basins. The basins subsequently flooded with lava (long since congealed) to create several dark, nearly circular plains easily visible to the naked eye. The Earth probably once bore such features, but erosion and plate motions have erased them. On the Moon's windless, rainless, airless surface, however, they remain—a record of events in the early Solar System that gives clues not only to the Moon's birth but to that of the Solar System as well.

In this chapter we will describe the Moon's surface and why astronomers believe so many of its features were carved by impact. We will see that lunar rocks differ significantly from terrestrial ones and how they point to the Moon's having been born in a cataclysmic event early in the Earth's history. We will also discuss how the Moon affects Earth today through tides and eclipses. But we will begin with a short physical description of the Moon to help us visualize this nearest world.

6.1 DESCRIPTION OF THE MOON

General Features

The Moon is about one fourth the diameter of the Earth, a barren ball of rock, possessing no air, water, or life. In the words of lunar astronaut Buzz Aldrin, the Moon is a place of "magnificent desolation." But you don't have to walk on the Moon to see its desolation. Even to the naked eye, the Moon is a world of grays without the vivid colors of the terrestrial landscape. Yet even gray has its variety, as you can see where the dark roughly circular areas stand out from the lighter background,* as shown in figure 6.1.

Surface Features

Through a small telescope or even a pair of binoculars you can see that the dark areas are smooth while the bright areas are covered with numerous large circular pits called **craters**, as illustrated in figure 6.2. Craters usually have a raised rim and range in size from tiny holes less than a centimeter across to gaping scars in the Moon's crust such as Clavius,† about 240 kilometers (150 miles) across. Some of the larger craters have mountain peaks at their center.

---

*These dark features create the face of the "Man in the Moon."
†Most lunar craters are named for famous scientists. For example, Cristoph Clavius (1537–1612) was a German astronomer and mathematician.

FIGURE 6.1
The Moon. (Courtesy Stephen E. Strom.)

FIGURE 6.2
Photograph showing the different appearance of the lunar highlands and maria. The highlands are heavily cratered and rough. The maria are smooth and have few craters. (© UCO/ Lick Obs image.)

A

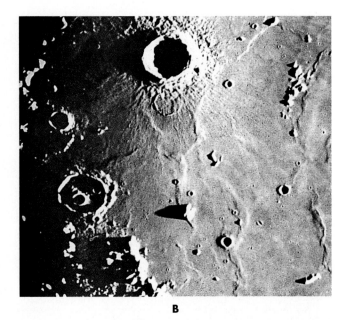

B

FIGURE 6.3
(A) Overlapping craters in the Moon's highlands. (B) Isolated craters in the smooth Mare. (©UC Regents; UCO/Lick Obs image.)

☀ Question: In (A) a small crater lies at the edge of a larger one. Which formed more recently, the small one or the large one?

If you pour a tiny drop of cranberry juice into a glass of milk, you will see an effect similar to the one that makes lunar crater central peaks. The cranberry juice drops into the milk, but is then pushed up again as the milk "rebounds," as shown in the high-speed photo above. (The Harold E. Edgerton 1992 Trust/Palm Press, Inc.)

The large, smooth, dark areas are called **maria** (pronounced MAR-ee-a), from the Latin word for "seas." However, these regions, like the rest of the Moon, are totally devoid of water. This usage comes from early observers who believed the maria looked like oceans and who gave them poetic names such as Mare (pronounced MAR-ay) Serenitatis (Sea of Serenity) or Mare Tranquillitatis (Sea of Tranquility), the site where astronauts first landed on the Moon.

The bright areas that surround the maria are called **highlands.** The highlands and maria differ in brightness because they are composed of different rock types. The maria are basalt, a dark, congealed lava rich in iron, magnesium, and titanium silicates. The highlands, on the other hand, are mainly anorthosite, a rock type rich in calcium and aluminum silicates. This difference has been verified from rock samples obtained by astronauts. Moreover, the samples also show that the highland material is generally less dense than mare rock and considerably older.

The highlands are not only brighter and their rocks less dense than the maria, they are also more rugged, being pitted with craters. In fact, highland craters are so abundant that they often overlap, as shown in figure 6.3A. Contrast this picture with the mare region shown in figure 6.3B, in which only a few, small craters are visible.

From many craters, long, light streaks of pulverized rock called **rays** radiate outward, as shown in figure 6.4. A particularly bright set spreads out from Tycho near the Moon's south pole and can be seen easily with a pair of binoculars when the Moon is full.

A small telescope reveals still other surface features. Lunar canyons known as **rilles**, perhaps carved by ancient lava flows, wind away from some craters, as shown in figure 6.5. Elsewhere, straight rilles gouge the surface, probably the result of crustal cracking. Drying mud and chocolate pudding left too long in the refrigerator show similar cracks.

Origin of Lunar Surface Features

Nearly all the surface features we see on the Moon—craters, maria, and lunar rays—were made by the impact of solid bodies on its surface. When such an object hits a solid surface at high speed, it disintegrates in a cloud of vaporized rock and fragments. The resulting explosion blasts a hole whose diameter depends on the mass and velocity of the impacting object. The hole's shape is circular, however, unless the impact is grazing.

As the vaporized rock expands from the point of impact, it forces surrounding rock outward, piling it into a raised rim. Pulverized rock spatters in all directions, forming rays.

FIGURE 6.4
The long, narrow, white streaks radiating away from the crater at the top are lunar rays. (©UC Regents; UCO/ Lick Obs image.)

☼ Can you see rays near other craters?

☼ Question: Some rays cross maria. What does this imply about the relative age of the rays and the maria?

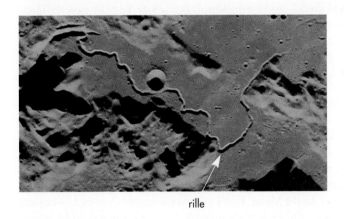

rille

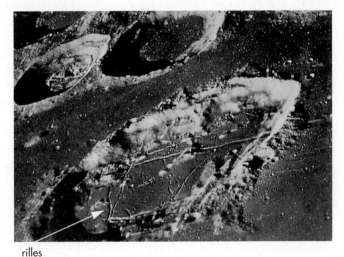
rilles

FIGURE 6.5
Photographs of some lunar rilles. (Courtesy NASA.)

Sometimes the impact compresses the rock below the crater sufficiently that it rebounds upward, creating a central peak, as shown in figure 6.6.

Astronomers believe the maria are also impact features, but to understand their formation, we must briefly describe the early history of the Moon. From the great age of the highland rocks (in some cases as old as 4.5 billion years), astronomers deduce that these rugged uplands formed shortly after the Moon's birth. At that time, the Moon was probably molten, allowing heavier (dense), iron-rich material to sink to its interior while lighter (less dense) material floated to the lunar surface. On reaching the surface, the less dense rock cooled and congealed, forming the Moon's crust. A similar process probably formed the Earth's continents. The highlands were then heavily bombarded by solid bodies from space, forming the numerous craters we see there.

FIGURE 6.6
Central peak in a crater and slumped inner walls. Apollo astronauts took this photograph of the crater Eratosthenes on the last manned flight to the Moon. This crater is 58 kilometers (approximately 36 miles) in diameter. (Courtesy NASA.)

FIGURE 6.7
Photograph of mountains along the edge of a mare. These mountains were probably thrown up by the impact that created the mare. (©UC Regents; UCO/Lick Obs image.)

As the Moon continued to cool, its crust thickened. But before the crust grew too thick, a small number of exceptionally large bodies—objects over 100 kilometers (about 60 miles)—struck the surface, blasting huge craters and pushing up mountain chains along their edges, as you can see in figure 6.7. Subsequently, molten material from deep within the Moon gradually flooded the vast crater and congealed to form the smooth, dark lava plains that we see now. Because the denser material sunk to the Moon's interior during its molten stage, the erupted lava from those depths was denser than crustal rock into which it flooded. Moreover, because it was molten more recently, the mare material is therefore younger than the highlands. By the time the maria formed, most of the impacting bodies were gone—collected into the Earth and Moon by earlier collisions. Thus, too few bodies remained to crater the maria, which therefore remain smooth, even to this day.

Our home planet Earth furnishes additional evidence that most lunar craters formed early in the Moon's history. Like the Moon, Earth too was presumably battered by impacts in its youth. Although the vast multitude of these craters have been obliterated by erosion and plate tectonics, a few remain in ancient rock whose measured age is typically hundreds of millions of years. From the scarcity of such craters, astronomers can deduce that the main bombardment must have ended billions of years earlier.

Craters and maria so dominate the lunar landscape that we might not notice the absence of folded mountain ranges and the great rarity of volcanic peaks, land forms common on Earth. Why have such features not formed on the Moon?

6.2 STRUCTURE OF THE MOON

The Moon's small size relative to the Earth explains the differences between the two bodies. We saw in chapter 4 that radioactive elements in the Earth's interior heat it. The Moon is also heated by radioactive material, but because its volume compared to its surface area is small relative to the Earth's, heat escapes far more easily from the Moon. Thus the Moon has cooled far more than the Earth. (Think of how a small baked potato cools much faster than a big one.) Moreover, because its mass is much smaller than the Earth's, the Moon contains much less radioactive material and so cannot generate as much heat. Thus, without a strong heat source, the Moon lacks the convection currents that drive plate tectonic activity on the Earth. Confirmation of this comes from studies of the Moon's interior.

(Courtesy Peter Arnold.)

Crust and Interior

The Moon's interior can be studied by seismic waves just as the Earth's can. One of the first instruments set up on the Moon by the Apollo astronauts was a seismic detector. Measurements from that and other seismic detectors placed by later Apollo crews show that the Moon's interior is essentially inactive and has a much simpler structure than the Earth's.

The Moon's surface layer is shattered rock that forms a **regolith**—meaning "blanket of rock"—tens of meters deep. The regolith consists of both rock chunks and fine powder, the result of successive impacts breaking rock into smaller and smaller pieces. This powdery nature is easily seen in the crispness of the astronauts' footprints. Samples of the regolith picked up by astronauts show that these surface rocks are typically the same type as the underlying rock. That is, the regolith on maria is generally basaltic, whereas that on the highlands is anorthositic. In places the regolith may extend several hundred meters below the surface. Analysis of the regolith shows that over time its rocks have been broken up by high-velocity impacts, supporting the interpretation that the surface has been bombarded by meteoritic bodies.

Below this surface layer of rocky rubble is the Moon's crust, about 100 kilometers (60 miles) thick, on the average. The crust is much thinner (about 65 kilometers) on the side of the Moon that faces the Earth than on the far side, but the reason for this difference is not clear. It may result from the Earth's gravity shifting the Moon's core slightly toward Earth billions of years ago when the Moon's interior was hotter. The crust on the near side—being slightly closer to the Moon's core because of that shift—might therefore have become hotter and as a result thinner than that on the far side. Subsequently the Moon cooled, leaving the crust thinner on one side than the other. The thinner crust on the near side made it much easier for maria to form, as you can see in figure 6.8.

The Moon's crust, like the Earth's, is composed of silicate rocks relatively rich in aluminum and poor in iron. Beneath the crust is a thick mantle of solid rock, extending down about 1000 kilometers (600 miles). The Moon's mantle is probably rich in olivine, the same type of dense, greenish rock that composes most of the Earth's mantle. Unlike the Earth's mantle, however, it appears too cold and rigid to be stirred by the Moon's feeble heat.

Footprint of an astronaut on the Moon. (Courtesy NASA.)

FIGURE 6.8
(A) The near side of the Moon and (B) the far side. Note how uncommon maria are on the far side. (Courtesy NASA/National Geographic Society.)

☀ Why can't we see the far side of the Moon from Earth?

A

B

FIGURE 6.9
An artist's impression of the Moon's interior. Notice the thinner near-side crust and the displacement (exaggerated for clarity) of the core toward the Earth.

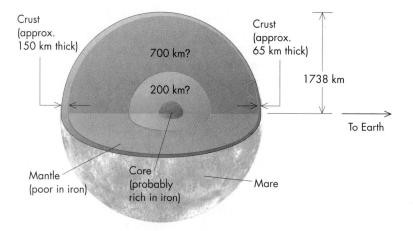

The Moon's low density (3.3 grams per cubic centimeter) tells us its interior contains little iron. Recall how in chapter 4 we saw that the Earth's high density (about 5 grams per cubic centimeter) was an indication that it had a large iron core. Some molten material may lie below the mantle, as illustrated in figure 6.9, but the Moon's core is smaller and contains far less iron and nickel than the Earth's. These factors, plus the Moon's slow rotation, lead astronomers to think that the Moon's core is unable to generate a magnetic field as the Earth does. Measurements made by the Apollo astronauts confirm that the Moon has essentially no magnetic field. Thus a compass would be of no use to an astronaut lost on the Moon.

The Absence of a Lunar Atmosphere

The Moon's surface is never hidden by lunar clouds or haze, nor does the spectrum of sunlight reflected from it show any trace of gases. With no atmosphere to absorb and trap heat, temperatures on the Moon soar during the day and plummet at night. Likewise, no wind blows, and so the lunar surface lies dead and silent under a black sky.

The Moon has no atmosphere for two reasons. First, its interior is too cool to cause volcanic activity, which as we saw in chapter 4 was probably the source of much of the Earth's early atmosphere. Second, even if volcanos had created an atmosphere in its youth, the Moon's small mass creates too weak a gravitational force for it to retain the erupted gas. In chapter 2 we learned that the Moon's escape velocity is only about a fourth that of the Earth's (2.4 kilometers per second versus 11 kilometers per second) and so atoms in the Moon's atmosphere would have found it easy to escape its gravity. With no atmosphere and no plate tectonics, the Moon has been essentially unchanged for billions of years. The footprints left on the Moon by the astronauts in 1969 will probably still be there a million years from now.

6.3 ORBIT AND MOTIONS OF THE MOON

By watching the Moon for a few successive nights, you can see it move against the background stars as it follows its orbit around the Earth. The Moon's orbit is elliptical, with an average distance from Earth of 380,000 kilometers (about 250,000 miles) and a period of 27.3 days.* Its distance can be measured by triangulation, radar, or laser beams (fig. 6.10). To triangulate its distance, astronomers observe the Moon from two different spots on the Earth. The distance between the locations, the angles to the Moon, and a little trigonometry give the Moon's distance.

---

*This is the time to complete an orbit around the Earth and is shorter than the cycle of the phases, as discussed in chapter 1.

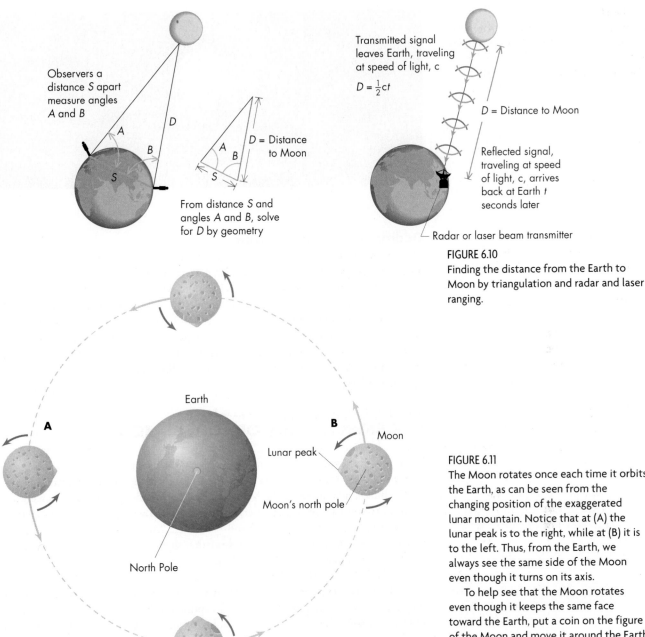

Observers a distance *S* apart measure angles *A* and *B*

A

D

B

S

D = Distance to Moon

A

B

S

From distance *S* and angles *A* and *B*, solve for *D* by geometry

Transmitted signal leaves Earth, traveling at speed of light, c

$D = \frac{1}{2}ct$

D = Distance to Moon

Reflected signal, traveling at speed of light, c, arrives back at Earth *t* seconds later

Radar or laser beam transmitter

FIGURE 6.10
Finding the distance from the Earth to Moon by triangulation and radar and laser ranging.

Earth

A

B

Moon

Lunar peak

Moon's north pole

North Pole

FIGURE 6.11
The Moon rotates once each time it orbits the Earth, as can be seen from the changing position of the exaggerated lunar mountain. Notice that at (A) the lunar peak is to the right, while at (B) it is to the left. Thus, from the Earth, we always see the same side of the Moon even though it turns on its axis.

To help see that the Moon rotates even though it keeps the same face toward the Earth, put a coin on the figure of the Moon and move it around the Earth so that the same edge of the coin always faces the Earth.

A more accurate method is to bounce either a radar pulse or a laser beam off the Moon. Half the time interval between the transmission and the return of the reflected radiation multiplied by the speed of light gives the Moon's distance to an accuracy of centimeters.

The Moon's Rotation

As it orbits, the Moon keeps the same side facing the Earth, as you can see by watching it through a cycle of its phases. You might think from this that the Moon does not rotate. Figure 6.11 shows, however, that the mountain on the side facing the Earth points to the right when the Moon is at A and to the left when the Moon is at B. Thus, the Moon *does* turn on its axis but with a rotation period exactly equal to its orbital period, a condition known as **synchronous rotation.** The Earth's gravity causes this locking of the Moon's spin to its orbital motion, as we will discuss in section 6.6.

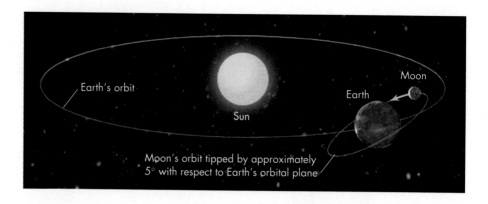

Oddities of the Moon's Orbit

The Moon's orbit is tilted by about 5° with respect to the Earth's orbit around the Sun, as illustrated in figure 6.12. It is also tilted with respect to the Earth's equator and is thus unlike most of the moons of Jupiter, Saturn, and Uranus, which lie nearly exactly in their planets' equatorial plane. These oddities indicate that our Moon formed differently from the moons of other planets, a conjecture supported by the odd mass ratio of the Earth and Moon.

Most moons are tiny compared with their planets. Even the largest of the moons of Jupiter and Saturn have masses less than 1/1000 that of their planet. But our Moon's mass is 1/81 that of the Earth's. Why is our Moon so different in its mass and orbit from the moons of other planets?

6.4 ORIGIN AND HISTORY OF THE MOON

Lunar rocks brought back to Earth by the Apollo astronauts have led astronomers to radically revise their ideas of how the Moon formed. Before the Apollo program, lunar scientists had three hypotheses of the Moon's origin. In one, the Moon was originally a small planet orbiting the Sun that approached the Earth and was captured by its gravity (capture theory). In another, the Moon and Earth were "twins," forming side by side from a common cloud of dust and gas (twin formation theory). In the third, the Earth initially spun enormously faster than now and formed a bulge that ripped away from the Earth to become the Moon (fission theory).

Each of these hypotheses led to different predictions about the composition of the Moon. For example, had the Moon been a captured planet, its composition might be very unlike the Earth's. If the Earth and Moon had formed as twins, their overall composition should be similar. Finally, if the Moon was once part of the Earth, its composition should be nearly identical to the Earth's crust. When the rock samples were analyzed, astronomers were surprised that for some elements the composition was the same, but for others it was very different. For example, the Moon has a relatively high abundance of high-melting-point materials such as gold, and an almost complete lack of low-melting-point materials, such as water. It also has much less iron than the Earth, as we pointed out when discussing its interior and low density.

The failure of evidence based on lunar surface samples to confirm any of the three hypotheses led astronomers to consider alternatives, and over the last decade a completely different picture of the Moon's origin has emerged. According to the new hypothesis, the Moon formed from debris blasted out of the Earth by the impact of a Mars-sized body, as shown in figure 6.13A. The great age of lunar rocks and the absence of any impact feature on the Earth indicate that this event must have occurred during the Earth's own formation, at least 4.5 billion years ago. The colliding body melted and vaporized millions of cubic kilometers of the Earth's surface rock and hurled it into space in an incandescent plume. As the debris cooled, its gravity gradually drew it together into what we now see as the Moon.

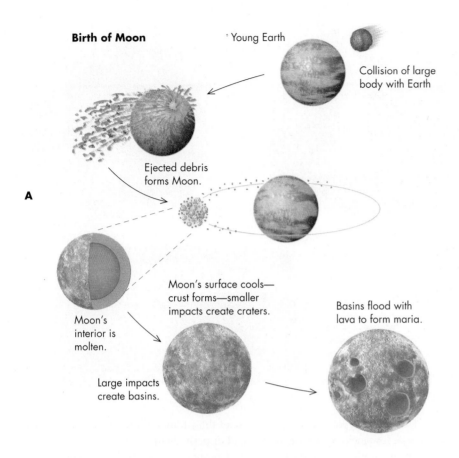

Birth of Moon

Young Earth

Collision of large body with Earth

Ejected debris forms Moon.

A

Moon's interior is molten.

Moon's surface cools— crust forms—smaller impacts create craters.

Basins flood with lava to form maria.

Large impacts create basins.

FIGURE 6.13
(A) The birth of the Moon. (B) This computer simulation 1–4 shows how the Moon might have formed when a Mars-sized body hit the young Earth and splashed out debris that later assembled into the Moon. The scale changes between the frames (1) is a close-up of the impact. (4) shows the newly assembled Moon (lower left) orbiting the Earth, which is still surrounded by debris. (Courtesy A. G. W. Cameron, Harvard Center for Astrophysics.)

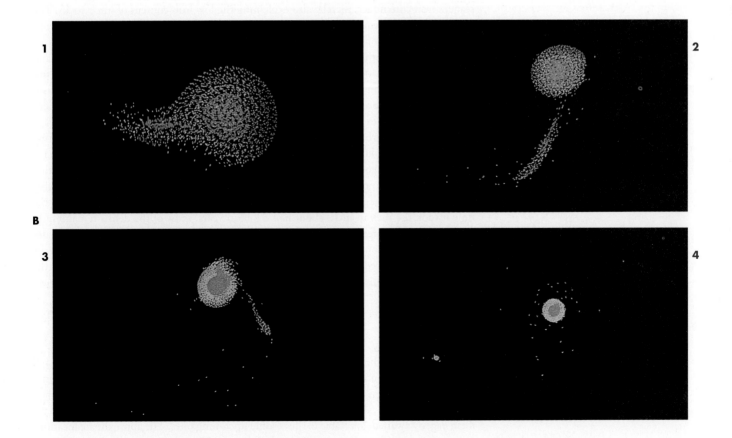

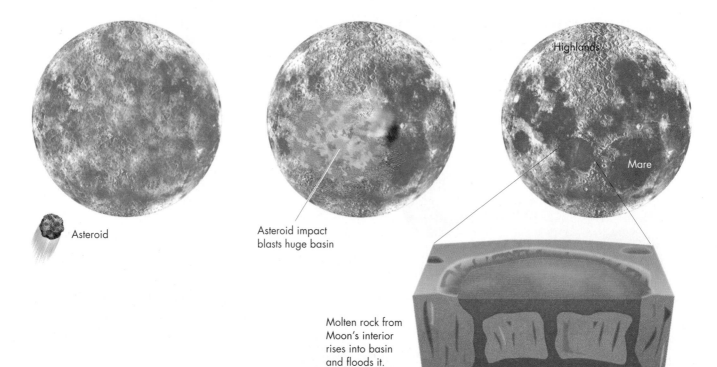

Asteroid

Asteroid impact
blasts huge basin

Highlands

Mare

Molten rock from
Moon's interior
rises into basin
and floods it.

FIGURE 6.14
A rain of debris creates craters on the
young Moon. Large impacts late in the
process form the maria basins. Lava floods
the basins to make the maria.

This violent birth hypothesis explains many of the oddities of the Moon. The impact
would vaporize low-melting-point materials and disperse them, leaving, for example, little water to be incorporated into the lunar body. Computer models (fig. 6.13B) of such an
event also show that only surface rock would be blasted out of the Earth, leaving our
planet's iron core intact, thereby also explaining the low iron content of lunar rocks. The
splashed-out rock would condense in an orbit whose shape and orientation were determined by the collision rather than by the orientation of the Earth's equator. Furthermore,
we would expect both similarities and differences in composition between the Earth and
Moon because the Moon was made partly from Earth rock and partly from rock of the impacting body. A bonus of the hypothesis is that it explains why the Earth's rotation axis is
tipped so much more than the rotation axis of Mercury, Venus, or the giant planets,
Jupiter and Saturn: the impact knocked the Earth part way over.

After the Moon's birth, stray fragments of the ejected rock pelted its surface, creating
the craters that blanket the highlands, as depicted in figure 6.14. A few huge fragments
plummeting onto the Moon later in its formation process blasted enormous holes that
later flooded with molten interior rock to become the maria. That rock was probably
melted in the Moon's interior by radioactive decay, as happened in the Earth. During the
time it took the rock to melt, about half a billion years, most of the debris remaining in
the Moon's vicinity fell onto its surface. Thus, by the time the maria flooded, little material was left to fall on them, and so they are only lightly cratered. Since that time, the
Moon has experienced no major changes. It has been a virtually dead world for all but the
earliest times in its history, but it nevertheless creates effects on Earth: eclipses and tides.

6.5 ECLIPSES

An **eclipse** occurs when one astronomical body casts its shadow on another. As discussed
in chapter 1, for us on the Earth two types of eclipses can occur: lunar and solar. A lunar
eclipse happens when the Earth's shadow falls on the Moon. A solar eclipse happens when
the Moon's shadow falls on the Earth. The great beauty of eclipses and their rarity make
them eagerly awaited, and table 6.1 lists some upcoming ones. The approximate locations
of the tabulated solar eclipses are shown in figure 6.15B.

TABLE 6.1   Some Upcoming Solar and Lunar Eclipses

Solar eclipses		Lunar eclipses	
Feb 26, 1998	Northwest South America and Caribbean	Sept 16, 1997	Australia, most of Asia, most of Africa and Europe
Aug 11, 1999	Greenland, Europe, Southern Asia	Jan 21, 2000	North and northwest Asia, Europe, North America, South America
June 21, 2001	Southern Africa/Madagascar	May 16, 2003	Antarctica, most of Africa and Europe, most of the Americas
Dec 4, 2002	Southern Africa, Australia	Nov 8/9, 2003	Most of Asia, Africa, Europe, The Americas except extreme SW Alaska

FIGURE 6.15
(A) Sketch of how the Moon's shadow travels across the Earth. (B) Location of some recent and upcoming total solar eclipses. (From Kuhn, Karl F. *Astronomy*, p. 204, St. Paul, Minn., 1989 West Publishing.)

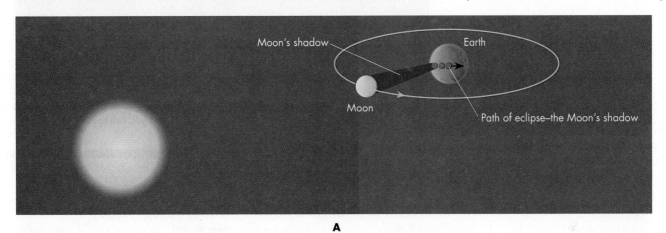

A

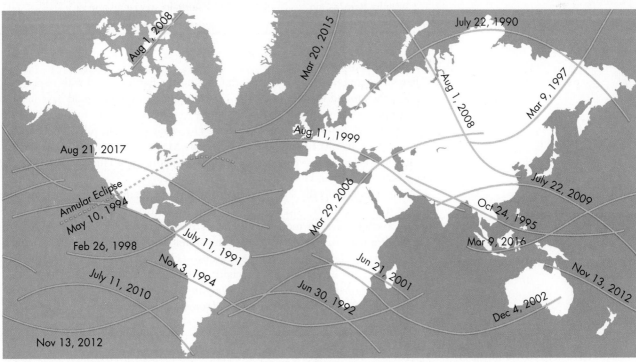

B

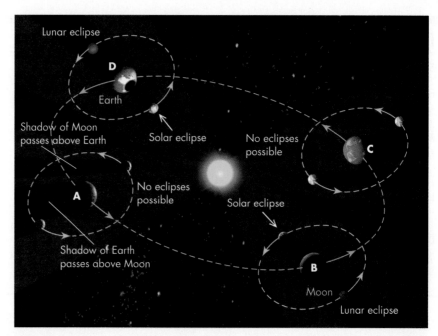

FIGURE 6.16
The Moon's orbit keeps approximately the same orientation as the Earth orbits the Sun. Because of its orbital tilt, the Moon generally is either above or below the Earth's orbit. Thus the Moon's shadow rarely hits the Earth, and the Earth's shadow rarely hits the Moon, as you can see in (A) and (C). Eclipse seasons occur when the Moon's orbital plane "points" at the Sun. A solar eclipse will then occur at new moon and a lunar eclipse at full moon, as you can see in (B) and (D).

TABLE 6.2   Pairing of Eclipses at Eclipse Seasons

March 9, 1997	Solar (total)	March 24, 1997	Lunar (partial)
East Asia, Japan (partial)		Africa, Europe, and Americas	
Sept 2, 1997	Solar (partial)	Sept 16, 1997	Lunar (total)
Australia, New Zealand, South Pacific		Australia, most of Asia, most of Africa, and Europe	

Rarity of Eclipses

Eclipses are rare because the Moon's orbit around the Earth is tilted with respect to the Earth's orbit around the Sun (see fig. 6.12). Without that tilt we would have lunar and solar eclipses every month, but with the tilt, the Moon lies above the Earth's orbit for half the month and below it for the other half. The result is that at full moon the Earth's shadow generally falls either above or below the Moon, and at new moon the Moon's shadow falls below or above the Earth, as shown in figure 6.16. Thus eclipses do not generally occur very often. How then are eclipses possible?

As the Earth orbits the Sun, the Moon's orbit keeps nearly the same direction of tilt, as shown in figure 6.16. This orbital tilt is kept fixed—like that of the spinning Earth—by a gyroscopic effect or more technically, by the conservation of angular momentum. The result is that twice each year the Moon's orbital plane points at the Sun, as shown in figure 6.16. At those times—**eclipse seasons**—eclipses will happen when the Moon crosses the Earth's orbital plane (the ecliptic*). Table 6.2 shows that in 1997 the eclipse seasons were in March and September. Only in those months could eclipses happen: at other times the shadows of the Earth and Moon always fell on empty space.

---

*As mentioned in chapter 1, this is the reason the ecliptic is so named.

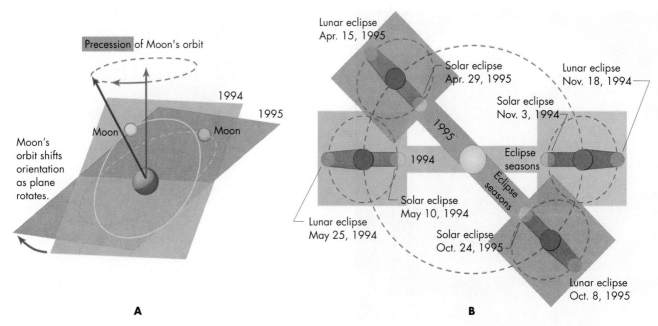

FIGURE 6.17
(A) Precession of the Moon's orbit. Notice its similarity to twisting a tilted book that has one edge resting on a table.
(B) Precession of the Moon's orbit causes eclipses to come at different dates in successive years.

You can also see from figure 6.16 that when a solar eclipse occurs at new moon, conditions are right for a lunar eclipse to happen at either the previous or the following full moon. Thus eclipses generally occur in pairs, with a solar eclipse followed approximately 14 days later by a lunar eclipse, or vice versa, as shown for the 1997 eclipses listed in table 6.2.

This simple pattern does not always work, because the tilt of the Moon's orbit is not exactly fixed. An imaginary line perpendicular to the orbit (shown in fig. 6.17A) slowly changes direction just as the Earth's rotation axis does. That is, the Moon's orbit precesses, swinging once around about every 18.6 years. This orbital precession makes the dates of the eclipse seasons shift by 1/18.6 year (about 20 days) each year. Thus, in 1995, eclipses occurred about 2 to 3 weeks earlier than in 1994, as shown in figure 6.17B.

If one of the eclipse seasons occurs in early January with the next in June, a third season may sometimes happen in late December. As a result, as many as five solar and five lunar eclipses can occur each year. No matter when the eclipse season falls, at least two solar and two lunar eclipses must happen each year, but that does not mean they will be visible to an observer at a given location, since the eclipse may be visible only from another part of the Earth. Because the Moon is so small compared with the Earth, its shadow is small, and therefore you can see a solar eclipse only from within a narrow band, as illustrated in figure 6.15. Lunar eclipses, however, are visible from anywhere the Moon is above the horizon at the time of the eclipse.

Appearance of Eclipses

Eclipses are beautiful and marvelous events and well worth watching. During a lunar eclipse, the Earth's shadow gradually spreads across the Moon's face, cutting an ever deeper dark semicircle out of it. The shadow takes about an hour to completely cover the Moon and produce totality. At totality, the Moon generally appears a deep ruddy color, almost as if dipped in blood. Sometimes it even disappears. After totality, the Moon again becomes lit, bit by bit, reverting to its unsullied, silvery light.

Be extremely careful when watching a Solar eclipse. Looking at the Sun through improper filters will blind you. A safer way is to not look directly at the Sun, but to use eyepiece projection to view the Sun. Hold a piece of paper about a foot from the eyepiece of a small telescope (or even binoculars), and a large image of the Sun will be visible on it. This method also allows many people to watch the eclipse simultaneously.

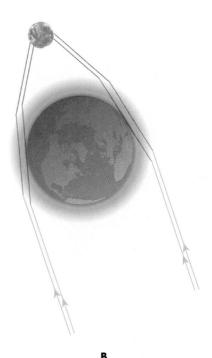

A   **B**

FIGURE 6.18
(A) Photograph of a total lunar eclipse. (Photo courtesy Dennis di Cicco.) (B) As sunlight falls on the Earth, some passes through the Earth's atmosphere and is slightly bent so that it ends up in the Earth's shadow. In its passages through our atmosphere, most of the blue light is removed, leaving only the red. That red light then falls on the Moon, giving it its ruddy color at totality.

A little light falls on the Moon even at totality because the Earth's atmosphere bends some sunlight into the shadow, as shown in figure 6.18. The light reaching the Moon is red because interactions with air molecules removes the blue light as it passes through our atmosphere and is bent, exactly as happens when we see the setting Sun.

A solar eclipse begins with a black "bite" taken out of the Sun's edge as the Moon cuts across its disk. Such partial eclipses are fun to watch but should be observed with care so as not to hurt your eyes. However, unless a large part of the Sun is covered by the Moon, you may not even notice an eclipse is happening. On the other hand, if you are fortunate enough to be at a location where the eclipse is total, you will see one of the most amazing sights in nature.

As the moment of totality approaches, the landscape takes on an eerie light. Shadows become incredibly sharp and black: even individual hairs on your head cast crisp shadows. Sunlight filtering through leaves creates tiny bright crescents on the ground. Seconds before totality, pale ripples of light sweep across the ground and to the west the deep purple shadow of the Moon hurtles down on you at more than 1000 miles an hour. In one heartbeat you are plunged into darkness. Overhead the sky is black, and stars may appear. The corona of the Sun—its outer atmosphere—gleams with a steely light around the Moon's black disk. Perhaps a solar prominence, a tiny glowing red flamelike cloud in the Sun's atmosphere, may protrude beyond the Moon's black disk (fig. 6.19). Birds call as if it were evening. A deep chill descends because for a few minutes the Sun's warmth is blocked by the Moon. The horizon takes on sunset colors: the deep blue of twilight with perhaps a distant cloud in our atmosphere glowing orange. As the Moon continues in its

FIGURE 6.19
(A) Photograph of a total solar eclipse. The bright halo of light is the Sun's corona, its outer atmosphere. (Photo courtesy Dennis di Cicco.) (B) The landscape is eerily lit during a total solar eclipse. The dark color of the sky is the Moon's shadow. (Courtesy Richard Wainscoat.)

orbit, it uncovers the Sun and instantly it is daylight again. Now the cycle continues in reverse. If you ever have the chance to see a total eclipse, do it. It is worth traveling hundreds of miles to see.

The rarity of eclipses may lead you to think that the astronomical effects of the Moon are uncommon. A day spent by the ocean, however, will reveal a far more common and more powerful lunar influence.

6.6 TIDES

Anyone who has spent even a few hours by the sea knows that the ocean's level rises and falls during the day. A blanket set on the sand 10 feet from the water's edge may be inundated an hour later, or a boat pulled ashore may be left high and dry. This regular change in the height of the ocean is called the **tides** and is caused mainly by the Moon.

Cause of Tides

Just as the Earth exerts a gravitational pull on the Moon, so too the Moon exerts a gravitational attraction on the Earth and its oceans and draws material toward it. The attraction is stronger on the side of the Earth near the Moon and weaker on the far side (see fig. 6.20) because the force of gravity weakens with distance (recall Newton's law of gravity, p. 83). The difference between the strong force on one side and the weaker force on the other is called a **differential gravitational force**.

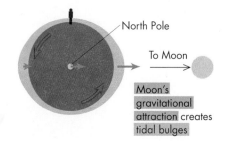

FIGURE 6.20
Tides are caused by the Moon's gravity creating tidal bulges.

FIGURE 6.21
(A) Arrows schematically show Moon's gravitation force at different points on the Earth. (B) Tidal forces from the point of view of an observer on the Earth. These arrows represent the difference between the Moon's gravitational force at a given point and its force at the Earth's center.

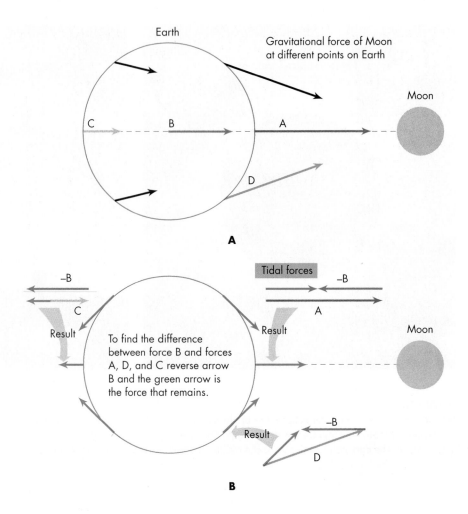

To find the difference between force B and forces A, D, and C reverse arrow B and the green arrow is the force that remains.

The differential gravity draws water in the oceans into a **tidal bulge** on the side of the Earth facing the Moon, as shown in figure 6.20. But curiously, it creates an identical tidal bulge on the Earth's far side. This second tidal bulge can be viewed as a result of the Moon's gravity pulling the Earth "out from under" the water on the far side. A better way to view it, however, is to examine the Moon's gravitational forces on the Earth and its oceans as seen by a person on the Earth, as shown in figure 6.21.

The arrows in figure 6.21A represent the Moon's gravitational force on the Earth. Points on the side of the Earth near the Moon (A) undergo a stronger pull toward the Moon than those on the far side (C), and so arrow A is longer than C. Likewise, arrow A is longer than B, which represents the Moon's pull on the center of the Earth, and B is longer than C.

Because the force at A is larger than the force at B, matter at A will be pulled away from B. This creates one tidal bulge. But B in turn is pulled away from C and creates the second tidal bulge. If we now draw a second set of arrows to represent the *difference* between the force at B and at every other point (the differential gravitational force), we find the forces illustrated in figure 6.21B. These drive the oceans into the bulges* that we see.

An analogy may help you better understand tidal forces. Imagine holding a child by its hands and swinging it around you in a circle. As you swing the child, its feet fly out away from you because of their inertia even though you are holding it by its hands. So too as

---

*Tides also occur in the atmosphere and solid ground, but the latter are very small because the ground is rigid and cannot move as easily as water or air.

the Earth and Moon swing around their common center of mass the Moon's gravity pulls water on the side of the Earth toward the Moon into a bulge while inertia makes water on the far side "fly out" into a second bulge.

In the above discussion we have ignored the Earth's rotation. The tidal bulges are aligned approximately with the Moon, but the Earth spins. Its rotation therefore carries us first into one bulge and then the next. As we enter the bulge, the water level rises, and as we leave it, the level falls. Because there are two bulges, we are carried into high water twice a day, creating two high tides. Between the times of high water, as we move out of the bulge, the water level drops, making two low tides each day (fig. 6.22).

This simple picture must be altered to account for the inability of the ocean to flow over land areas. Thus water tends to pile up at coast lines when the tidal bulge reaches shore. In most locations the tidal bulge has a depth of about 2 meters (6 feet), but it may reach 10 meters (30 feet) or more in some long narrow bays (as you can see in the photographs of high and low tides along the Maine Coast) and may even rush upriver as a tidal bore—a cresting wave that flows upstream. On some rivers, surfers ride the bore upstream on the rising tide.

The motion of the Moon in its orbit makes the tidal bulge shift slightly from day to day. Thus high tides come about 50 minutes later each day, the same delay as in moonrise, discussed in chapter 1.

Solar Tides

The Sun also creates tides on the Earth, but although the Sun is much more massive than the Moon, it is also much farther away. The result is that Sun's tidal force on the Earth is only about one-third the Moon's, Nevertheless, it is easy to see the effect of their tidal cooperation in spring tides, which are abnormally large tides that occur at new and full moon. At those times the lunar and solar tidal forces work together, adding their separate tidal bulges, as illustrated in figure 6.23A. Notice that spring tides have nothing to do with the seasons, rather they refer to the "springing up" of the water at new and full moon.

It may seem odd that spring tides occur at both new and full moon because the Moon and Sun pull together when the Moon is new but in opposite directions when it is full.

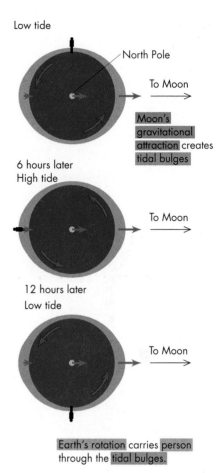

FIGURE 6.22
As the Earth rotates, it carries points along the coast through the tidal bulges. Because there are two bulges where the water is high and two regions where the water is low, we get two high tides and two low tides each day.

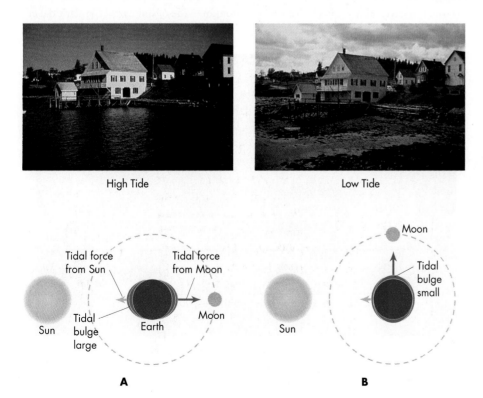

High Tide   Low Tide

A   B

FIGURE 6.23
The Sun's gravity creates tides too, though its effect is only about one-third that of the Moon. (A) The Sun and Moon each create tidal bulges on the Earth. When the Sun and Moon are in line, their tidal forces add together to make larger than normal tides. (B) When the Sun and Moon are at 90° as seen from Earth, their tidal bulges are at right angles and partially nullify each other, creating smaller than normal tidal changes.

However, both the Sun and Moon create two tidal bulges, and the bulges add regardless of whether the Sun and Moon are on the same or opposite sides of the Earth. On the other hand, at first and third quarters, the Sun and Moon's tidal forces work at cross-purposes, creating tidal bulges at right angles to one another as shown in figure 6.23B. The so-called neap tides that result are therefore not as extreme as normal high and low tides.

Tidal Braking

Tides create forces on the Earth and Moon that slow their rotation, a phenomenon known as **tidal braking**. Figure 6.24 shows how the Moon tidally brakes the Earth. As the Earth spins, friction between the ocean and the solid Earth below drags the tidal bulge ahead of the imaginary line joining the Earth and Moon, as depicted in figure 6.24. The Moon's gravity pulls on the bulge, as shown by the arrow in the figure, and holds it back. The resulting drag is transmitted through the ocean to the Earth, slowing its rotation the way a brake shoe on a car or your hand placed on a spinning bicycle wheel slows the wheel.

As the Earth's rotation slows, the Moon accelerates in its orbit, moving farther from the Earth, as required by the need to conserve angular momentum. The Moon accelerates because the tidal bulge it raises on the Earth exerts a gravitational force back on the Moon (as predicted by Newton's third law of motion), which pulls the Moon ahead in its orbit, as shown by the arrows in figure 6.24. That acceleration makes the Moon move away from the Earth at about 3 centimeters (roughly 1 inch) per year, a tiny increase in the Earth-Moon distance, but nevertheless detectable with laser rangefinders. Thus the Moon was once much closer to the Earth, and the Earth spun much faster, perhaps as rapidly as once every 5 hours several billion years ago. Over that immense period of time, the Moon has receded to its present distance and the Earth's rotation has slowed to 24 hours. These processes occur even now: tidal braking lengthens the day by about 0.002 seconds each century.

Tidal braking is also the reason the Moon always keeps the same face to the Earth. Just as the Moon raises tides, which slow the Earth, the Earth raises tides on the Moon, which slow it. These lunar tides distort the Moon's crust and have braked the Moon severely, locking it into synchronous rotation. The Moon's braking of the Earth will eventually make the Earth rotate synchronously with the Moon's orbital motion. Billions of years from now the Earth and Moon will orbit so that each constantly presents the same face to the other: the Moon will then be visible only from one side of the Earth! Similar tidal ef-

FIGURE 6.24
Tidal braking slows the Earth's rotation and speeds up the Moon's motion in its orbit.

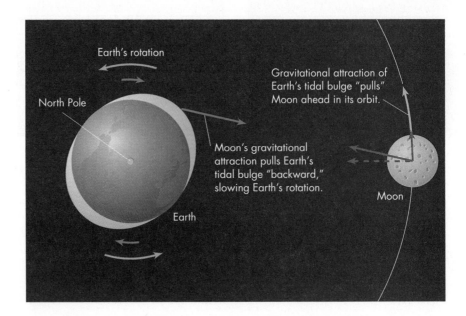

fects have locked some of the moons of other planets into synchronous rotation, but the planets themselves have not been noticeably slowed. On the other hand, tidal braking by the Sun probably slowed the rotation of Mercury and Venus.

The Moon's gravitational pull on the Earth may also stabilize our climate. Astronomers have recently discovered with computer simulations that the tilt of a planet's rotation axis may change erratically by many tens of degrees if the planet has no moon. Because the tilt causes seasons, changes in the tilt will alter the severity of the seasons. Our Moon is large enough that its gravitational attraction on Earth's equatorial bulge helps hold the Earth's tilt relatively fixed, sparing us catastrophically large climate changes.

6.7 MOON LORE

The Moon figures prominently in folklore around the world. Most stories concerning its powers are false. For example, people often claim that the full moon triggers antisocial behavior, hence the term "lunatic." All studies to look for such effects have found nothing. Automobile accidents, murders, admissions to clinics, and so forth, show no increase when the Moon is full.

On the other hand, "once in a blue moon," indicating a rare event, is a phrase with a basis in fact because, on rare occasions, the Moon may look blue. This odd coloration comes from particles in the Earth's atmosphere. Normally, our atmosphere filters the blue colors from light better than the red ones. For example, light from the rising or setting Sun passes through so much atmosphere that little blue light remains by the time it reaches us. Therefore the Sun looks red when it is low in the sky. However, if the atmosphere contains particles whose size falls within a very narrow range, the reverse may occur. Dust from volcanic eruptions or smoke from forest fires may have just the right size to filter out the red light, allowing mainly the blue colors to pass through. Under these unusual circumstances, we may therefore see a "blue Moon."

A different meaning for "blue moon" has appeared within the last few years. This new meaning applies to months with two full moons. Because the cycle of phases is 29.5 days, unless the Moon is full on the first day of the month, the next full moon will fall in the following month. The odds of the Moon being full on the first of the month are about 1 in 30, and so two full moons in a given month happens about every 2 1/2 years. Why such occasions should be referred to as "blue moons" is unexplained.

The harvest moon, the full moon nearest the time of the autumn equinox, is another well-known phrase. As it rises in the east at sunset, the light from the harvest moon helps farmers see to get in the crops. Full moons in other months also have special names, but only the harvest and hunter's moon are widely accepted. Other names occasionally used in American folklore are listed in table 6.3.

TABLE 6.3   Names Used for Full Moons

January	old moon	July	thunder or hay moon
February	hunger moon	August	grain or green corn moon
March	sap or crow moon	September	harvest moon
April	egg or grass moon	October	hunter's moon
May	planting moon	November	frost or beaver moon
June	rose or flower moon	December	long night moon

SUMMARY

The Moon is the Earth's satellite. It is much smaller than the Earth: about one-fourth the Earth's radius and about 1/81 its mass. Its small size has allowed its internal heat to escape, keeping its core cool, thereby preventing plate tectonic motions. The Moon has no atmosphere because it is too cool to create one by volcanic outgassing and too small for its low gravity to retain gases even if an atmosphere had formed.

With neither atmosphere nor tectonic activity, the Moon's surface is unaltered except by impact features: craters, rays, and the maria. Maria are enormous lava flows that have flooded into basins made by large impacting bodies late in the Moon's formation.

The Moon may have formed when a Mars-sized body collided with the Earth and splashed material from the Earth into orbit. That debris, drawn together by its own gravity, would then have reassembled into the Moon.

The Moon's shadow sometimes falls onto the Earth, causing a solar eclipse. When the Earth's shadow falls onto the Moon, we see a lunar eclipse.

The Moon's gravity creates tides, and as the Earth rotates beneath the tidal bulge of the ocean, our planet's rotation is slowed. Similar tidal braking exerted by the Earth on the Moon has slowed the Moon's spin, making it synchronous with its orbital motion around the Earth.

QUESTIONS FOR REVIEW

1. What are the Moon's mass and radius compared with the Earth's?
2. Describe a crater and how it is formed.
3. What are lunar rilles? What are rays?
4. How do the maria differ from the highlands?
5. What formed the maria? Why are they smooth?
6. Why has the Moon no atmosphere?
7. Why is the Moon's surface cratered but the Earth's not?
8. How do astronomers believe the Moon formed? What supports this theory? How does the theory explain why the Earth and Moon have such different densities?
9. How are tides formed on the Earth?
10. Why does the Moon form two tidal bulges on the Earth?
11. Why aren't there eclipses each month?
12. What is an eclipse season?
13. How does the Moon rotate? Why does it spin in this manner?

THOUGHT QUESTIONS

1. Why has the Moon's interior cooled more than the Earth's?
2. Bergmann's rule states that individuals of a given species—for example, bears—will be larger in cold climates than in warmer climates. How is an explanation of this rule similar to an explanation of the temperature difference between the Earth's and the Moon's interior?
3. If the day were 12 hours long, what would be the approximate time interval between high and low tide?
4. Highway surfaces develop "potholes" over time. How can you use the number of potholes as an indication of the "age" of the paving? How is this like using craters to estimate the age of the Moon's surface?
5. Why will an astronaut's footprint on the Moon last so long?

PROBLEMS

1. A laser pulse takes 2.56 seconds to travel from Earth to the Moon and return. Given that the speed of light is 300,000 kilometers per second, how far away is the Moon?
2. A lunar crater has an angular diameter of 1 minute of arc. What is its diameter in kilometers?
3. A spacecraft orbits the Moon 100 kilometers above its surface and with an orbital period of 114 minutes. What is the Moon's mass?
4. The Moon's orbital period is 27.2122 days. Its synodic period, the period of the phases, is 29.5306 days. Show that 242 orbital periods very nearly equals 223 synodic periods. How long is this in years? What does this suggest about eclipses and why? (This match of cycles is called the *saros* and was used by ancient astronomers to predict eclipses.)

TEST YOURSELF

1. The large number of craters on the lunar highlands compared to those on the maria is evidence that
   (a) the surface of the maria is liquid and craters quickly disappear there.
   (b) the material composing the highlands is very soft and easily cratered.
   (c) the bodies that struck the Moon and made the craters were clumped in such a manner that they missed hitting mare areas.
   (d) the maria are much younger than the highlands.
   (e) the maria are much older than the highlands.
2. Figure 6.19 shows an eclipse of the Sun. The black circle in the middle is
   (a) the Earth's shadow on the Sun.
   (b) the Sun's shadow on the Moon.
   (c) the Moon covering the Sun.
   (d) the Earth's shadow on the Moon.
   (e) a dark cloud in our atmosphere.
3. The photographs above fig. 6.23 were taken at high tide and the next low tide. About how much time elapsed between the pictures?
   (a) 3 hours.
   (b) 24 hours.
   (c) 12 hours.
   (d) 6 hours.
   (e) 1 month.
4. What evidence indicates that the Moon lacks a large iron core?
   (a) The Moon has a very strong magnetic field.
   (b) The Moon always keeps the same side facing the Earth.
   (c) The Moon has no atmosphere.
   (d) The Moon's average density is about 3.3 grams per cubic centimeter, similar to that of rock.
   (e) The Moon has so many volcanos that all the iron in its core has been erupted onto its surface.
5. Eclipses do not occur each month because
   (a) the Moon's orbit is so elliptical.
   (b) the Moon's orbit is tilted with respect to the Earth's orbit around the Sun.
   (c) the Moon takes 6 months to complete its orbit around the Earth.
   (d) the Earth's rotation axis is tilted.
   (e) the statement is false. Eclipses do occur each month, but they are generally visible only in very remote parts of the Earth.

FURTHER EXPLORATIONS

Benningfield, D. "Mysteries of the Moon." *Astronomy* 19 (December 1991): 50.

Brueton, Diana. Many Moons: the Myth and Magic, Fact and Fantasy of our Nearest Heavenly Body. Englewood Cliffs, NJ, Prentice-Hall, 1991.

Cadogan, Peter H. "The Moon's Origin." *Mercury* 12 (March/April 1983): 2.

Comins, N. "What If? The Earth Without the Moon." *Astronomy* 19 (February 1991): 48.

Goldman, Stuart J. "Clementine Maps the Moon." *Sky & Telescope* (August 1994): 20.

Goldreich, Peter. "Tides and the Earth-Moon System." *Scientific American* 226 (April 1972): 42.

Hartmann, W. K. "Birth of the Moon." *Natural History* (November 1989): 68.

Maran, Stephen P. "Quakes on the Moon." *Natural History* (February 1982): 82.

Nicholson, T. D. "The Moon Illusion." *Natural History* (August 1991): 66.

Runcorn, S. K. "The Moon's Ancient Magnetism." *Scientific American* 257 (December 1987): 60.

Wood, John A. "The Moon." *Scientific American* 233 (September 1975): 93.

Web sites

"The Nine Planets: A Multimedia Tour of the Solar System," by Bill Arnett.
www.seds.org/billa/tnp
Excellent pictures, glossary, and other info. One of the best astronomy web sites.

"Views of the Solar System" by Calvin Hamilton.
bang.lanl.gov/solarsys/moon.htm
Nice pictures, lots of text, fact sheets, and glossary. Essentially an on-line text.

"Welcome to the Planets"
pds.jpl.nasa.gov/planets
Nice Pictures, fact sheets, and other info.

National Space Science Data Center.
nssdc.gsfc.nasa.gov/photo_gallery/photo_gallery-moon.html
A NASA site with lots of pictures, fact sheets, glossary.

All the above have many additional links to other sites.

KEY TERMS

craters, 186
differential gravitational force, 201
eclipse, 196
eclipse seasons, 198
highlands, 188
maria, 188
rays, 188

regolith, 191
rilles, 188
synchronous rotation, 193
tidal braking, 204
tidal bulge, 202
tides, 201

 PROJECTS

Look at the Moon on an evening when it is nearly full. Make a sketch of the light and dark markings that you see on its surface with the naked eye. Then observe the Moon with binoculars and make an enlarged sketch that shows more detail. Mark a few of the craters you can see. Estimate the diameter of these craters from your knowledge that the Moon's radius is about 1000 miles (1700 km). How big is the largest crater you can see compared to the size of your town? Can you see any lunar rays? If so, sketch them on your drawing. How long are the rays?

The Solar System

The Solar System consists of the Sun, its nine planets (including the Earth) and their moons, and innumerable smaller objects such as comets and asteroids (fig. OV4.1). Shaped like a thin disk with all the planets orbiting the Sun in the same direction, the Solar System is held together by the Sun's gravitational force.

FIGURE OV4.1
The Solar System, illustrating its flattened, disklike shape.

Mercury
diameter = 4880 km

Venus
diameter = 12,100 km

Earth
diameter = 12,800 km

Mars
diameter = 6,800 km

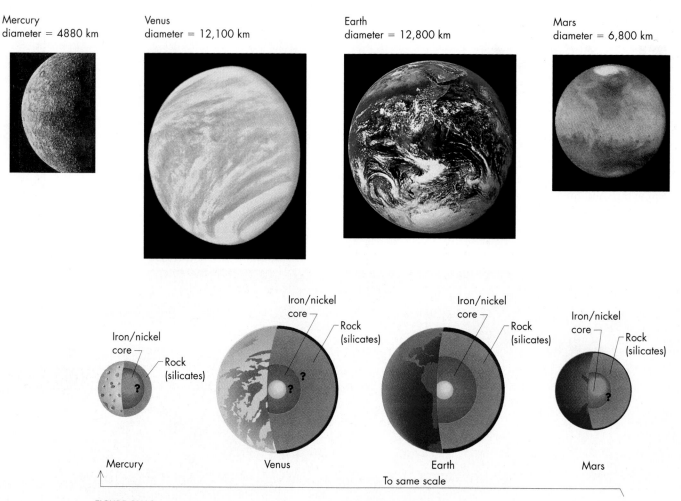

FIGURE OV4.2
The inner planets: Mercury, Venus, Earth, and Mars. (Courtesy NASA/JPL.)

The nine planets form two very different families: the inner planets—those nearest the Sun (Mercury, Venus, Earth, and Mars)—are made of rock (silicates) with iron-rich cores (fig. OV4.2). The four largest outer planets—Jupiter, Saturn, Uranus, and Neptune (fig. OV4.3)—are far more massive than the inner planets. Although they have inner cores with approximately the same rock and iron composition as the inner planets, about 70% of their vast bulk is hydrogen and its compounds, such as methane, ammonia, and water. The rest is helium. Beneath their atmospheres, these huge worlds have no solid surface. Instead, squeezed by the immense gravity generated by their huge mass, their gas liquifies to form deep hydrogen oceans.

Pluto differs dramatically from the other planets. It is far tinier than any of the other planets. Being so far from the Sun, it is always bitterly cold and is probably made throughout of a mix of rock and ice. Several of the moons of the outer planets have a similar composition.

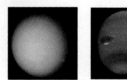

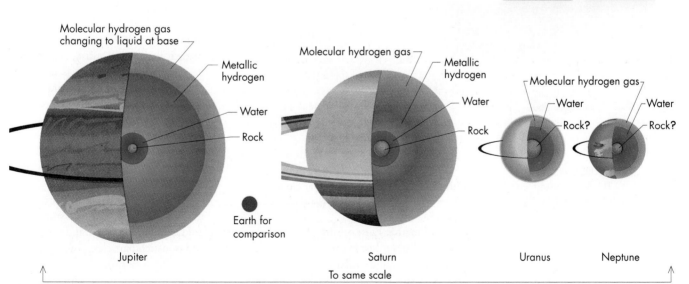

FIGURE OV4.3
The outer planets: Jupiter, Saturn, Uranus, Neptune. (Courtesy NASA/JPL.)

The Sun—our star—dwarfs the other objects in the Solar System (fig. OV4.4). It has nearly the same composition as Jupiter and Saturn (hydrogen and helium). The Sun's huge bulk squeezes the gas in its core so that the atoms there move at furious speed, with a temperature of almost 15 million Kelvin. At such a high temperature, hydrogen atoms fuse when they collide and form helium atoms. The fusion of hydrogen supplies energy to keep the Sun's core hot and support it against its crushing gravity. The Sun also contains heavier atoms such as carbon, oxygen, silicon, and iron, but these atoms are rare in our star and, because of the high temperature, exist there only as gas.

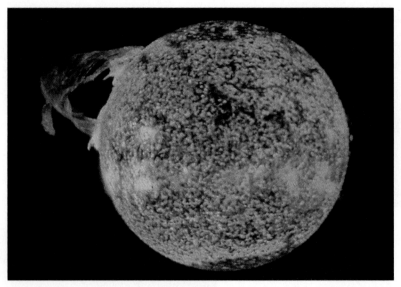

FIGURE OV4.4
The Sun. (Courtesy NASA.)

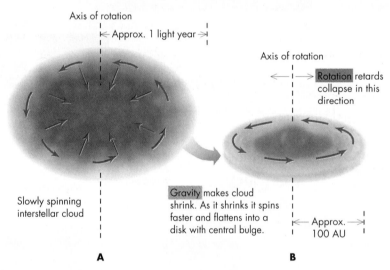

Axis of rotation
← Approx. 1 light year →

Axis of rotation
← Rotation retards collapse in this direction →

Slowly spinning interstellar cloud

Gravity makes cloud shrink. As it shrinks it spins faster and flattens into a disk with central bulge.

← Approx. 100 AU →

A   B

FIGURE OV4.5
The Solar System's birth began with the collapse of an interstellar cloud of gas and dust.

Astronomers believe the Solar System formed about 4.5 billion years ago when a slowly spinning interstellar cloud of gas and dust collapsed (fig. OV4.5). The cloud's gas was mostly hydrogen. The dust was composed mostly of silicates, compounds of iron, and water (rock, rust, and ice, in the simplest terms). The cloud's spin made it flatten into a disk, and at its center, gravity drew gas and dust into a central core. As gravity squeezed and thereby heated the core, it began to glow and eventually became hot enough for hydrogen atoms to fuse into helium atoms. At this stage the core became a star—our Sun.

Meanwhile, dust particles within the disk around the core stuck together and grew into larger pieces of matter. Dust orbiting far from the Sun was cold enough that water and gases such as ammonia could freeze on it, so distant dust became a mixture of ice, rock, and iron-rich particles. Dust near the Sun was too hot for ice to form, so it became a mixture of rock and iron compounds (fig. OV4.6).

Over about one hundred thousand years, the dust gradually grew into chunks ranging in size from tens of meters to hundreds of kilometers that astronomers call planetesimals. Planetesimals growing near the Sun were made mostly of rock and iron compounds. Planetesimals far from the Sun were rock and iron compounds plus ordinary ice. Because water was so much more abundant in the cloud than rocky or iron material, the ability to add ice to themselves allowed the planetesimals far from the Sun to grow much larger than those forming near the Sun.

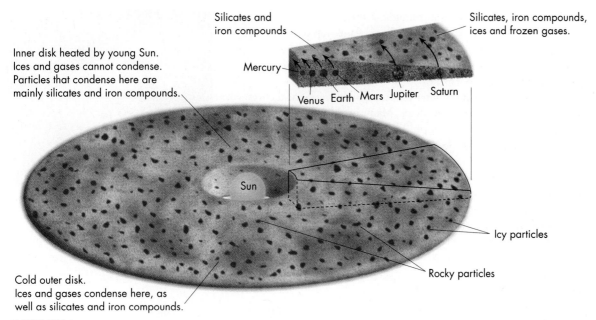

Inner disk heated by young Sun.
Ices and gases cannot condense.
Particles that condense here are
mainly silicates and iron compounds.

Silicates and
iron compounds

Silicates, iron compounds,
ices and frozen gases.

Mercury

Venus   Earth   Mars   Jupiter   Saturn

Sun

Icy particles

Rocky particles

Cold outer disk.
Ices and gases condense here, as
well as silicates and iron compounds.

FIGURE OV4.6
Heat from the young Sun prevented ice from condensing in the inner parts of the Solar Nebula. The planetesimals—and ultimately the planets—that formed there are therefore composed mainly of rock and iron.

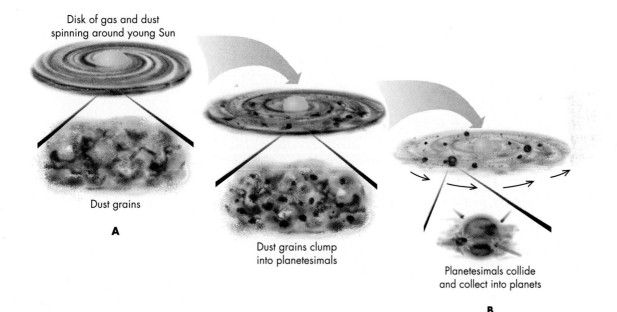

Disk of gas and dust
spinning around young Sun

Dust grains

A

Dust grains clump
into planetesimals

Planetesimals collide
and collect into planets

B

FIGURE OV4.7
Collisions between planetesimals led to their growth into the planets we see today.

Circling the Sun in a flattened disk-shaped cloud, the planetesimals gradually collided and grew more massive yet. Aided by their gravity as they grew in size, they gradually drew in all the solid material near them to become the nine planets (fig. OV4.7). Those that formed near the Sun grew from rock and iron-rich planetesimals and became the inner planets, Mercury, Venus, Earth, and Mars. Those that formed farther from the Sun grew from the planetesimals rich in ice and gases and became the outer planets, Jupiter, Saturn, Uranus, and Neptune. These giant worlds had large masses, so that their gravitational attraction was strong enough to draw in hydrogen and helium gas from their surroundings. This gas became the deep, hydrogen-rich atmospheres that we observe today on these cold and distant planets. The planetesimals near the Sun were too small to attract gas from their surroundings, so they contained little or no hydrogen or helium.

Rocky planetesimals a few astronomical units from the Sun may have begun to collect into a planet, but the huge planet Jupiter, forming a little farther out, constantly disturbed them and prevented them from gathering into a single world. We see these survivors of failed planet formation as the asteroids (fig. OV4.8).

As Jupiter and the other outer planets orbited the Sun, their gravity drew icy planetesimals toward them. Many of these planetesimals collected around the planets to form disks from which their moons grew. Other planetesimals, though initially drawn inward toward a planet, had enough orbital speed of their own to avoid capture. Some of these were flung inward toward the Sun to perish in its heat. Others struck the inner planets, delivering gases and water to these rocky worlds. Yet others were flung into the outer fringes of the Solar System, where they orbit in long slow loops. These remote icy objects form the Oort cloud and the Kuiper belt (fig. OV4.9), reservoirs for comets.

FIGURE OV4.8
The asteroid Ida and its tiny "moon," Dactyl. (Courtesy NASA/JPL.)

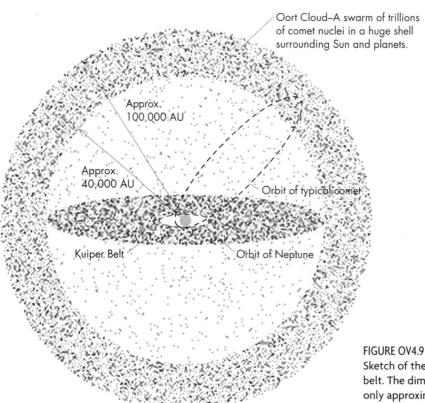

Oort Cloud–A swarm of trillions of comet nuclei in a huge shell surrounding Sun and planets.

Approx. 100,000 AU

Approx. 40,000 AU

Orbit of typical comet

Kuiper Belt

Orbit of Neptune

FIGURE OV4.9
Sketch of the Oort cloud and the Kuiper belt. The dimensions shown are known only approximately. Orbits and bodies are not to scale.

A few objects from the Kuiper belt strayed into orbits closer to the Sun and came under the influence of Neptune's gravity. One may have been captured to form Neptune's moon, Triton. Two became locked in an orbital resonance with Neptune, making three orbits for every two Neptune makes. We see these semi-captives of Neptune today as the planet Pluto and its moon Charon. The orbits of smaller residents of the Oort cloud and Kuiper belt occasionally shift under the influence of gravitational disturbances to carry these icy survivors in close to the Sun. There, our star's heat melts and vaporizes their surface layers, which then stream away from the Sun to form a long tail. We see these visitors to the inner Solar System as comets (fig. OV4.10).

FIGURE OV4.10
Comet Ikeya-Seki. (Courtesy Richard Cromwell/Steward Observatory, University of Arizona.)

The gravitational force that drew planetesimals together into planets generated lots of heat. The smaller worlds—Mercury and Mars—have cooled since their birth and are now inactive, with no signs of current volcanic activity. Earth and Venus, being larger, have retained much of their heat, and we see evidence of this in volcanic eruptions (fig. OV4.11) and in the slow creep of the rocky crust of our planet. Venus may be similarly active, but astronomers are still not certain.

FIGURE OV4.11
An erupting volcano. (Courtesy Novosti Pree Agency/Science Photo Library.

CHAPTER SEVEN

SURVEY OF THE SOLAR SYSTEM

An artist's view of the Solar System

The planets, to correct relative size

The Solar System consists of the Sun and the bodies orbiting in its gravitational field: the nine planets, their moons, and swarms of asteroids and comets. In this chapter we will survey the general properties of this system and describe its birth, reserving for later chapters more detailed descriptions of its components. Although earthlings have not walked on any Solar System objects except the Earth and Moon, we have detailed photographs sent to us from spacecraft of most of the planets and their satellites. Some are naked spheres of rock; others are mostly ice. Some have thin, frigid atmospheres so cold that ordinary gases crystalize as snow above their cratered surfaces; others have thick atmospheres the consistency of wet cement and no solid surface at all. On Mars, a dead volcano over twice the height of Mt. Everest looms above a red desert. On Uranus, an ocean thousands of miles deep heaves beneath poisonous clouds. Yet despite such diversity the Solar System possesses an underlying order, an order from which astronomers attempt to read the story of how our Solar System came to be.

As long ago as the eighteenth century, astronomers recognized that the highly ordered pattern in which the planets move is a clue to their origin. All the planets circle the Sun in the same direction, and all lie within a region of space shaped like a flattened disk, as we can directly infer from their motion within the zodiac. That pattern suggested to those early astronomers that the planets formed from a vast, spinning disk of gas and dust surrounding the Sun. As they learned more about the planets from more powerful telescopes and from the spectrum analysis of their light, astronomers discovered a second underlying pattern. Despite major visible surface differences, the planets form two main families: solid, rocky bodies composed mainly of silicates (like the Earth) and lying near the Sun, and liquid and gaseous bodies rich in hydrogen (like Jupiter) and lying far from the Sun.

From these and other regularities, astronomers hypothesize that the Sun and planets formed from the collapse of a huge, slowly spinning cloud of gas and dust. Most of the cloud's material fell inward and ended up in the Sun, but in response to rotation, some settled into a swirling disk around it. Then, within that disk, dust particles coagulated—perhaps aided by electrostatic effects such as those that make lint cling to your clothes—to form pebble-sized chunks of material, which in turn collided and sometimes stuck, growing ever larger to become the planets we see today. The type of planet formed—rocky or gaseous—depended strongly on the temperature and hence the place in the disk where it condensed.

Near the Sun's fiery surface, only dust with a high melting point could form from the original gases, and so planets there became masses of rock and metal like the Earth. But in the bitterly cold regions far from the Sun, ice and frozen gas could condense as well as rock and metals. Thus planets like Jupiter are rich in these lighter materials. However, regardless of their composition, the objects that formed in the disk retained the motion of the original gas and dust, and so we see them today, moving in a flattened system, all orbiting the Sun in the same direction. How long ago did all this happen? From the amount of radioactivity in meteorites, astronomers estimate the Solar system formed about 4.5 billion years ago. With this overview in mind, we can now look in more detail at what our Solar System is like and how astronomers think it formed.

7.1 COMPONENTS OF THE SOLAR SYSTEM

The Sun

The Sun is a star, a ball of incandescent gas whose light and heat are generated by nuclear reactions in its core. It is by far the largest body in the **Solar System**: more than 1000 times the mass of all the other bodies put together, and its gravitational force holds the planets and other bodies in the system in their orbital patterns about it. This gravitational domination of the planets by the Sun justifies our calling the Sun's family the Solar System.

The Sun is mostly hydrogen (about 71%) and helium (about 27%), but it also contains very small proportions of nearly all the other chemical elements (carbon, iron, uranium, and so forth) in vaporized form, as we can tell from the spectrum of the light it emits.

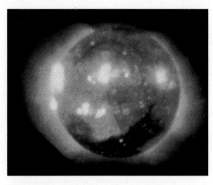

An x-ray image of the Sun (Courtesy NASA.)

The Planets

The planets are much smaller than the Sun and orbit about it. They emit no visible light of their own but shine by reflected sunlight. In order of increasing distance from the Sun, they are Mercury, Venus, Earth, Mars, Jupiter, Saturn, Uranus, Neptune, and Pluto. The planets move around the Sun in approximately circular* orbits all lying in nearly the same plane, as shown in the top view in figure 7.1 and the side view in figure 7.2. Thus the Solar System is like a spinning pancake, with the planets traveling around the Sun in the same direction: counterclockwise, as seen from above the Earth's north pole.

As the planets orbit the Sun, each also spins on its rotation axis. The spin is generally in the same direction as each planet's orbital motion around the Sun (again, counterclockwise, as seen from above the Earth's north pole), and the tilt of the rotation axes relative to the plane of planetary orbits is generally not far from the perpendicular. However, there are three exceptions: Venus, Uranus, and Pluto. Uranus and Pluto have extremely large tilts to their rotation axes: their rotation axes lie nearly in their orbital planes (fig. 7.2). Venus's rotation axis has only a small tilt, but it spins backward, a motion technically called "retrograde rotation." However, despite this backward spin, Venus orbits the Sun in the same direction as the rest of the planets.

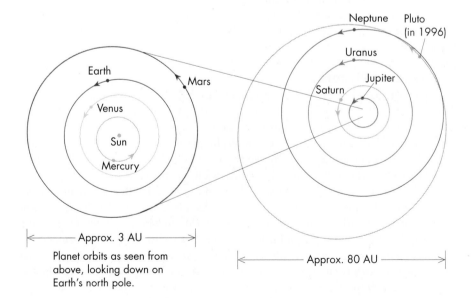

Approx. 3 AU

Planet orbits as seen from above, looking down on Earth's north pole.

Approx. 80 AU

FIGURE 7.1
An artist's view of the Solar System from above. The orbits are shown in the correct relative scale in the two drawings.

---

*The circularity of the orbits is only approximate because we know from Kepler's laws that in reality the orbits are ellipses. However, the amount of ellipticity is, in general, very small.

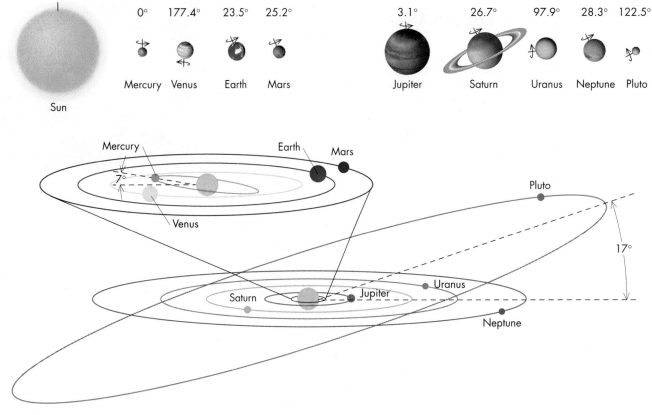

FIGURE 7.2
Planets and their orbits from the side. Sketches also show the orientation of the rotation
axes of the planets and Sun. Orbits and bodies are not to the same scale.*

The flattened structure and the orderly orbital and spin properties of the Solar System
are two of its most fundamental features, and any theory of the Solar System must explain them. But a third and equally important feature is that the planets fall into two
families called inner and outer planets based on their size, composition, and location in
the Solar System.

Two Types of Planets

The **inner planets,** Mercury, Venus, Earth, and Mars, are small rocky bodies with relatively
thin or no atmospheres. The **outer planets,** Jupiter, Saturn, Uranus, Neptune, and Pluto, are
gaseous, liquid, or icy. Except for Pluto, they are much larger than the inner planets and have
deep, hydrogen-rich atmospheres. For example, Jupiter is over 10 times the Earth's diameter and has 318 times its mass. These differences can be seen in figure 7.3, which also shows
a small part of the edge of the Sun to illustrate how the Sun drawfs even the large planets.

In describing the planets we have used the terms *rock* and *ice.* By rock we mean material composed of silicon and oxygen (SiO_2) with an admixture of other heavy elements
such as aluminum (Al), magnesium (Mg), sulfur (S), and iron (Fe). By ice we mean
frozen liquids and gases such as ordinary water ice (H_2O), frozen carbon dioxide (CO_2),
frozen ammonia (NH_3), frozen methane (CH_4), and so on. If we consider the Solar System as a whole, rock is rare because the silicon atoms that compose it are outnumbered
more than 25,000 to 1 by hydrogen. However, in the warmth of the inner Solar System,
rock dominates because intrinsically more abundant materials such as hydrogen,
methane, and ammonia cannot condense to mingle with it. Thus the inner planets are
composed mainly of rock.

---

*Technically, Venus's tilt angle is about 177°, but it can also be described as 3° (180° − 177°),
with a "backwards", or retrograde spin.

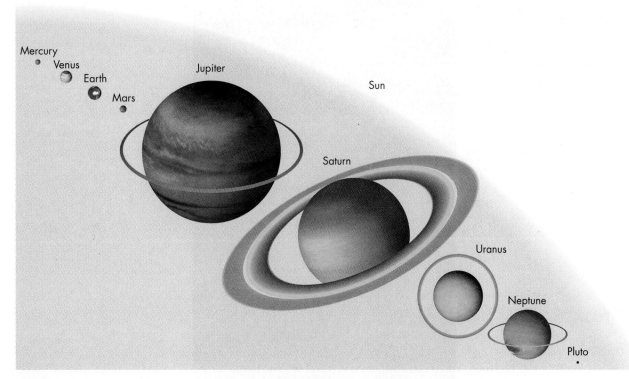

FIGURE 7.3
The planets and Sun to scale.

The outer planets have no true "surface"; rather their atmospheres thicken with depth and eventually liquefy. There is no distinct boundary between "atmosphere" and "crust" as we have on the Earth. In the deep interior, the liquid may be compressed into a solid, as happens in the Earth's inner core, but the transition from liquid to solid is also probably not sharply defined. Thus we can never "land" on Jupiter because we would simply sink ever deeper into its interior.

Instead of "inner" and "outer" planets, astronomers sometimes use "terrestrial" and "Jovian" to describe the two types of planets. The **terrestrial planets** (Mercury to Mars) are so named because of their resemblance to the Earth. The **Jovian planets** (Jupiter to Neptune) are named for their resemblance to Jupiter. Pluto is not included as a Jovian planet because of its small size.

Satellites

As the planets orbit the Sun, most are themselves orbited by satellites. Jupiter, Saturn, and Uranus have large families of 16, 22, and 15 clearly identified moons, respectively.* Neptune has 8, Mars has 2, and Earth and Pluto each only 1. Only Mercury and Venus are moonless.

The moons generally move along approximately circular paths that are roughly in the planet's equatorial plane and so in general lie in the same plane as the planet's orbit around the Sun. Only the moons of Uranus and Pluto deviate, having their orbits tilted like the planets themselves. Thus each planet and its moons resemble a miniature Solar System—an important clue to the origin of the satellites.

*Astronomers have recently discovered two more moons orbiting Uranus. Because these new moons are very small (about 80 and 160 km in diameter) and move in elongated and highly tilted orbits, astronomers think they are captured asteroids.

Evening sky looking west showing the crescent Moon and the planets Venus and Jupiter. Note that these objects lie in approximately a line along the Zodiac, showing the flatness of the Solar System. (Courtesy Milton Hayes.)

Asteroids and Comets

Asteroids and comets are far smaller than planetary bodies. The **asteroids** are rocky or metallic bodies with diameters that range from a few meters up to about 1000 km (about one-tenth the size of the Earth). The **comets,** on the other hand, are icy bodies about 10 km (about 6 miles) or less in diameter that grow huge tails of gas and dust as they near the Sun and are partially vaporized by its heat. Thus these minor bodies exhibit the same split into two families that we see for the planets, that is, rocky and icy bodies.

Asteroids and comets differ not only in their composition, but also in their location within the Solar System. Most asteroids circle the Sun in the large gap between the orbits of Mars and Jupiter, a region called the **asteroid belt.** They are probably material that failed—perhaps as a result of disturbance by Jupiter's gravity—to aggregate into a planet. Most comets, on the other hand, orbit far beyond Pluto in a region of the Solar System called the Oort cloud, and only rarely do they move into the inner Solar System.

The **Oort cloud,** named for the Dutch astronomer who proposed its existence, is thought to be a spherical region that completely surrounds the Solar System and extends from about 40,000 to 100,000 AU from the Sun (fig. 7.4). Although the majority of comets probably originate in the Oort cloud, some come from a disklike swarm of icy objects that lies just beyond the orbit of Neptune, a region called the Kuiper belt. We will discuss more details of the Oort cloud and Kuiper belt in chapter 10, but for now we simply note that together they probably contain more than 1 trillion (10^{12}) comet nuclei.

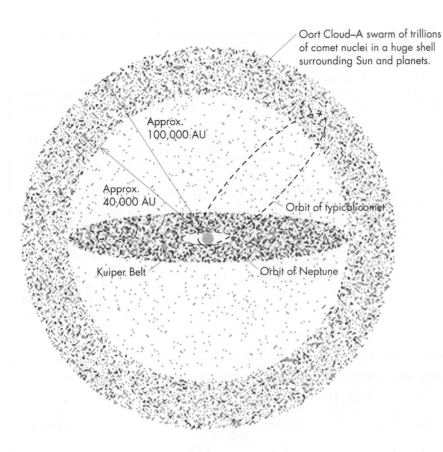

Oort Cloud—A swarm of trillions of comet nuclei in a huge shell surrounding Sun and planets.

Approx. 100,000 AU

Approx. 40,000 AU

Orbit of typical comet

Kuiper Belt

Orbit of Neptune

FIGURE 7.4
Sketch of the Oort cloud and the Kuiper belt. The dimensions shown are known only approximately. Orbits and bodies are not to scale.

Composition Differences between the Inner and Outer Planets

We stated earlier that the inner planets are rocky and the outer planets are hydrogen-rich. These composition differences are so important to our understanding of the history of the Solar System that we should look at them more closely and see how they are determined.

Astronomers can deduce a planet's composition in several ways. From its spectrum, they can measure its atmospheric composition and get some information about the nature of its surface rocks. However, spectra give no clue as to what lies deep inside a planet where light cannot penetrate. To learn about the interior, astronomers must therefore use alternative methods.

We saw in chapter 4 how earthquake waves could reveal what lay inside the Earth, but this method has not yet been used for other planets. Although quake detectors were landed on Mars, they did not work properly, and such detectors would require very special modification to work on the Jovian planets because they have no surface on which to land! Thus we must try other means to study the interior of planets. One such technique uses the planet's density.

Density as a Measure of a Planet's Composition

The average density of a planet is its mass divided by its volume. Both these quantities can be measured relatively easily. For example, we showed in chapter 2 how to determine a body's mass from its gravitational attraction on a second body orbiting around it by applying Newton's modification of Kepler's third law. Thus from this law, we can calculate a planet's mass by observing the orbital motion of one of its moons or a passing spacecraft. We can determine a planet's volume (V) from the formula $V = 4\pi R^3/3$, where R is the planet's radius. We can measure R in several ways, for example, from its angular size and

EXTENDING OUR REACH

BODE'S LAW: THE SEARCH FOR ORDER

A curious and as yet unexplained feature of the orbits of the planets is their regular spacing. Very roughly, each planet is about twice as far from the Sun as its inner neighbor. This progression of distance from the Sun can be expressed by a simple mathematical relation known as **Bode's law,** which works as follows: write down 0, 3, and then successive numbers by doubling the preceding number until you have nine numbers. That is, 0, 3, 6, 12, 24, and so on. Next add 4 to each, and divide the result by 10, as shown in table 7.1. The resulting numbers, with two exceptions, are *very* close to the distances of the planets from the Sun in astronomical units.

Bode's law was worked out before the discovery of Uranus, Neptune, and Pluto, and when Uranus was discovered and found to fit the law, interest was focused on the "gap" at 2.8 AU. Astronomers therefore began to search for a body in the gap, and, as we will see in chapter 10, Giuseppi Piazzi, a Sicilian astronomer, soon discovered the asteroid Ceres, which fitted the rule splendidly.

Ironically, the next planet to be found, Neptune, did not fit the rule at all, though Pluto does, at least approximately! These irregularities make astronomers unsure how to treat Bode's law. Is it just chance, or is it telling us something profound about the Solar System that we have yet to understand? Perhaps it merely shows the human fascination with patterns and our tendency to see order where none may actually exist.

TABLE 7.1   Bode's Law

Bode's Law	Number	Planet	True distance
(0 + 4)/10 =	0.4	Mercury	0.39
(3 + 4)/10 =	0.7	Venus	0.72
(6 + 4)/10 =	1.0	Earth	1.00
(12 + 4)/10 =	1.6	Mars	1.52
(24 + 4)/10 =	2.8	?	?
(48 + 4)/10 =	5.2	Jupiter	5.2
(96 + 4)/10 =	10.0	Saturn	9.5
(192 + 4)/10 =	19.6	Uranus	19.2
		Neptune	30.1
(384 + 4)/10 =	38.8	Pluto	39.5

distance, a technique we used in chapter 1 to measure the radius of the Moon. With the planet's mass, M, and volume, V, known, we can calculate its average density straightforwardly by dividing M, by V, (fig. 7.5).

Once the planet's average density is known, we can compare it with the density of abundant, candidate materials to find a likely match. For example, we saw in chapter 4 that the average density of the Earth (5.5 grams per cubic centimeter) was intermediate between silicate rock (about 3 grams per cubic centimeter) and iron (7.8 grams per cubic centimeter). Therefore, we inferred that the Earth has an iron core beneath its rocky crust, a supposition which was verified from studies with earthquake waves.

Although density comparison is a powerful tool for studying planetary composition, it also has drawbacks. First, there may be several different substances that will produce an equally good match to the observed density. Second, the density of a given material can be affected by the planet's gravitational force. For example, a massive planet may crush rock whose normal density is 3 grams per cubic centimeter to a density of 7 or 8 grams per cubic centimeter. Thus, in making a match to determine the composition, we must take into account compression by gravity.

Mass

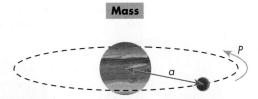

Observe motion of a satellite orbiting planet. Determine satellite's distance (a) from planet and orbital period, P. Use Newton's form of Kepler's third law (discussed in chapter 2)

$$M = \frac{4\pi^2 a^3}{GP^2}$$

Insert measured values of a and P, and value for constant G. Solve for M.

Radius

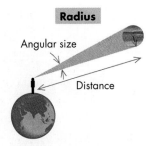

Measure angular size of planet and use relation between angular size and distance to solve for R.

Average Density

Average density (ρ) is mass (M) divided by volume (V)

$$\rho = \frac{M}{V}$$

Volume

$$V = \frac{4\pi R^3}{3}$$

For a spherical planet of radius R

FIGURE 7.5
Measuring a planet's mass, radius, and average density.

All the terrestrial planets have an average density similar to the Earth's (3.9 to 5.5 grams per cm^3). On the other hand, all the Jovian planets have a much smaller average density (0.71 to 1.67 grams per cm^3), similar to that of ice. After correcting for the above gravitational compression, we conclude that all the inner planets contain large amounts of rock and iron and the iron has sunk to the core, as shown in figure 7.6. Likewise, the outer planets contain mainly light materials, as borne out by their spectra, which show them to be mostly hydrogen, helium, and hydrogen-rich material such as methane (CH_4), ammonia (NH_3), and water (H_2O).

The outer planets probably have cores of iron and rock about the size of the Earth beneath their deep atmosphere, as illustrated in figure 7.6. Astronomers deduce the existence of these cores in two ways. First, if the outer planets have the same relative amount of heavy elements as the Sun, they should contain several Earth masses of iron and silicates, and because these substances are much denser than hydrogen, they must sink to the planet's core. Secondly, mathematical analysis of these planets' rotation rates shows that the shape of their equatorial bulges can be best explained if they have small, dense cores.

Our discussion of the composition of the planets not only underlines the difference between the two families of planets, but also furnishes another clue to their origin: the planets and Sun were all made from the same material. Astronomers think this because Jupiter and Saturn have a composition almost identical to that of the Sun, and the inner planets have a similar composition *if we were to remove the Sun's hydrogen and helium.**
Thus, we can explain the compositional difference between the inner and outer planets by proposing a process that would keep the inner planets from collecting and capturing the light gases.

Age of the Solar System

An important clue to the origin of the Solar System comes from its age. Despite great differences in size, structure, and composition, the planets, asteroids, and comets all seem to have formed at nearly the same time. We can directly measure that date for the Earth, Moon, and some asteroids from the radioactivity of their rocks, and we find that none are

---

*Carbon, nitrogen, neon, and other elements normally in gaseous compounds are also relatively rare in the inner planets.

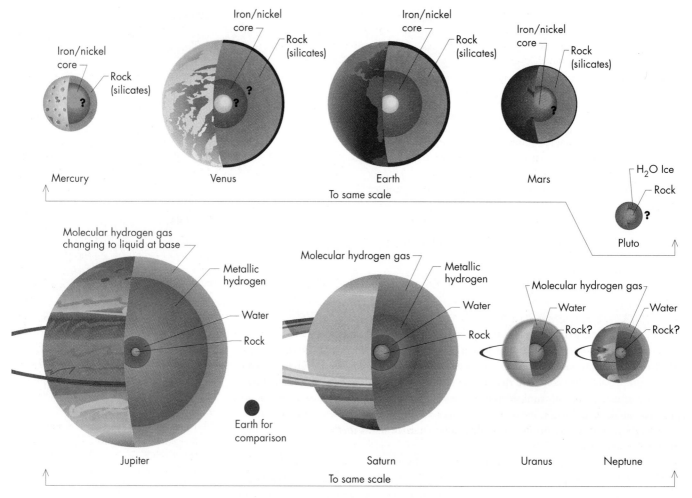

FIGURE 7.6

Sketches of the interiors of the planets. Details of sizes and composition of inner regions are uncertain for many of the planets. (Pluto is shown with the terrestrial planets for scale reasons only.)

more than about 4.5 billion years old. (Rocks from asteroids are the origin of many of the meteors that have reached Earth and are thus available for study, as we will discuss in chapter 10). Likewise, we find a similar age for the Sun based on its current brightness and temperature and its presumed rate of nuclear fuel consumption.

7.2 ORIGIN OF THE SOLAR SYSTEM

Since we were not around 4.5 billion years ago to witness the birth of the Solar System, our best explanation of its origin must be a reconstruction based on observations that we make now, billions of years after the event. Those observations, discussed above, are summarized below. Each must be explained by whatever theory we devise.

1. The Solar System is flat, with all the planets orbiting in the same direction.
2. There are two types of planets, inner and outer, with the rocky ones near the Sun and the gaseous or liquid ones further out.
3. The composition of the outer planets is similar to the Sun's, while that of the inner planets is like the Sun's minus the gases that condense only at low temperatures.
4. All the bodies in the Solar System whose ages have so far been determined are less than about 4.5 billion years old.

We have listed only the most important observed features that our theory must explain. There are many additional clues from the structure of asteroids, the number of craters on planetary and satellite surfaces, and the detailed chemical composition of surface rocks and atmospheres.

The currently favored theory for the origin of the Solar System derives from theories proposed in the eighteenth century by Immanuel Kant, the great German philosopher, and Pierre Simon Laplace, a French mathematician. Kant and Laplace independently proposed what is now called the **solar nebula hypothesis** that the Solar System originated from a rotating, flattened disk of gas and dust, with the outer part of the disk becoming the planets and the center becoming the Sun. This theory offers a natural explanation for the flattened shape of the system and the common direction of motion of the planets around the Sun.

The modern form of the theory proposes that the Solar System was born 4.5 billion years ago from an **interstellar cloud,** an enormous rotating aggregate of gas and dust like the one shown in figure 7.7. Such clouds are common between the stars in our galaxy even today, and astronomers now think all stars have formed from them. Thus, although our main concern in this chapter is with the birth of the Solar System, we should bear in mind that our theory applies more broadly and implies that most stars could have planets, or at least surrounding disks of dust and gas from which planets might form.

Interstellar Clouds

Because interstellar clouds are the raw material of the Solar System, we need to describe them more fully. Although such clouds are found in many shapes and sizes, the one that became the Sun and planets was probably a few light years in diameter and contained about twice the present mass of the Sun. If it was like typical clouds we see today, it was made mostly of hydrogen (71%) and helium (27%) gas, with tiny traces of other chemical elements, such as gaseous carbon, oxygen, and silicon. In addition to the gases, interstellar clouds also contain tiny dust particles called **interstellar grains.**

Interstellar grains range in size from large molecules to micrometers or larger and are believed to be made of a mixture of silicates, iron compounds, carbon compounds, and water frozen into ice. Astronomers deduce the presence of these substances from their spectral lines that are seen in starlight that has passed through dense dust clouds. Moreover, a few hardy interstellar dust grains, including tiny diamonds, have been found in ancient meteorites. This direct evidence from grains and the data from spectral lines shows that the elements occur in proportions similar to those we observe in the Sun. This is additional evidence that the Sun and its planets could have formed from an interstellar cloud.

The cloud began its transformation into the Sun and planets when the gravitational attraction between the particles in the densest parts of the cloud caused it to collapse inward, as shown in figure 7.8A. The collapse may have been triggered by a star exploding

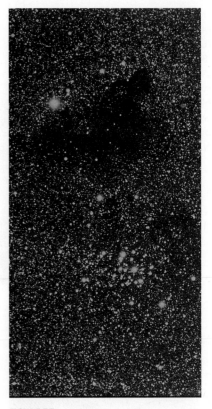

FIGURE 7.7
Photograph of an interstellar cloud (the dark region at top) which may be similar to the one from which the Solar System formed. (Anglo-Australian Telescope Board, photo by David Malin.)

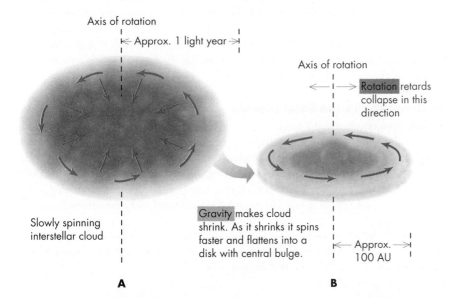

Axis of rotation
Approx. 1 light year

Slowly spinning interstellar cloud

Gravity makes cloud shrink. As it shrinks it spins faster and flattens into a disk with central bulge.

Axis of rotation
Rotation retards collapse in this direction

Approx. 100 AU

A   **B**

FIGURE 7.8
Sketch illustrating the (A) collapse of an interstellar cloud and (B) its flattening.

FIGURE 7.9
(A) Disk of dust and gas around forming star. (Courtesy NASA.)
(B) Photograph in false color of a disk of dust around the young star, β Pictoris. The dark circle blots out the star's direct light, which would otherwise overexpose the image. (Courtesy NASA/JPL.)

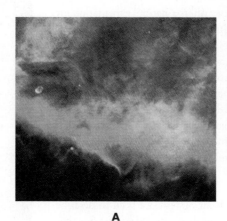

A

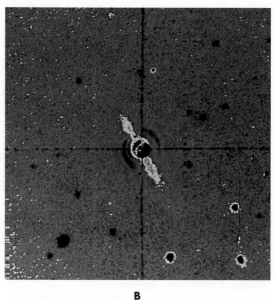

B

nearby or by a collision with another cloud. But regardless of its initial cause, the infall was not directly to the center. Instead, because the cloud was rotating, it flattened, as shown in figure 7.8B.

Flattening occurred because rotation retarded the collapse perpendicular to the cloud's rotation axis. A similar effect happens in an old-fashioned pizza parlor where the chef flattens the dough by tossing it into the air with a spin.

Formation of the Solar Nebula

It took a few million years for the cloud to collapse and become a rotating disk with a bulge in the center. The disk is called the **solar nebula**, and it eventually condensed into the planets while the bulge became the Sun. This explains the first obvious property of the Solar System—its disklike structure—which we mentioned at the beginning of the section.

The solar nebula was probably about 200 AU in diameter and perhaps 10 AU thick. Its inner parts were hot, heated by the young Sun and the impact of gas falling on the disk during its collapse, but the outer parts were cold, far below the freezing point of water. We are fairly certain of these dimensions and temperatures because we can observe disks around other stars and, in a few cases, can even detect other planets. For example, figure 7.9A shows a picture made with the Hubble Space Telescope of gas and dust disks near the Orion Nebula. The stars at the centers of these disks have not yet become hot enough to emit much visible light. Figure 7.9B, on the other hand, shows a disk (seen nearly edge on) in which the star has reached full brilliance. Although the picture is grainy and in false color in order to emphasize the limited detail, you can see the disk edge on.

Condensation in the Solar Nebula

Condensation occurs when a gas cools and its molecules stick together to form liquid or solid particles.* For condensation to happen, the gas must cool below a critical temperature (the value of which depends on the substance condensing and the surrounding pressure). For example, suppose we start with a cloud of vaporized iron at a temperature of 2000 K. If we now cool the iron vapor to about 1300 K, tiny flakes of iron will condense from it. Likewise, if we cool a gas of silicates to about 1200 K flakes of rocky material will condense.

---

*Technically, condensation is the change from gas to liquid, and deposition is the change from gas to solid. However, we will not make that distinction here.

At lower temperatures, other substances will condense. For example, water can condense at room temperature, as you can see as steam escapes from a boiling kettle. Here, water molecules in the hot steam come into contact with the cooler air of the room. As the vaporized water cools, its molecules move more slowly, so that when they collide, electrical forces can bind them together, first into pairs, then into small clumps, and eventually into the tiny droplets that make up the cloud we see at the spout.

An important feature of condensation is that when a mixture of vaporized materials cools, the materials with the highest vaporization temperatures condense first. Thus, as a mixture of gaseous iron, silicate, and water cools, it will make iron grit when its temperature reaches 1300 K, silicate grit when it reaches 1200 K, and finally water drops when it cools to only a few hundred degrees K. It is a bit like putting a jar of chicken soup in the freezer. First the fat freezes, then the broth, and finally the bits of chicken and celery.

However, the condensation process stops if the temperature never drops sufficiently low. Thus, in the example above, if the temperature never cools below 500 K, water will not condense and the only solid material that forms from the gaseous mixture will be iron and silicates.

This kind of condensation sequence occurred in the solar nebula as it cooled after its collapse to a disk. But because the temperature within the disk from the Sun to almost the orbit of Jupiter never dropped low enough, water and other substances with similar condensation temperatures could not condense there. On the other hand, iron and silicate, which condense even at relatively high temperatures, could condense everywhere within the disk. Thus the nebula became divided into two regions: an inner zone of silicate/iron particles, and an outer zone of similar particles on which ices also condensed, as illustrated schematically in figure 7.10. Water, hydrogen, and other easily vaporized substances were present as gases in the inner solar nebula, but they could not form solid particles there. However, some of these substances combined chemically with silicate grains so that the rocky material from which the inner planets formed contained within it small quantities of water and other gases.

Accretion and Planetesimals

In the next stage of planet formation, the tiny particles that condensed from the nebula must have begun to stick together into bigger pieces in a process called **accretion.**

The process of accretion is a bit like building a snowman. You begin with a handful of loose snowflakes and squeeze them together to make a snowball. Then you add more snow

(Courtesy Fundamental Photograph/Diane Schiumo.)

FIGURE 7.10
Artist's depiction of the condensation of dust grains in the solar nebula and the formation of rocky and icy particles.

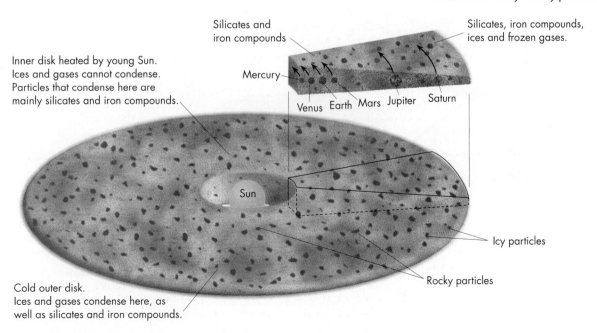

Silicates and iron compounds

Silicates, iron compounds, ices and frozen gases.

Mercury

Venus Earth Mars Jupiter Saturn

Inner disk heated by young Sun. Ices and gases cannot condense. Particles that condense here are mainly silicates and iron compounds.

Sun

Icy particles

Rocky particles

Cold outer disk. Ices and gases condense here, as well as silicates and iron compounds.

by rolling the ball on the ground. As the ball gets bigger, it is easier for snow to stick to it, and it rapidly grows in size.

Similarly in the solar nebula, tiny grains stuck together and formed bigger grains that grew into clumps, perhaps held together by electrical forces similar to those that make lint stick to your clothes. Subsequent collisions, if not too violent, allowed these smaller particles to grow into objects ranging in size from millimeters to kilometers (fig. 7.11A). These larger objects are called **planetesimals** (that is, small, planetlike bodies).

Because the planetesimals near the Sun formed from silicate and iron particles, while those farther out were cold enough that they could incorporate ice and frozen gases, there were two main types of planetesimals: rocky/iron ones near the Sun and icy/rocky/iron ones farther out. This then explains the second observation we described at the beginning of the section—that there are two types of planets—as described below.

Formation of the Planets

As planetesimals moved within the disk and collided with one another, planets formed. Computer simulations show that some collisions led to the shattering of both bodies, but gentler collisions led to merging, with the planetary orbits gradually becoming approximately circular. Moreover, in some such simulations, the distance between orbits is similar to that given by Bode's law.

Merging of the planetesimals increased their mass and thus their gravitational attraction. That in turn helped them grow even more massive by drawing planetesimals into clumps or rings around the Sun. Within these clumps, growth went even faster, so that over a time lasting about 100,000 years larger and larger objects formed, as depicted in figure 7.11B.

Planet growth was especially rapid in the outer parts of the solar nebula. Planetesimals there had more material from which to grow because ice was about 10 times more abundant than silicate and iron compounds. Thus planetesimals in the outer solar nebula could in principle become 10 times larger than those in the inner nebula.

Additionally, once a planet grew somewhat larger than the diameter and mass of the Earth, it was able to attract and retain gas by its own gravity. Because hydrogen was overwhelmingly the most abundant material in the solar nebula, planets large enough to tap that reservoir could grow vastly larger than those that formed only from solid material. Thus Jupiter, Saturn, Uranus, and Neptune probably began as Earth-sized bodies of ice and rock, but their gravitational attraction resulted in their becoming surrounded by the

FIGURE 7.11
(A) Sketches illustrating how dust grains may have grown into planetesimals. (B) Sketches of how the planetesimals may have grown into planets.

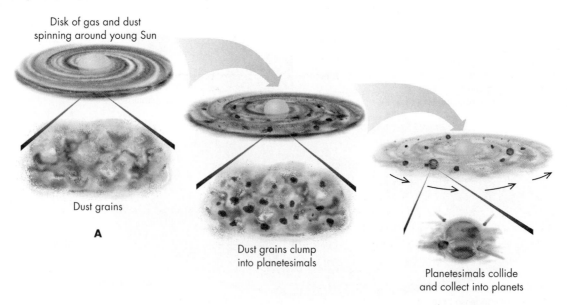

Disk of gas and dust
spinning around young Sun

Dust grains

A

Dust grains clump
into planetesimals

Planetesimals collide
and collect into planets

B

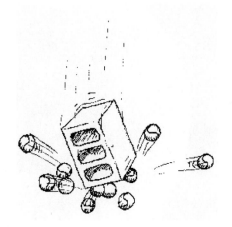

huge envelopes of hydrogen-rich gases that we see today. The smaller and warmer bodies of the inner Solar System could not capture hydrogen, and therefore remained small and lack that gas. This explains the third observation we mentioned at the beginning of the section—that the outer planets have a composition similar to the Sun's.

As planetesimals struck the growing planets, their impact released gravitational energy that heated both the planetesimal and the planet. Gravitational energy is liberated whenever something falls. For example, when a cinderblock falls onto a box of tennis balls, the impact scatters the balls in all directions giving them kinetic energy—energy of motion. In much the same manner, planetesimals falling onto a planet's surface give energy to the atoms in the crustal layers, energy that appears as heating. You can easily demonstrate that motion can generate heat by hitting a steel nail a dozen or so times with a hammer and then touching the nail to your lip: the metal will feel distinctly hot. Imagine now the vastly greater heating created as mountain-sized masses of rock plummet onto a planet. The heat so liberated, in combination with the radioactive heating described earlier, melted the planets and allowed matter with high density (such as iron) to sink to their cores, while matter with lower density (such as silicate rock) "floated" to their surfaces. We saw in chapter 4 that the Earth's iron core probably formed by this process, and astronomers believe that the other terrestrial planets formed their iron cores and rocky crusts and mantles the same way.* A similar process probably occurred for the outer planets when rock and iron material sank to their cores.

Formation of Moons

The moons of the outer planets probably were formed from planetesimals orbiting the growing planets. Once a body grew massive enough so that its gravitational force could draw in additional material, it became ringed with debris. Thus moon formation was a scaled-down version of planet formation, and so the satellites of the outer planets have the same regularities as the planets around the Sun.

All four giant planets have flattened satellite systems in which the satellites (with few exceptions) orbit in the same direction. Many of these satellites are about as large as Mercury, and they would be considered full-fledged planets were they orbiting the Sun. A few of these bodies even have atmospheres, but they have too little mass (and thus too weak a gravitational attraction) to have accumulated large quantities of hydrogen and other gases from the solar nebula as their parent planets did. Thus these moons are composed mainly of rock and ice, giving them solid surfaces—surfaces that are generally cratered and that, in a few cases, show signs of volcanic activity. These distant moons might in the future be ideal bases for studying those planets with no surface to land on.

Final Stages of Planet Formation

The last stage of planet formation was a rain of planetesimals that blasted out the huge craters such as those we see on the Moon and on all other bodies in the Solar System with solid surfaces. Figure 7.12 shows some of the planets and moons bearing this vivid testimony to the role of planetesimals in planet building.

---

*Although the above scenario accurately explains much of what we know of our Solar System—that is, there is not as yet any evidence to disprove it—some astronomers suggest a slightly modified theory of planet building. They propose that the solar nebula was initially so hot that iron condensed into dust long before silicate rock did and that the first planetesimals were therefore mostly made of iron.

These iron-rich bodies then clumped to form what we see today as the cores of the planets. Only later, after the nebula had cooled further, could rocky material condense and accumulate around the existing iron cores. Which scenario is correct, accumulation followed by differentiation or the reverse, we simply do not yet know.

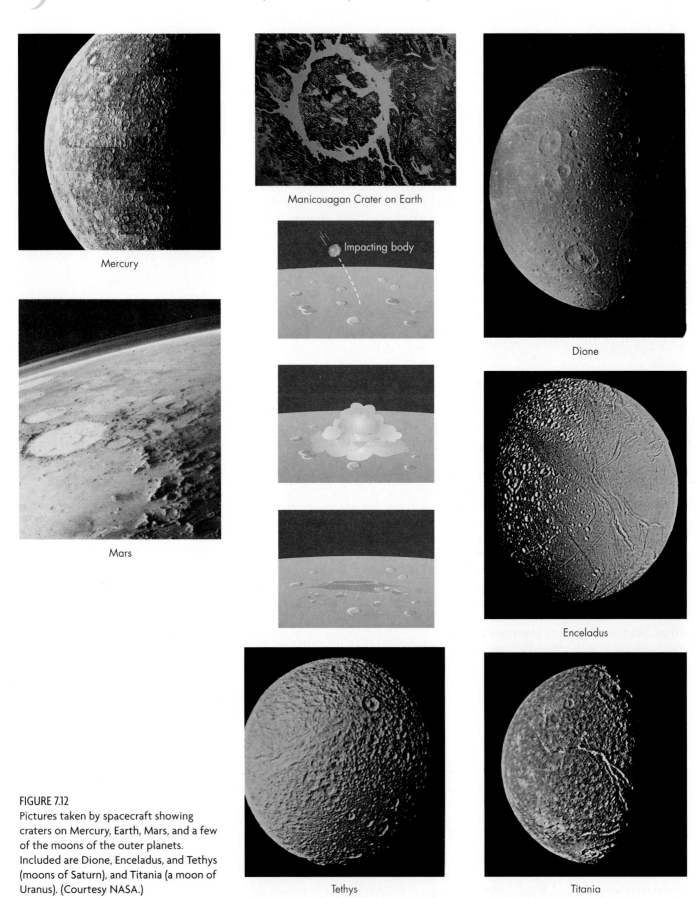

Mercury

Manicouagan Crater on Earth

Impacting body

Mars

Dione

Enceladus

Tethys

Titania

FIGURE 7.12
Pictures taken by spacecraft showing
craters on Mercury, Earth, Mars, and a few
of the moons of the outer planets.
Included are Dione, Enceladus, and Tethys
(moons of Saturn), and Titania (a moon of
Uranus). (Courtesy NASA.)

Occasionally an impacting body was so large that it did more than simply leave a crater. For example, we saw in chapter 6 that the Moon may have been created when the Earth was struck by a Mars-sized body. Likewise, as we will discuss in chapter 8, Mercury may have suffered a massive impact that blasted away its crust. The extreme tilt of the rotation axes of Uranus and Pluto may also have arisen from planetesimal collisions. In short, planets and satellites were brutally battered by the remaining planetesimals.

Although planet building consumed most of the planetesimals, some survived to form small moons, the asteroids, and comets. Rocky planetesimals and their fragments remained between Mars and Jupiter where, stirred by Jupiter's gravitational force, they were unable to assemble into a planet. We see them today as the asteroid belt. Jupiter's gravity (and that of the other giant planets) also disturbed the orbits of icy planetesimals, tossing some in toward the Sun and others outward in elongated orbits to form the swarm of comet nuclei we call the Oort cloud. The few that remain in the disk from Neptune's orbit out to beyond Pluto's we call the Kuiper Belt.

Formation of Atmospheres

Atmospheres were the last part of the planet-forming process. The inner and outer planets are thought to have formed atmospheres differently, a concept that explains their very different atmospheric composition. The outer planets probably captured most of their atmospheres from the solar nebula as mentioned above; because the nebula was rich in hydrogen, so are their atmospheres.

The inner planets were not massive enough and were too hot to capture gas from the solar nebula (as mentioned earlier) and are therefore deficient in hydrogen and helium. Venus, Earth, and Mars probably created their original atmospheres by volcanic eruptions, though some of their atmospheric gases may have come from infalling comets and icy planetesimals that vaporized on impact. In fact, as a general rule, bodies too small to have captured atmospheres directly but that show clear signs of extensive volcanic activity (now or in the past) have atmospheres. More quiescent ones do not. Moreover, small bodies such as Mercury and our Moon cannot keep any atmosphere at all because their weak gravitational force means that their escape velocity is rather small, and any atmospheric gases tend to escape easily from them.

Cleaning up the Solar System

Only a few hundred thousand or million years were needed to assemble the planets from the solar nebula, though the rain of infalling planetesimals may have lasted longer. Such a time is long in the human time frame, but short in the Solar System's. All the objects within the Solar System are about the same age—the fourth property of the Solar System mentioned at the beginning of the section.

One process still had to occur before the Solar System became what we see today: the residual gas and dust must have been removed. Just as a finished house is swept clean of the debris of construction, so too was the Solar System. In the sweeping process, the Sun was probably the cosmic broom, with its intense heat driving a flow of tenuous gas outward from its atmosphere. As that flow impinges on the remnant gas and dust around the Sun, the debris is pushed away from the Sun to the fringes of the Solar System. Such gas flows are seen in most young stars, and astronomers are confident the Sun was no exception. Even today some gas flows out from the Sun, but in its youth the flow was more vigorous.

The above theory for the origin of the Solar System explains many of its features, but astronomers still have many questions about the Sun and its family of planets and moons. For example, could some planetary systems have many more planets than our own, while others have fewer? Will all planetary systems have both terrestrial and Jovian types of planets? Even our own system has many remaining mysteries. For instance, astronomers recently discovered several small icy objects beyond Pluto's orbit. Are they the first of a new family of objects at the outer edge of the Solar System or just stray comets or odd asteroids?

RE-MODELING

OTHER PLANETARY SYSTEMS

Over the last few years, astronomers have discovered evidence for planets around some dozen nearby stars. These planets are not detected by their light but by the slight gravitational tug they exert on their parent stars. As a planet orbits its star, the planet exerts a gravitational tug that makes the star's position wobble slightly, just as you wobble a little if you swing a heavy weight around you. The wobble creates a Doppler shift in the star's light that astronomers can measure. From that shift and its change in time, they can deduce the planet's orbital period, mass, and distance from the star. So far, the method can only detect relatively massive planets (Jupiter-size), and figure 7.13 shows models of some of these newly discovered systems.

Astronomers have been both delighted and surprised by these discoveries. They are delighted to confirm that planets have formed around other stars, but they are suprised that several of these systems have huge planets very close to their parent stars. According to the solar nebula theory, large planets should only form far from their star, so what is going on?

One possibility is that these giant planets did form far from their stars but their orbits subsequently shrank. Friction with gas and dust in a dense disk might allow this to happen. Unfortunately, because astronomers cannot yet determine the diameter of these extrasolar planets, they cannot deduce their density and thereby tell what they are made of. Thus, although we know that our Solar System is not unique, our inability to explain these other systems implies that we have yet to fully understand our own system's origin.

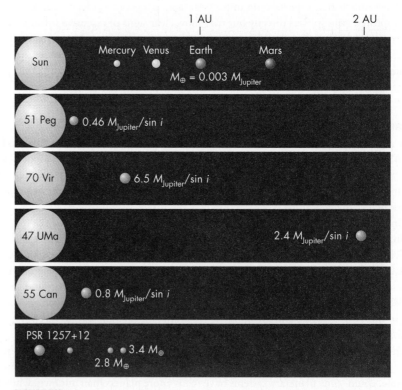

FIGURE 7.13
Sketches of several newly discovered planetary systems compared to our Solar System. (Courtesy G. Marcy at San Francisco State University.)

SUMMARY

The Solar System consists of a star (the Sun) and planets, asteroids, and comets, which orbit it in a broad, flat disk. All the planets circle the Sun in the same direction and, with a few exceptions, spin in the same direction. Their moons also form flattened systems, generally orbiting in the same direction. The planets fall into two main categories: small, high-density bodies (the inner planets) and large, low-density bodies (the outer planets). The former are rich in rock and iron; the latter are rich in hydrogen and ice.

These features of the Solar System can be explained by the solar nebula hypothesis. In this hypothesis, the Solar System was born from a cloud of interstellar gas that collapsed to a disk called the solar nebula. The center of the nebula became the Sun, and the disk became the planets. This explains the compositional similarities and the common age of the bodies in the system.

The flat shape of the system and the common direction of motion around the Sun arose because the planets condensed within the nebula's rotating disk. Planet growth occurred in two stages: dust condensed and clumped to form planetesimals; and then later the planetesimals aggregated to form planets and satellites. Two kinds of planets formed because lighter gases and ice could condense easily in the cold outer parts of the nebula, but only rocky and metallic material could condense in the hot inner parts. Impacts of surviving planetesimals late in the formation stages cratered the surfaces and may have tilted the rotation axes of some planets. Some planetesimals survive to this day as the asteroids and comets.

QUESTIONS FOR REVIEW

1. Name the nine planets in order of increasing distance from the Sun.
2. Make a table listing separately the inner and outer planets.
3. What properties apart from position distinguish the inner and outer planets?
4. Make a sketch of the Solar System showing top and side views.
5. What is the Oort cloud? Where is it located, and what kind of objects come from it?
6. What is Bode's law?
7. What is an interstellar cloud? What does it have to do with the Solar System?
8. What is the solar nebula? What is its shape and why?
9. Why are there two main types of planets?
10. How do we know the composition of Jupiter?
11. What are planetesimals?
12. What is the difference between condensation and accretion?
13. Describe the planetesimal theory of planet formation.
14. How does the planetesimal theory of planet formation explain the asteroids?
15. How did the craters we see on many of the planets form?
16. Describe a theory of how planets may have formed their atmospheres.
17. How would you describe the formation of the Solar System to a little brother or sister?

PROBLEMS

1. Calculate the densities of Venus and Jupiter, given the following data: The mass and radius of Venus are 4.87×10^{27} grams and 6051 kilometers. The mass and radius of Jupiter are about 1.9×10^{30} grams and 71,492 kilometers. How do these numbers compare with the density of rock (about 3 grams per cm^3) and water (1 gram per cm^3)? (Note: Be sure to convert kilometers to centimeters if you are expressing your answer in grams per cm^3.)

2. Look up the mass and radius of Mercury and Jupiter and calculate their escape velocity, using the expression in chapter 2. Does this help you see why the one body has an atmosphere but the other doesn't? (Note: Be sure to convert kilometers to meters, or the appropriate unit.)

3. At what distance beyond Pluto does Bode's law predict the next planet should be?

TEST YOURSELF

1. Which of the following planets are primarily rocky with iron cores?
   (a) Venus, Jupiter, and Neptune.
   (b) Mercury, Venus, and Pluto.
   (c) Mercury, Venus, and Earth.
   (d) Jupiter, Uranus, and Neptune.
   (e) Mercury, Saturn, and Pluto.

2. One explanation of why the planets near the Sun are composed mainly of rock and iron is that
   (a) the Sun's magnetic field attracted all the iron in the young Solar System into the region around the Sun.
   (b) the Sun is made mostly of iron. The gas ejected from its surface is therefore iron, so that when it cooled and condensed it formed iron-rich planets near the Sun.
   (c) the Sun's heat made it difficult for other substances such as ices and gases to condense near it.
   (d) the statement is false. The planets nearest the Sun contain large amounts of hydrogen gas and subsurface water.
   (e) the Sun's gravitational attraction pulled iron and other heavy material inward and allowed the lighter material to float outward.

3. Which of the following features of the Solar System does the solar nebula hypothesis explain?
   (a) All the planets orbit the Sun in the same direction.
   (b) All the planets move in orbits that lie in nearly the same plane.
   (c) The planets nearest the Sun contain only small amounts of substances that condense at low temperatures.
   (d) All the planets and the Sun, to the extent that we know, are the same age.
   (e) All of the above.

4. The numerous craters we see on the solid surfaces of so many Solar System bodies are evidence that
   (a) they were so hot in their youth that volcanos were widespread.
   (b) the Sun was so hot that it melted all these bodies and made them boil.
   (c) these bodies were originally a mix of water and rock. As the young Sun heated up, the water boiled, creating hollow pockets in the rock.
   (d) they were bombarded in their youth by many solid objects.
   (e) all the planets were once part of a single, very large and volcanically active mass that subsequently broke into many smaller pieces.

5. Why is Pluto not considered a Jovian planet?
   (a) Its mass and radius are so small, and it lacks the thick atmosphere of hydrogen seen on the other Jovian planets.
   (b) It is so far out in the Solar System.
   (c) Its interior is mostly rock and iron.
   (d) Its atmosphere is rich in oxygen, making it more like the Earth.
   (e) It is not really orbiting the Sun but is simply drifting through the outer edge of the Solar System.

FURTHER EXPLORATIONS

Beatty, J. Kelly, and Andrew Chaikin. *The New Solar System,* 3rd ed. New York: Cambridge University Press, 1990.

Cameron, A. G. W. "The Origin and Evolution of the Solar System." *Scientific American* 233 (September 1975): 32.

Dent, W. R. F. "Observing the Formation of Stars and Planets." *Endeavour* 16 (No. 3, 1992): 139.

Falk, S. W., and D. N. Schramm. "Did the Solar System Start with a Bang?" *Sky and Telescope* 58 (July 1979): 18.

Ferris, T. "A Plumb Line to the Sun: Finding the Scale of the Solar System." *Mercury* 18 (May/June 1989): 66.

Hartmann, W. K. "In the Beginning." *Astronomy* 4 (June 1976): 6.

———. "Cratering in the Solar System." *Scientific American* 236 (January 1977): 84.

Jaki, S. I. "The Titius-Bode Law: A Strange Bicentenary." *Sky and Telescope* 43 (May 1972): 280.

Maran, Stephen P. "Where Do Comets Come From?" *Natural History* 91 (May 1982): 80.

Reeves, H. "The Origin of the Solar System." *Mercury* 6 (March/April 1977): 80.

Stern, A. "Where Has Pluto's Family Gone?" *Astronomy* 20 (September 1992): 40.

Wetherill, G. W. "The Formation of the Earth from Planetesimals." *Scientific American* 244 (June 1981): 163.

Web Sites

"The Nine Planets: A Multimedia Tour of the Solar System," by Bill Arnett.
www.seds.org/billa/tnp
Excellent pictures, glossary, and other info. One of the best astronomy web sites.

"Views of the Solar System," by Calvin Hamilton.
bang.lanl.gov/solarsys/earth.htm
Nice pictures, lots of text, fact sheets, and glossary. Essentially an on-line text.

"Welcome to the Planets."
pds.jpl.nasa.gov/planets
Nice pictures, fact sheets, and other info.

KEY TERMS

accretion, 229
asteroid belt, 222
asteroids, 222
Bode's law, 224
comets, 222
condensation, 228
inner planets, 220
interstellar cloud, 227
interstellar grains, 227

Jovian planets, 221
Oort cloud, 222
outer planets, 220
planetesimals, 230
solar nebula, 228
solar nebula hypothesis, 227
Solar System, 219
terrestrial planets, 221

CHAPTER EIGHT

THE TERRESTRIAL PLANETS

Mercurian mare

MERCURY

VENUS

Volcanic peak on Venus

Ocean on Earth

MARS

EARTH

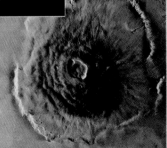

75,000'

The Four Inner Planets: Mercury, Venus, Earth, and Mars (all Courtesy of NASA)

Terrestrial planets, as their name suggests, have a size and structure similar to Earth's. Within our Solar System, Mercury, Venus, Earth, and Mars fit in this category. Orbiting in the inner part of the Solar System, close to the Sun, these rocky worlds are too small and too warm to have captured massive hydrogen envelopes such as those that cloak the outer planets. Nor have they the array of moons possessed by their cold, giant brethren. In fact, of the terrestrial planets, only Earth has a large moon, although Mars is orbited by two tiny captured asteroids.

8.1 PORTRAITS OF THE TERRESTRIAL PLANETS

The terrestrial planets roughly resemble one another in size, as you can see in figure 8.1, which shows them in correct proportion. These four planets have markedly different atmospheres and surfaces, as illustrated in the insets below each one. For example, Mercury, the smallest, looks like the Moon: a gray, bare, cratered body with essentially no atmosphere. Venus, on the other hand, is covered with deep clouds of sulfuric acid droplets suspended in a dense atmosphere of carbon dioxide. This thick atmosphere traps sunlight by the greenhouse effect, heating Venus to a fairly constant 750 K (about 900° F) and making it hotter even than Mercury. From this baking surface, immense volcanic peaks tower above desolate plains. One Venusian peak, Maxwell Montes, rises to 11 kilometers (approximately 6.8 miles) above its surroundings, making it much higher than Earth's highest peak, Mount Everest, which reaches only 8.8 kilometers (approximately 5.5 miles) above sea level. Compared with Earth, with its gleaming blue seas, white clouds and ice caps, red deserts, and green jungles, Mercury looks dead and Venus resembles medieval visions of hell.

Mars, on the other hand, looks more familiar. Caps of ice and frozen carbon dioxide gleam at its poles, surrounded by vast red deserts pocked with craters and furrowed with dunes. Canyons and river beds, now dry but once filled with flowing water, slope toward an immense gash in the Martian crust, the Valles Marineris, which runs for about 5000 kilometers (3000 miles) roughly along the Martian equator. To the northwest of this great rift lies a highland dotted with ancient volcanic peaks. The tallest,* Olympus Mons, rises approximately 25 kilometers (about 15.5 miles)—nearly three times the height of Mount Everest—into the cold, thin carbon dioxide atmosphere of Mars.

Our goal in this chapter is to learn more about Mercury, Venus, and Mars so that we can better understand how these neighboring planets came to be so different from Earth. We will discover that size plays a major role. For example, because Mercury is so small, it generates little internal heat to create surface activity, and so its crust is essentially unchanged from its birth. Mars and Venus, on the other hand, are large enough to have hot interiors rather like Earth's, and so they have active surfaces with mountains and volcanic peaks.

Size, coupled with distance from the Sun, creates the great atmospheric differences between these terrestrial worlds. Mercury, for example, is too small and its surface is too hot to retain an atmosphere, while Mars, only slightly larger but farther from the Sun and therefore cooler, has retained one. Venus and Earth are both large enough to have sizeable atmospheres, but Earth's slightly greater distance from the Sun has made it cool enough to have liquid water in its atmosphere. That simple fact has led to the profound difference between the atmospheres of Earth and Venus because liquid water can remove carbon dioxide from air. Moreover, liquid water has allowed life to form and flourish here, and life has not only removed additional carbon dioxide from our air, but has also added oxygen.

---

*Astronomers have recently remeasured these peaks and think Ascraeus Mons may be slightly higher.

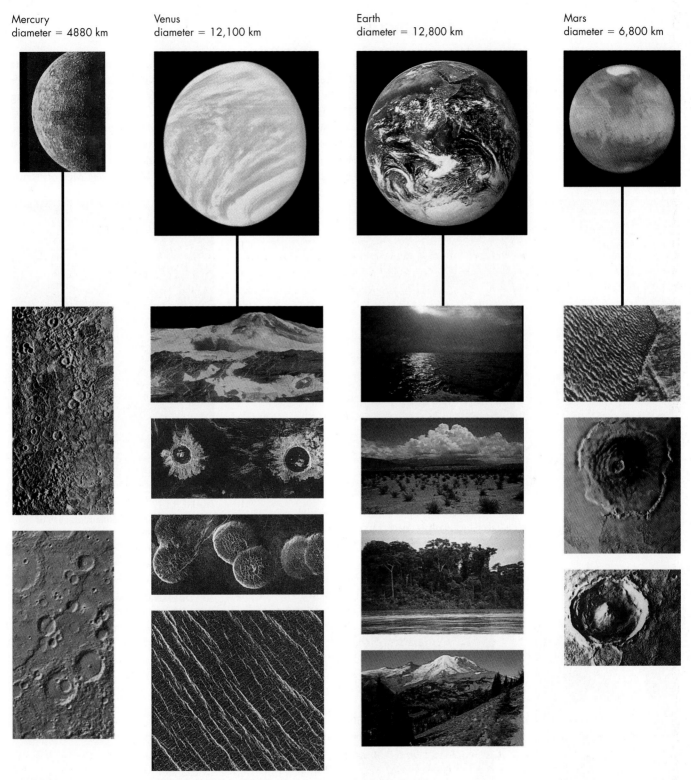

Mercury
diameter = 4880 km

Venus
diameter = 12,100 km

Earth
diameter = 12,800 km

Mars
diameter = 6,800 km

FIGURE 8.1
Pictures (left to right) of Mercury, Venus, Earth, and Mars showing representative surface features. Planets are shown approximately to correct relative scale. (Courtesy NASA/JPL/USGS.)

8.2 MERCURY

Mercury is the smallest terrestrial planet and is named for the Roman deity who was the messenger of the gods because it changes its position on the sky faster than any other planet. Mercury resembles our Moon in both size and appearance, with its radius being about-one third and its mass about $\frac{1}{20}$ that of the Earth. Astronomers can determine Mercury's radius from its angular size and distance, as we discussed in the previous chapter (see fig. 7.5). They can calculate its mass from its gravitational attraction on the Mariner 10 spacecraft, which flew past the planet in 1974 and 1975, taking pictures of its surface. Those images show that circular craters like those on the Moon cover Mercury's sur-

FIGURE 8.2
Picture of Mercury's cratered surface (taken by a passing spacecraft). (Courtesy NASA.)

FIGURE 8.3
Caloris Basin is the semicircular set of rings here. Only half the basin is shown here. This picture of the Basin, a Mercurian mare, was taken by the passing Mariner spacecraft. (Courtesy NASA.)

☀ Does Caloris Basin look like lunar maria? Compare figure 8.3 (above) to figure 6.7. How are they similar? How are they different?

Mercury (on the left) and the Moon (on the right) shown to approximately correct relative size. Note that both these airless objects are heavily cratered. (Courtesy NASA/JPL)

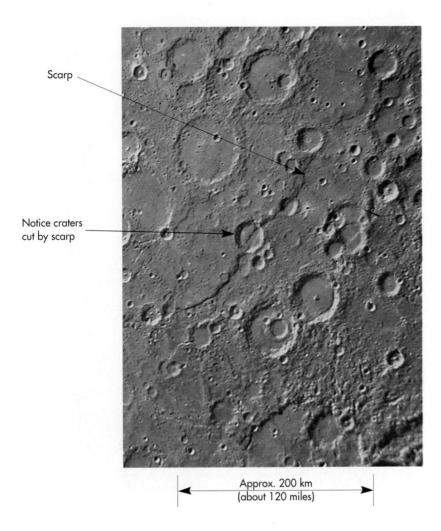

Scarp

Notice craters
cut by scarp

Approx. 200 km
(about 120 miles)

FIGURE 8.4
Photograph of a scarp, a cliff, running across Mercury's surface. (Courtesy NASA.)

Which formed first, the scarp or the craters it passes through? On what do you base your answer?

face, as illustrated in figure 8.2. The largest of these craters by far is the vast Caloris Basin, shown in figure 8.3. With a diameter of 1300 kilometers (about 800 miles), this mountain-ringed depression is reminiscent of lunar maria. Moreover, its circular shape and surrounding hills indicate that, like the maria, it was formed by impact.

Mercury's surface is not totally Moonlike, however. Congealed lava flows flood many of Mercury's old craters and pave much of its surface. On our Moon, such flows are found almost exclusively within the maria. In addition, enormous scarps—cliffs formed where the crust has shifted—run for hundreds of kilometers across Mercury's surface, as seen in figure 8.4. The scarps may have formed as the planet cooled and shrank, wrinkling like a dried apple. Mercury also possesses the curious landscape illustrated in figure 8.5, which lies exactly on the far side of the planet from Caloris Basin. Astronomers call this hummocky, jumbled surface "weird terrain" and think it was churned up by earthquake waves generated by the impact that created Caloris. As the waves traveled around Mercury, they converged on its far side, heaving up the rock, much as dropping cream into coffee creates a tiny splash as the ripples reconverge.

Mercury's Temperature and Atmosphere

Mercury's surface is one of the hottest places in the Solar System. At its equator, noon temperatures reach approximately 700 K (about 800° F). On the other hand, nighttime temperatures are among the coldest in the Solar System, dropping to approximately 100 K (about −280° F). These extremes result from Mercury's closeness to the Sun and from its lack of atmosphere. No atmospheric gas moderates the inflow of sunlight during the day or retains heat during the night. Despite a high equatorial temperature, however, Mercury's

FIGURE 8.5
Picture of odd terrain opposite Caloris
Basin. (Courtesy NASA.)

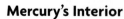

Approx. 260 km
(about 160 miles)

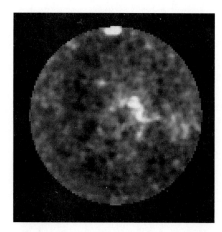

FIGURE 8.6
Radar map of Mercury showing the region
thought to be a polar ice cap — the white
blob at the top. (Courtesy Cal
Tech/JPL/NASA.)

poles are very cold. Sunlight shines so indirectly on them that they receive little heat, and
with no atmosphere to distribute warmth, the poles have grown very cold—so cold, in fact,
that features thought to be small ice caps have formed there, as the radar map of Mercury
(see fig. 8.6) illustrates.

Mercury lacks an atmosphere for the same reason the Moon does. Although traces of
gas, perhaps captured from interplanetary space, have been detected spectroscopically,
Mercury's small mass makes its gravitational attraction too small to retain much gas
around it. Moreover, its proximity to the Sun makes keeping an atmosphere difficult because the resulting high temperature in the equatorial regions causes molecules to move so
fast that they readily escape into space. In fact, Mercury probably never had an extensive
atmosphere because it lacks volcanos (thought to be the source of much of our atmosphere, as discussed in chapter 4). Where then has the ice at its poles come from? One theory suggests the ice came from comets that have struck its surface. Comets occasionally
plunge into the Sun, and so some must occasionally hit Mercury. Such impacts vaporize
the comet, creating a wispy and ephemeral atmosphere, most of which quickly escapes
into space. Some gas, however, may drift toward the cold polar regions and freeze there,
much as frost condenses on automobile windshields on a subfreezing morning. Over time,
such frost deposits might build up the caps seen on this otherwise hot planet.

Mercury's Interior

Mercury probably has an iron core beneath its silicate crust, but astronomers have little
proof because no spacecraft has landed there to deploy seismic (earthquake) detectors.

Their conclusion is therefore based on Mercury's density and gravitational field. A massive planet's gravity can compress its interior to high density, but Mercury is too small for this effect to be significant. Thus its high density (5.4 grams per cm³) indicates an iron-rich interior with only a thin rock (silicate) mantle, as depicted in figure 8.7.

Why Mercury is so relatively rich in iron but poor in silicates is unclear. One possibility is that silicates did not condense as easily as iron compounds in the hot, inner solar nebula where Mercury formed. Another possibility is that Mercury once had a thicker rocky crust but that it was blasted off by the impact of an enormous planetesimal, as the computer simulation in figure 8.8 illustrates.

Whether Mercury has a liquid (molten) inner core like the Earth's is unknown. We infer from its small mass that Mercury has probably been less heated than the Earth by radioactivity, and its small radius would have allowed heat to escape readily. Thus the core may not be hot enough to be molten, or the molten part may be very small. We discussed in chapter 4 that circulating motions in the Earth's molten iron core when combined with our planet's spin may generate its magnetic field. The small size or solidity of Mercury's core may therefore explain why it has such a weak magnetic field (it is only about 1% that of the Earth's). Moreover, as we shall see below, Mercury spins very slowly. It thus lacks both the attributes that scientists think are needed to create a strong magnetic field. If Mercury is mostly solid, its iron-nickel core could be a huge permanent magnet, but one much weaker than the dynamo-created magnetic field of Earth.

Mercury's Rotation

Mercury spins very slowly. Its rotation period is 58.646 Earth days, exactly two thirds its orbital period around the Sun of 87.969 Earth days, as you can verify with a pocket calculator. This means that it spins exactly three times for each two trips it makes around the Sun. How did this proportion occur, or is it just coincidence?

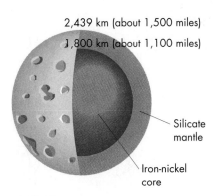

FIGURE 8.7
Artist's depiction of Mercury's interior.

2,439 km (about 1,500 miles)
1,800 km (about 1,100 miles)
Silicate mantle
Iron-nickel core

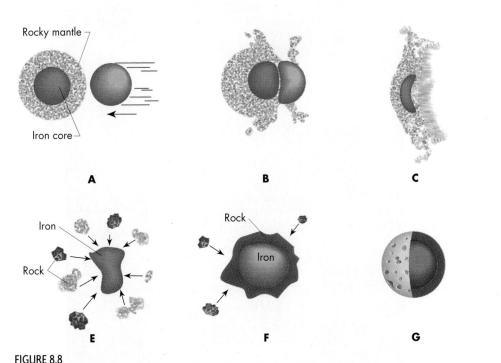

FIGURE 8.8
Computer simulation of collision between Mercury and a large planetesimal. The impact strips away most of the rocky crust and leaves a highly distorted iron core surrounded by some rocky debris. Gravity eventually reshapes the planet into a sphere. (Based on a computer simulation, courtesy W. Benz, University of Arizona; W. Slattery, Los Alamos; A. G. W. Cameron, CFA.)

The Sun has affected Mercury's rotation by tidal forces, just as the Earth has affected the Moon's. That is, the Sun's gravity exerts a force on Mercury, which tends to twist the planet and make it rotate with the same period in which it orbits the Sun. Mercury's orbit, however, is very elliptic. Thus, in accordance with Kepler's second law of planetary motion, Mercury's orbital speed changes as it moves around the Sun. Because of that changing speed, the Sun cannot lock Mercury into a purely synchronous spin; the closest to synchrony it can get is three spins for each two orbits (fig. 8.9). Such an integer ratio of periods is called a *resonance*.

Resonance occurs when a force that acts repeatedly on a body causes its motion to grow ever larger. For example, pushing a child on a swing is a resonance. If you push just as the swing starts to move forward, the child will swing higher and higher, and the pushing force will be in resonance with the motion of the swing. On the other hand, if you push before the backward motion is stopped, the swinging motion will decrease, and no resonance occurs. Likewise, applying power resonantly to a car stuck in a ditch may "rock" it out. A similar resonance exists between the Sun's changing gravitational tug on Mercury as it moves along its elongated orbit and its rotation. The result is the 2:3 relation between its orbital and spin periods.

Mercury's odd rotation gives it an extremely long solar day (the time between successive sunrises) of 176 Earth days. During that time, the Sun sometimes changes its direction of motion across the sky. For example, if sunset occurs when Mercury is at the point of its orbit nearest the Sun, the Sun will set and then briefly rise again before setting a second time!

FIGURE 8.9
Mercury's odd rotation. The planet spins three times for each two orbits it makes around the Sun.

☀ What resemblance do you see between Mercury's motions and those of our Moon?

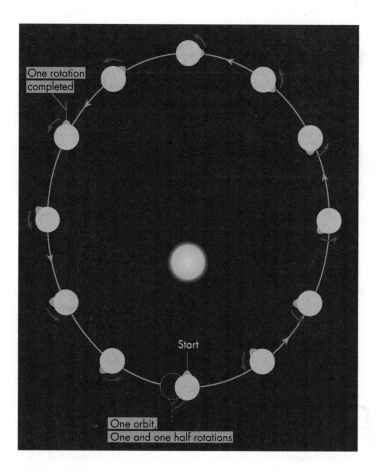

8.3 VENUS

Of all the planets, Venus is most like the Earth in diameter and mass. It is named for the Roman goddess of love. Because Venus is so similar to the Earth in size, we might therefore expect it to be like the Earth in other ways. However, Venus and the Earth have radically different surfaces and atmospheres. Many of these differences we have only just discovered because dense clouds perpetually cloak Venus, as shown in figure 8.10. Nevertheless, astronomers know that its surface looks nothing like the Earth's, and its atmosphere is much hotter and denser than ours and has a very different composition.

The Venusian Atmosphere

The atmosphere of Venus is mainly (96%) carbon dioxide. Astronomers know the atmospheric composition from its spectrum and from measurements with space probes. Gases in its atmosphere absorb some of the sunlight falling on the planet and create absorption lines in addition to those of the Sun itself. These lines then reveal the composition and

FIGURE 8.10
Photograph through ultraviolet filter of the clouds of Venus. The picture is artificially colored and enhanced to show the clouds clearly. (Courtesy NASA.)

Why does Venus not look round in this photograph? Hint: think of Galileo's observations of Venus.

The Moon and Venus, over Ottawa Lake, Wisconsin. (Courtesy Alan Dyer.)

density of the gas. Moreover, spacecraft have descended through the atmosphere to the surface and have sampled its atmosphere. Thus we have learned that, in addition to carbon dioxide, Venus's atmosphere contains about 3.5% nitrogen and very small amounts of water vapor and other gases.

Spectra also reveal the nature of the Venusian clouds: they are composed of sulfuric acid droplets that formed when sulfur compounds—perhaps ejected from volcanos—combined with the traces of water in the atmosphere. These clouds permanently cover the planet and are very high and thick, beginning at about 30 kilometers (19 miles) above the surface and extending upward to about 60 kilometers (37 miles). In fact, the clouds are so thick that no surface features can be seen through them with ordinary telescopes. This deep cloud layer strongly reflects sunlight falling on it, making Venus very bright as seen from Earth—so bright that if the air is very clear, you can see Venus in broad daylight. Below the clouds, the Venusian atmosphere is relatively clear, and some sunlight penetrates to the surface. The light is tinged orange, however, because the blue wavelengths are absorbed in the deep cloud layer.

Venus's atmosphere is extremely dense. It exerts a pressure roughly 100 times that of the Earth's, equivalent to the pressure you would feel under 3000 feet of water. We will discuss later in this chapter why Venus has such a dense atmosphere, but there are other features of the planet that we need to understand first. One of the most important of those features is that its lower atmosphere is extremely hot. Observations made with radio telescopes from Earth show that the surface temperature is more than 750 K (about 900° F), a value confirmed by spacecraft landers.* What makes the atmosphere of a planet so similar to the Earth in size, and only slightly nearer the Sun, so very hot?

The Greenhouse Effect

Venus's carbon dioxide atmosphere creates an extremely strong **greenhouse effect**. We discussed in chapter 4 how gases in a planet's atmosphere may allow sunlight to enter and warm the surface but prevent the heat so generated from escaping to space. That is, certain gases may trap a planet's heat by hindering or blocking its infrared radiation from escaping to space, as was illustrated in figure 4.16. In chapter 4, we discussed how carbon dioxide in the Earth's atmosphere is especially effective at trapping heat, creating a weak, but necessary, greenhouse effect here. Venus, however, has about 300,000 times more carbon dioxide than Earth, and so its greenhouse effect is correspondingly much stronger. In fact, it is so effective at trapping heat that the surface of Venus is hotter than Mercury, even though Venus is farther from the Sun.

The high temperature and density of Venus's atmosphere create a high atmospheric pressure on the planet. That high pressure pushes the Venusian atmosphere upward just as raising the pressure in a balloon makes it expand. Near the top of the atmosphere, the gas is much less dense and does not trap heat as readily as the gas near the surface. The upper atmosphere is therefore much cooler than the surface, with a temperature of about 300 K (approximately room temperature here on Earth) at an altitude of 40 kilometers (about 25 miles). These upper regions were surveyed by a weather balloon, dropped from a Russian spacecraft, that floated through the Venusian atmosphere for about 2 days, measuring temperature, wind speed, and direction.

---

*Such landers do not survive long, however, because Venus is hot enough to melt the lead in their electronic instrumentation.

The Surface of Venus

The surface of Venus is hidden beneath its thick clouds, but planetary scientists can map its ground features with radar both from Earth and from spacecraft orbiting Venus. Just as radar penetrates terrestrial clouds to show an aircraft pilot a runway even through fog, so too radar penetrates the Venusian clouds, revealing the planet's surface. Figure 8.11 shows a radar map of Venus made by the *Pioneer* Venus orbiter, the first satellite to observe Venus with radar. Such maps show that Venus is less mountainous and rugged than Earth, with most of its surface being low, gently rolling plains. Only two major highland regions, Ishtar and Aphrodite, rise above the lowlands to form land masses similar to terrestrial continents, but much smaller in both size and elevation. Ishtar, named for the Babylonian goddess of love, is about the size of Greenland and is studded with volcanic peaks, the highest of which, Maxwell Montes, rises more than 11 kilometers (about 6.8 miles) above the average level of the planet. (Notice that because no oceans exist on Venus, "sea level" has no meaning as a reference height.) The other highland region, Aphrodite, bears the

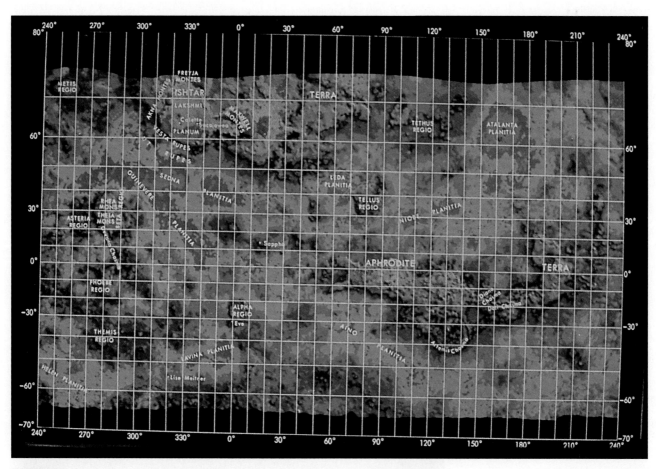

FIGURE 8.11
Global radar map of Venus made by the *Pioneer* Venus orbiting satellite. Colors indicate relative height of surface features. Lowlands are blue; mountain peaks are red. (Courtesy NASA/JPL.)

ancient Greek name for Venus and is about the size of South America. Together, Ishtar and Aphrodite comprise only about 8% of Venus's surface, a far smaller fraction than for Earth, where continents and their submerged margins cover about 45% of the planet.

The best pictures of the Venusian surface are the radar maps made by *Magellan,* a United States spacecraft previously orbiting the planet. *Magellan* transmitted a radar beam downward to the surface of Venus where the beam reflected back to the satellite and was recorded. The recorded information was stored and transmitted to Earth, where it was processed to create pictures of surface features such as figure 8.12, which shows a view of the entire planet with its numerous volcanic peaks, channels, and wrinkled crust.

Figure 8.13 shows close-ups of some of the more intriguing pictures that *Magellan* has transmitted. Details as small as 100 meters (100 yards) are visible. Thanks to spacecraft, we have better maps of Venus than we had of the Earth itself 40 years ago.

The many odd and unique structures seen in the radar maps have proved puzzling to astrogeologists. Venus is so similar in diameter and mass to the Earth that they expected to see landforms there similar to those on the Earth. For example, some astrogeologists predicted that there would be evidence of plate tectonics, such as continental blocks, crustal rifts, and trenches at plate boundaries. But few such features are visible. Instead,

Magellan met a deliberatedly engineered fiery doom in 1994. Its orbit was altered so that it plunged into Venus's atmosphere. Analysis of its final tumblings gave astronomers data on the density of Venus's upper atmosphere.

FIGURE 8.12
Global radar map of Venus made by the spacecraft *Magellan* in orbit around Venus. The picture is artificially colored to simulate lighting conditions observed by the Russian *Venera* lander. (Courtesy NASA/JPL.)

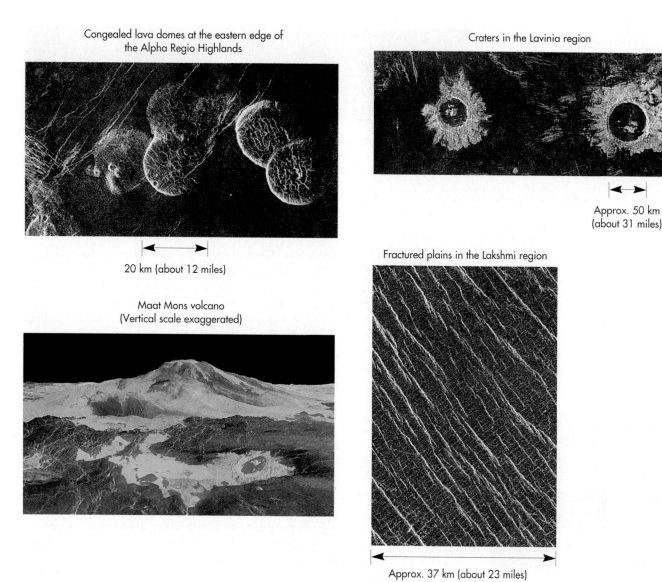

Congealed lava domes at the eastern edge of the Alpha Regio Highlands

20 km (about 12 miles)

Maat Mons volcano (Vertical scale exaggerated)

Craters in the Lavinia region

Approx. 50 km (about 31 miles)

Fractured plains in the Lakshmi region

Approx. 37 km (about 23 miles)

FIGURE 8.13
Gallery of *Magellan* radar pictures. The orange color is artificially added to match the color of the landscape observed by the Russian *Venera* lander. Features are also exaggerated vertically for clarity. (Courtesy NASA/JPL.)

Venus has a surface almost totally unlike the Earth's. Although Venus has some craters (often weirdly distorted, however) and crumpled mountains, volcanic landforms dominate. These include peaks with immense lava flows, "blisters" of uplifted rock, grids of long narrow faults, and peculiar lumpy terrain. All these features indicate a young and active surface, a deduction borne out by the scarcity of impact craters. From the small number of craters, astrogeologists have concluded that virtually all of Venus's original surface has been destroyed by volcanic activity. The surface we see is probably at most half a billion years old, much younger than Earth's, and some regions may be less than 10 million years old. Such estimates of crustal age are difficult to make, however, because the Venusian atmosphere is so dense that all but the largest infalling bodies (bigger than a few hundred meters) are broken up in its dense atmosphere.

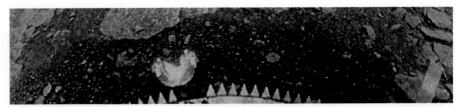

Picture made on the surface of Venus by the Russian spacecraft *Venera* that landed there. Sunlight filtering through the thick clouds gives the landscape its orange color. (Courtesy Vernadsky Institute, USSR Academy of Sciences.)

Are the Venusian volcanos such as Maxwell Mons still active? Eruptions have not been seen directly, but some lava flows appear very fresh. Moreover, electrical discharges, perhaps lightning, have been detected near some of the larger peaks. On Earth, volcanic eruptions frequently generate lightning, and some astronomers think the electrical activity indicates that Venus's volcanos are still erupting. Such eruptions might also explain brief increases in sulfur content detected in the Venusian atmosphere, changes similar to those produced on Earth by eruptions here.

The numerous volcanic peaks, domes, and uplifted surface regions suggest to some astrogeologists that heat flows less uniformly within Venus than within the Earth. Although some locations on Earth (Yellowstone Park and the Hawaiian Islands for example) are heated anomalously by "plumes" of rising hot rock, such plumes seem to dominate on Venus. As hot rock wells upward, it bulges the crust, stretching and cracking it. We may be viewing on our sister planet what Earth looked like as its crust began to form and before smooth heat flows were established.

Astrogeologists do not yet understand enough about how a planet forms a crust to explain why Venus differs so much from Earth. Some differences, however, may result from Venus's hotter surface and its lack of water. A hotter surface may make the Venusian crust thinner and therefore more easily penetrated by hot upwelling matter from its interior. The lack of water may affect tectonic processes because water acts like a lubricant in lavas. On Earth, some lavas are more "sticky" than others, creating volcanic peaks with very different shapes. For example, the volcano Mt. Fuji has steep slopes, while the great Hawaiian volcanos with their more fluid lava have shallow slopes.

Although most of what astronomers know of the Venusian surface comes from radar maps, several Russian *Venera* spacecraft have landed there and transmitted pictures back to Earth from the Venusian surface itself. The pictures show a barren surface covered with flat, broken rocks and lit by the pale orange glow of sunlight diffusing through the deep clouds. These robotic spacecraft have also sampled the rocks, showing them to be volcanic.

Mt. Fuji in Japan and one of the Hawaiian Islands. The lava that built these peaks differed greatly in its "stickiness." Fuji's lava was thicker and thus its slope is steeper. (Courtesy Photo Researchers/Maso Hayashi and Photo Researchers/Douglas.)

Interior of Venus

The deep interior of Venus is probably like the Earth's, an iron core and rock mantle. However, astronomers have no seismic information to confirm this conjecture and, as with Mercury, must rely on deductions from its gravity and density, which is similar to the Earth's.

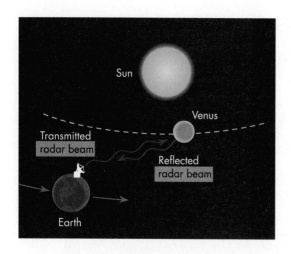

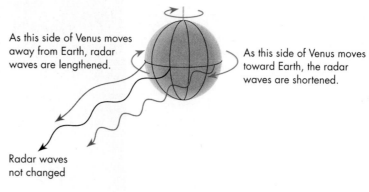

FIGURE 8.14
Measuring Venus's rotation by radar.

Rotation of Venus

Venus spins on its axis more slowly than any other planet in the Solar System, taking 243 days to complete one rotation. Moreover, its spin is retrograde ("backward") compared with the direction of rotation of the other terrestrial planets. Interestingly, these properties of its spin were first learned from radar observations. As a planet spins, a radar signal bounced from its surface will have its wavelength altered by the Doppler shift, as shown in figure 8.14. From the Doppler shift, the planet's rotational speed can be found, just as police radar can reveal the speed of your automobile. In turn, from the rotation speed and the planet's radius, its rotation period can easily be found. This slow and retrograde spin has led some astronomers to hypothesize that Venus was struck shortly after its birth by a huge planetesimal; the impact slowed Venus and set it spinning backward. A less dramatic explanation of the spin is that Venus has been affected by a combination of tidal forces exerted by the Sun—and perhaps the Earth—so that the tilt angle of its rotation axis may have shifted in time.

Venus rotates so slowly that it cannot generate a strong magnetic field as the Earth does. The slow rotation also makes the Solar day there very long, approximately 117 Earth days. Furthermore, because the planet's spin is retrograde (backward), the Sun rises in the west and sets in the east.

8.4 MARS

Mars is named for the Roman god of war, presumably because of its distinctly reddish color. Compared with Mercury and Venus, Mars seems positively Earthlike. Although its diameter is only about half and its mass about one-tenth the Earth's, its surface and atmosphere are less alien. For example, on a warm day, the temperature at the Martian equator reaches about 50° F, and although winds sweep dust and patchy clouds of ice crystals through its sky, the Martian atmosphere is generally clear enough for astronomers on Earth to view its surface clearly. Such views from here and from spacecraft orbiting Mars show a world of familiar features. Polar caps of sparkling white contrast with the reddish

FIGURE 8.15
Picture of Mars made by the Hubble space telescope orbiting Earth. (Courtesy HST.)

How big across is the polar cap in this picture? (Estimate its size from the radius of Mars, which you can look up in the appendix).

color of most of the planet and are visible from Earth, as depicted in figure 8.15. But it is space-based pictures, the legacy of the spacecrafts *Mariner* and *Viking,* that reveal the true marvels of Mars.

Along the equator runs a rift—Valles Marineris—that stretches 5000 kilometers (3000 miles) long, 100 kilometers (62 miles) wide, and 10 kilometers (6 miles) deep, as shown in figure 8.16. This canyon, named for the *Mariner* spacecraft whose pictures led to its discovery, dwarfs the Grand Canyon and would span the continental United States.

Figure 8.17 shows the Martian polar caps. These frozen regions change in size during the cycle of the Martian seasons, a cycle resulting from the tilt of Mars' rotation axis in the same way that our cycle of seasons is caused by the tilt of Earth's rotation axis. The Martian seasons are more extreme than terrestrial ones because the Martian atmosphere is much less dense than Earth's, and therefore it does not retain heat as well. Because Mars' seasonal changes are so extreme, its polar caps vary greatly in size, shrinking during the Martian summer and growing again during the winter. The southern cap is frozen carbon dioxide—dry ice—and in winter its frost extends in a thin layer across a region some 5900 kilometers (about 3660 miles) in diameter, from the south pole to latitude 40°, much as snow cover extends to middle latitudes such as New York in our winters. But because the frost is very thin over most of this vast cap, it shrinks in the summer to a diameter of about 350 kilometers (approximately 217 miles). The northern cap shrinks to

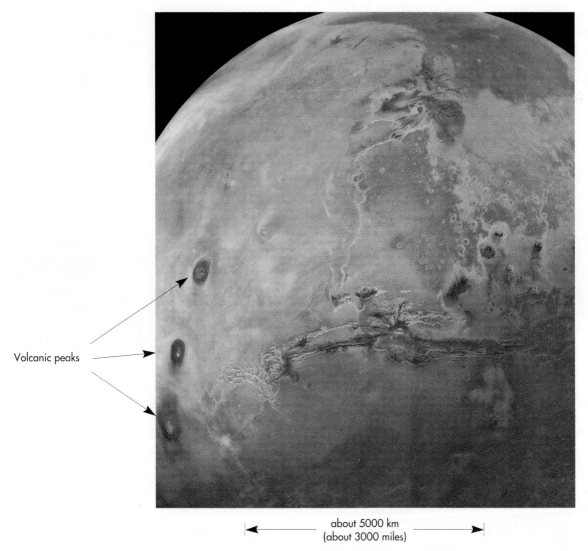

Volcanic peaks

about 5000 km
(about 3000 miles)

FIGURE 8.16

Photograph of Valles Marineris, the Grand Canyon of Mars. This enormous gash in Mars' crust may be a rift that began to split apart the Martian crust but that failed to open farther. The canyon is about 5000 kilometers (approximately 3000 miles) long. Were it on Earth, it would stretch from California to Florida. (Courtesy A. S. McEwen, USGS.)

a diameter of about 1000 kilometers (approximately 620 miles) and, although it has a surface layer of CO_2, the bulk of the frozen material there is ordinary water ice, as deduced from its temperature. Despite their great extent during the Martian winter, the polar caps contain very little water, far less, for example, than the ice caps of our own planet.

The northern cap consists of numerous separate layers, as can be seen in figure 8.17C. These strata indicate that the Martian climate changes cyclically. Thus Mars may have "ice ages" similar to those on Earth.

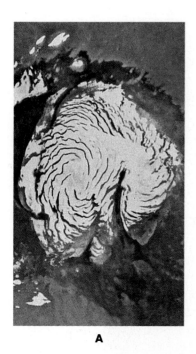

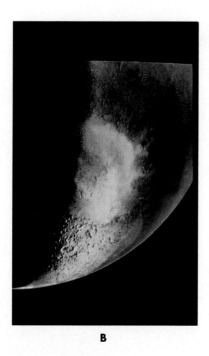

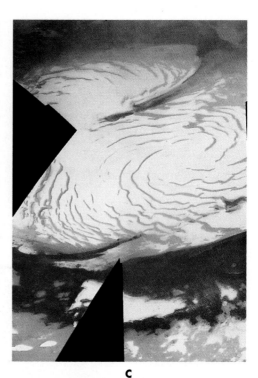

A   **B**   **C**

FIGURE 8.17
Pictures of (A) the north Martian polar cap (Courtesy A. S. McEwen, USGS.) and (B) the south Martian polar cap. (C) Note the layered structure visible in this enlarged view of the north polar cap. (Courtesy NASA/JPL)

FIGURE 8.18
Picture of dune fields in the Martian desert. (Courtesy NASA/JPL.)

The Martian poles are bordered by immense deserts with dunes blown into parallel ridges by the Martian winds, as illustrated in figure 8.18. At midlatitudes a huge upland, Tharsis bulge, is dotted with volcanic peaks, two of which can be seen in figure 8.19A. Another peak on Tharsis, Olympus Mons, rises about 25 kilometers (about 16 miles) above its surroundings, nearly three times the height of Earth's highest peaks, and is illustrated in figure 8.19B. If ever interplanetary parks are established, Olympus Mons should lead the list.

Astrogeologists believe that the Tharsis region formed as hot material rose from the deep interior of the planet and forced the surface upward as it reached the crust. The hot matter then erupted through the crust to form the volcanos, some of which appear relatively young. For example, the small number of impact craters in its slopes implies that Olympus Mons is no older than 250 million years and that it may in fact have been active much more recently. Some planetary geologists think the Tharsis bulge may also have created the gigantic Valles Marineris, which lies to the southeast. According to this theory, Valles Marineris formed as the Tharsis region swelled, stretching and cracking the crust. Other planetary scientists believe that this vast chasm is evidence for plate tectonic activity, like that of Earth, and that the Martian crust began to split, but the motion ceased as the planet aged and cooled.

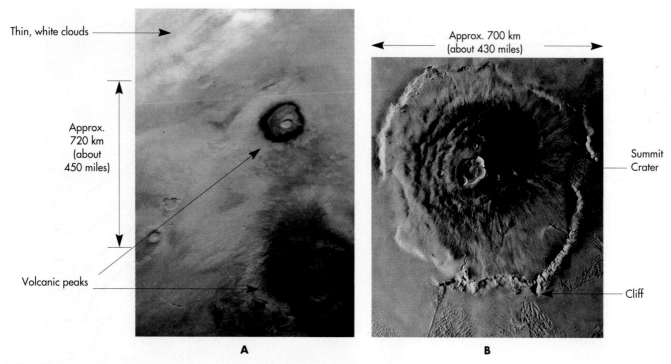

Thin, white clouds

Approx. 720 km (about 450 miles)

Volcanic peaks

Approx. 700 km (about 430 miles)

Summit Crater

Cliff

A   **B**

FIGURE 8.19
Photographs of (A) two volcanos on the Tharsis bulge and (B) Olympus Mons, the second largest known volcano (probably inactive) in the Solar System. (Courtesy A. S. McEwen, USGS.)

Perhaps the most surprising features revealed by the Viking spacecraft are the huge channels and dry river beds, such as those seen in figure 8.20. We infer from these features, which wind across the Martian surface and often contain "islands," that liquid water once flowed on Mars, even though no surface liquid is present now. In fact, many astronomers now believe that huge lakes and small oceans once existed on Mars. The strongest evidence for these ancient bodies of water is smooth terraces that look like old beaches around the inner edges of craters and basins, as you can see in the lower left part of figure 8.21. Moreover, some of these features are cut by narrow canyons that breach their rims and appear to drain into lowland areas (fig. 8.21). But where has the water gone? To understand, we need to look at the properties of the Martian atmosphere.

The Martian Atmosphere

Clouds and wind-blown dust are visible evidence that Mars has an atmosphere. Spectra confirm this and show that the atmosphere is mostly (95%) carbon dioxide with small amounts (3%) of nitrogen, and traces of oxygen and water. From the strength of the atmosphere's spectral lines, astronomers can measure the density of the gases, which turns out to be very low—only about 1% the density of Earth's. This composition and low density have been verified by spacecraft that have landed on Mars and made numerous observations not only of its atmospheric properties but also of its weather.

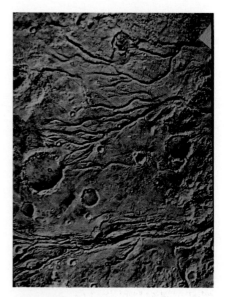

FIGURE 8.20
Picture of channels probably carved by running water on Mars. (Courtesy NASA.)

FIGURE 8.21
A Martian crater thought to have once been a "Crater Lake" on our neighboring planet. The crater is roughly 50km (about 30 miles) across. Note the inflow channel on the lower left and the outflow channel on the right. The smooth floor (apart from a few small craters) suggests that the crater bottom is covered with sediment left behind as the lake dried out. (Courtesy NASA/JPL).

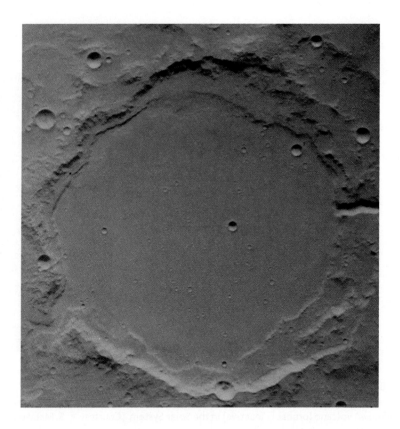

Although the Martian atmosphere is mostly carbon dioxide, its density is so low that the carbon dioxide creates only a very weak greenhouse effect. The consequent lack of heat trapping and Mars' greater distance from the Sun make the planet very cold. Temperatures at noon at the equator may reach a little above the freezing point of water, but at night they plummet to far below zero on the Fahrenheit scale. The resulting average temperature is a frigid 218 K (−67° F). Thus although water exists on Mars, most of it is frozen, locked up either below the surface in the form of permafrost or in the polar caps as solid water ice.

Clouds of dry ice (frozen CO_2) and water-ice crystals (H_2O) drift through the Martian atmosphere carried by the Martian winds. These winds, like the large-scale winds on Earth, arise because air that is warmed near the equator rises and moves toward the poles. This flow from equator to poles, however, is deflected by the Coriolis effect arising from the planet's rotation. The result is winds that blow around the planet approximately parallel to its equator. The Martian winds are generally gentle, but seasonally, and near the poles, they become gales, which sometimes pick up large amounts of dust from the surface. The resulting vast dust storms occasionally cover the planet completely and turn its sky pink.

No rain falls from the Martian sky, despite its clouds, because the atmosphere is too cold and contains too little water. In fact, there is so little water in the Martian atmosphere that even if all of it were to fall as rain, it would make a layer only about 12 micrometers deep (less than 1/2000 inch). For comparison, Earth's atmosphere holds enough water to make a layer a few centimeters (inches) deep. Despite such dryness, however, fog sometimes forms in some Martian valleys, and frost condenses on the ground on cold nights, as seen in figure 8.22. In addition, during the Martian winter, CO_2 "snow" falls on the Martian poles.

Mars has not always been so dry; the numerous channels in its highlands show that water once flowed freely there. But for a planet to have liquid water, it must have a warm atmosphere with a pressure similar to that of Earth's. If the pressure on a liquid is very low, molecules can break free from its surface, evaporating easily because no external force restrains them. On the other hand, if the pressure is high, molecules in a liquid must be heated strongly to turn them into gas. For example, at normal atmospheric pressure, water boils at 100 degrees Celsius. If the pressure is reduced, however, the boiling point drops, an effect used by food producers to make "freeze-dry" foods, such as instant coffee. Coffee is brewed normally, and then frozen and placed in a chamber from which the air is pumped out. The reduced pressure makes the liquid "boil" without heating and evaporate, leaving only a powder residue—instant coffee. Similarly, on Mars, any liquid water on its surface today would evaporate.

The existence of channels carved by liquid water on Mars is therefore strong evidence that in its past Mars was warmer and had a denser atmosphere. But that milder climate must have ended billions of years ago. From the large number of craters on Mars, astronomers deduce that its surface has not been significantly eroded by rain or flowing water for about 3 billion years. Why has Mars dried out, and where has its water and atmosphere gone?

Some water may lie buried below the Martian surface as ice. If the Martian climate was once warmer and then cooled drastically, water would condense from its atmosphere and freeze, forming sheets of surface ice. Wind might then bury this ice under protective

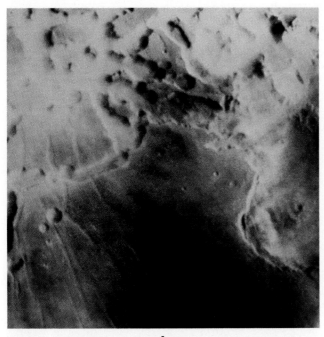

A

B

FIGURE 8.22
(A) Fog in Martian valleys and (B) frost on surface rocks near the Viking Lander. (Courtesy NASA/JPL.)

layers of dust, as happens in polar and high mountain regions of Earth. In fact, figure 8.23A shows indirect evidence of such buried ice. Astrogeologists think that subsurface ice melted, perhaps from volcanic activity, and drained away, causing the ground to collapse and leaving the lumpy, fractured terrain visible at the broad end of the channel. The water from the melted ice then flowed downstream, carving the 20-kilometer (12 mile)-wide canyon. Similarly, figure 8.23B shows a crater from which partially melted matter "squished" out on impact. Thus Mars's water may now mostly be subsurface ice. But why did the planet cool so?

If Mars had a denser atmosphere in the past, as deduced from the higher pressure needed to allow liquid water to exist, then the greenhouse effect might have made the planet significantly warmer than it is now. The loss of such an atmosphere would then weaken the greenhouse effect and plunge the planet into a permanent ice age. Such a loss could happen in at least two ways. According to one theory, Mars may have been struck by a huge asteroid whose impact blasted its atmosphere off into space. Such impacts, although rare, do occur, and our own planet may have been struck about 65 million years ago with results nearly as dire, as we shall learn in chapter 10.

A less dramatic explanation for how Mars lost most of its atmosphere is that Mars' low gravity allowed gas molecules to escape over the first 1 to 2 billion years of the planet's history. Regardless of which explanation is correct, the loss of its atmosphere would have cooled the planet and locked its remaining water up as permafrost. But why have the Martian volcanos not replenished its atmosphere, keeping the planet warm? Astronomers believe that the blame lies with Mars' low level of tectonic activity, a level set by conditions in its interior.

The Martian Interior

Astronomers believe that the interior of Mars is differentiated like the Earth's into a crust, mantle, and iron core. However, because Mars is so small compared with the Earth, its interior is probably cooler. Mars' smaller mass supplies less heat, and its smaller radius allows the heat to escape more rapidly. Unfortunately, astronomers have no direct confirmation

FIGURE 8.23
Pictures of (A) a channel cut by water released as subsurface ice melted. Note the lumpy terrain at the broad end of the channel where the surface collapsed as the water drained away. (B) Crater with surrounding flow patterns. Heat released by the impact that formed the crater has melted subsurface ice. The thawed material has oozed out. (Courtesy NASA.)

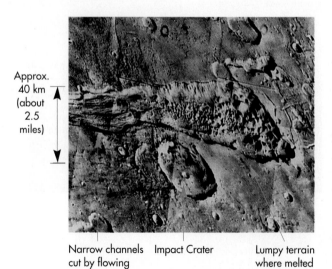

Approx. 40 km (about 2.5 miles)

Approx. 20 km (about 12 miles)

Material squished out by impact

Narrow channels cut by flowing water   Impact Crater   Lumpy terrain where melted water has flowed out

A

B

of Mars' interior structure because the seismic detectors landed there by the *Viking* spacecraft failed. Thus, as is the case for Mercury and Venus, astronomers must rely on indirect evidence from its density and gravitational field to learn about the interior of Mars. Having a mass between that of dead Mercury and lively Earth and Venus implies that Mars should be intermediate in its tectonic activity. Such seems to be the case, and additional evidence for this moderate activity comes from its weak magnetic field.

The Earth's magnetic field is generated by a combination of our planet's spin and motions driven by heat in its interior. Mars rotates as rapidly as the Earth does, yet it has only a very weak field, implying that Mars probably has only a small liquid iron-nickel core.

Although it possesses numerous volcanic peaks and uplifted highlands, implying that it had an active crust, at least in the past, Mars bears no evidence of large-scale crustal motion like the Earth's. For example, it has no folded mountains. Astronomers therefore think that Mars has cooled and its crust thickened to perhaps twice the thickness of the Earth's crust. As a result, its now weak interior heat sources can no longer break through to the surface or drive tectonic motions. The thick Martian crust may also explain why Mars has a small number of very large volcanos, while the Earth has a large number of small ones. Mars' immense volcanos are thus mute testimony to a more active past.

Mars' current low level of tectonic activity is also demonstrated by the many craters that cover its older terrain, far more than are seen on either Earth or Venus. The number of those craters implies that Mars has been geologically quiet for billions of years. Mars is probably not dead, however, because some regions (for example, the slopes of Olympus Mons and other volcanos) are essentially free of craters. Thus these immense peaks may still occasionally erupt. They do not erupt often enough, however, to replace the gas lost to space because of the planet's low gravity. Thus Mars may have entered a phase of planetary senescence.

The Martian Moons

Mars has two tiny moons, Phobos and Deimos, which are named for the demigods of Fear and Panic. These bodies are only about 10 kilometers across, and are probably captured asteroids. They are far too small for their gravity to have pulled them into spherical shapes. Both moons are cratered, implying bombardment by smaller objects. Phobos (see fig. 8.24) has cracks, suggesting that it may have been struck by a body large enough to split it nearly apart.

Phobos and Deimos were discovered in 1877, but by chance they appeared in literature nearly two centuries earlier in Jonathan Swift's book, *Gulliver's Travels*. Gulliver stops at the imaginary country Laputa whose inhabitants include numerous astronomers.

Approx.
16 km
(about
10 miles)

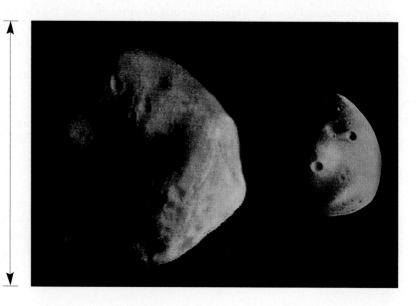

FIGURE 8.24
Picture of Phobos and Deimos, the moons of Mars. These tiny bodies are probably captured asteroids. (Courtesy NASA/JPL.)

What does the irregular shape of these bodies tell you about the strength of their surface gravity? Is it likely these moons have any atmosphere of their own?

Among the accomplishments of these people is the discovery of two tiny moons of Mars. Even earlier, Kepler guessed that Mars had two moons because the Earth has one moon and Jupiter, at least in Kepler's time, was known to have four. Mars, lying between these two bodies should therefore have a number of moons lying between 1 and 4, and he chose 2 as the more likely case.

Life on Mars?

Scientists have long wondered whether living organisms developed on Mars. Much of that interest grew from a misinterpretation of observations made in 1877 by the Italian astronomer, Giovanni Schiaparelli. Schiaparelli saw what he took to be straight-line features on Mars and called them *canali,* by which he meant "channels". In English-speaking countries, the Martian *canali* became canals, with the implication of intelligent beings to build them. The interest in these canals had become so great by 1894 that the wealthy Bostonian Percival Lowell built an observatory in northern Arizona to study Mars and search for signs of life there.

Most astronomers could see no trace of the alleged canals, but they did note seasonal changes in the shape of dark regions, changes that some interpreted as the spread of plant life in the Martian spring. By the early 1970s, scientists were excited by satellite photographs of water-carved canyons and old river beds because water—at least on Earth—is so important for life. Therefore, to further the search for life on Mars, the United States landed two *Viking* spacecraft on the planet in 1976. These craft carried instruments to search for signs of carbon chemistry in the soil and to look for metabolic activity of soil samples that were put in a nutrient broth carried on the lander. All tests either were negative or ambiguous.

Then, in 1996, an American and an English group of scientists reported possible signs of life in rocks from Mars. These were not samples returned to Earth by a spacecraft, but meteorites found on Earth. They arrived here after being blasted off the surface of Mars by the impact of a small asteroid. Such impacts are not uncommon, but most fragments are scattered in space or fall back to Mars, and of those that are shot into space, only a tiny fraction have just the right combination of speed and direction to reach Earth. Even these may first go into orbit around the Sun, where they may remain for millions of years.

How then do the fragments end up on Earth so long after their violent departure from Mars? Scientists can thank Jupiter because its gravitational attraction gradually shifts the orbit of the fragments so that about 13,000 years ago, a few struck Earth. Most of those fragments were lost—burnt up in the atmosphere, sunk in the ocean, or scattered and buried on the land. But a few landed in Antarctica.

Over the last few decades, scientists have discovered that Antarctica is a great place to look for meteorites. The dark fragments are easy to see on the snow, and the movement of the glaciers sweeps up fragments deposited over a huge area and concentrates them at the glacier's end. There, these astronomical relics are relatively easy to find and analysis of them shows that the vast majority are fragments of asteroids. Of the remaining fragments, a few are pieces blasted off the Moon by impact, but very few—roughly a dozen—appear to be Martian rocks.

Scientists can identify a meteorite's origin from its composition. Lunar ones contain very little trapped water molecules and are generally very poor in iron. The fragments believed to come from Mars contain tiny bubbles of trapped gas whose composition matches closely the atmosphere of Mars, as measured by the Viking Mars landers. Moreover, chemical analysis of the minerals in the meteorites show some of them to be unlike minerals on the Earth or Moon.

Chemical analysis of the radioactive elements in the meteorites can also reveal their age and how long they have orbited in space. One of the rocks that shows possible signs of past life probably formed on Mars about 3.5 billion years ago and was blasted into space about 16 million years ago. Other chemical signatures reveal it arrived on Earth about 13,000 years ago, during the last ice age.

Although the meteorites' Martian origin seems well established, it is the possible evidence for fossilized, primitive, bacterialike organisms within the rocks that make them so exciting and controversial. For one of the rocks, that evidence comes from three findings. First, scientists found small globs of carbonate deep within its cracks. Such globs might have formed if the rock spent some time underwater on Mars. Second, the carbonate globs are stained with magnetite, an iron oxide produced by many bacteria here on Earth. Moreover, near the magnetite-stained globs are many tiny, rod-shaped structures (fig. 8.25). These look very much like ancient terrestrial bacteria, but are much smaller. To some scientists they look like "fossilized" primitive Martian life. Finally, the carbonate globs contain traces of organic chemicals known as polycyclic aromatic hydrocarbons (PAHS, for short). Terrestrial bacteria make such chemicals when they die and decay, but PAHS can also form spontaneously given the proper mix of chemicals. In fact they have been found in a number of non-Martian meteorites and have also been detected by their spectrum lines in the radio emission from interstellar gas and dust clouds.

Any one of the these pieces of evidence (organic chemicals, carbonate crystals, and rod-shaped structures) would not itself make a very convincing case for life on ancient Mars. Taken together, however, they suggest its possibility. But given how cautious scientists want to be about making claims for life on another planet, many other explanations will be tested. For example, some scientists have already claimed that the carbonate blobs were formed by chemical reactions in the rocks, and others believe the PAHS are simply contaminants from the Antarctic ice cap. It will take years of further analysis of these and other Martian meteorites or perhaps a robot Mars explorer to tell us for sure whether life exists or once existed on that remote red world.

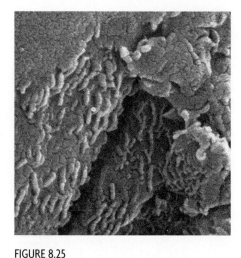

FIGURE 8.25
Fossils of ancient Martian life? The tiny rod-shaped structures look similar to primitive fossils found in ancient rocks on Earth. However, some scientists think these structures formed chemically. (Courtesy NASA.)

8.5 WHY ARE THE TERRESTRIAL PLANETS SO DIFFERENT?

We have seen above that the four terrestrial planets have little in common apart from being rocky spheres. They have different surfaces, atmospheres, and interiors. Astronomers think these differences arise from their different masses, radii, and distance from the Sun.

Role of Mass and Radius

As discussed earlier, a planet's mass and radius affect its interior temperature and thus its level of tectonic activity, with low-mass, small-radius planets being cooler inside than larger bodies. We see, therefore, a progression of activity from small, relatively inert Mercury, to slightly larger and once-active Mars, to the larger and far more active surfaces of Venus and Earth. Mercury's surface still bears the craters made as it was assembled from planetesimals. Mars has some craters, from which we infer that much of its surface is very old, but being larger and more tectonically active, it also has younger surface features such as volcanos, canyons formed by surface cracking as hot material rose inside it, and erosional features such as canyons and river beds carved by running water. In contrast, Earth and Venus retain essentially none of their original crust; their surfaces have been enormously modified by activity in their interiors over the lifetimes of these planets.

Role of Internal Activity

Internal activity, as we have seen, also affects a planet's atmosphere. In fact, the atmospheres of the terrestrial planets, though now greatly modified, are probably mostly volcanic gases vented as the result of their internal activity. Thus small, inactive Mercury probably never had much atmosphere, and Mars, active once but now quiescent, likewise could create only a thin atmosphere. Moreover, these planets have so little mass and consequently such a small surface gravity relative to Venus and Earth that they have difficulty retaining what little gases they might once have had. As a result, Mercury is virtually

Volcanos even today modify Earth's surface and atmosphere. (Science Photo Library/Mathew Shipp.)

without an atmosphere today, and Mars has only a vestige of its original atmosphere. On the other hand, highly active Venus and Earth have extensive atmospheres.

Astronomers think that the atmospheres that remain on Venus, Earth, and Mars have changed their composition appreciably over time from chemical processes. The atmospheres of all three bodies were probably originally nearly the same composition: primarily CO_2, but with small amounts of nitrogen and water. But these original atmospheres have been modified by sunlight, tectonic activity, and, in the case of the Earth, life.

Role of Sunlight

Sunlight affects a planet's atmosphere in several ways. First, of course, it warms a planet by an amount that depends on the planet's distance from the Sun, and so we expect that Venus will be warmer than Earth, and Earth will be warmer than Mars. These expected temperature differences are increased, however, by the atmospheres of these bodies. Moreover, even relatively small differences in temperature can lead to large differences in physical behavior and chemical reactions within an atmosphere. For example, Venus is just enough nearer the Sun than the Earth that even without a strong greenhouse effect, most of its atmosphere would be so warm that water would have difficulty condensing and turning to rain. Moreover, the Venusian atmosphere being warmer throughout, water vapor can rise to great heights in the Venusian atmosphere, whereas on cooler Earth, water vapor condenses to ice at about 30,000 feet, making our upper atmosphere almost totally devoid of water. You may have noticed this extreme dryness on a long plane trip because after such a flight your skin may feel itchy and the inside of your nose may feel cracked or stuffy. That dryness is the result of bringing upper atmospheric air into the cabin and compressing it to make it breathable but not adding moisture.

Role of Water Content

The great difference in the water content of the upper atmosphere of Earth and Venus has led to a drastic difference between their atmospheres at lower levels. At high altitudes, ultraviolet light from the Sun is intense enough to break apart any water molecules present into their component oxygen and hydrogen atoms, a process called **photo-dissociation**. Being very light, the hydrogen atoms so liberated escape into space, while the heavier oxygen atoms remain. Because water can rise to great heights in the Venusian atmosphere, over billions of years it has steadily been dissociated there and almost completely lost from our sister planet. In our atmosphere, however, water has survived.

Water makes possible chemical reactions that profoundly alter the composition of our atmosphere. For example, CO_2 dissolves in liquid water and creates carbonic acid, which in dilute form we drink as "soda water." In fact, the bubbles in soda water are just carbon dioxide that is coming out of solution. As rain falls through our atmosphere it picks up CO_2, making it slightly acidic, even in unpolluted air. As the rain falls on the ground, it reacts chemically with silicate rocks to form carbonates, locking some CO_2 into the rock.

Role of Biological Processes

Biological processes also remove some CO_2 from the atmosphere. For example, plants use it to make the large organic molecules such as cellulose, of which they are composed. This CO_2 is usually stored for only short periods of time, however, before decay or burning releases it back into the atmosphere. More permanent removal occurs when rain carrying dissolved CO_2 runs off into the oceans where sea creatures use it to make shells of calcium carbonate. As these creatures die, they sink to the bottom where their shells form sediment that eventually is changed to rock. Thus carbon dioxide is swept from our atmosphere and locked up both chemically and biologically in the crust of our planet. With most of the carbon dioxide removed from our atmosphere, mostly nitrogen is left. In fact, our atmosphere contains roughly the same total amount of nitrogen as the atmosphere of Venus.

Our atmosphere is also rich in oxygen, a gas found in such relative abundance nowhere else in the Solar System. Our planet's oxygen is almost certainly the product of green plants breaking down the H_2O molecule during photosynthesis, as we discussed in chapter 4. Thus the cause of the great difference between the atmospheres of Earth and Venus may simply be that life was able to start on Earth and thereby, with the aid of liquid water, not only sweep our atmosphere nearly free of carbon dioxide (0.03% of the atmosphere) but also produce oxygen in the process. In fact, if all the buried carbonate rock released its CO_2, the Earth's atmosphere, apart from its oxygen, would closely resemble Venus's.

If water is so effective at removing carbon dioxide from our atmosphere, why does any CO_2 remain? We add small amounts of CO_2 by burning wood and fossil fuels, but the major contribution is from natural processes. Atmospheric chemists hypothesize that tectonic activity gradually releases CO_2 from rock back into our atmosphere. At plate boundaries, sedimentary rock is carried downward into the mantle where it is melted. Heating breaks down the carbonate rock into carbon dioxide, which then rises with the heated rock to the surface and reenters the atmosphere. A similar process may once have occurred on Mars to remove carbon dioxide from its atmosphere, locking it up in rock there. Mars' lower level of tectonic activity, however, prevents its CO_2 from being recycled. Thus with so little of its original carbon dioxide left, Mars has grown progressively colder. Our Earth, because it is active, has retained enough CO_2 in its atmosphere to maintain a moderate greenhouse effect, making our planet habitable. Thus, poised between one planet that is too hot and another that is too cold, Earth has been blessed with a relatively stable atmosphere, one factor in the complex web of our environment to which we owe our existence.

Plants created most of the oxygen in our atmosphere by photosynthesis. (Courtesy Photo Researchers/Gregory G. Dimijiani.)

SUMMARY

The terrestrial planets—Mercury, Venus, Earth, and Mars—are alike in being rocky bodies of comparable size and internal structure, but they nevertheless differ dramatically from one another (table 8.1 and fig. 8.26). Their slight difference in mass and diameter has led to great differences in their internal heating and surface activity. In fact, when ordered by mass, the terrestrial planets are also ordered by activity. Mercury, the least massive, is least active and has the oldest surface. The many impact craters created late in its formation make Mercury look much like our Moon.

TABLE 8.1   Comparison of the Terrestrial Planets

	Mercury	Venus	Earth	Mars
$R_{equator}$ (Earth units)	0.382	0.949	1.0	0.533
(km)	2439	6052	6378	3398
Mass (Earth units)	0.055	0.815	1.0	0.107
(kg)	3.30×10^{23}	4.87×10^{24}	5.98×10^{24}	6.42×10^{23}
Density (grams/cm³)	5.43	5.25	5.52	3.94
Atmospheric composition	None	CO_2 (96%) N_2 (3%)	N_2 (78%) O_2 (21%)	CO_2 (95%) N_2 (2.7%)
Pressure (bars)	0.0	about 90	1.0	about 0.007
Surface features	Craters, maria, scarps	Some craters, mountains, volcanic peaks, congealed lava plains	Oceans, mountains, volcanic peaks	Deserts, canyons, volcanic peaks
Sidereal day	58.65 Earth days	243.02 Earth days	23.9345 hours	24.62 hours
Solar day	176 Earth days	116.8 Earth days	24 hours	24.66 hours
Satellites	None	None	Moon	Phobos and Deimos
Distance from Sun	0.387 AU	0.723 AU	1 AU	1.524 AU
Orbital period	87.969 Earth days	224.70 Earth days	365.26 Earth days	686.98 Earth days
Axial tilt	7°	177.4°	23.45°	23.98°

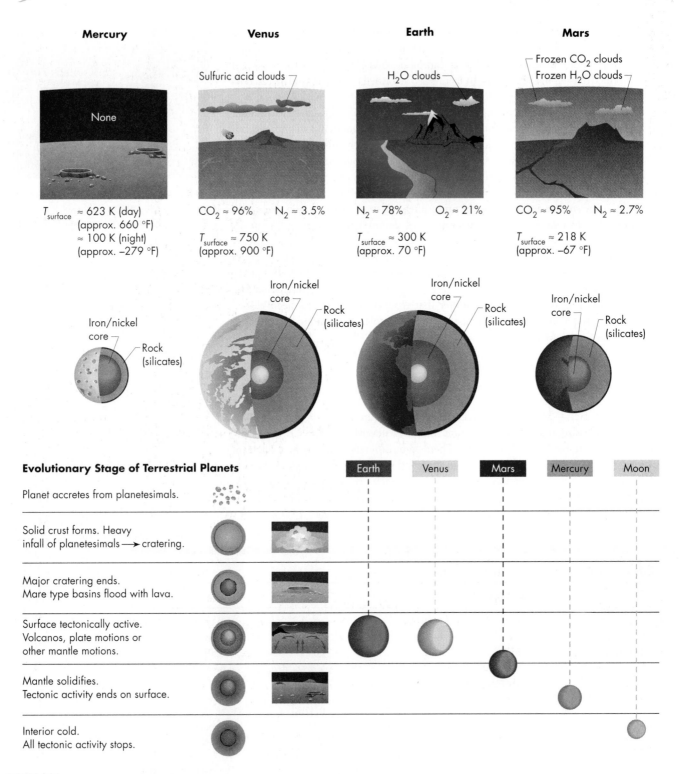

FIGURE 8.26
Gallery comparing interiors, atmospheres, and surfaces of the terrestrial planets.

Mars, intermediate in mass and radius between Mercury and Earth, shows a surface with an intermediate level of activity. Much of it is cratered, implying a great age. Some regions, however, show volcanic uplift like that seen on Venus, implying that its interior was once much hotter than now. Moreover, its surface shows an amazing variety of landforms: polar caps of ordinary ice and frozen carbon dioxide, immense deserts with dune fields, and canyons created both by crustal cracking and erosion from running water.

Venus and Earth have young surfaces, extensively altered by volcanic activity. On Earth, the flow of heat from its core has led to plate motion. On Venus, perhaps because of its hotter surface, the flow of heat to its surface is less uniform. As a result its surface is marked by isolated regions of intense volcanic uplift. In fact, Venus is probably as active geologically as the Earth.

Mercury, Venus, and Mars have very different atmospheres. Mercury was too small to either make or retain an atmosphere and is now essentially airless. Venus and Mars both have carbon dioxide atmospheres, but Venus' is about 100 times denser than Earth's, while Mars' is about 100 times less dense. The great thickness of the Venusian atmosphere creates a strong greenhouse effect, heating it to about 900° F. High in the Venusian atmosphere, thick clouds of sulfuric acid droplets block our view of the Venusian surface. Astronomers can nevertheless study the Venusian surface with radar telescopes.

Mars' atmosphere is too thin now for a strong greenhouse effect, but the old river channels, cut presumably by running water, imply that Mars once had a denser and warmer atmosphere, perhaps even resembling Earth's.

The great differences between the atmospheres of Venus, Earth, and Mars can probably be explained by the differences in their water content and the fact that life evolved on Earth.

QUESTIONS FOR REVIEW

1. How does Mercury's radius and mass compare with the Earth's?
2. Does Mercury have an atmosphere? Why or why not?
3. What is the surface of Mercury like?
4. What is peculiar about Mercury's rotation? What causes this oddity?
5. How does Venus compare with the Earth in mass and diameter?
6. Can we see the surface of Venus? Why or why not?
7. How do astronomers know what the surface of Venus is like?
8. What is the dominant gas in Venus's atmosphere? How do astronomers know?
9. What are the clouds of Venus made of?
10. Why is Venus so hot?
11. What sort of features are seen on Venus's surface? Is the surface young or old? How do astronomers know?
12. How does Mars compare with the Earth in mass and diameter?
13. Describe some of the surface features seen on Mars.
14. What is the Martian atmosphere like?
15. What explanation(s) have been offered for why the atmospheres of the terrestrial planets are so different?
16. What are the Martian polar caps composed of?
17. What is the evidence that Mars once had running water on its surface?
18. How do astronomers explain why the Earth's atmosphere ended up with so little CO_2 compared with Mars and Venus?
19. What is the evidence that leads some scientists to believe life may have existed on Mars?

THOUGHT QUESTIONS

1. Some scientists have proposed "terraforming" Venus and Mars. That is, they propose introducing organisms that would use carbon dioxide to photosynthesize oxygen and make both planets more suitable for humans. Do you think this is a good idea? Why or why not?
2. What evidence do we see from the terrestrial planets that supports the theory for the origin of the Solar System discussed in chapter 7?
3. What are some examples of resonances in everyday life?
4. What role might planetesimal impacts have played in the history of the terrestrial planets?
5. How does the surface temperature of Venus relate to concern about global warming of the Earth?
6. Suppose scientists discover simple life forms on Mars. Would that alter the way you look at life here on Earth?

TEST YOURSELF

1. Why does Mercury have so many craters and the Earth so few?
   (a) Mercury is far more volcanically active than the Earth.
   (b) Mercury is much more massive than the Earth and therefore attracted more impacting bodies.
   (c) The Sun has heated Mercury's surface to the boiling point of rock and the resulting bubbles left craters.
   (d) Erosion and plate tectonic activity have destroyed most of the craters on the Earth.
   (e) Mercury's iron core and its resulting strong magnetic field have attracted impacting bodies.

2. Which body in the inner Solar System has the densest atmosphere?
   (a) Mercury.
   (b) Venus.
   (c) Earth.
   (d) Mars.
   (e) Our Moon.

3. Why is Venus's surface hotter than Mercury's?
   (a) Venus rotates more slowly, so it "bakes" more in the Sun's heat.
   (b) Clouds in Mercury's atmosphere reflect sunlight back into space and keep its surface cool.
   (c) Carbon dioxide in Venus's atmosphere traps heat radiating from its surface, thereby making it warmer.
   (d) Venus is closer to the Sun.
   (e) Venus's rapid rotation generates strong winds that heat the ground by friction as they blow.

4. What are the Venusian clouds made of?
   (a) Mostly water droplets.
   (b) Frozen carbon dioxide.
   (c) Sodium chloride.
   (d) Sulfuric acid droplets.
   (e) Ice crystals.

5. Mercury's average density is about 1.5 times greater than the Moon's, even though the two bodies have similar radii. What does this suggest about Mercury's composition?
   (a) Mercury's interior is much richer in iron than the Moon's.
   (b) Mercury contains proportionately far more rock than the Moon.
   (c) Mercury's greater mass has prevented its gravitational attraction from compressing it as much as the Moon is compressed.
   (d) Mercury must have a uranium core.
   (e) Mercury must have a liquid water core.

FURTHER EXPLORATIONS

Barlow, Nadine G., Janes, Daniel, and Mark A. Bullock. "The Prodigal Sister [Geology of Venus]." Mercury 24 (September/October 1995): 23.

Beatty, J. K., and A. Chaikin. *The New Solar System,* 3rd edition. Cambridge, Mass.: Sky Publishing Corporation; and New York: Cambridge University Press, 1990.

Beatty, J. Kelly. "Working Magellan's Magic." *Sky and Telescope* 86 (August 1993): 16.

Esposito, L. W. "Does Venus Have Active Volcanoes?" *Astronomy* 18 (July 1990): 42.

Goldman, S. J. "Venus Unveiled." *Sky and Telescope* 83 (March 1992): 258.

Gore, Rick. "The Planets—Between Fire and Ice." *National Geographic* 167 (January 1985): 4.

Kargel, Jeffrey S., and Robert G. Strom. "Global Climatic Change on Mars." *Scientific American* 275 (November 1996): 80.

Kasting, J. F. "How Venus Lost Its Oceans." *Oceanus* 32 (Summer 1989): 54.

Kaula, W. M. "Venus: A Contrast in Evolution to Earth." *Science* 247 (9 March 1990): 1191.

Plaut, Jeffrey J. "Venus in 3-D." *Sky and Telescope* 86 (August 1993): 32.

Robinson, M. S. "Surveying the Scars of Ancient Martian Floods." *Astronomy* 17 (October 1989): 38.

Saunders, S. "The Exploration of Venus: a Magellan Progress Report." *Mercury* 20 (September/October 1991): 130.

Saunders, R. S. "The Surface of Venus." *Scientific American* 263 (December 1990): 60.

Schubert, G. and C. Covey. "The Atmosphere of Venus." *Scientific American* 245 (July 1981): 66.

Stofan, Ellen R. "The New Face of Venus." *Sky and Telescope* 86 (August 1993): 22.

Strom, R. G. "Mercury: The Forgotten Planet." *Sky and Telescope* 80 (September 1990): 256.

Tennesen, M. "Mars: Remembrance of Life Past." *Discover* 10 (July 1989): 82.

Web Sites
"Welcome to the Planets" (many pretty pictures, fact sheets on the planets and the other Solar System objects, and a glossary). Prepared by the Jet Propulsion Laboratory.
www.pds.jpl.nasa.gov/planets
"NSSDC Photo Gallery" (National Space Science Data Center—NASA) Good pictures and fact sheets.
www.nssdc.gsfc.nasa.gov/photo_gallery/

"Views of the Solar System," by Calvin Hamilton. Excellent pictures, fact sheets, and glossary. Lots of links to other sites.
www.bang.lanl.gov/solarsys/
"The Nine Planets," by Bill Arnett. "A Multimedia Tour . . ." One of the best astronomy sites. Lots of excellent links, pictures, text, and glossary.
www.seds.org/billa/tnp/

VIDEOS

Flying by the Planets: The Videos (available from The Astronomical Society of the Pacific, 390 Ashton Ave., San Francisco, CA 94122.)

KEY TERMS

greenhouse effect, 248
photo dissociation, 264
resonance, 246

CHAPTER NINE

THE OUTER PLANETS

JUPITER

SATURN

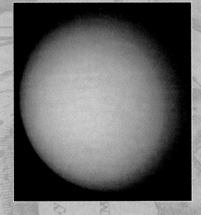

URANUS

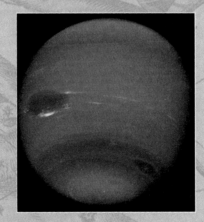

NEPTUNE

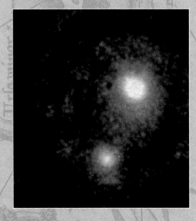

PLUTO

The Outer Planets: Jupiter, Saturn, Uranus, Neptune, and Pluto (all Courtesy of NASA)

Beyond Mars, the Solar System is a realm of ice and frozen gas. In this frigid zone, far from the Sun, where solar heat is only a vestige of what we receive on Earth, the giant planets formed. The low temperature—typically about 150 Kelvin (over 100 degrees below zero on the Fahrenheit scale)—allowed bodies condensing there, within the solar nebula, to capture hydrogen and hydrogen-rich gases, such as methane, ammonia, and water. Because these gases were far more abundant in the young Solar System than the silicate- and iron-rich material from which the terrestrial planets condensed, planets that formed in this cold environment had more material available for their growth. As a result, these cold planets became vastly larger than those near the Sun, and they developed very different structure and composition. The four largest planets—Jupiter, Saturn, Uranus, and Neptune—are composed mainly of gaseous and liquid hydrogen and its compounds. Although these giant bodies may have cores of molten rocky matter, they lack solid surfaces and, consequently, have no surface features, such as mountains and valleys. Rather, it is their atmospheric features that give them such different appearances. Pluto, however, is an exception to these rules. By far the smallest planet in the Solar System, Pluto has little in common with the giant planets and more closely resembles their larger moons, which are composed of ice and rock.

The moons of the outer Solar System range in size from very small bodies to those as large as Mercury. As they orbit their parent planets, they form families rather like miniature Solar Systems. Some of the larger moons have brightly colored surfaces, others have numerous craters, and a few have surface features unlike anything seen elsewhere in the Solar System. A few moons even have active volcanoes. In fact, astronomers consider these diverse bodies, virtually unknown before the space age, to be some of the most interesting members of the Sun's family.

9.1 JUPITER

To the ancient Romans, Jupiter was the king of the gods. Although they did not know how immense this planet is, they nevertheless chose its name appropriately.

Appearance and Physical Properties

Jupiter is the largest planet both in radius and in mass. In fact, its mass is larger than that of all other planets in the Solar System combined. It is slightly more than 10 times the Earth's diameter and more than 300 times its mass. Dense, richly colored parallel bands of clouds cloak the planet, as shown in figure 9.1. Spectra of the sunlight reflected from these clouds show that Jupiter's atmosphere consists mostly of hydrogen, helium, and hydrogen-rich gases such as methane (CH_4), ammonia (NH_3), and water (H_2O). These gases were also directly detected in December 1995 when the *Galileo* space probe parachuted into Jupiter's atmosphere. The clouds themselves are harder to analyze, but theoretical calculations of the chemistry of Jupiter's atmosphere suggest they are particles of water, ice, and ammonia compounds. Their bright colors may come from complex organic molecules whose composition is still uncertain. Time-lapse pictures show that its clouds move swiftly, sweeping around the planet in jet streams that are far faster than those of Earth. Moreover, Jupiter itself rotates once every 10 hours, spinning so fast that its equator bulges significantly.

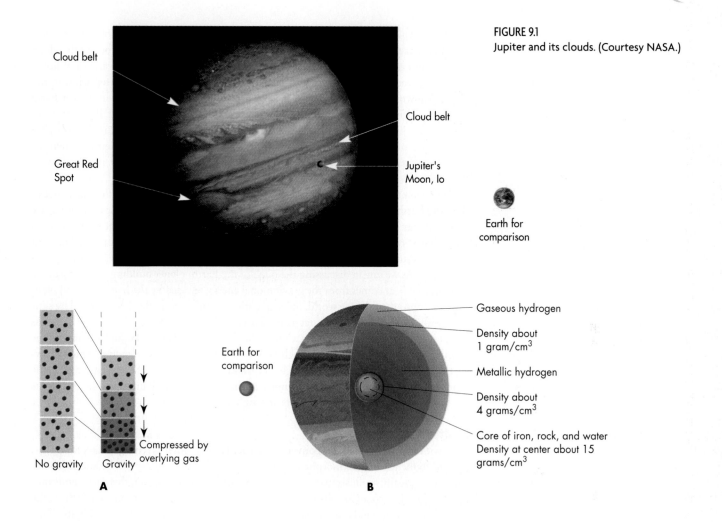

FIGURE 9.1
Jupiter and its clouds. (Courtesy NASA.)

Cloud belt

Great Red
Spot

Cloud belt

Jupiter's
Moon, Io

Earth for
comparison

No gravity   Gravity   Compressed by
overlying gas

A

Earth for
comparison

Gaseous hydrogen

Density about
1 gram/cm³

Metallic hydrogen

Density about
4 grams/cm³

Core of iron, rock, and water
Density at center about 15
grams/cm³

B

FIGURE 9.2
(A) The density of gas increases with depth because the overlying gas compresses the matter. (B) A sketch of what astronomers think Jupiter's interior is like.

Jupiter's Interior

Astronomers cannot see through Jupiter's cloud layers to its interior, nor can they probe its interior with seismic detectors. Instead, they must rely on theory to tell them what lies inside this giant planet. For example, despite its great mass, Jupiter is much less dense, on the average, than the Earth. To determine a planet's average density, astronomers divide its mass by its volume. The mass can be found by calculating the planet's gravitational attraction on one of its moons, using the method described in chapter 7 (see fig. 7.5). We also described in that figure how to find the planet's radius, and hence its volume. When such calculations are made for Jupiter, we find that its average density is only slightly greater than that of water—1.3 grams per cubic centimeter—showing that the bulk of the planet, not just its atmosphere, must be composed mainly of very light elements, such as hydrogen. Calculations based on Jupiter's low density, its shape, and its gravitational attraction on its satellites and passing spacecraft all give astronomers information on what lies below its clouds.

Jupiter's immense mass exerts a tremendous gravitational force on its atmosphere and interior, holding the planet together and compressing its gas. Near the cloud tops, this compression is only slight, so the gas density there is low. Deep in the interior, however, the weight of thousands of kilometers of gas compresses the matter to a high density. Thus the density of the gas increases with depth, as illustrated in figure 9.2A. Everywhere in the planet, however, the forces of gravity and pressure are nearly in balance, just as they are in our own atmosphere.

Deep within Jupiter, the compression created by its gravity presses molecules so close together that the gas changes to liquid. Thus, about 10,000 kilometers below the cloud tops—about one sixth of the way into the planet—Jupiter's interior is a vast sea of liquid hydrogen. Deeper still, the weight of the liquid hydrogen compresses matter below it into a state known as liquid metallic hydrogen, a form of hydrogen that scientists on Earth have created in tiny high-pressure chambers. Astronomers theorize that if the material from which Jupiter formed has the same overall composition as the Sun's, then Jupiter must also contain heavy elements such as silicon and iron. Because of their high density, these elements have probably sunk to Jupiter's center, forming a core of iron and rocky material a few times larger than the Earth (see fig. 9.2B).

Jupiter's interior is extremely hot, perhaps 30,000 K, which is about five times hotter than the Earth's core. This heat rises slowly to the planet's surface and escapes into space as low-energy infrared radiation. Astronomers have long known from measurements of this infrared radiation that Jupiter emits more energy than it receives from the Sun, but they are still uncertain what supplies that heat. Much of it may be left over from Jupiter's formation. As we saw in discussing the birth of the Earth, planet building is a hot process. Most of that heat comes from the gravitational energy released by the hail of gas and planetesimals onto the forming world. Subsequently, additional heat may be generated by slow but steady shrinkage of the planet as it adjusts to the greater gravity created by the mass added in the last stages of its formation. In fact, giant gas planets like Jupiter may still be shrinking (and therefore heating) slightly. Yet more heat is released as matter slightly heavier than hydrogen, such as helium, sinks toward Jupiter's core. Whatever its source, the heat generates convection currents similar to those in the Earth's interior. These currents stir both Jupiter's deep interior and its atmosphere.

Jupiter's Atmosphere

Heat within Jupiter generates convection currents in its outer layers that carry warm gas upward to the top of its atmosphere. Here, the gas radiates heat to space, becoming cooler and sinking again, as illustrated in figure 9.3A. Because of its motion, this rising and sinking gas is subject to a Coriolis effect* created by Jupiter's rapid rotation that deflects the gas into powerful winds called jet streams. We see these winds as the cloud belts, illustrated in figure 9.3B. Winds in these jets streams can have velocities of up to 300 kilometers per hour (nearly 200 miles per hour) with respect to the planet's overall rotation. Yet despite such high speeds, winds in adjacent regions may blow in opposite directions, as shown in figure 9.3C. Such reversals of wind direction (wind shear) from place to place also occur on Earth, where equatorial winds generally blow from east to west, while mid-latitude winds generally blow from west to east.

These reversals occur because the Coriolis effect deflects northern hemisphere winds to the right and southern hemisphere winds to the left of the direction they would travel in the absence of the Coriolis effect. As a result, winds moving toward the equator are deflected to the west, while winds moving toward the poles are deflected to the east, as shown in figure 9.3B. The cumulative effect of these deflections is a set of jet streams blowing in opposite directions with respect to Jupiter's overall rotation.

As the various jet streams circle Jupiter, gas between them is spun into huge whirling atmospheric vortices, much as a pencil between your palms twirls as you rub your hands together. Some of these spinning regions are brightly colored, as figure 9.4 shows. Brown and shades of white dominate, but one exceptionally large vortex—bigger across than the Earth—is nearly brick red. Known as the Great Red Spot, this vortex was discovered in the seventeenth century. Since then, it has changed slightly in appearance and location,

*The Coriolis effect is a consequence of the conservation of angular momentum. It is not some new force.

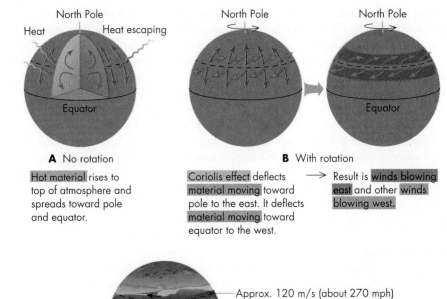

A No rotation

Hot material rises to top of atmosphere and spreads toward pole and equator.

B With rotation

Coriolis effect deflects material moving toward pole to the east. It deflects material moving toward equator to the west.

⟶ Result is winds blowing east and other winds blowing west.

FIGURE 9.3
(A) Rising gas from Jupiter's hot interior cools near the top of the atmosphere and sinks. (B) The Coriolis effect, arising from Jupiter's rotation, deflects the gas, creating winds that blow as narrow jet streams. (C) The wind varies widely in speed and direction from region to region.

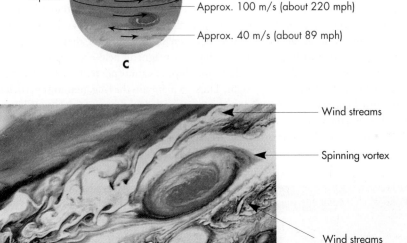

Equator

— Approx. 120 m/s (about 270 mph)

— Approx. 100 m/s (about 220 mph)

— Approx. 40 m/s (about 89 mph)

C

Wind streams

Spinning vortex

Wind streams

FIGURE 9.4
Vortices form between atmospheric streams of different velocities much the way a pencil twirls between rubbed palms. Colors enhanced for clarity. (Courtesy NASA/JPL.)

but both direct observations of the spot and theoretical models show that it is an essentially permanent feature of Jupiter's atmospheric circulation.

Matter also circulates in Jupiter's deep interior. There, convection in the metallic liquid hydrogen combines with the planet's rapid rotation to generate a magnetic field by a natural dynamo process similar to that which generates the Earth's magnetic field. Jupiter's dynamo process is far more powerful than Earth's, however, and creates the strongest magnetic field of any planet in the Solar System. Measurements taken by passing spacecraft show that the field near the top of Jupiter's cloud layer is about 20 times the strength of the magnetic field on the Earth's surface. When Jupiter's larger radius is taken into account, its field is some 20,000 times as powerful as Earth's, as was deduced even before it was directly measured. For example, astronomers inferred that Jupiter had a strong magnetic field from its auroral activity—just barely detectable from Earth—and from the intense radio emission from the giant planet.

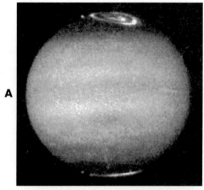

FIGURE 9.5
(A) Jupiter's auroral zone as seen by the Hubble Space Telescope. (Courtesy STSCI) (B) Aurora and lightning on Jupiter, as observed by the *Voyager* spacecraft as it flew over Jupiter's nightside. (Courtesy NASA/JPL.)

FIGURE 9.6
Jupiter's ring as viewed from the *Voyager* spacecraft. In this picture, Jupiter blocks the Sun's light. Note how small and thin its ring is compared to those of Saturn. (Courtesy NASA/JPL.)

The Earth's auroral activity is linked to its magnetic field, which steers incoming energetic particles from the Sun into our upper atmosphere, where the particles trigger the lovely pale glow of the northern and southern lights. Our planet's magnetic field also traps such particles in the Van Allen radiation belts. Jupiter's magnetic field does much the same: particles trapped magnetically in radiation belts far above the planet emit radio waves, making Jupiter a powerful source of radio emission. Some of those particles descend into Jupiter's upper atmosphere where they create an aurora, as you can see in figure 9.5A, a picture made with the Hubble Space Telescope. Jupiter's aurora also shows in figure 9.5B as the pale glow along the edge of the planet in a picture taken by the passing *Voyager* spacecraft. This figure also depicts another familiar phenomenon in Jupiter's atmosphere: lightning. Cloud particles carried up and down by the rising and sinking atmospheric motions collide and generate atmospheric electricity—thunderstorms—just as such motions do on Earth. In the picture, you can see many storms lighting up the Jovian night.

Jupiter's Ring

For centuries, astronomers believed that the only Solar System planet with rings was Saturn. But in 1977, thin rings were detected around Uranus, leading astronomers to wonder if similar rings might surround Jupiter. The opportunity to look for such rings came with the *Voyager I* spacecraft, which flew by Jupiter in 1979. Pictures taken from the craft clearly show that Jupiter has a ring, although only a very thin one, as illustrated in figure 9.6. Jupiter's ring is thought to be made of tiny particles of rock dust held in orbit by Jupiter's immense gravitational attraction. These particles, are so tiny, however, that radiation from the Sun and collisions with gas trapped in Jupiter's magnetic field exert frictional forces on them, making them gradually drift down into Jupiter's atmosphere. There they mingle with the swirling gas and are lost from sight. Thus, to maintain the ring, new dust particles must constantly be added to it. Where does this new ring material come from? According to current theories, the tiny satellites of Jupiter's system occasionally collide and fragment, creating new dust to replenish the rings.

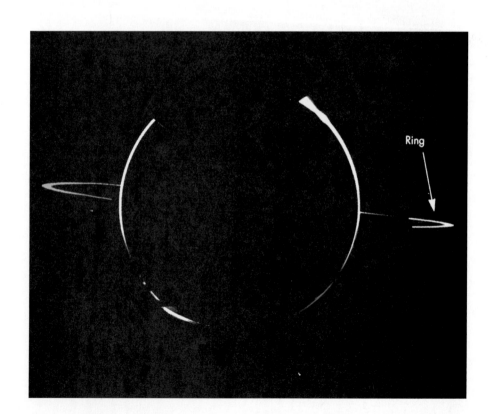

Jupiter's Moons

When Galileo first viewed Jupiter with his telescope, he saw four moons orbiting the planet, bodies now called the "Galilean satellites." Over the past three centuries, astronomers have found additional Jovian moons, with the total now reaching 16. Most of these are too small to be readily seen from Earth and were discovered by examining pictures taken by the *Voyager* spacecraft as it flew past Jupiter.

The Galilean satellites—Io, Europa, Ganymede, and Callisto—are very large: all but Europa are larger than our Moon, and Ganymede has a diameter bigger than Mercury's, making it the largest moon in the Solar System. They orbit along with most of the rest of Jupiter's moons, approximately in Jupiter's equatorial plane, forming a flattened disk, rather like a miniature Solar System. In fact, the Galilean moons most likely formed by a scaled-down version of the process that created the Solar System. That is, they probably aggregated from planetesimals and gas that collected around Jupiter during its formation. Some astronomers believe that Jupiter heated this orbiting debris, affecting the composition and density of the moons that formed from it. Such heating would melt ices and partially evaporate them, so that the moons nearest Jupiter would have less ice and gas and therefore be denser, much as the Sun heated the solar nebula, thereby creating the difference between the terrestrial and outer planets. Evidence that Jupiter heated its moons before or during their formation comes from their density: the densest Galilean satellites are those nearest Jupiter. In order of increasing distance, the densities (in units of grams per cm^3) are 3.53 for Io, 2.99 for Europa, 1.94 for Ganymede, and 1.85 for Callisto. Further evidence of heating comes from their strange, often colorful surface features, as illustrated in figure 9.7, and in the case of Io, from its active volcanos.

Io (pronounced* /eye-oh/) is named for a mythological maiden with whom Jupiter fell in love and whom he changed into a heifer (a young cow) so that his wife Hera would not suspect his infidelity. Io is the nearest to Jupiter of the Galilean moons and therefore is subject to a strong tidal force created by Jupiter's gravity. That tidal force locks Io's spin to its orbital motion the way our Moon's spin is locked to its motion around the Earth. But Io also undergoes a strong gravitational attraction from Europa, the Galilean satellite next

Earth's Moon for comparison

FIGURE 9.7
Photographs of the four Galilean satellites of Jupiter. (Courtesy NASA/JPL.)

☀ Why are these moons called the Galilean satellites?

*American astronomers tend to say "eye-oh," but European ones prefer "ee-oh."

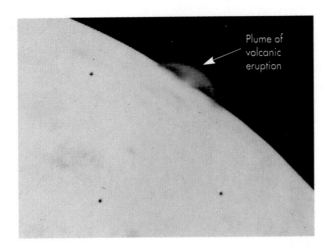

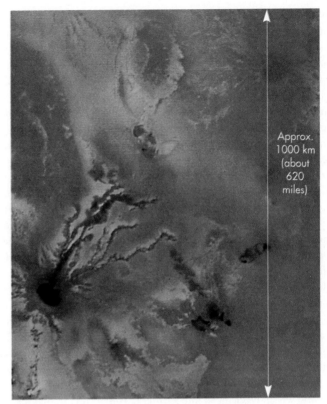

Approx. 1000 km (about 620 miles)

FIGURE 9.8
A volcanic eruption and lava flow on Io. The "lava" is believed to be mostly molten sulfur. These pictures were taken by the *Voyager* spacecraft as it flew past Jupiter. (Courtesy NASA/JPL.)

☀ What explanation has been offered for the source of heat in Io?

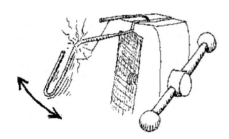

closest to Jupiter. and Europa's gravitational tug twists Io from side to side. Moreover, Europa's pull forces Io into an orbit whose shape constantly changes. The outcome of these two effects is that Io is subject to a strong and changing gravitational force from Jupiter that distorts its shape. This deformation heats it by internal friction, much as bending a paperclip until it breaks heats the wire (touch the freshly broken end and it will feel hot). Over billions of years, the heating has melted not only the ice but also the rocky matter in Io's interior. As molten matter oozes to the surface, it erupts, creating volcanic plumes and lava flows, as shown in figure 9.8. Such activity has driven most of Io's water into space, where it is lost because of Io's weak gravitational attraction. Sulfur, also common in terrestrial volcanoes, is now the major component of Io's volcanic outpourings, and the erupted sulfur and its compounds give Io its rich red, yellow, and orange colors. A recent sulfur lava flow can be seen in figure 9.8 where the molten material has snaked down the slope from the dark volcanic peak. (The peak's darker color indicates that the sulfur was molten there.)

Europa, the smallest of the Galilean moons, is named for another maiden whom Jupiter pursued. According to the legend, Jupiter disguised himself as a bull and carried her on his back across the Hellespont from Asia to Europe, thereby giving the continent its name. Despite this romantic myth, Europa looks rather like a cracked egg. Long, thin lines score its surface, as shown in figure 9.9. The white material is probably a crust of ice, while the red material is probably mineral-rich water that has oozed to the surface through the cracks and frozen. The absence of large craters on its surface suggests that some process has eradicated them, for Europa's surface—like that of Ganymede and Callisto—must

FIGURE 9.9
Picture of cracks on Europa. (Courtesy NASA/JPL.)

☀ Look up the radius of Europa in appendix 5 and estimate the width of a large crack.

surely have been cratered during and following its formation. To explain the lack of craters, astronomers think that Europa may be heated by Jupiter's gravitational forces, although not as strongly as Io is. That heat, in combination with a small amount from radioactive decay of rocky material in its core, may be sufficient to keep a layer of water melted beneath Europa's crust. Moreover, the heat may soften the surface ice, allowing it to "flow," glacier-like, and obliterate craters as they form.

Ganymede and Callisto look somewhat like our own Moon, being basically grayish brown and covered with craters made during the late stages of their formation, much as our own Moon was pockmarked by infalling debris. But these similarities to our Moon are only superficial, because the surfaces of Ganymede and Callisto are probably mostly ice, the white of the craters being similar to the white you see on ice cubes when they are shattered. Ganymede's surface is less heavily cratered than Callisto's, implying that craters in the smoother areas have been destroyed, perhaps much as happened on our own Moon. That is, astronomers think that after Ganymede's surface solidified and was cratered by infalling debris, a few large bodies may have hit it. These large bodies created basins that subsequently flooded with water and then froze into a relatively smooth surface with few craters. However, the smooth regions do show curious parallel ridges, possibly created by tectonic forces as the water in them froze.

From the average density of the Galilean satellites, astronomers deduce that their interiors are composed mainly of rocky material. Heating of these moons during their youth may have allowed iron in them to sink to their centers and form cores, as happened with our own planet. Indirect evidence for just such a dense core in Ganymede comes from the *Galileo* spacecraft orbiting Jupiter. As the craft swings past that moon, its trajectory is slightly altered by Ganymede's gravity in a way suggesting that Ganymede has an ice mantle surrounding an iron-rich core. Such a core also explains why this moon has a magnetic field. The rest of Jupiter's moons are much smaller than the Galilean satellites, but they, too, are heavily cratered. The orbits of the outermost of these moons are steeply tilted relative to the others, suggesting that these objects may be captured asteroids.

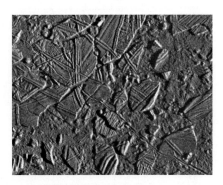

Ice floes (?) on Europa (Courtesy NASA/JPL).

9.2 SATURN

Saturn is the second largest planet and lies about 10 AU from the Sun. Surrounded by its lovely rings (illustrated in fig. 9.10), Saturn bears the name of an ancient Roman harvest god. In later mythology, Saturn came to be identified with Cronus (also spelled Kronos), whom the ancient Greeks considered the father of the gods.

Saturn's Appearance and Physical Properties

Saturn's diameter, like Jupiter's, is about 10 times larger than the Earth's diameter. Saturn and Jupiter therefore have similar volumes. Their masses differ considerably, however: Saturn's mass is about 95 times that of the Earth, or about one-third Jupiter's mass. Because Saturn's mass is so much smaller than Jupiter's but its volume is about the same, Saturn's average density is very small—only 0.7 grams per cubic centimeter, which is less than the density of water. Such a low density suggests that Saturn, like Jupiter, is composed mostly of hydrogen and hydrogen-rich compounds. Spectra of the planet bear this out, and astronomers believe these two giant planets have very similar compositions and internal structures, as depicted in figure 9.11.

Saturn radiates more energy than it gains from the Sun, implying that, like Jupiter, it has an internal heat source. Astronomers think that Saturn's heat does not come from its slow gravitational contraction, however, as most of Jupiter's does. Instead, they hypothe-

FIGURE 9.10
Saturn as pictured by the *Voyager* spacecraft. (Courtesy NASA/JPL.)

☀ What is the dark spot on the lower portion of Saturn's disk? Why does the ring look broken as it goes behind Saturn at the upper right?

FIGURE 9.11
Internal structure of Saturn.

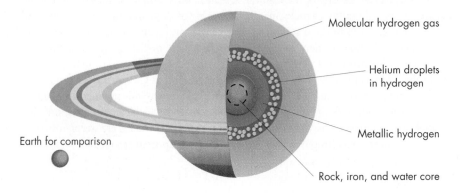

Molecular hydrogen gas

Helium droplets in hydrogen

Metallic hydrogen

Earth for comparison

Rock, iron, and water core

size that deep beneath Saturn's cold clouds, helium droplets condense in its atmosphere, much as water droplets condense in Earth's atmosphere. As the helium droplets fall toward Saturn's core, they release gravitational energy that heats the planet's interior.

If both Jupiter and Saturn have hot interiors and similar compositions, why do they look so different externally? In particular, why does Saturn show only faint cloud belts and markings compared with the striking patterns seen on Jupiter? Saturn's greater distance from the Sun and its consequently lower temperature provide an answer. Saturn's atmosphere is cold enough for ammonia gas to freeze into cloud particles that veil its atmosphere's deeper layers, making markings below the clouds indistinct.

Saturn's Rings

Saturn's spectacular rings, illustrated in figure 9.12, were first seen by Galileo. Through his small, primitive telescope, however, they looked like "handles" on each side of the planet, and it was not until 1659 that Christiaan Huygens, a Dutch scientist, observed that the rings were detached from Saturn, and encircled it.

The rings are very wide but very thin. The main band extends from about 30,000 kilometers above the top of Saturn's atmosphere to a little more than twice the planet's radius (136,000 kilometers, or about 84,000 miles), as illustrated in figure 9.12. Some faint inner rings can be seen even closer to Saturn, and faint outer rings extend considerably further from the planet. Yet despite the rings' immense breadth, they are probably less than a few hundred meters thick—thin enough to allow stars to be seen through them.

The British physicist James Clerk Maxwell, a pioneer in the study of electromagnetism, demonstrated that the rings must be a swarm of particles. He showed mathematically that no material could plausibly be strong enough to hold together in a solid sheet of such vast size. Spectra of the rings support Maxwell's theory: the inner and outer parts orbit Saturn at different velocities, obeying Kepler's third law, as shown by their Doppler shift. Thus the rings must be a swarm of individual bodies.

Astronomers have since discovered that the ring particles are relatively small, only a few centimeters to a few meters across. Although these particles are far too small to be seen individually with telescopes, they reflect radar signals bounced off them. From the strength of the radar "echo," astronomers can estimate the particle sizes. Recently, more precise measurements were made with radio signals from the *Voyager* spacecraft. As it transmitted data to Earth, the signals passed through the rings, scattering slightly from the particles. From the amount of scattering of the *Voyager* signal and the radar waves, astronomers can deduce not only the size of the ring particles but also their composition. Better information about the composition of the rings, however, comes from analyzing the spectrum of

FIGURE 9.12
Rings of Saturn. (Courtesy NASA/JPL.)

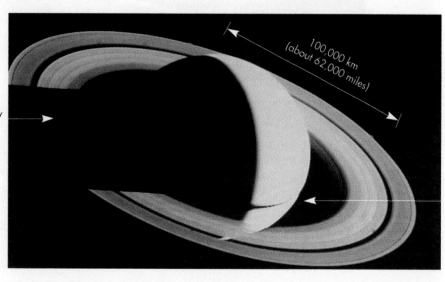

Saturn's shadow on rings

Shadow of rings on Saturn

their light. Such analyses show that the rings are composed primarily of water and ice. However, the *Voyager* spacecraft pictures, such as figure 9.13, show that some parts of the rings are much darker than others, implying that the composition of the rings is not the same everywhere. Particles in the darker ring segments may be rich in carbon compounds similar to those found in some asteroidal material. Figure 9.13 also shows that the rings are not uniformly filled: they consist of numerous separate ringlets. Large gaps in the rings had been seen from Earth, but the many narrow gaps in the rings came as a surprise. What causes these gaps that create the multitude of ringlets?

As long ago as 1866, Daniel Kirkwood, the astronomer who discovered the gaps in the distribution of asteroids, noticed that the largest gap—known as Cassini's division—occurs where ring particles orbit Saturn in exactly one third the time (⅓ the orbital period) of its moon, Enceladus. Thus, any ring particle that attempted to orbit in the gap would undergo a strong and repeated gravitational force from Enceladus every third orbit. Kirkwood concluded that, over long periods of time, the cumulative effect of Enceladus's force would pull particles from the gap. He therefore hypothesized that Enceladus's gravitational attraction creates Cassini's division, just as Jupiter's creates the gaps in the asteroid belt. A few years later, Kirkwood revised his theory of the gap to account for the action of the four largest Saturnian moons. Today, astronomers believe that Mimas, whose period is twice that of particles in Cassini's division, causes that large gap, but they think the many narrow gaps apparent in figure 9.13 have a different cause.

Narrow gaps in the rings probably arise from a complex interaction between the ring particles and the tiny moons that orbit within the rings. As these moonlets—only tens of kilometers or less in size—orbit Saturn, their gravitational attraction on the ring particles generates waves. These waves spread through the rings much like ripples in a cup of coffee that is lightly tapped. Such ripples are circular in a cup or on a still pond, but in a planetary ring system, they take a different form. Because the inner part of the ring orbits faster than the outer part—a consequence of Kepler's laws—the spreading waves wrap into a tightly wound pattern called "spiral density waves." The crests of these density waves form the narrow rings.

Moonlets may generate gaps within the rings in yet another way. If two moonlets move along orbits that lie very close together, their combined gravitational force may

FIGURE 9.13
False-color photographs of the ringlets, showing the substructure of Saturn's rings. (Courtesy NASA/JPL.)

deflect ring particles into a narrow stream between them. Such "**shepherding satellites**" occur not only in Saturn's rings but also in Uranus's. But what created planetary rings in the first place?

Origin of Planetary Rings

Earlier in this century, some astronomers thought that planetary rings were material left over from a planet's formation, perhaps matter that had failed to condense into a satellite. But they now realize that rings are short-lived because they are subject to forces in addition to gravity. For example, gas trapped in a planet's magnetic field may exert a force on the ring particles, gradually causing them to spiral into the planet's atmosphere, as may happen to the material in Jupiter's ring. Thus new material must be added to the rings from time to time, for without such replenishment, they would disappear in a few million years. One source of new material is the satellites orbiting the planet. A moon in a satellite system as complex as those of Jupiter and Saturn is subject not only to the gravitational forces of the planet, but also to that of the other moons. The cumulative effect of such forces alters the satellites' orbits and may lead to collisions between them. Alternatively, a moon may drift so close to its planet that tidal forces break the moon apart, a theory first suggested more than a century ago.

The Roche Limit

In 1849, the French scientist M. E. Roche (pronounced /rohsh/), while studying the problem of a planet's gravitational effect on its moons, demonstrated mathematically that if a moon gets too close to its planet, the planet's gravity could rip the moon apart. This disruption occurs because the planet pulls harder on one side of the satellite than the other. If the difference in this pull exceeds the moon's own internal gravitational force, the moon will be pulled apart, as shown in figure 9.14. Thus if a moon—or any body held together by gravity—approaches a planet too closely, the planet raises a tide so large it pulls the encroaching object to pieces. Roche calculated the distance at which the tide becomes fatally large and showed that for a moon and planet of the same density, breakup occurs if the moon comes nearer to its planet than 2.44 planetary radii, a distance now called the **Roche limit**. All planetary rings lie near their planet's Roche limit, suggesting that most rings are caused by satellite disruption. But moons are not the only bodies that can stray into the danger zone. Asteroids or comets may occasionally pass too close to a planet, and shattered fragments of such bodies may help keep rings filled.* The existence side-by-side

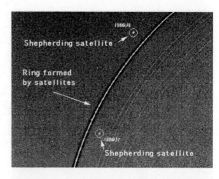

Two shepherding satellites (the dots inside the small circles) and portion of the narrow ring they created. The picture's contrast is strongly enhanced to show these faint moons. (Courtesy NASA/JPL.)

Comet Shoemaker-Levy 9. This image shows the 20 or so fragments into which it was broken by Jupiter's tidal force. (Courtesy STSCI/HST.)

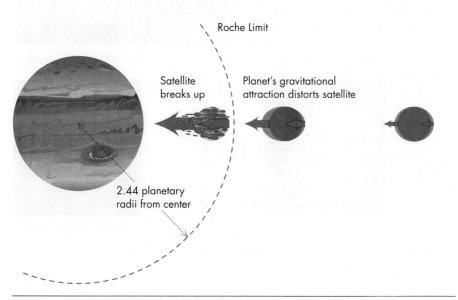

Roche Limit

Satellite breaks up

Planet's gravitational attraction distorts satellite

2.44 planetary radii from center

FIGURE 9.14
The Roche limit. A planet's gravity pulls more strongly on the near side of a satellite than on its far side, stretching it and, if strong enough, pulling it apart.

---

*In 1992 Comet Shoemaker-Levy 9 passed extremely close to Jupiter and was broken into at least 22 large fragments. These pieces struck Jupiter in July 1994.

An infrared picture of Titan that penetrates its clouds and shows surface features. The light areas may be a "continent" surrounded by seas of liquid nitrogen or ethane. (Courtesy P. Smith and M. Lemmon U. Ariz: EONA/STSCI.)

of ringlets with different compositions (some rich in ice, others rich in carbon) is additional evidence that the rings formed by the breakup of many different small objects.

The Roche limit applies only to bodies held together by gravity, however. Artificial satellites or small bodies bonded together by chemical forces can pass safely through the Roche limit without effect.

Saturn's Moons

Saturn has several large moons and about a dozen known smaller ones. Most of these bodies, like Jupiter's moons, orbit in a flat "mini–Solar System" aligned with Saturn's equator. Saturn's moons have a smaller average density than the Galilean satellites of Jupiter, from which astronomers deduce that their interiors must be mostly ice. Moreover, all are about the same density, implying that they were not strongly heated by Saturn as they formed. (Recall that the Galilean satellites nearest Jupiter have a higher density than those farther away, presumably a result of Jupiter's heat driving away much of the icy material from the nearer bodies, leaving them with a higher proportion of rocky material.)

Titan, the largest Saturnian moon, has a diameter of about 5000 kilometers (3000 miles), making it slightly bigger in diameter than the planet Mercury and comparable in

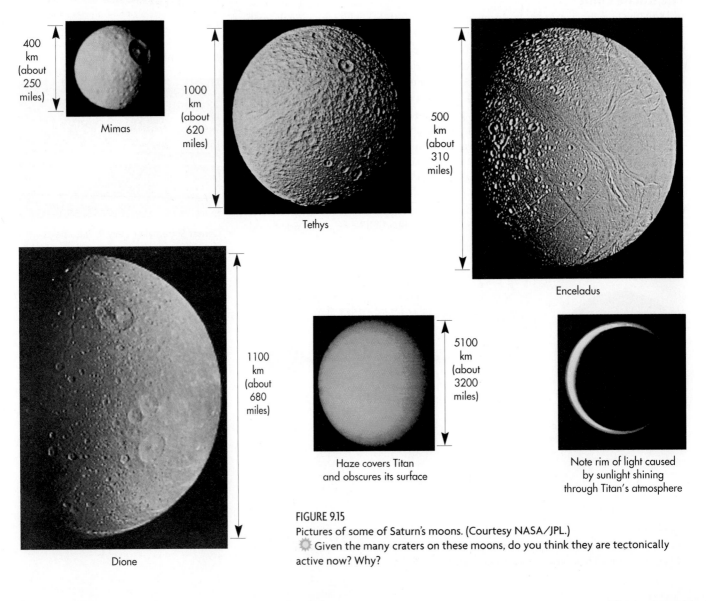

400 km (about 250 miles)
Mimas

1000 km (about 620 miles)
Tethys

500 km (about 310 miles)
Enceladus

1100 km (about 680 miles)
Dione

5100 km (about 3200 miles)
Haze covers Titan and obscures its surface

Note rim of light caused by sunlight shining through Titan's atmosphere

FIGURE 9.15
Pictures of some of Saturn's moons. (Courtesy NASA/JPL.)
☀ Given the many craters on these moons, do you think they are tectonically active now? Why?

mass and radius with Jupiter's large moons Ganymede and Callisto. Because Titan is farther from the Sun than these bodies, it is much colder. Thus, as gas molecules leak from its interior, they move relatively slowly and are unable to escape Titan's gravitational attraction. This immense moon therefore possesses its own atmosphere, which spectra show to be mostly nitrogen. Clouds in Titan's atmosphere perpetually hide its surface. On the basis of chemical calculations, however, some astronomers believe that Titan's surface is covered with oceans of liquid nitrogen or of the hydrocarbon ethane (C_2H_6), or both, and that ethane "rain" may fall from its clouds.

Figure 9.15 is a picture gallery of Saturn's moons taken by the *Voyager* spacecraft. Notice that most of these bodies are heavily cratered, implying that they have been extensively bombarded by infalling bodies. The smoother surface seen, for example, on Enceladus suggests that water has erupted from its interior, flooding old craters and drowning them as it freezes. The white markings surrounding many of the craters are probably shattered ice. Extremely dark material, however, coats one side of Iapetus, as illustrated in figure 9.16. The dark side faces the direction in which Iapetus travels, leading some astronomers to think that as it orbits Saturn, Iapetus runs into debris that has coated its leading face with carbon-rich material. Alternatively, the dark material may be dust left behind when tiny bodies strike it and vaporize the ice on its surface. Yet another possibility is that Iapetus has erupted some dark substance from its interior. Thus these distant and alien moons still have many unsolved mysteries.

9.3 URANUS

Uranus, although small compared with Jupiter and Saturn, is much larger than the Earth. Its diameter is about four times that of the Earth, and its mass is about 15 Earth masses. Lying approximately 19 AU from the Sun (more than twice Saturn's distance), Uranus is difficult to study from Earth, visible only as a blue but featureless disk. Even pictures of it taken by the *Voyager* spacecraft show few details (fig. 9.17), although computer processing of the images shows that it has faint cloud bands.

Uranus was unknown to the ancients, even though it is just visible to the naked eye. It was discovered by Sir William Herschel, a German émigré to England. Herschel, a musician, was at the time only an amateur astronomer interested in hunting comets, a task in

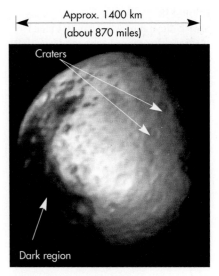

Approx. 1400 km
(about 870 miles)

Craters

Dark region

FIGURE 9.16
Iapetus, Saturn's third largest satellite. The huge dark area on one side of it may be dust from Phoebe, an extremely dark Saturnian moon that orbits near Iapetus. Such dust might be debris ejected from Phoebe when small asteroids or meteorites strike its surface. (Courtesy NASA/JPL.)

Earth for comparison

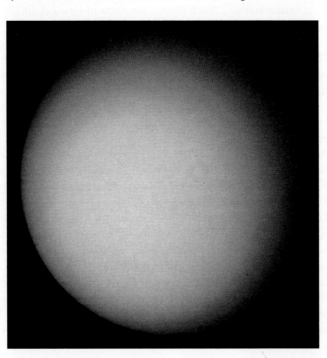

FIGURE 9.17
Uranus as pictured from the *Voyager* spacecraft. Because of Uranus's odd tilt (recall fig. 7.2), we are viewing it nearly pole-on. Note the lack of clearly defined cloud belts. (Courtesy NASA/JPL.)

FIGURE 9.18
Sketch illustrating why Uranus is blue. Methane absorbs red light, removing the red wavelengths from the sunlight that falls on the planet. The surviving light— now missing its red colors—is therefore predominantly blue. As that light scatters off cloud particles in the Uranian atmosphere and returns to space, it gives the planet its blue color.

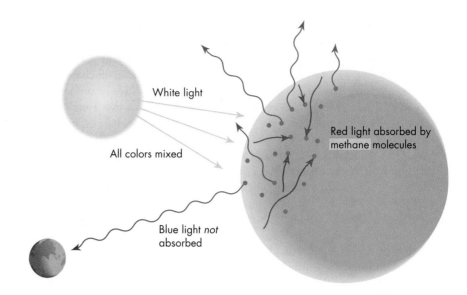

White light

All colors mixed

Red light absorbed by methane molecules

Blue light *not* absorbed

which he collaborated with his sister Caroline. In 1781, he observed a pale blue object whose position in the sky changed from night to night. Herschel at first thought he had discovered a comet, but observations over several months showed that the body's orbit was nearly circular, and he therefore concluded that he had found a new planet. For this discovery, King George III named Herschel his personal astronomer, and to honor the king, Uranus was briefly known as "Georgium Sidus," or 'George's Star.' In ancient Greek mythology, Uranus was the father of Cronus and was identified as the God of the Heavens.

Uranus's Atmosphere
Uranus's spectrum shows that its atmosphere is rich in hydrogen and methane. In fact, methane gives the planet its deep blue color. When sunlight falls on Uranus's atmosphere, the methane gas strongly absorbs the red light. The remaining light, now blue, scatters from cloud particles in the Uranian atmosphere and is reflected into space, as depicted in figure 9.18. The cloud particles that cause the scattering are thought to be primarily crystals of frozen methane. Such crystals can form in Uranus's atmosphere because, being so far from the Sun, it is extremely cold.

Some astronomers speculate that in the deep interior of Uranus, the pressure is large enough to compress methane to a state in which it releases some of its carbon. That carbon, itself under high pressure, may then form diamonds within Uranus.

Uranus's Interior
Astronomers must rely on indirect methods to study the interior of Uranus, using, for example, its density and shape. From its mass and radius, astronomers can calculate that Uranus has an average density of about 1.2 grams per cubic centimeter, somewhat greater than that of Jupiter or Saturn, implying that Uranus contains proportionately less hydrogen than either of these larger bodies. On the other hand, the density is too low for Uranus to contain much rock or iron material. Astronomers therefore believe it must be composed of material that is light and abundant, such as ordinary water mixed with methane and ammonia. This mix satisfactorily explains both the density and the spectrum of Uranus. Confirmation of the abundance of water also comes from studies of the planet's shape. Uranus, like Jupiter and Saturn, rotates moderately fast, with its equator rotating faster than its poles. At its equator, Uranus spins once every 17 hours, bulging the planet's equator. The size of such a bulge depends in part on the planet's gravitational attraction and therefore on how the mass generating that attraction is distributed inside the planet. Thus astronomers can deduce the density and composition deep inside Uranus from the size of its bulge. Such studies are consistent with the hypothesis that Uranus is composed of mostly water and hydrogen-rich gases and that it may have a core of rock and iron-rich material, as illustrated in figure 9.19. It is not known whether the core formed first and then attracted the lighter gases and ices that condensed around it, or whether

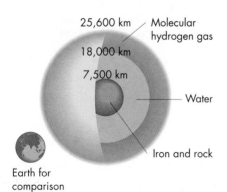

25,600 km — Molecular hydrogen gas

18,000 km

7,500 km

Water

Iron and rock

Earth for comparison

FIGURE 9.19
Artist's view of a recently suggested model for the interior of Uranus. Note how different it is from Jupiter and Saturn, which both contain large regions of liquid and metallic hydrogen.

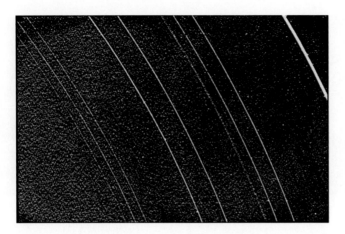

FIGURE 9.20
Photograph of a section of the rings of Uranus. (Courtesy NASA/JPL.)

Cliffs
approx. 20 km–more than 12 miles–high

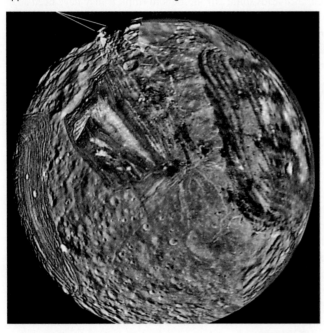

Enlargement
of cliff region

FIGURE 9.21
Miranda, an extremely puzzling moon, as observed by the *Voyager* spacecraft. Note the enormous cliffs glinting in the sunlight at the top of the picture. (Courtesy NASA/JPL.)

the core formed by heavy material sinking to the center after the planet formed. In fact, some astronomers think Uranus's core may be simply highly compressed ices with little rocky material.

Uranus's Rings and Moons

Uranus is encircled by a set of narrow rings, as illustrated in figure 9.20. The rings, like those of Saturn, are composed of a myriad of small particles, perhaps a meter or so in diameter, moving in individual orbits. The Uranian rings are very dark, however, implying that they are not made of, or coated with, ice like the bright rings of Saturn; instead, they may be rich in carbon particles or organic-like material. Uranus's rings are also extremely narrow compared with those of Saturn. According to one theory, the ring particles are held in such narrow zones by "shepherding satellites," as discussed for Saturn's rings.

Uranus has five large moons and about 10 smaller ones. Like the moons of Jupiter and Saturn, they form a regular system and are probably composed mainly of ice and rock. Many of the Uranian moons are heavily cratered, but Miranda, the smallest of the five large moons, has a surface totally unlike that of any other Solar System body (fig. 9.21).

The five large Uranian moons are named for characters in English literature. Titania and Oberon are the queen and king of the fairies in Shakespeare's A Midsummer Night's Dream. *Ariel and Miranda are characters in Shakespeare's* The Tempest. *Umbriel is a sprite in Alexander Pope's poem "The Rape of the Lock," in which Ariel also appears. The smaller Uranian moons are also named for characters from Shakespeare, such as Puck and Cordelia.*

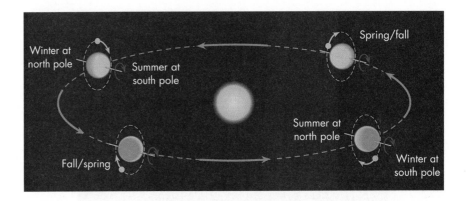

The surface is broken into distinct areas that seem to bear no relation to one another. One region is wrinkled, while an adjacent one has small hills and craters, rather like our Moon. Miranda's patchwork appearance leads some astronomers to think it may have been shattered by impact with another large body. The pieces were subsequently drawn back together by their mutual gravity, giving this peculiar moon its jumbled appearance. Alternatively, the curious surface might have been caused by rising and sinking motions in Miranda's interior. Regardless of the cause, Miranda has some extremely curious and unexplained surface features such as a set of cliffs, visible in the top corner of figure 9.21, that are twice the height of Mt. Everest.

Uranus's Odd Tilt

Uranus's rotation axis is tipped about 90° with respect to its orbital plane, so that its equator is nearly perpendicular to its orbit. That is, it spins nearly on its side, as illustrated in figure 9.22. Moreover, the orbits of Uranus's moons are similarly tilted. They orbit Uranus in its equatorial plane and, as a result, their orbits are also tilted at approximately 90° with respect to the planet's orbit. Some astronomers therefore hypothesize that during its formation, Uranus was struck by an enormous planetesimal whose impact tilted the planet and splashed out material to create its family of moons.

The strong tilt of its axis gives Uranus an odd pattern of day and night. For part of its orbit, one pole is in "perpetual" day and the other pole is in "perpetual" night. Thus sunlight heats the planet very unevenly, perhaps explaining why Uranus lacks the cloud bands seen on the other giant, gaseous planets. The lack of cloud bands may be temporary, however. Recent observations with an infrared telescope on Earth reveal an odd dark marking in Uranus's atmosphere, perhaps a feature that fades and reappears.

9.4 NEPTUNE

Neptune is the outermost of the large planets and is very similar to Uranus in size, with a diameter about 3.9 times that of the Earth and a mass about 17 times the Earth's. Through a telescope on the Earth, it looks like a chip of sapphire, a lovely blue color, but because of its great distance from the Sun—about 30 AU—it is difficult to study. Pictures taken of it by passing spacecraft show it to be a deep blue world, reminiscent of Jupiter, as shown in figure 9.23. Cloud bands encircle it, and it even has briefly shown a Great "Dark" Spot, a huge dark blue atmospheric vortex similar to Jupiter's Great Red Spot.

Neptune, named for the Roman god of the sea, was discovered in the 1840s from predictions made independently by a young English astronomer, John Couch Adams, and a French astronomer, Urbain Leverrier. Adams and Leverrier both noticed that Uranus was not precisely following its predicted orbit, and they therefore inferred that its motion was being disturbed by the gravitational force of an as yet unknown planet. From the size of these orbital disturbances, Adams and Leverrier predicted where the unseen body must lie.

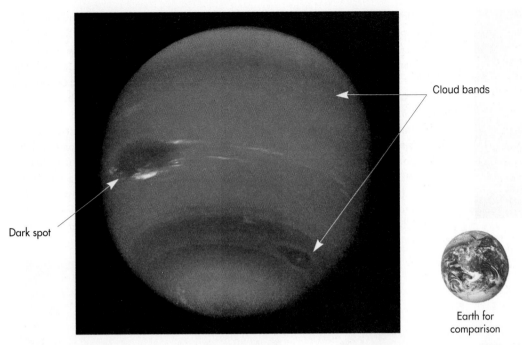

Cloud bands

Dark spot

Earth for
comparison

FIGURE 9.23
Photograph of Neptune taken by the
Voyager spacecraft. Notice the dark blue
oval, an atmospheric vortex whose origin
is probably similar to that of Jupiter's Red
Spot. (Courtesy NASA/JPL.)

Approximately how big is this spot
compared to the Earth?

Adams completed his calculations in 1845, but when he reported his results, the astronomer royal, Sir George Airy, was unconvinced and gave a low priority to the search for the unseen planet. In 1846, however, Airy was startled to read a paper by Leverrier detailing calculations nearly identical to those made by Adams. This spurred Airy to begin a search in earnest, but by then it was too late: Leverrier had given his predicted positions to Johann Galle, a German astronomer who that same night pointed his telescope to the predicted location and saw Neptune. Assignment of credit for the discovery of the new planet led to a rancorous dispute tinged with national pride that lasted decades. The discovery is now credited equally to Adams and Leverrier. Ironically, Galileo had seen Neptune in 1613 while observing Jupiter's moons. His observation notes record a dim object whose position changed with respect to the stars, as would be expected for a planet. Galileo failed, however, to appreciate the significance of that motion, so Neptune eluded discovery for another two centuries.

Neptune's Structure
Neptune's structure is probably similar to Uranus's. That is, the planet is composed mostly of ordinary water surrounded by a thin atmosphere rich in hydrogen and hydrogen compounds, such as methane. Our understanding of the atmosphere's composition is derived from the planet's spectrum. The composition of the interior is inferred from the planet's low average density, which is 1.67 grams per cm^3. As we have seen for the other giant planets, such a low density implies that the planet must be composed mostly of light atoms. But during its formation, Neptune must also have collected heavy elements such as silicon and iron. These denser materials probably sank to the center of Nepture (just as they did in the other giant planets), where they now form a core of rock and iron a little smaller than the Earth.

Neptune's Atmosphere
Neptune's blue color is caused, like Uranus's, by methane in its atmosphere. Unlike Uranus, however, Neptune has distinctive cloud belts. Why do these two so similar bodies have such different cloud formations? Infrared observations offer a partial explanation. Neptune, like Jupiter and Saturn, radiates more energy than it gains from the Sun. That

Curiously, recent pictures of Neptune taken by the Hubble Space Telescope show that the spot is gone! Does it grow and disappear like terrestrial storms or did Voyager's camera just happen to catch a rare event? Yet another mystery about these outer worlds! (Courtesy STSCI).

excess energy (also like Jupiter's and Saturn's) is probably left over from the planet's birth or is supplied by heavier material sinking even now to the planet's core. Whatever its source, the heat deep within Neptune generates convection currents that rise to its outer atmosphere. There, the rising gas is deflected into a system of winds by the Coriolis effect caused by Neptune's spin—one rotation every 16 hours. The resulting winds create cloud bands similar to those seen on Jupiter and Saturn, but tinted deep blue by Neptune's methane-rich atmosphere.

Neptune's winds are extremely fast. For example, near its equator, strong easterly winds blow opposite to the planet's direction of rotation, much like the trade winds on Earth. But these Neptunian winds reach speeds of nearly 2200 kilometers per hour (about 1300 miles per hour). As these gales sweep around Neptune, gas sandwiched between adjacent streams spins, as described in our discussion of Jupiter's atmosphere. Such localized spinning motion creates the Great Dark Spot, as the large, dark blue spot illustrated in figure 9.24 is called. Why are such winds absent on Uranus? One explanation is that being nearer the Sun

Earth for
comparison

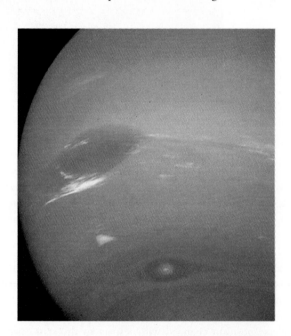

FIGURE 9.24
Photograph of the Great Dark Spot of Neptune, showing some lighter clouds in its vicinity. (Courtesy NASA/JPL.)

FIGURE 9.25
Rings of Neptune. The black band blocks the light reflected from Neptune that would otherwise overwhelm and make invisible the much fainter light of the rings in the picture. (Courtesy NASA/JPL.)

than Neptune, Uranus's upper atmosphere is warmer than Neptune's. Being warmer outside, Uranus thus has a smaller temperature difference between its core and surface than Neptune does. With less temperature difference, Uranus has weaker convection, making it a duller-looking planet.

Neptune's Rings and Moons

Neptune, like the other giant planets, has rings, but they are very narrow, more like those of Uranus than those of Saturn, as shown in figure 9.25. They are probably composed of debris from small satellites or comets that have collided and broken up, and they contain proportionally more dust than the rings of either Saturn or Uranus. Moreover, photographs of the rings show that in some places, the ring particles are not distributed uniformly around the ring but are gathered into arcs in some places. This clumping—probably the result of "shepherding satellites"—leads astronomers to think that Neptune's rings may be relatively short-lived and that they may dissipate in a million years or perhaps even faster unless new ring material is added from the breakup of other small bodies.

Neptune has six small moons orbiting close to the planet in roughly circular orbits and two other moons at much greater distances. One of these two, Triton, is immense and orbits "backwards" (counter to Neptune's rotation) compared with the motion of most other satellites). Moreover, its orbit is highly tilted with respect to Neptune's equator (most large moons follow orbits approximately parallel to their planet's equator). These orbital peculiarities lead many astronomers to think that Triton is a surviving icy planetesimal from the inner Kuiper belt. When Neptune captured it, the encounter destroyed or expelled any mid-size moons that Neptune originally possessed. In fact, Neptune's outermost moon, Nereid, follows such a highly elliptical path far beyond the other moons that it just barely missed escaping from its parent world.

Triton intrigues astronomers for more than its orbital oddity. It is massive enough that its gravity, in combination with its low temperature, allows it to retain gases around it. Triton is thus one of the few moons in the Solar System with an atmosphere.

Triton's surface also has many unusual features, as shown in figure 9.26. Wrinkles give much of the surface a texture that looks like a cantaloupe. Craters pock the surface elsewhere, and dark streaks extend from some of them. Some astronomers believe that these streaks are matter ejected by geysers of nitrogen, ice, and carbon compounds. Sunlight, pale though it is at Triton's immense distance from the Sun, may warm gases trapped below the surface, making them expand and burst through surface cracks. The erupted material cools and condenses in Triton's cold, thin atmosphere, where winds carry and deposit it as a black "soot."

Neptune's moons are named for mythological sea deities, in keeping with the planet's own name.

FIGURE 9.26
The surface of Triton. (Courtesy NASA/JPL.)

Wrinkled terrain

Windblown volcanic debris

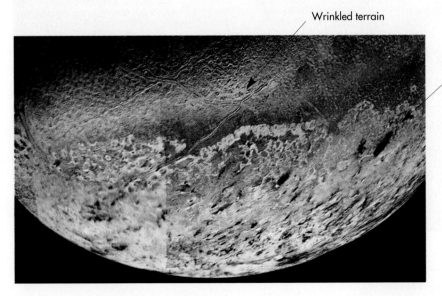

A

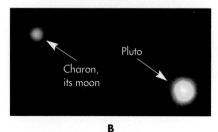

Charon,
its moon

Pluto

B

FIGURE 9.27
(A) Pluto looks like merely another dim star in this photograph taken at Lick Observatory in California. Only its change in position from night to night shows it to be a planet. (© UC Regents; UCO/Lick Obs image.)
(B) A picture of Pluto and its moon Charon obtained with the Hubble Space Telescope and therefore free of the blurring effects of our atmosphere. (Courtesy NASA.)

☀ Why does the position of Pluto change but not that of the stars?

FIGURE 9.28
Maps of Pluto and its moon Charon, generated by computer analysis of Hubble Space Telescope images. (Courtesy STSCI.)

9.5 PLUTO

Pluto, named for the Greek and Roman god of the underworld, was the last of the nine planets to be discovered. It was found in 1930 by Clyde Tombaugh, an astronomer at Lowell Observatory in Flagstaff, Arizona. Tombaugh painstakingly examined pairs of photographs of the sky for over a year—scanning millions of star images—searching for objects whose position changed between the exposures, the telltale motion that distinguishes a planet from a star.

Pluto's great distance from the Sun combines with its small size to make it very dim. Even in the largest telescopes on the ground Pluto looks like a dim star, as illustrated in figure 9.27. For years after its discovery, astronomers knew only that Pluto was smaller than the Earth. Then, in 1978, James Christy of the U.S. Naval Observatory discovered a moon orbiting Pluto, a moon that now makes it possible for astronomers to measure Pluto's radius and mass.

Pluto's moon, Charon, is named for the boatman who, in mythology, ferries dead souls across the river Styx to the underworld. Circling Pluto in a tiny orbit whose average radius is 19,600 kilometers, Charon takes only 6.4 days to complete a trip around the planet. From this orbital data, Pluto's mass can be calculated using Newton's modified form of Kepler's third law of planetary motion (see fig. 7.5). Such a calculation shows that Pluto's mass is about 0.002 times the Earth's, making Pluto by far the least massive planet. Observations of Charon's motion around Pluto show that its orbit is tilted steeply with respect to Pluto's orbital plane and that Pluto's rotation axis is similarly tipped, reminiscent of Uranus's odd tilt.

As Charon orbits Pluto, it occasionally* eclipses the planet from our view. From these eclipses, the diameters of both Pluto and Charon can be measured. Pluto's diameter turns out to be a little less than one fifth the Earth's, and Charon, though smaller, turns out to be surprisingly large. Satellites are generally dwarfed by their planets, but Charon is slightly more than half Pluto's diameter. Moreover, its orbit is locked in synchrony with Pluto's rotation, suggesting a tight gravitational coupling between the bodies. Such eclipse studies are not the only way to study Pluto and Charon, however. Observations of them made with interferometers (see chapter 5) and from the Hubble Space Telescope show Pluto's and Charon's tiny disks, nicely confirming the eclipse data.

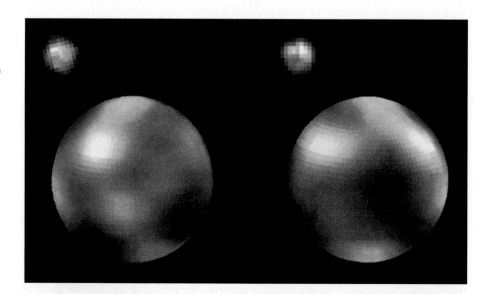

---

*Every 124 years!

From Pluto's mass and radius, astronomers can deduce its density to be about 2.1 grams per cm^3, a value suggesting that the planet must be a mix of water, ice, and rock. They know little, however, about its surface. Although astronomers have photographed Pluto and its moon, Charon, with the Hubble Space Telescope, the best "pictures" of this distant system come from computer analysis of its changing brightness during eclipses by Charon. These images (fig. 9.28) show that Pluto's south pole is much brighter than its equator, implying the presence of a polar cap there. Spectra suggest that the cap is frozen methane.

Astronomers have also detected a very tenuous methane atmosphere on Pluto. In the bitter cold (40 K, or about $-387\ °F$) of this remote planet, molecules move so slowly that, despite its tiny mass, Pluto's gravity is strong enough to retain a thin atmosphere.

Pluto's small size and peculiar orbit (it crosses Neptune's orbit, as shown in fig. 9.29) once led some astronomers to hypothesize that it was originally a satellite of Neptune that escaped and now orbits the Sun independently.

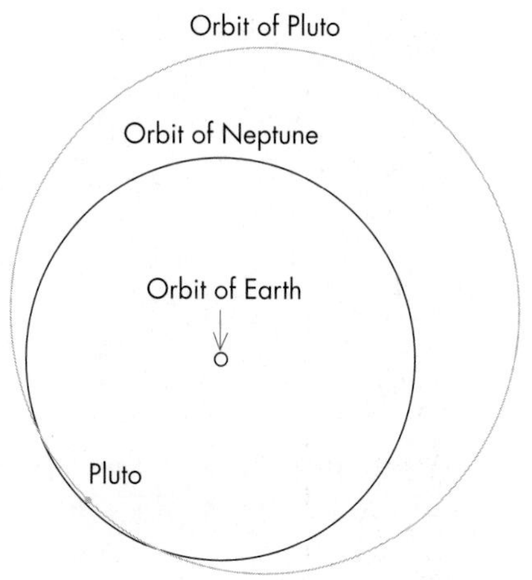

 RE-MODELING

Today however, astronomers believe almost the reverse—that Neptune has "captured" Pluto. Pluto's orbital period is 247.7 years, very close to $^3/_2$ of Neptune's. Thus, Pluto makes two orbits around the Sun for every three made by Neptune. This match of orbital periods may have created a cumulative gravitational attraction on Pluto that "tugged" it into its current orbit. In fact, some dozen other much smaller objects, presumably icy objects a few hundred kilometers in diameter, orbit at nearly the same distance from the Sun as Pluto. As we will see in the next chapter, these remote icy worlds are probably remnant planetesimals, survivors from the birth of the Solar System.

Orbit of Pluto

Orbit of Neptune

Orbit of Earth

Pluto

FIGURE 9.29
Pluto's odd orbit is highly tilted and highly eccentric (elongated)—so much so that it crosses Neptune's orbit.

If Pluto's orbit crosses Neptune's, why don't they collide? Referring to figure 7.2 may help you answer the question.

SUMMARY

The outer planets formed in the solar nebula far from the Sun, where it was cold enough for them to accumulate hydrogen-rich material. Because of hydrogen's abundance relative to other elements, Jupiter, Saturn, Uranus, and Neptune grew much larger than the terrestrial planets in both diameter and mass and are composed mostly of hydrogen and its compounds, such as methane (CH_4), ammonia (NH_3), and water (H_2O). These gases form a deep atmosphere, often richly colored, that grows denser with depth and eventually becomes liquid. The interiors of Jupiter and Saturn are composed mainly of hydrogen—liquid just below the atmosphere and metallic deeper down. Each is believed to have a rocky core whose diameter—within a factor of 2 or 3—is comparable to the Earth's. Uranus and Neptune may have some liquid hydrogen below their hydrogen- and methane-rich atmospheres, but they are probably mostly water, ammonia, and methane surrounding a rocky core. Because they are smaller bodies, their weaker gravity has not allowed them to capture or retain as much hydrogen and helium, but they have retained the heavier gases.

The giant planets are probably heated by continued gravitational contraction and settling of heavier matter toward their cores. As this heat flows outward, it generates convective motions that stir them. The planets' rotation creates a Coriolis effect on the rising gas, drawing it into the cloud belts seen on Jupiter, Saturn, and Neptune.

Rotation also creates an obvious equatorial bulge on Jupiter and Saturn, which spin in about 10 hours. The spinning motion, when combined with convection in their metallic hydrogen cores, generates powerful magnetic fields that in turn contribute to the formation of auroras and energetic, charged particle belts like the Van Allen belts of our Earth.

Each of the four giant planets has a ring system composed of small orbiting particles. The rings are probably debris from small objects that have broken up either as the result of collisions or of tidal forces exerted by the planet. Collisions continue to break up small particles, replacing ring material that falls into the planets' atmospheres.

Many satellites orbit each of the four giant planets. Some of these moons are comparable in diameter to Mercury. They are thought to be made mostly of water, ice, and rock, although Io, an exceptionally active moon of Jupiter, may be mostly sulfur and rock. Many of the moons are heavily cratered, but a few are smooth, indicating surface activity (Io's volcanos of sulfur, for example) that has filled in old craters.

Pluto, despite being an outer planet, is more like a moon in size and mass. It is the smallest planet—probably mostly ice and rock—with a thin methane atmosphere. Many astronomers now think Pluto is a surviving icy planetesimal.

QUESTIONS FOR REVIEW

1. How do Jupiter's mass and radius compare with the Earth's? How do they compare with those of the other outer planets?
2. What are the major gaseous substances that make up Jupiter and Saturn?
3. Why are the outer planets so large?
4. What is the interior structure of Jupiter and Saturn thought to be?
5. What does Jupiter look like?
6. What is the Great Red Spot?
7. Do Jupiter and Saturn have solid surfaces?
8. What sort of atmospheric motion and activity are observed in Jupiter?
9. What sort of activity has been seen on Io? What is Io's heat source thought to be?
10. How do astronomers know what lies inside the outer planets?

11. What are the rings of Saturn made of? How do astronomers know?
12. What creates the gaps between the rings?
13. How might the rings have formed?
14. What is the Roche limit? Why does such a limit exist?
15. How do Uranus and Neptune differ inside from Jupiter?
16. Why are Uranus and Neptune so blue?
17. What are the satellites of the outer planets thought to be composed of?
18. What evidence makes some astronomers believe that Pluto might once have been a moon of Neptune?
19. What is unusual about Uranus's rotation axis? What might explain this peculiarity?
20. How did the discovery of a moon orbiting Pluto help astronomers better understand this planet?

THOUGHT QUESTIONS

1. Why do some astronomers think that Pluto should not properly be considered a planet? Do you agree?
2. If Jupiter were moved closer to the Sun, what do you think would happen to it?
3. Why is it not surprising that Uranus and Neptune have proportionally less hydrogen than their more massive companions, Jupiter and Saturn?

4. Why do the relatively uncratered surfaces of Europa and Enceladus imply that these moons may have been "active" recently?

PROBLEMS

1. Use the modified form of Kepler's third law, illustrated in figure 7.5 (and discussed in detail in chapter 2), to calculate Pluto's mass from the orbital data for Charon that is given in the text. Be sure to convert the orbital period to seconds and the orbital radius to meters before putting those numbers into the formula.
2. How long does it take sunlight to reach Pluto?

3. Calculate the density of Charon, given that its radius is approximately 593 kilometers and its mass is about 1.1×10^{24} grams. (Be sure to convert kilometers to centimeters or meters.) Is it likely that Charon has a large iron core? Why?
4. Show that Pluto's orbital period is very close to 3/2 of Neptune's. Use the data in Table 4 of the appendix.

TEST YOURSELF

1. The low average densities of Jupiter and Saturn compared with the Earth's suggest
   (a) they are hollow.
   (b) their gravitational attraction has compressed their cores into a rare form of iron.
   (c) they contain large quantities of light elements, such as hydrogen and helium.
   (d) they are very hot.
   (e) volcanic eruptions have ejected all the iron that was originally in their cores.
2. What makes some astronomers think that Uranus was hit by a large body early in its history?
   (a) It goes around the Sun in a direction opposite to the other planets.
   (b) Its rotation axis has such a large tilt.
   (c) Its composition is so different from that of Neptune, Jupiter, and Saturn.
   (d) It has no moons.
   (e) All of the above.
3. Why would it be foolish to send astronauts to land on Jupiter?
   (a) It has no solid surface for them to land on.
   (b) Its clouds are so hot that any spacecraft getting near it would burn up.
   (c) Its gravitational attraction is so weak that they would float off.
   (d) All of the above.
   (e) The idea is perfectly reasonable.
4. What is the Roche limit?
   (a) The mass a planet must exceed to have satellites.
   (b) The smallest mass a planet can have and still be composed mainly of hydrogen.
   (c) The greatest distance from a planet that its satellites can orbit without falling into the Sun.
   (d) The distance at which a moon held together by gravity will be broken apart by the planet's gravitational attraction.
   (e) The distance astronomers can see into a planet's clouds.
5. Astronomers think that the inner core of Jupiter is composed mainly of
   (a) hydrogen.
   (b) helium.
   (c) uranium.
   (d) rock and iron.
   (e) water.

FURTHER EXPLORATIONS

Beatty, J. K., and A. Chaikin. *The New Solar System,* 3rd edition. Cambridge, Mass.: Sky Publishing Corporation and Cambridge University Press, 1990. (This fine book covers the Solar System in great detail.)

Beatty, J. Kelly, and Stuart J. Goldman. "The Great Crash of 1994: A First Report." *Sky & Telescope* (October 1994): 18.

Beebe, Reta. Jupiter: The Giant Planet. 2nd ed. Washington, DC: Smithsonian Institution Press, 1997.

Binzel, Richard. "Pluto." *Scientific American* 262 (June 1990): 50.

Burnham, Robert. "Into the Maelstrom." *Astronomy* 24 (April 1996): 42.

Cuzzi, Jeffrey N., and Larry W. Esposito. "The Rings of Uranus." *Scientific American* 257 (July 1987): 52.

Gore, Rick. "Neptune: *Voyager's* Last Picture Show." *National Geographic* 178 (August 1990): 35.

Johnson, Torrence V. , Robert Hamilton Brown , and Laurence A. Soderblom. "The Moons of Uranus." *Scientific American* 256 (April 1987): 48.

Kinoshita, June. "Neptune." *Scientific American* 261 (November 1989): 82.

Lunine, Jonathan I. "Neptune at 150." *Sky and Telescope* 92 (September 1996): 38.

Owen, Tobias. "Titan." *Scientific American* 246 (February 1986): 98.

Sagan, Carl. "The First New Planet." *Astronomy* 23 (March 1995): 34.

Soderblom, Laurence A. and Torrence V. Johnson. "The Moons of Saturn." *Scientific American* 246 (January 1982): 101.

Talcott, Richard. "Galileo's Dazzling Flyby." *Astronomy* 24 (November 1996): 56.

Web Sites

"Welcome to the Planets" (many pretty pictures, fact sheets on the planets and the other Solar System objects, and a glossary). Prepared by the Jet Propulsion Laboratory.
www.pds.jpl.nasa.gov/planets/

"NSSDC Photo Gallery" (National Space Science Data Center—NASA) Good pictures and fact sheets.
www.nssdc.gsfc.nasa.gov/photo_gallery/

"Views of the Solar System," by Calvin Hamilton. Excellent pictures, fact sheets, and glossary. Lots of links to other sites.
www.bang.lanl.gov/solarsys/

"The Nine Planets," by Bill Arnett. "A Multimedia Tour . . ." One of the best astronomy sites. Lots of excellent links, pictures, text, and glossary.
www.seds.org/billa/tnp/

VIDEOTAPES

Voyager: Missions to Jupiter and Saturn
Voyager: Neptune Encounter Highlights

Uranus: I Will See Such Things
Miranda: The Movie

(The above are available from the Astronomical Society of the Pacific, 390 Ashton Ave., San Francisco, CA 94112.)

KEY TERMS

Roche limit, 283

shepherding satellites, 283

CHAPTER TEN

METEORS, ASTEROIDS, AND COMETS

A crater in northern Arizona, made by an asteroid that struck Earth (Courtesy Meteor Crater, Northern Arizona)

Comet Ikeya-Seki (Richard Cromwell, Steward Obs., U. Arizona)

An Asteroid (Courtesy of NASA)

Orbiting the Sun and scattered throughout the Solar System are numerous bodies much smaller than the planets—the asteroids and comets. The asteroids are generally rocky objects in the inner Solar System. The comets are icy bodies and spend most of their time in the outer Solar System. These small members of the Sun's family, remnants from the formation of the Solar System, are of great interest to astronomers because they are our best source of information about how long ago and under what conditions the planets formed. In fact, some asteroids and comets may be planetesimals—the solid bodies from which the planets were assembled—that have survived nearly unchanged from the birth of the Solar System.

Apart from their scientific value, asteroids and comets merit study because they can be both beautiful and deadly. A comet in the dawn sky with its tail a shining plume is a sight not to miss. But "miss" is precisely what we hope will happen if a large comet or asteroid were on a collision course with Earth. Such a collision in the past may have exterminated much of Earth's ancient life, and such an event in the future could well have equally disastrous effects on today's living things.

In this chapter we will see why astronomers think asteroids and comets are related to planetesimals. We will also discover why meteorites, fragments of these bodies that by chance fall into our atmosphere, are such important clues to the time of birth and the structure of the ancient Solar System. Likewise, we will study how a comet changes from a 10-kilometer diameter ball of ice into a beautiful banner of light in the night sky and why an asteroid may be the reason you have hair rather than scales.

10.1 METEORS AND METEORITES

If you have spent even an hour looking at the night sky, you have probably seen a "shooting star," a streak of light that appears in a fraction of a second and as quickly fades (fig. 10.1). Astronomers call this brief but lovely phenomenon a **meteor**.

A meteor is the glowing trail of hot gas and vaporized debris left by a solid object heated by friction as it moves through the Earth's atmosphere. Most of the heating occurs between about 100 and 50 kilometers in the outer fringes of the atmosphere. The solid body, while in space and before it reaches the atmosphere, is called a **meteoroid**.

Heating of Meteors

Meteors heat up on entering the atmosphere for the same reason a reentering spacecraft does. When an object plunges from space into the upper layers of our atmosphere, it collides with atmospheric molecules and atoms. These collisions convert some of the body's energy of motion (kinetic energy) into heat, as shown in figure 10.2. In a matter of seconds, the outer layer of the meteor reaches thousands of degrees Kelvin and glows. Given that reentry speeds are typically at least 10 kilometers per second and are often 30 to 40 kilometers per second, the collisions with air molecules are extremely violent and tear atoms off the body, vaporizing the surface layers. The trail of hot evaporated matter and atmospheric gas emits light, making the glow that we see.

FIGURE 10.1
A time exposure photograph captures a "shooting star" (meteor) flashing overhead. The curved streaks are star trails. (Courtesy Ronald A. Oriti, Santa Rosa Junior College, Santa Rosa, Calif.)

☀ Why is the meteor's track straight whereas the star images are curved?

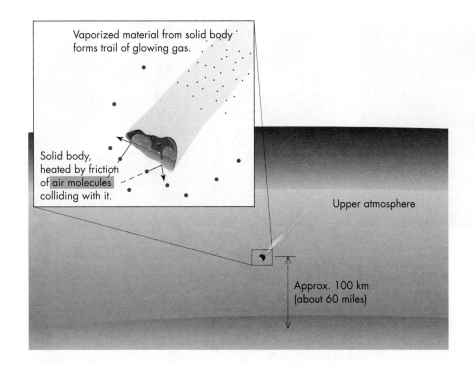

Vaporized material from solid body forms trail of glowing gas.

Solid body, heated by friction of air molecules colliding with it.

Upper atmosphere

Approx. 100 km (about 60 miles)

FIGURE 10.2
Sketch depicting how a meteor is heated by air friction in the upper atmosphere.

If the meteoroid is larger than a few centimeters, it creates a ball of incandescent gas around it and may leave a luminous or smoky trail. Such exceptional meteors, sometimes visible in daylight, are called "fireballs."

Meteoroids bombard the Earth continually: a hail of solid particles that astronomers estimate amounts to hundreds of *tons* of material each day. Slightly more strike between midnight and dawn than in the evening hours, and so it is best to watch for meteors in the early morning. This difference arises for the same reason that if you run through rain your front will get wetter than your back. That is, the dawn side of our planet advances into the meteoritic debris near us in space, while the night side moves away from it.

Most meteors that we see last only a few seconds and are made by meteoroids the size of a raisin or smaller. These tiny objects are heated so strongly that they completely vaporize. Larger pieces, though heated and partially vaporized, are so drastically slowed by air resistance that they may survive the ordeal and reach the ground.* We call these fragments found on the Earth **meteorites**.

Meteorites

Astronomers classify meteorites into three broad categories based on their composition: iron, stony (that is, composed mainly of silicate compounds), and stony/iron. Most stony meteorites are composed of smaller rounded chunks of rocky material stuck together. The grains are called **chondrules**,[†] and meteorites that have this lumpy structure are called **chondritic meteorites**. Chondrules appear to have been rapidly melted and cooled in the

Some stony meteorites have a smooth structure with no sign of chondrules. Astronomers think that these so-called achondritic meteorites were melted and resolidified, probably in a large asteroid.

---

*People are sometimes hit by meteorites, but only rarely. There is one well documented case of a person being killed, and another case is known where a meteor punched a hole through a roof and hit a woman, badly bruising her. See section 10.4 for other reports.
†The *ch* is pronounced as a *k* sound. Thus "chondritic" is pronounced /kon-DRIT-ic/ and means "coarse grained" in Greek.

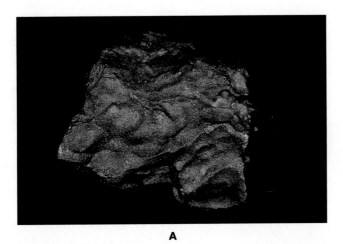

A

B

FIGURE 10.3

Meteorites. (A) An iron meteorite. (From the collection of Ronald A. Oriti, Santa Rosa Junior College, Santa Rosa, Calif.)

☀ What has caused the pits and irregular shapes of the meteorite's surface?

(B) A chondritic meteorite. (Courtesy John A. Wood, Harvard-Smithsonian Center for Astrophysics.) Notice the small round lumps (chondrules) compared with the smooth texture of the meteorite in A.

Amino acids occur in two molecular forms— right handed and left handed. All living things on Earth contain only left-handed amino acids. The amino acids found in meteorites occur in both forms, however. Their presence indicates that they are not contamination by terrestrial organisms but were formed in space by nonbiological processes.

solar nebula. The cause of the heating is not known. Figure 10.3 shows a meteorite of uniform composition and a chondritic meteorite.

Chondrules contain traces of radioactive material, which can be used to measure their age, as described in chapter 4. They are extremely old, 4.5 billion years, and are believed to be the first solid material that condensed within the solar nebula. Many chondrules contain even older material: dust grains that have survived from before the birth of the Solar System. Thus, chondritic meteorites offer us valuable information about the early history of the Solar System.

In some chondritic meteors, the chondrules are embedded in (or are themselves) a black, carbon-rich, coal-like substance and are therefore called **carbonaceous chondrites.** This carbonaceous matter contains organic compounds, including amino acids, the same complex molecules used by living things for the construction of their proteins and genetic material. Thus the presence of amino acids in meteoritic matter indicates that the raw material of life can form in space and that it might therefore have been available right from the start within the Solar System. Irrespective of the existence of amino acids in meteorites, we can still ask how and where these bodies formed, and what brought them to Earth. Astronomers think that most of them are fragments of asteroids* and comets.

10.2 ASTEROIDS

Asteroids are small, generally rocky, bodies that orbit the Sun. Most lie in the **asteroid belt,** a region between the orbits of Mars and Jupiter, stretching from about 2 to 4 AU from the Sun, as shown in figure 10.4. Giuseppe Piazzi discovered the first of this swarm of bodies in 1801 during his search for the "missing planet," which, according to Bode's law, should have been at 2.8 AU from the Sun. He named the asteroid that he found Ceres in honor of the patron goddess of Sicily, his home. Thousands of other asteroids have subsequently been found, but Ceres remains the largest. Despite the huge number of asteroids, their combined mass is very small, amounting to probably less than $\frac{1}{1000}$ the mass of the Earth.

Size and Shape

The diameter of an asteroid is difficult to measure because nearly all are so small that they appear in ground-based telescopes merely as points of light. Moreover, the amount of light reflected from an asteroid is not a good clue to its size because a large, poorly reflective ob-

---

*Although most meteorites are fragments of asteroids or comets, some are chunks of rock from the surface of the Moon and Mars, blasted into space by the collision of an asteroid with these bodies.

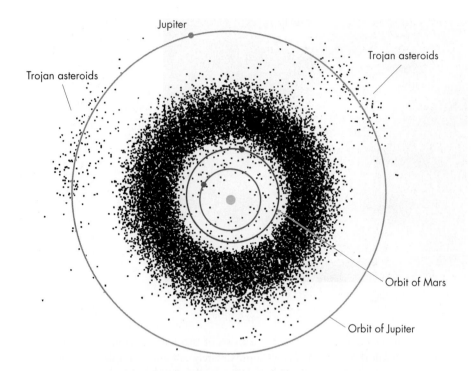

FIGURE 10.4
Diagram showing the distribution of 21,785 asteroids. Notice that most lie between Mars and Jupiter, but a small number form two loose clumps—the Trojan asteroids—located on Jupiter's orbit. (Courtesy E. L. G. Bowell, Lowell Observatory.) Making the plotted points large enough to see causes them to appear far more closely packed than they really are.

☀ With a ruler, estimate how far (in AU) the center of the Trojan swarms are from the Sun and Jupiter.

How does this compare with Jupiter's distance from the Sun? Does this help you understand why the Trojan asteroids are located where they are?

FIGURE 10.5
The asteroid Gaspra, as photographed by the spacecraft *Galileo*. (Courtesy NASA/JPL.)

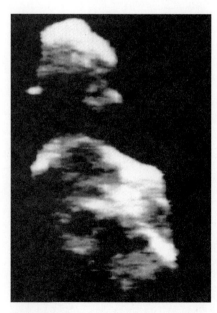

The asteroid Toutatis. (Courtesy NASA/JPL.)

ject will look as bright as a small, highly reflective one. For this reason the emitted infrared radiation is a better measure of diameter; bigger bodies emit more than smaller ones of the same temperature. From such measurements astronomers have found that asteroids range tremendously in diameter, from Ceres—about 1000 kilometers (less than 1/10 the Earth's size) across—down to kilometer-sized bodies and even smaller. For example, the diameter of the tiny asteroid 1991 BA, which passed about 170,000 kilometers (less than half the distance to the Moon) from Earth in January 1991 is probably less than 9 meters (approximately 30 feet).

Most asteroids—such as Gaspra (fig. 10.5), whose picture was taken by the *Galileo* spacecraft as it traveled toward Jupiter—are irregularly shaped. A more extreme example

Dactyl

Courtesy NASA/JPL.

is Toutatis, which is dumbbell-shaped. It consists of two separate chunks, about 4 kilometers and 2.5 kilometers in diameter (approximately 2.5 and 1.6 miles) held together weakly by their gravity. Astronomers imaged this small asteroid with radar when it passed within 3.5 million kilometers (approximately 2.2 million miles) of Earth. Only Ceres and a few dozen other large asteroids are approximately spherical. Their gravitational force is strong enough to crush their material into a sphere. Small bodies with weak gravities remain irregular and are made more so by collisions blasting away pieces. Collisions leave the parent body pitted and lumpy, and the fragments become smaller asteroids in their own right.

Astronomers cannot directly photograph most asteroids, but they can nevertheless infer an asteroid's shape from the brightness of sunlight reflected off its surfaces and the infrared radiation it emits. As an irregularly shaped asteroid orbits the Sun, it slowly tumbles, so that sometimes we see a small end and sometimes a larger side. Thus its reflected brightness will vary. Larger bodies that are nearly spherical do not exhibit such fluctuations as they spin.*

Composition

When sunlight falls on an asteroid, the minerals in its surface create absorption features in the spectrum of the reflected light, from which we can determine the asteroid's composition. Such spectra show that asteroids belong to three main compositional groups similar to those of meteorites: carbonaceous bodies, silicate bodies, and metallic iron-nickel bodies. The groups are not mixed randomly throughout the asteroid belt: inner-belt asteroids tend to be silicate-rich, and outer-belt ones tend to be carbon-rich.

Origin of Asteroids

The properties of asteroids that we have discussed above (composition, size, and their location between Mars and Jupiter) give us clues to their origin and support the solar nebula hypothesis for the origin of the Solar System. As we described in chapter 7, the asteroids are probably fragments of planetesimals, the bodies from which the planets were built.

---

*Dark areas on some large asteroids may make their light vary slightly as they spin.

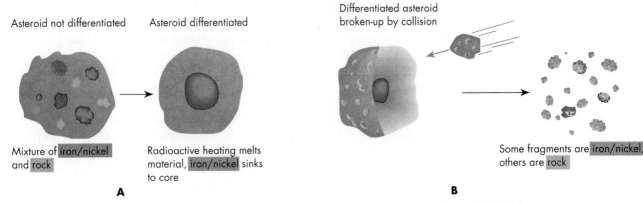

Asteroid not differentiated   Asteroid differentiated

Differentiated asteroid broken-up by collision

Mixture of iron/nickel and rock

Radioactive heating melts material, iron/nickel sinks to core

A

Some fragments are iron/nickel, others are rock

B

FIGURE 10.6
Sketch depicting (A) differentiation in asteroids and (B) their subsequent break-up by collision to form iron and stony bodies.

According to the solar nebula hypothesis, bodies that condensed in the inner asteroid belt have a different composition from those that condensed farther out. The inner belt, being warmer, is richer in the easy-to-condense silicate and iron materials and contains less of the hard-to-condense carbon-rich materials, as is indeed observed.

The existence of stony and iron asteroids might at first seem to be evidence *against* the solar nebula theory. How could a swirling mass of gas and dust have separated so as to form some bodies of rock and some bodies of iron? Such separation would be a bit like shaking a piece of cake and having it disintegrate into eggs, flour, sugar, and milk.

We saw in chapter 4, however, that chemical elements *can* be separated by differentiation. That is, astronomers think that the Earth's rocky crust and iron core formed as a result of melting, with the iron then sinking to the core and the lighter rock floating to the top. Asteroids also differentiated (fig. 10.6A), and some even appear to have had volcanic eruptions. This activity, now long dormant, is deduced from their spectra, which show the presence of basalt, a volcanic rock.

After differentiation, collisions with neighboring asteroids broke up most of the large bodies (fig. 10.6B). The fragments are what we see today: pieces of crust became stony asteroids, while pieces of core became iron asteroids. But for a body to differentiate, it must be large enough to be able to melt from the heat liberated within it by radioactivity. Thus the existence of stony and iron asteroids is strong evidence that the early Solar System contained intermediate-sized bodies, that is, planetesimals.

The solar nebula hypothesis also offers an explanation for the reason the asteroids are concentrated between Mars and Jupiter. Any small planet there would have to compete for material with Jupiter, whose immense gravity would disturb the accretion process and prevent the planet's growth.*

---

*This is a less dramatic view than the one that was common 50 years ago, when asteroids were believed to be fragments of a terrestrial planet destroyed by collision. This no longer seems likely, given the tiny amount of material in the asteroid belt, but it may have given birth to the Superman story. Superman is a refugee from Krypton, an imaginary planet destroyed in just that fashion.

Unusual Asteroids

Even today, Jupiter affects the asteroid belt. Figure 10.7 shows a partial census of the number of asteroids found at each distance from the Sun within the belt. A clear gap can be seen at 2.5 AU and slight concentrations occur at 3.4 and 4.0 AU.

The seemingly empty regions in the asteroid belt are called **Kirkwood gaps,** and they are caused by the same process that creates the gaps in Saturn's rings: gravitational forces of an outlying body. Saturn's moons, particularly Mimas, create the gaps in the rings. Jupiter creates the Kirkwood gaps. As it orbits the Sun, Jupiter exerts a gravitational force on asteroids in the belt that slightly alters their orbits. If an asteroid has an orbital period that when multiplied by an integer equals Jupiter's period, the asteroid may be subject to a cumulative force that makes it drift to a new orbit. The gap at 2.5 AU arises because, according to Kepler's third law, an asteroid at that distance would have an orbital period exactly one-third that of Jupiter. Every third trip around the Sun, the asteroid would undergo exactly the same tug from Jupiter. Over time it will be shifted to a new orbit, leaving a gap.

Not all asteroids are found in the main belt. A few, the so-called Trojan asteroids, travel along Jupiter's orbit in two loose swarms, 60° ahead and 60° behind it, as illustrated in figure 10.4. Mars too has a similar family.

Farther from the main belt are the **Apollo asteroids** whose orbits carry them into the inner Solar System across the Earth's orbit. Fortunately for us, there are only about 700 such bodies, and so the chance of collision is slim. Nevertheless, on the average, one such body hits the Earth about every 10,000 years. The Apollo asteroids have diameters of about 1 kilometer or less. They may be related to comets, but "dead" ones, shifted to their peculiar orbit by Jupiter's gravitational force and stripped of ice and gas by their repeated passage around the Sun. We will describe how this happens in more detail later in this chapter.

Chiron is another odd "asteroid." Its orbit stretches from just inside that of Saturn almost to Uranus, putting it far outside the main asteroid belt. In addition to this odd orbit, Chiron changes brightness oddly; sometimes it flares up and ejects gas. Such behavior is more like that of a comet than a "normal" asteroid. But if Chiron is a comet, it is not a

An asteroid 1 kilometer in diameter striking the Earth will release on impact about the same amount of energy as a 40,000-megaton nuclear explosion. Some astronomers have therefore urged the construction of an international monitoring system to warn of the approach of small asteroids. If detected early enough, such bodies might be deflected away from Earth by nuclear bombs launched toward them and exploded on their surface.

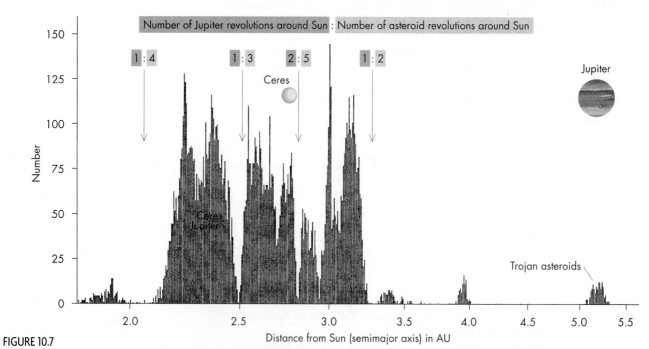

FIGURE 10.7

The number of asteroids at each distance from the Sun within the asteroid belt. Notice the conspicuous gaps where there are few if any objects. These empty zones are called Kirkwood gaps. (Courtesy E. L. G. Bowell, Lowell Observatory.)

☼ What feature in Saturn's rings is similar to these gaps?

"normal" one. From its brightness, astronomers deduce that Chiron has a diameter of about 180 kilometers (approximately 112 miles), making it much larger than most comets that enter the inner Solar System. Perhaps Chiron is neither asteroid nor comet, but a surviving icy planetesimal.

10.3 COMETS

A bright comet is a stunning sight, as you can see in figure 10.8. But such sights are now sadly rare because light pollution from our cities drowns the view for most people. Comets have long been held in fear and reverence, and their sudden appearance and equally sudden disappearance after a few days or so have added to their mystery. In fact Halley's comet was allegedly excommunicated by Pope Calixtus III when its appearance in 1456 coincided with a major assault by the Turks on southeastern Europe.

Light pollution not only spoils our view of the night sky, but it also wastes money. Groups such as the International Dark-Sky Association have shown that better designed street and other outdoor lights make the ground brighter, the sky darker, and use less electricity.

Structure of Comets

Comets consist of two main parts, as illustrated in figure 10.9. The largest part is the long **tail**, a narrow column of dust and gas that may stretch across the inner Solar System for as much as 100 million kilometers (nearly an AU!).

FIGURE 10.8
Photograph of Comet West in the dawn sky. (Courtesy STSCI.)

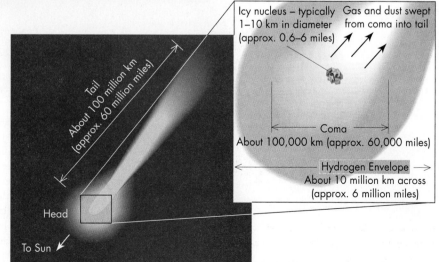

FIGURE 10.9
Artist's depiction of the structure of a comet, showing the tiny nucleus, surrounding coma, and the long tail.

The tail emerges from a cloud of gas called the **coma,** which may be some 100,000 kilometers in diameter (10 times or so the size of the Earth). However, despite the great volume of the coma and the tail, these parts of the comet contain very little mass. The gas and dust are extremely tenuous, and so a cubic centimeter of the gas contains only a few thousand atoms and molecules. By terrestrial standards this would be considered a superb vacuum. This extremely rarified gas is matter that the Sun's heat has boiled off the heart of the comet, its nucleus.

The comet **nucleus** is a block of ice and gases that have frozen in the extreme cold of interplanetary space into an irregular mass whose approximate diameter is 10 kilometers. The nucleus of a comet has been described as a giant "iceberg" or "dirty snowball," and it contains most of the comet's mass. Our best information about the nucleus comes from studies of Comet Halley* made by the *Giotto* spacecraft.

Giotto,[†] a spacecraft launched by the European space community as part of an international study of Halley, approached to within 600 kilometers of Halley's nucleus and sent pictures back to Earth of the nucleus. From these its size could be measured. From the size and estimates of the mass of the nucleus, its density can be calculated. This turns out to be about 0.2 grams per cubic centimeter, a value implying that the icy material of the nucleus is "fluffy," like snow, not hard and compacted like pure ice. Unfortunately, the mass estimates are not very accurate, and so the density we infer from them is uncertain.

FIGURE 10.10
Photograph made by the *Giotto* spacecraft of the nucleus of Halley's comet. The spacecraft was approximately 1000 kilometers (about 620 miles) from the nucleus—deep inside Halley's coma of gas—during the time this picture was made. (Copyright 1986, Max Planck Institute for Aeronomy; courtesy Harold Reitsema, Ball Aerospace.)

*Comets are generally named for their discoverer or for the year they were first seen. Halley's gets its name because Sir Edmund Halley was the first to propose that some comets move around the Sun like planets, and he predicted that the great comet he saw in 1682 would reappear in 1759. It did, but Halley did not live to see his prediction verified. There is controversy about how to pronounce "Halley". Most astronomers say it to rhyme with /Sally/, although there is some evidence that Halley himself pronounced his name as /Haw-lee/.

[†]The name *Giotto* was given to the spacecraft in honor of the Italian artist who painted a portrait of a comet as part of a Christmas scene for a church altarpiece. Some astronomers believe his painting depicts Halley's comet as it appeared in 1301. Other astronomers believe that Giotto based it on a different comet.

Despite its icy composition, the nucleus is extremely dark, as you can see in figure 10.10, which is one of the pictures made by the *Giotto* spacecraft. Astronomers think that the dark color comes from dust and carbon-rich material (similar to that of the carbonaceous chondritic meteorites) coating the surface of the nucleus. Other visible features of the nucleus are its irregular shape and the jets of gas erupting from the frozen surface. The jets form when sunlight heats and vaporizes the icy material. The irregular shape is probably the outcome of uneven melting of the nucleus during passage by the Sun on previous orbits.

Composition of Comets

The escaped gas from the comet offers astronomers a way to probe the comet's composition. Spectra of gas in the coma and tail show that comets are rich in water, CO_2, CO, and small amounts of other gases that condensed from the primordial solar nebula. Evaporating water is broken up by solar ultraviolet radiation to create oxygen and hydrogen gas, and most comets are surrounded by a vast cloud of hydrogen created in this way.

If it passes by the Sun too often, the escape of gas from the comet eventually erodes it away. Also, some comets literally fall into the Sun. Since new comets show up every few years, there must be a source to replace those devoured by the Sun, and it is to their origin that we now turn.

Origin of Comets

Astronomers think that most comets come from the **Oort cloud,** the swarm of trillions of icy bodies* believed to lie far beyond the orbit of Pluto, as we discussed in chapter 7. You may recall that astronomers think the Oort cloud formed from planetesimals that originally orbited near the giant planets and were tossed into the outer parts of the Solar System by the gravitational force of those planets. There, they form a spherical shell that completely surrounds the Solar System and extends to perhaps as much as 150,000 AU from the Sun, as shown in figure 10.11. Astronomers deduce this shape for the Oort cloud from the many comet orbits that are highly tilted with respect to the main plane of the

Photograph of comet Hyakutake. Many people saw this beautiful comet in the Spring of 1996. (Courtesy M. Skrutskie, Univ. of Massachusetts.)

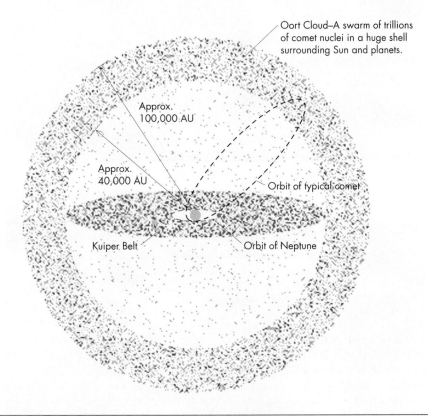

Oort Cloud–A swarm of trillions of comet nuclei in a huge shell surrounding Sun and planets.

Approx. 100,000 AU

Approx. 40,000 AU

Orbit of typical comet

Kuiper Belt

Orbit of Neptune

FIGURE 10.11

Schematic drawing of the Oort cloud, a swarm of icy comet nuclei orbiting the Sun at 40,000 to 100,000 AU. Also shown is the Kuiper Belt, another source of comet nuclei.

*Astronomers call these cold and inert bodies "comets," even though they do not have tails.

Solar System. However, as we will discuss later, some comets also seem to come from a flatter, less remote region—the so-called Kuiper Belt, also shown schematically in figure 10.11. The Kuiper Belt begins about the orbit of Neptune and extends from there to about 10,000 AU, where it may merge with the inner part of the Oort cloud.

Each comet nucleus moves along its own path, and those in the Oort cloud take millions of years to complete an orbit. With orbits so far from the Sun, these icy bodies receive essentially no heat from the Sun, and calculations indicate that their temperature is a mere 3 K, or about −454 °F. Thus the gases and ices remain deeply frozen.

Such cold and distant objects are invisible to us on Earth; so, if we are to see a comet, its orbit must somehow be altered to carry it closer to us and the Sun. Astronomers think that such orbital changes may arise from the chance passage of a star far beyond the outskirts of the Solar System or from tidal forces exerted on the Oort cloud by the Milky Way. Such gravitational effects disturb the orbits of the comet nuclei in the Oort cloud, altering their paths and making some drop in toward the inner Solar System, as shown in figure 10.12. A single disturbance may shift enough orbits to supply comets to the inner Solar System for tens of thousands of years.

As the comet falls inward toward the inner Solar System, the Sun's radiation heats it and begins to melt the ices. At a distance of about 5 AU from the Sun (Jupiter's orbit), the heat is enough to vaporize the ices, forming gas that escapes to make a coma around the comet nucleus. The escaping gas carries tiny dust grains that were frozen into the nu-

FIGURE 10.12
Sketch of how a passing star alters the orbit of a comet nucleus. On its new orbit the comet will pass by the Sun and be visible from Earth.

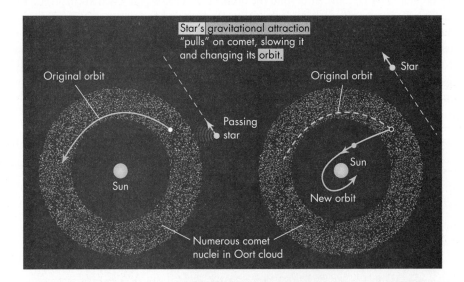

FIGURE 10.13
Sketch of how radiation pressure pushes on dust particles. Photons hit the dust, and their impact drives the dust away from the Sun, forming a dust tail.

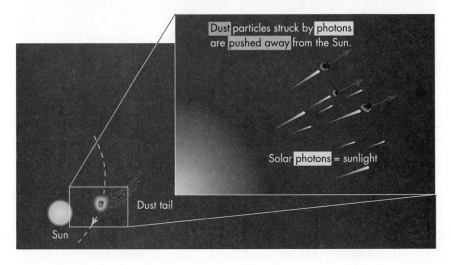

cleus with it. The comet then appears through a telescope as a dim, fuzzy ball. As the comet falls ever nearer the Sun, its gas boils off even faster, but now the Sun begins to exert additional forces on the cometary gas and dust.

Formation of the Comet's Tail

Sunlight striking dust grains imparts a tiny force to them, a process known as **radiation pressure.** We don't feel radiation pressure when sunlight falls on us because the force is tiny and the human body is far too massive to be shoved around by solar photons. However, the microscopic dust grains in the coma do respond to radiation pressure and are pushed away from the Sun, as shown in figure 10.13. Because all the grains move in the same direction, away from the Sun, a tail begins to form.

The tail pushed out by radiation pressure is made of dust particles, but figure 10.14 shows that comets often have a second tail. That tail is created by an outflow of gas that streams from the Sun into space, a flow called the **solar wind.**

The solar wind blows away from the Sun at about 400 kilometers per second. It is very tenuous, containing only a few atoms per cubic centimeter. But the material in the comet's coma is tenuous too, and the solar wind is dense enough by comparison to blow it into a long plume.

Magnetic fields carried along by the solar wind enhance its effect on the comet's tail, helping to drag matter out of the coma and channel its flow, just as magnetic fields in the Earth's atmosphere channel particles to form the aurora. Thus two forces, radiation pressure and the solar wind, act on the comet to drive out a tail. Because those forces are directed away from the Sun, the comet's tail always points away from the Sun, and the tail may even point out ahead of the comet as it moves away from the Sun (fig. 10.15).

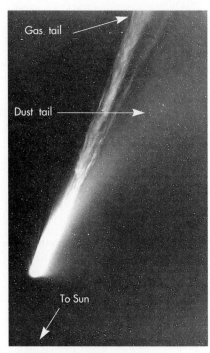

FIGURE 10.14
Photograph of Comet Mrkos clearly illustrating the two tails, one of dust, one of gas. (Courtesy Cal Tech/Palomar Observatory.)

FIGURE 10.15
Sketch illustrating how radiation pressure and the solar wind make a comet's tail always point away from the Sun.

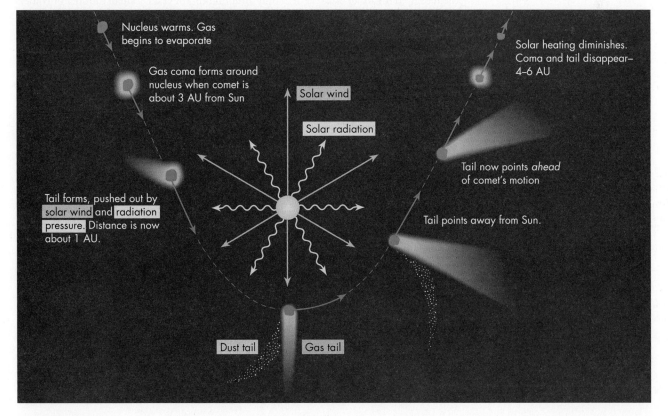

Some laundry detergents have "whiteners" in them that fluoresce. They convert ultraviolet radiation into visible light and thereby make a white shirt or blouse look brighter. You can see this effect strongly if you shine a black light on a freshly washed shirt in a darkened room.

It might help you understand this seemingly odd phenomenon if you think of a runner carrying a torch. If the air is still, the smoke from the torch will of course trail behind the runner. However, if a strong wind is blowing at, say, 40 miles per hour, the smoke will be carried along in the direction of the wind regardless of which way the runner moves. Likewise, the high velocity of the solar wind (400 kilometers per second versus about 40 kilometers per second for the comet) carries the tail of a comet outward from the Sun regardless of the comet's motion.

Light from the Comet's Tail

The gases and dust swept out into the tail are lit up by the Sun. The dust particles reflect sunlight, and the gases emit light of their own by a process called **fluorescence.** Fluorescence is produced when light at one wavelength is converted to light at another wavelength. A familiar example is the so-called black light that you may have seen for illuminating posters.

Black light is really ultraviolet radiation that we have difficulty seeing because of its short wavelength. When such ultraviolet radiation falls on certain paints or dyes, the chemicals in the pigment absorb the ultraviolet radiation and convert it into visible light.

A major part of a comet's light is created by fluorescence. A photon of ultraviolet—and thus energetic—radiation from the Sun lifts electrons in the atoms of the comet's gas molecules to an upper, excited level in a single leap. The electron then returns to its original level in two or more steps, emitting a photon each time it drops. The combined energy of these photons must equal that of the absorbed ultraviolet photon to conserve energy. Thus the energy of each emitted photon must be less than that of the original ultraviolet one. That smaller energy then gives them a longer wavelength, which we can see with our eyes. Thus fluorescence creates the soft glow of the comet's light. In addition, the spectrum of the fluorescing gas tells us of what the comet is made.

Short-Period Comets

Although most comets that we see from Earth swing by the Sun on orbits that will bring them back to the inner Solar System only after millions of years, a small number of comets reappear at time intervals less than 200 years. These **short-period comets** include Halley's, which has a period of 76 years.

The origin of short-period comets is still under study. At one time it was thought that they came from the Oort cloud, but as they moved through the region of the Solar System containing the giant planets, their orbits were shifted by a close encounter with one of the planets into smaller orbits with periods of centuries rather than millenniums. However, astronomers now think that short-period comets come from the icy nuclei, orbiting beyond Neptune, in the Kuiper belt. Support for this origin comes from the detection of several dozen small, presumably icy, bodies orbiting near and somewhat beyond Pluto. In fact, astronomers estimate that the Kuiper belt contains well over 30,000 icy objects bigger than 100 km in diameter, and its total mass is hundreds of times larger than the asteroid belt between Mars and Jupiter. These frozen objects are probably survivors of the Solar System's birth—icy planetesimals still orbiting in the disk—but they too far apart to form additional planets.

Fate of Short-Period Comets

A comet is a bit like the Schulz cartoon character Pigpen, who trails dirt wherever he goes.

Repeated orbits of a short-period comet past the Sun gradually whittle it away: all the ices and gases evaporate, and only the small amount of solid matter, dust and grit, remains. This fate is like that of a snowball made from snow scooped up alongside the road, where small amounts of gravel have been packed into it. If such a snowball is brought inside, it melts and evaporates, leaving behind only the grit accidentally incorporated in it. So too, the evaporated comet leaves behind in its orbit grit that continues to circle the Sun. The material left by the comet produces a delightful benefit: it is a source of meteors.

A

FIGURE 10.16
Sketch showing how (A) in mid-August, at the time of the Perseid meteor shower, the Earth is moving along its orbit (B) When the Earth crosses the debris strewn along a comet's orbit, the scattered material plunges into our atmosphere, producing the diverging pattern of meteors characteristic of a meteor shower. (Bodies and orbits are not to scale.)

Meteor Showers

If you go outside on a clear night and have an unobscured view of the sky, you will see on the average one meteor every 15 or so minutes.* Most of these meteors are stray fragments of asteroids that arrive at the Earth randomly.

At some times of year, instead of one meteor per quarter hour or so, you may observe one every few minutes. Furthermore, if you watch such meteors carefully, you will see that they appear to come from the same general direction in the sky. Meteors of this type are called "shower meteors," and the events that produce them are called **meteor showers.**

The most famous meteor shower occurs each year in mid-August. From August 11 to 13, meteoroids rain into our atmosphere from a direction that lies toward the constellation Perseus. The meteors themselves have no association with Perseus. Rather, they are following an orbit around the Sun that happens to lie roughly in that direction (fig. 10.16A), and the Earth happens to cross their orbits in mid-August. Thus at that time we encounter far more meteoroids than usual.

This encounter creates an effect similar to what you observe when you drive at night through falling snow: the flakes seem to radiate from a point in front of you the location of which depends on a combination of the direction and speed of both the wind and your car. Thus during the time that the Earth crosses the path followed by the meteoroids, they seem to diverge from a common point (fig. 10.16B) called the **radiant.** Meteor showers are generally named for the constellation from which they appear to diverge, and appendix table 6 lists several of the brighter and more impressive showers and their dates. Each shower therefore marks when the Earth crosses the path of a still-active comet or even a "dead" one.

B

*Returning from work or late parties, I often wait outside looking at the sky until I see a meteor. I seldom have to wait more than 20 minutes.

As a comet orbits the Sun and its icy and gaseous material evaporates, it leaves in its path a trail of dust and small bits of solid material ejected from its nucleus. When we cross or near such a trail, our planet is blasted by this microscopic debris that rains into our atmosphere, burning up, and creating a meteor shower.

Be sure to check the November 17 shower the next few years. Astronomers predict that these so-called Leonids may put on good shows in 1997, 1998, and 1999.

On rare occasions the Earth will pass through a particularly dense clump of material left by the comet. If that happens, thousands of meteors per hour may spangle the sky. Such a display happened in November 1966 when dawn observers on the west coast of the United States and Canada saw literally dozens of meteors per second! The sky looked as if someone had lit a celestial Fourth of July sparkler.

Spectacles of this kind are one of the delights of astronomy. However, on even rarer occasions, far more sinister meteoritic events may occur.

10.4 GIANT IMPACTS

Every few thousand years, the Earth is hit by a huge meteoroid, a body tens of meters or more in size. Such bodies will produce not only a spectacular glare as they pass through the atmosphere, but also an enormous blast on impact.

As we discussed earlier in this section, meteoroids may have a very large kinetic energy, that is, energy of motion. If the meteoroid does not burn up on passage through the atmosphere, its remaining kinetic energy is released when it hits the ground or when it breaks up in the atmosphere.

The energy so released can be huge, as we can easily show from the formula for a body's kinetic energy, E, given by the expression

E = Kinetic energy of an object
v = Velocity of the object
m = Mass of the object

$$E = \frac{mv^2}{2}$$

where m is its mass and v is its velocity. For a meteoroid weighing 100 kilograms (about 200 pounds) and traveling at 30 kilometers per second (= 3×10^4 meters per second), the kinetic energy of impact will be $100 \times (3 \times 10^4)^2/2 = 4.5 \times 10^{10}$ joules, which is about the same as that released when a hundred tons of dynamite are exploded. Such a body would make a crater roughly 30 meters (100 feet) in diameter. A body 10 meters in diameter, the size of a small house, would have on impact the explosive power of a thermonuclear bomb and make a crater nearly a kilometer in diameter. Were such a body to hit a heavily populated area, it would have catastrophic results.

Such a catastrophe sets the stage for Arthur C. Clarke's science-fiction novel Rendezvous with Rama.

Fortunately we have been spared such disasters recently, but there have been some close calls and some truly horrific impacts in the distant past.

Giant Meteor Craters

One of the most famous meteor impacts was the event that formed the giant crater in northern Arizona. About 50,000 years ago, a meteoroid estimated to have been some 50 meters in diameter hit the Earth about 40 miles east of what is now Flagstaff. Its impact vaporized tons of rock, which expanded and peeled back the ground, creating a crater about 1.2 kilometers across and 200 meters deep (fig. 10.17).

More recently, in 1908, an asteroid broke up in our atmosphere over a largely uninhabited part of north-central Siberia. This so-called Tunguska event, named for the region where it hit, leveled trees radially outward from the blast point to a distance of some 30 kilometers. The blast was preceded by a brilliant fireball in the sky and was followed by clouds of dust that rose to the upper atmosphere. Sunlight reflected off this dust gave an eerie glow to the night sky for several days. According to some accounts, the blast killed two people. Casualties were few because the area was so remote.

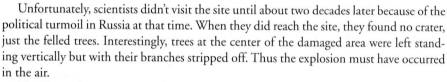

A

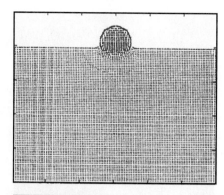

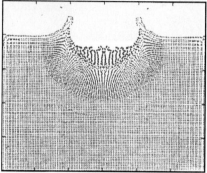

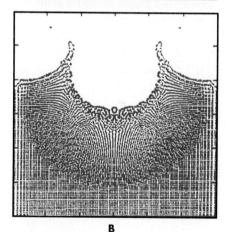

B

FIGURE 10.17
(A) Photograph of the Arizona meteor crater. (Courtesy Meteor Crater, northern Arizona.) (B) A computer simulation of a crater's formation. (Courtesy Helen Pongracic, University of Sydney.)

Unfortunately, scientists didn't visit the site until about two decades later because of the political turmoil in Russia at that time. When they did reach the site, they found no crater, just the felled trees. Interestingly, trees at the center of the damaged area were left standing vertically but with their branches stripped off. Thus the explosion must have occurred in the air.

The Tunguska event is just the most recent giant impact. More ancient impact scars occur in many places on our planet. The huge, ring-shaped Manicouagan Lake, about 70

 RE-MODELING

GHOST CRATERS
OR NO TELL-TALE FRAGMENTS

The evidence of an air blast but no crater or meteoritic fragments at one time led to the idea of a comet's being the cause. On entering the atmosphere, the cometary ices would have heated and expanded explosively, and if any pieces did survive to reach the ground, they would have melted long before anyone explored the area.*

Recently, astronomers have calculated that stony or iron asteroids can also create devastation without leaving a crater or telltale fragments. A typical asteroid approaches Earth at a speed of 10 to 30 km/s. When it hits our atmosphere, it compresses and heats the air ahead of it. The hot compressed air obeys Newton's law of action-reaction and exerts a tremendous force

back on the asteroid that may shatter it. The resulting fragments plunge deeper into our atmosphere, where air resistance heats them further until they vaporize, creating a fireball at a height of 20 kilometers (12 miles) or so above the ground. No trace of the asteroid survives to form a crater or fragments at the ground. All that remains is the 6000-K incandescent air of the fireball that blasts downward and out, crushing and igniting whatever lies below it, exactly the way a nuclear bomb burst would.

---

*Stories of unusual radioactivity at the site are totally false but persistent, especially in the kinds of supermarket tabloids that feature articles about 165-pound babies and "How Aliens Abducted My Wife."

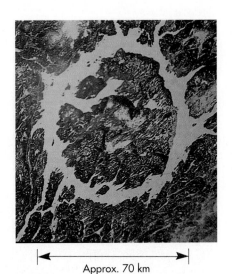

|←——— Approx. 70 km ———→|
(about 43 miles)

FIGURE 10.18
Picture (from Earth orbit) of the Manicouagan Crater in Quebec. (Courtesy Landsat/EOS.)

☀ What might explain why this crater is so much less sharply defined than the Arizona meteor crater?

kilometers (approximately 43 miles) in diameter and pictured in figure 10.18, is a meteor crater, as is Wolf Creek Crater in northwestern Australia.

Astronomers have found even larger craters but are not sure they are impact features. Two such craters are the vast arc (nearly 500 kilometers across) on the east edge of Hudson Bay and a basin (about 300 kilometers across) in central Europe. Still other craters may lie hidden under sediment on land or beneath the oceans.

Mass Extinction and Asteroid/Comet Impacts

At the end of the Cretaceous period, about 65 million years ago, an asteroid or comet hit the Earth. Its impact and the subsequent disruption of the atmosphere is blamed for exterminating the dinosaurs and many less conspicuous but widespread creatures and plants. In fact, the sudden disappearance of large numbers of life forms at the end of the Cretaceous period defines the end of the Mesozoic era.

The evidence that an extraterrestrial body caused this devastation comes from the relatively high abundance of the otherwise rare element iridium found in sediments from that time. Iridium, a heavy element similar to platinum, is very rare in terrestrial surface rocks because, according to geological theory, most of it sank to the Earth's deep interior at the time our planet formed its core. On the other hand, samples of meteoritic material contain moderate amounts of iridium because most of these bodies have not formed iron cores. Thus the presence of so much iridium in a layer of clay laid down 65 million years ago is suggestive of a link to meteoritic material. The amount of iridium in the earth sediments at that layer is the quantity that would be dispersed from a piece of meteoritic material 10 kilometers in diameter. Therefore many astronomers and paleontologists believe the Earth was hit by an asteroid of that size.

A 10-kilometer asteroid hitting the Earth would produce an explosion on impact equivalent to that of several billion nuclear weapons. The impact would not only make an immense crater, but it would also blast huge amounts of dust and molten rock into the air. The molten rock raining down would raise the surface temperature as high as that under an electric broiler and ignite global wildfires. The hot fragments and blast would also create nitrogen oxides, which would combine with water to form a rain of highly concentrated nitric acid. This devastating combination of heat, acid rain, and blast would then be followed by months of darkness and intense cold caused by the dust shroud blotting out the Sun. It seems likely that the biosphere would be devastated, leading to mass extinctions, just as the fossil record shows.

This frightening picture is supported not only by the iridium layer and the sudden disappearance of dinosaurs, but also by a layer of soot, as well as a layer of tiny quartz pellets believed to have been created by the melting and blast of a violent impact. Moreover, some astronomers believe that a faint, circular depression about 65 million years old near Chicxulub* (/cheek-shoo-loob/) in the Yucatán region of Mexico may be the crater formed by this impact.

The Cretaceous mass extinction may have played an especially important role in our own evolution. Before that event, reptiles were the largest animals on Earth. Subsequently, mammals have assumed that niche. Small mammals may have escaped the fury of the heat and acid rain by remaining in burrows, and they may have survived the subsequent cold by virtue of their fur. You may be running your fingers through your hair, rather than your claws across your scales, because of that impact.

Other mass extinctions have occurred earlier and later than the Cretaceous event, and a few scientists believe that our planet is bombarded roughly every 26 million years. On the other hand, many geologists believe that massive volcanic eruptions or drastic changes in sea level might have the same effect. Thus, like so many of the most interesting issues in science, this story has no definitive explanation at this time.

---

*"Flea of the devil."

RE-MODELING

METEORITES CAN BE DEADLY

As astronomers have studied the details of the impact believed responsible for the Cretaceous extinction, they have come to realize that meteor impacts may be more common (and deadly!) than once believed. Although recent, well-documented cases of people being killed by meteorites are rare, they do occur. Careful searches through newspapers have also turned up many very near misses. Moreover, reports of meteorites hitting buildings and even cars are surprisingly common. Searching old records from Europe and China, astronomers have found many instances of fatalities. One of the grimmer of these reports is from a fifteenth-century Chinese chronicle that describes "stones that fell like rain" that killed 10,000 people. Even allowing for exaggeration in such old records, we can conclude that meteor falls can be far from harmless. In fact, astronomers conclude that although meteoritic disasters are rare, when they occur they have the potential to kill enormous numbers of people.

One well-documented event that proved harmless only because of its remoteness was a brilliant flash detected in February 1994 over the south Pacific ocean by a spy satellite. Analysis of the flash suggests that it marked the breakup of a 7-meter (20-foot) diameter meteoroid. The resulting blast had an energy between 10 and 20 kilotons of TNT, comparable to the bomb that destroyed Hiroshima. Had the object burst over a city, the casualties might have been in the hundreds of thousands.

SUMMARY

Our Solar System contains numerous small bodies: asteroids, meteors, and comets. They are important astronomically because they give us information about the time of formation, composition, and physical conditions in the solar nebula.

Asteroids are rocky, metallic, or carbon-rich objects found mainly in the asteroid belt between the orbits of Jupiter and Mars. Comets are icy bodies found mainly in the Oort cloud, far beyond the orbit of Pluto.

A comet becomes visible if a passing star or some other event disturbs the comet's orbit so that it drops in toward the inner Solar System. There solar heating thaws the frozen nucleus and evaporates gases. Radiation pressure and the solar wind then sweep the liberated gas and dust into a tail.

Some comets are caught in short-period orbits where they may be melted away to a collection of dust and grit. If the Earth crosses or nears the path of such a skeletal comet, the debris falling into the atmosphere causes a meteor shower.

The Earth is hit now and then by large asteroidal or cometary bodies, producing craters or, in very rare instances, mass extinctions.

QUESTIONS FOR REVIEW

1. What is the difference between a meteor, a meteoroid, and a meteorite?
2. How do we know what asteroids are made of?
3. Where are most asteroids found?
4. What are Apollo asteroids?
5. What shape are typical asteroids and how do we know? Why is Ceres not so shaped?
6. What do asteroids tell us about the formation of the Solar System?
7. What makes a "shooting star"?
8. How is a meteor heated?
9. What creates meteor showers? When do some occur?
10. What is the Oort cloud?
11. What is the life history of a comet from the Oort cloud to an object we see?
12. What creates a comet's tail?
13. Why are there two tails to some comets?
14. How do we know asteroids have a composition similar to that of some meteorites?
15. What evidence is there that the Earth has been hit by asteroids or comets?
16. What was the Tunguska event?
17. Why do some scientists believe that asteroids and comets play a role in mass extinctions?

PROBLEMS

1. Use Kepler's third law to find the semimajor axis of Halley's comet, given that its orbital period is 76 years.
2. With the result of problem 1 and the fact that for a very elliptical orbit the distance furthest from the Sun is roughly twice the semimajor axis, estimate how far from the Sun Halley's comet gets. What planet is about that same distance from the Sun? How does this distance compare with the distance to the Kuiper belt?
3. Use the formula for the kinetic energy of a moving body to estimate the energy of impact of a 1000-kilogram (roughly 1-ton) object hitting the Earth at 30 kilometers per second. Express your answer in kilotons of TNT, using the conversion that 1 kiloton is about 4×10^{12} joules. Note: Be sure to convert kilometers/second to meters/second.
4. Use Kepler's third law to determine the period of a comet whose orbit extends to 50,000 AU, within the inner Oort cloud.
5. Given that the temperature of a body decreases as the square root of its distance from the Sun, estimate the temperature of a comet nucleus in the Oort cloud. Take the temperature at 1 AU to be 300 Kelvin.

TEST YOURSELF

1. The tail of a comet
   (a) is gas and dust pulled off the comet by the Sun's gravity.
   (b) always points away from the Sun.
   (c) trails behind the comet, pointing away from the Sun as the comet approaches it, and toward the Sun as the comet moves out of the inner Solar System.
   (d) is gas and dust expelled from the comet's nucleus by the Sun's heat and radiation pressure.
   (e) both (b) and (d).
2. The bright streak of light we see as a meteoroid enters our atmosphere is caused by
   (a) sunlight reflected from the solid body of the meteoroid.
   (b) radioactive decay of material in the meteoroid.
   (c) a process similar to the aurora that is triggered by the meteoroid's disturbing the Earth's magnetic field.
   (d) frictional heating as the meteoroid speeds through the gases of our atmosphere.
   (e) the meteoroid's disturbing the atmosphere so that sunlight is refracted in unusual directions.
3. Meteor showers such as the Perseids in August are caused by
   (a) the breakup of asteroids that hit our atmosphere at predictable times.
   (b) the Earth passing through the debris left behind by a comet as it moves through the inner Solar System.
   (c) passing asteroids triggering auroral displays.
   (d) nuclear reactions in the upper atmosphere triggered by an abnormally large meteoritic particle entering the upper atmosphere.
   (e) None of the above.
4. Astronomers think that most comets come from
   (a) interstellar space.
   (b) material ejected by volcanic eruptions on the moons of the outer planets.
   (c) condensation of gas in the Sun's hot outer atmosphere.
   (d) small icy bodies in the extreme outer parts of the Solar System that are disturbed into orbits that bring them closer to the Sun.
   (e) luminous clouds in the Earth's upper atmosphere created when a small asteroid is captured by the Earth's gravitational force.
5. The asteroid belt lies between the orbits of
   (a) Earth and Mars.
   (b) Saturn and Jupiter.
   (c) Venus and Earth.
   (d) Mars and Jupiter.
   (e) Pluto and the Oort cloud.

FURTHER EXPLORATIONS

Brandt, John C., and Robert D. Chapman. *Rendezvous in Space: the Science of Comets:* W. H. Freeman and Co., New York: 1992.

Binzel, Richard P., M. Antonietta Barucci, and Marcello Fulchignoni. "The Origin of the Asteroids." *Scientific American* 265 (October 1991): 88.

Chapman, Clark R. "Worlds between Worlds." *Astronomy* 24 (June 1996): 46.

Cowen, Ron. "The Day the Dinosaurs Died." *Astronomy* 24 (April 1996): 34.

"Debate: What Caused the Mass Extinction?" *Scientific American* 263 (October 1990): 76.

Alvarez, Walter, and Frank Asaro. "An Extraterrestrial Impact."

Courtillot, Vincent E. "A Volcanic Eruption."

Delsemme, A. H. "Whence Come Comets?" *Sky and Telescope* 77 (March 1989): 260.

Dodd, R. T. *Thunderstones and Shooting Stars.* Cambridge, Mass.: Harvard University Press, 1986.

Gehrels, Tom. "Collision with Comets and Asteroids." *Scientific American* 274 (March 1996): 54.

Grieve, Richard A. "Impact Cratering on the Earth." *Scientific American* 262 (April 1990): 60.

Lewis, John S. *Rain of Iron and Ice.* Reading, Mass.: Addison-Wesley, 1996.

Luu, Jane X., and David C. Jewitt. "The Kuiper Belt." *Scientific American* 274 (May 1996): 46.

Marsden, Brian G. "Comet Swift-Tuttle: Does It Threaten Earth?" *Sky and Telescope* 85 (January 1993): 16.

Morrison, D., and C. R. Chapman. "Target Earth: It Will Happen." *Sky and Telescope* 79 (March 1990): 261.

Rao, Joe. "The Leonids: King of the Meteor Showers." *Sky and Telescope* 90 (November, 1995): 24.

"Special Issue: Comets." *Mercury* 25 (November–December, 1996).

Stern, Alan. "The Sun's Fab Four [Ceres, Pallas, Juno, and Vesta]." *Astronomy* 23 (June 1995): 30.

Verschuur, Gerrit L. *Impact! The Threat of Comets and Asteroids.* New York: Oxford University Press, 1996.

Weissman, Paul R. "Comets at the Solar System's Edge." *Sky and Telescope* 85 (January 1993): 26.

Web Sites

"Welcome to the Planets." Many pretty pictures, fact sheets on the planets and the other Solar System objects, and a glossary. Prepared by the Jet Propulsion Laboratory.
www.pds.jpl.nasa.gov/planets

"Views of the Solar System," by Calvin Hamilton. Excellent pictures, fact sheets, and glossary. Lots of links to other sites.
www.bang.lanl.gov/solarsys/

"The Nine Planets," by Bill Arnett. "A Multimedia Tour . . ." One of the best astronomy sites. Lots of excellent links, pictures, text, and glossary.
www.seds.org/billa/tnp/

KEY TERMS

Apollo asteroids, 304
asteroid belt, 300
asteroids, 300
carbonaceous chondrites, 300
chondritic meteorites, 299
chondrules, 299
coma, 306

fluorescence, 310
Kirkwood gaps, 304
meteor, 298
meteor showers, 311
meteorites, 299
meteoroid, 298
nucleus, 306

Oort cloud, 307
radiant, 311
radiation pressure, 309
short-period comets, 310
solar wind, 309
tail, 305

 PROJECTS

1. Pick a clear night and watch the night sky with a friend for half an hour. Count the number of meteors you see. Note the direction they come from. From your knowledge of these bodies, were the ones you saw (if any!) cometary or asteroidal?

2. Watch a meteor shower and count the number you see. Note the position the meteors seem to radiate from and mark it on a star map. If you are in a summer course, try for the Perseids in mid-August (August 10–14), but check with your instructor, a Web site, or some popular astronomy magazine for more precise dates. For autumn courses, try for the Orionids (October 18–23) or better yet, the Leonids (early morning November 14–19). The Leonids promise to be very good for 1997–1999. Later in the season you might try for the Geminids (as early as 10 P.M. December 12–14). For spring courses, try for the Lyrids (April 20–22) or the η Aquarids (May 2–7). What was the average time interval between meteors? Does your estimate of the radiant match the predicted position?

Stars

We owe our existence to stars. Our own star—the Sun—warms us and drives our weather. Some other star, long dead, created the atoms of which we and our planet are made. But beyond such practical considerations, stars are simply beautiful (fig. OV5.1).

Stars are almost unimaginably far away. The nearest is more than 20 trillion miles (40 trillion km) from the Earth, or in more convenient units, 4 light-years. Measuring such vast distances is difficult, and astronomers use two main techniques. For "nearby" stars, they observe the star at two times of year separated by 6 months so that the Earth is on opposite sides of its orbit (fig. OV5.2). From these widely spaced points, astronomers can get a "stereo" view of nearby stars, much as our two eyes give us a stereo view. From that stereo view, the star's distance can be found mathematically. The distance to more remote stars is measured by comparing their brightness to that of stars of known distance, just as you can estimate the distance to a street light by comparing its brightness to that of a nearby light.

FIGURE OV5.1
Photograph of the Pleiades (Courtesy Anglo-Australian Observatory/Photographs by David Malin.)

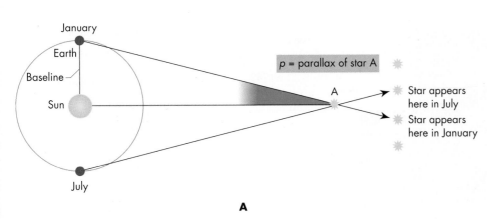

FIGURE OV5.2
Measuring the distance to a star by parallax.

The immense distance to stars creates many barriers to knowledge of them. Even though the most powerful telescopes, most stars are merely points of light. Studying stars with remotely controlled spacecraft is also hopeless. Present spacecraft would take tens of thousands of years to reach even the nearest star. Thus astronomers must use indirect means for studying stars. For example, we saw in chapter 3 how astronomers analyze a star's light to reveal its composition (fig. OV5.3) and how they deduce its temperature from its color (fig. OV5.4). They can measure its mass with a modification to Kepler's third law if a star has a companion star orbiting it (a fairly common occurrence) (fig. OV5.5).

Many of the things that astronomers would like to know about a star, such as its age, how it was born, and how it will die, they cannot learn by any observational technique. To answer questions such as these, astronomers use mathematical models. The results of such calculations are summarized below.

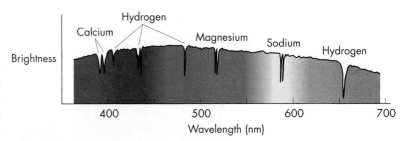

FIGURE OV5.3
Sketch of the spectrum of a star.

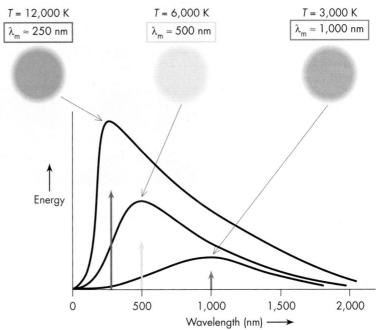

FIGURE OV5.4
Deducing the temperature of a hot object from its color.

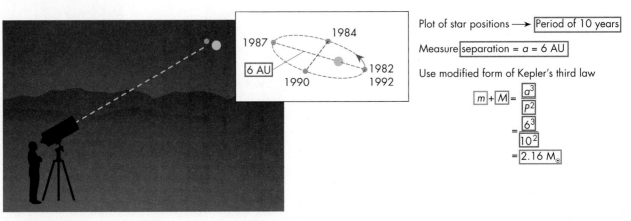

Plot of star positions ⟶ Period of 10 years

Measure separation = a = 6 AU

Use modified form of Kepler's third law

$$m + M = \frac{a^3}{P^2}$$
$$= \frac{6^3}{10^2}$$
$$= 2.16 \, M_\odot$$

FIGURE OV5.5
Measuring the mass of a binary star pair from its orbit.

The outward pressure force balances the inward gravitational force everywhere inside a star.

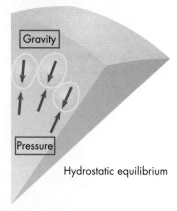

Hydrostatic equilibrium

FIGURE OV5.6
Inside a star, pressure outward must balance the gravitational pull inward.

A star is a massive ball of gas held together by the collective gravity of its innumerable atoms. A star's great mass creates—according to the law of gravity—a tremendous inward crushing force. That force is counterbalanced by the pressure created by the random careening of the star's atoms off one another (fig. OV5.6). At the same time, gravity's inward pull also heats the star by compressing its gas.

That heating is immense, and it needs to be to create a pressure large enough to support the star against its own gravity. Because gravity's force increases with mass, as a rule, the more massive a star is, the hotter and brighter it is, a result known as the mass-luminosity law (fig. OV5.7).

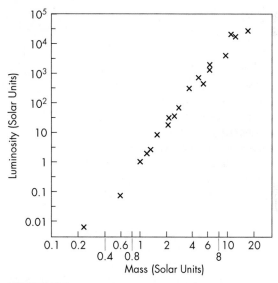

FIGURE OV5.7
The more massive a star, the more luminous it is (the Mass–Luminosity law).

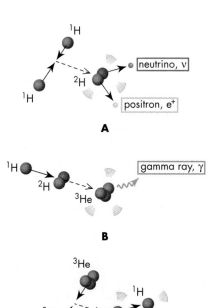

FIGURE OV5.8
Sketch of how hydrogen atoms fuse into helium to supply a star's energy.

Some of the heat a star needs to support its mass leaks outward and escapes into space. We see the heat lost from the Sun as sunlight and the heat lost from other stars as starlight. Stars in general need to replenish the energy they lose as starlight, or they will collapse. The process that generates that energy in most stars, including the Sun, is the conversion of hydrogen atoms into helium atoms (fig. OV5.8). Normal stars like the Sun are made mostly of hydrogen and helium (roughly 75% and 25% respectively), and they have enough fuel to burn for millions of years or more. Our Sun has enough fuel to burn at its present rate for approximately another 5 billion years.

But even billions of years is not forever, and so a star will eventually use up its fuel. As its supply of hydrogen dwindles, a star adjusts its structure to maintain a balance between gravity and pressure. The first adjustment is that the star's core shrinks. This compresses the core and makes it hotter. As it heats, the core releases more energy, which pushes the surface layers outward, expanding the star. As the star grows larger, its surface cools and grows redder (fig. OV5.9). The star thus becomes what astronomers call a red giant. Our Sun will become a red giant some 5 billion years from now.

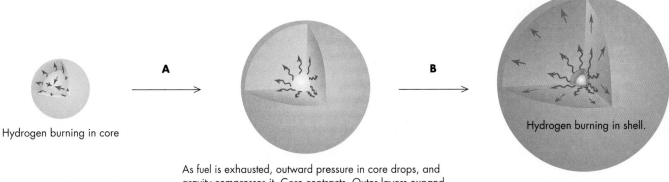

Hydrogen burning in core

A

As fuel is exhausted, outward pressure in core drops, and gravity compresses it. Core contracts. Outer layers expand. Rising heat in contracting core creates pressure that causes outer layers to expand. Shell of H outside core ignites.

B

Hydrogen burning in shell.

Outer layers expand and cool. A red giant forms.

FIGURE OV5.9
Sketch illustrating the evolution of a star as it ages.

Becoming a red giant doesn't mark the end of a star's life. When a star exhausts its hydrogen fuel supply, it may be able to use other fuels. For example, the Sun will be able to use helium as a fuel (converting it to carbon) for another few billion years, but eventually the helium will be used up, and the Sun will again adjust its structure, growing far brighter, larger, and cooler. In fact it will becomes so cool that tiny dust particles of carbon and silicates will condense in its outer layers, where they will be swept outward by the flood of radiation pouring from the Sun's core. The dust will drag gas along with it, peeling off the Sun's outer layers to form a slowly expanding shell surrounding its hot, collapsed core (fig. OV5.10). With no fuel left to burn, the core will cool and shrink to about the size of the Earth, becoming what astronomers call a white dwarf.

FIGURE OV5.10
Sketch of how a low mass star ejects its outer layers to become planetary nebula as it nears the end of its life.

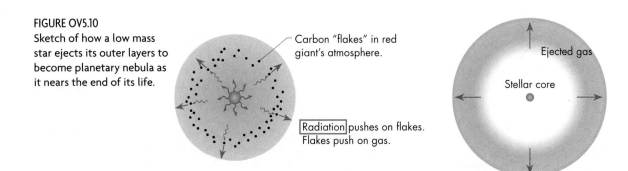

Carbon "flakes" in red giant's atmosphere.

Radiation pushes on flakes. Flakes push on gas.

Ejected gas

Stellar core

The expanding shell of gas is called a planetary nebula (fig. OV5.11), and it glows from the heat it receives from the star's exposed core. The gas gradually mingles with other gas in space to eventually become a new generation of stars.

Stars much more massive than the Sun are hot enough in their cores that they can fuse the carbon in their cores into heavier elements, starting with oxygen, and building up to iron. Such "nucleosynthesis" in stars that have long been dead created the chemical elements that compose our planet, the air we breath, and our very bodies.

The formation of an iron core marks the end for a massive star. Iron atoms cannot fuse to release energy, and the star's huge mass crushes its core and triggers a collapse and explosion—what astronomers call a supernova (fig. OV5.12). These final stages in a star's life may leave a dim remnant of the star's core. It may collapse into a ball about 20 kilometers across and made of neutrons, or if it is sufficiently massive, it may collapse and become a black hole. A black hole gets its name because its gravity's pull is so strong that light itself cannot escape from it.

How can we make such claims about the lives of stars? We obviously can't watch an individual star age. However, by observing groups of stars, we can deduce the stages in a star's life, just as by looking at a group of people (infants, teenagers, young adults, and old people) we can infer how an individual will age. Studies of star groups show that the above sketches of how stars form and grow old are at least essentially correct, although few astronomers will be surprised if our knowledge of the details of the lives of these amazing structures changes.

FIGURE OV5.11
Photograph of a planetary, gas ejected from a dying star (Courtesy Anglo-Australian Observatory/Photographs by David Malin.)

FIGURE OV5.12
Photograph of a supernova remnant, the debris blown out of a high mass star as it explodes at the end of its life. (Courtesy Anglo-Australian Observatory/Photographs by David Malin.)

CHAPTER ELEVEN

THE SUN, OUR STAR

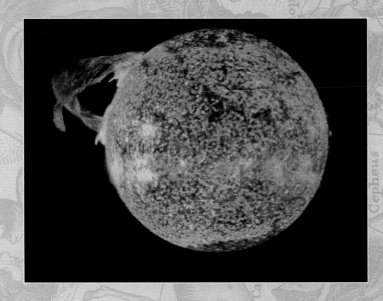

The Sun as viewed from space through a special filter (Courtesy of NRL).

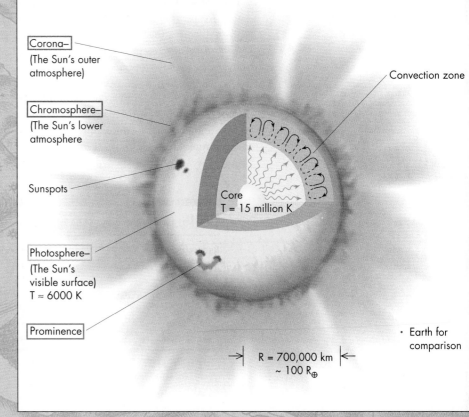

Corona–
(The Sun's outer atmosphere)

Chromosphere–
(The Sun's lower atmosphere

Sunspots

Photosphere–
(The Sun's visible surface)
T ≈ 6000 K

Prominence

Convection zone

Core
T = 15 million K

· Earth for comparison

R = 700,000 km
~ 100 $R_\oplus$

A cutaway sketch of the Sun

The Sun is a star, a dazzling luminous ball of gas, more than 100 times bigger in diameter than the Earth. Although to the naked eye it gleams peacefully, a telescope reveals the Sun's surface to be violently agitated, with rising fountains of incandescent gas and twisted magnetic field. Even greater violence wracks its core. There, a nuclear "furnace" every *second* burns 600,000,000 tons of hydrogen into helium, producing in one heartbeat the energy of 100 billion nuclear bombs. How that energy is released, and how the Sun manages to hold itself together is the main theme of this chapter. We will begin by describing the Sun: its radius, mass, and so on. Then we will discuss how the crushing force of its gravity balances the explosive power in its core. Finally we will see how energy escaping from the core stirs its atmosphere into the inferno we see.

11.1 SIZE AND STRUCTURE

The Sun is immense, dwarfing the Earth and even Jupiter, and its immensity is what makes it shine. With a radius over 100 times that of the Earth, and a mass 300,000 times that of the Earth, the Sun has an enormous gravity that crushes the material in its interior. To offset that crushing force and prevent its own collapse, the Sun must be extremely hot. But hot objects always lose energy, and the Sun is no exception. We see that lost energy as sunshine, and welcome it as the source of our life. But sunshine is a death warrant for the Sun because the energy it carries off must be replenished, or the Sun will collapse. Fortunately for us, the Sun does replace its lost energy, but only at the cost of consuming itself—a dilemma that is not unique to the Sun but is shared by most stars.

Before we discuss how the Sun replaces its lost energy, we need to understand better some of its overall properties. How much mass must the Sun support? How rapidly does it lose energy? What resources has it available to supply its energy needs? Table 11.1 lists some of these vital statistics, and figure 11.1 shows the Sun's portrait, illustrating its main features. It will be helpful many times in this chapter to refer to this picture.

Astronomers can measure some of the Sun's properties, such as its diameter and surface temperature. However, most of its other properties, for example, its internal temperature and density, they can only deduce. Such deductions are based on computer models that

TABLE 11.1	Properties of the Sun
Radius ($R_\odot$*)	7×10^8 m, or about 109 R_{Earth}
Mass ($M_\odot$*)	2×10^{30} kg, or about 300,000 M_{Earth}
Distance	1.5×10^8 km, or 1 AU
Temperature of surface	5780 Kelvin (approx 9900 °F)
Temperature of core	15 million Kelvin (approx 27 million °F)
Composition	71% hydrogen, 27% helium, 2% heavier elements
Power output	4×10^{26} watts

* Astronomers use the symbol $\odot$ to stand for the Sun. Thus $R_\odot$ is the Sun's radius and $M_\odot$ is its mass. The symbol is the ancient Egyptian hieroglyph for "Sun."

FIGURE 11.1
Cutaway picture of the Sun, showing its
interior and atmosphere.

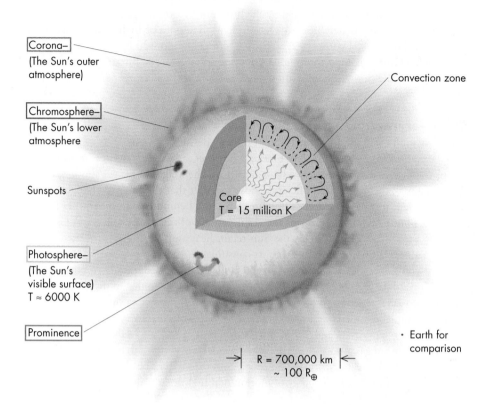

Corona–
(The Sun's outer
atmosphere)

Chromosphere–
(The Sun's lower
atmosphere

Sunspots

Photosphere–
(The Sun's
visible surface)
T ≈ 6000 K

Prominence

Convection zone

Core
T = 15 million K

Earth for
comparison

R = 700,000 km
~ 100 $R_\oplus$

use the laws of physics to predict the Sun's properties. The correctness of the model's predictions is then judged by whether the predictions agree with observable properties. For
example, if a model predicted the Sun's surface temperature to be 10,000 K, it would be
rejected as incorrect. Thus our understanding of the Sun's properties comes from a combination of theory and measurement.

Measuring the Sun's Properties

The Sun is about 150 million kilometers (1 AU, or 93 million miles) from Earth. Astronomers originally measured its distance by triangulation, but they now use radar, either
bounced directly from the Sun or from other bodies whose distance is known in AU from
Kepler's third law.

Once the Sun's distance is known, we can find its radius from its angular size, as we
showed in chapter 1. We also need to know the Sun's distance if we are to measure its mass
with the help of Kepler's third law, as we showed in chapter 2. From its mass and radius
we can calculate that its surface gravity is about 30 times the Earth's. It is against this
crushing force that the Sun must struggle.

The Sun's internal heat balances the crushing force of its gravity. Its surface temperature can be found from its color and Wien's law using the methods of chapter 3. Astronomers cannot measure its interior temperature directly, but from calculations and indirect measurements (discussed in section 11.3), they deduce that its core temperature is
about 15 million K (approximately 27 million °F).

From measurements of the amount of solar energy that reaches the Earth, astronomers calculate that the Sun radiates a total of 4×10^{26} watts of power into space from its surface, energy that must be replenished by the fusion of hydrogen in its core, as we will see in section 11.2. Fortunately, the Sun has a plentiful supply of hydrogen: its spectrum shows it is about 71% hydrogen, 27% helium, and 2% vaporized heavier elements such as carbon and iron, a composition similar to Jupiter's and Saturn's. But unlike these mostly liquid bodies, the Sun is gaseous throughout because its high temperature breaks most molecular bonds, vaporizing even iron, and allowing the atoms to move freely as a gas.

The Solar Interior

When we look at the Sun, we see through the low density, tenuous gases of its outer atmosphere. Our vision is ultimately blocked, however, as we peer deeper into the Sun. There, the material is compressed to high density by the weight of the gas above it. In this dense material, the atoms are sufficiently close together that they absorb enough light to limit our view, just as frosted glass obscures what lies behind it. This obscuring layer is called the **photosphere**, and it forms the visible surface of the Sun. The blocking is a nuisance to astronomers, but it helps the Sun retain heat and, like a well-insulated house, thereby reduces the amount of fuel it must consume.

If we could see into the Sun, we would find that its density and temperature rise steadily inward. In the photosphere, the density is comparable to that of the air around us; deeper, the material above pushes down on that below, compressing the gas like a pile of pillows. A similar compression occurs in the atmosphere of the Earth and other planets, as we discussed in chapters 4 and 9. But the greater mass of the Sun leads to a vastly greater compression of its gas, and so near the core, the density is more than 100 times that of water. Despite this great density, the Sun is gaseous throughout because its high temperature gives the atoms so much energy of motion that they are unable to bond with one another to form a liquid or solid substance.

The temperature also rises as we plunge into the Sun's interior. The photosphere is 6000 K*, but below it, heat is partially trapped and the temperature soars to about 15 million K at the core. Figure 11.2, based upon theoretical calculations, illustrates how the temperature and density change through the Sun. No spacecraft has or is ever likely to make the measurements directly, but we are confident that they are correct because the Sun needs such high temperatures and densities to keep it from collapsing under its own gravity.

FIGURE 11.2
Plots of how density and temperature change through the Sun.

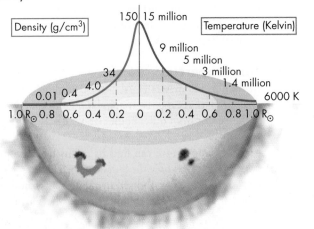

*More precisely, the temperature of the Sun's photosphere is 5780 K.

Energy Transport

Your own experience and experiments in the laboratory show that heat always flows from hot to cold. Applying this principle to the Sun, we can therefore infer that because its core is hotter than its surface, heat will flow outward from its center, as illustrated in figure 11.3. Near the core, the energy moves by radiation carried by photons through what is called the **radiative zone**. Because the gas there is so dense, a photon travels less than an inch before it is absorbed by an atom and stopped. The photon is eventually re-emitted, but it will be almost immediately reabsorbed. The constant absorption and re-emission slows photons like cars caught in stop-and-go traffic. Even though photons travel at the speed of light between absorptions, it takes them nearly a million years to move from the core to the surface. Thus, today's sunshine was born in the Sun's core before the birth of human civilization!

The flow of energy from the core toward the Sun's surface is slowed in the region just below the photosphere where the gas is cooler and less transparent. Here, photons are even less effective in moving energy, and convection currents like those in the Earth and giant planets carry the energy to the surface. The rising and sinking gas occupies the **convection zone,** and we can infer the gas's motion there from the numerous tiny, bright regions surrounded by narrow darker zones, called **granulation** (fig. 11.4). The bright areas are

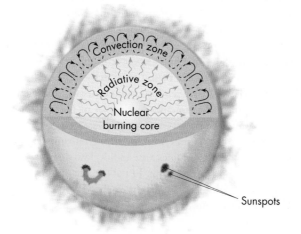

Sunspots

FIGURE 11.3
Sketch of how energy flows from the Sun's core to its surface. In the deep interior, radiation carries the energy. Near the surface, convection carries the energy.

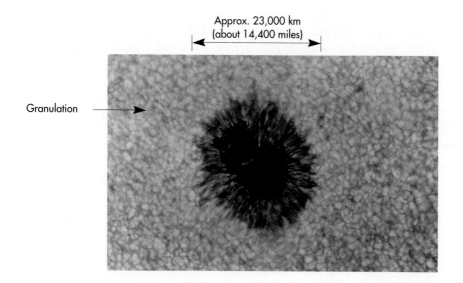

Approx. 23,000 km
(about 14,400 miles)

Granulation

FIGURE 11.4
Photograph of solar granulation near a sunspot. The bright patches are immense bubbles of hot gas rising from deep within the Sun. The darker areas around them are cooler gas sinking back into the Sun's interior. The sunspot is about 23,000 kilometers in diameter. (Courtesy William Livingston/NOAO.)

☀ How big across is one of the rising bubbles that forms the granulation? (You can use the sunspot to estimate the size of the granulation.)

FIGURE 11.5
Photograph of a portion of the solar chromosphere at a total solar eclipse. (Courtesy Dennis di Cicco.)

FIGURE 11.6
Photograph of spicules in the chromosphere. The spicules are the thin, stringy features that look like tufts of grass. (Courtesy NOAO.)

bubbles of hot gas many hundreds of kilometers across, rising from deep within the Sun. They are brighter because they are hotter than the gas around them. On reaching the surface, these hot bubbles radiate their heat to space and cool. That cooler matter then sinks back toward the hotter interior where it is reheated and rises again to radiate away more heat. Astronomers can measure the speed of these up and down motions using the Doppler effect and find that the bubbles rise at hundreds of kilometers per second.

The Solar Atmosphere

Astronomers refer to the extremely low density gases that lie above the photosphere as the Sun's atmosphere. This region marks a gradual change from the dense gas of the photosphere to the extremely low density gas of interplanetary space. A similar transition occurs in our own atmosphere, where the gas density decreases steadily with altitude and eventually merges with the near-vacuum of space.

Although the gas density in the Sun's atmosphere decreases above the photosphere, the gas temperature behaves very differently. Immediately above the photosphere the temperature decreases, but at higher altitudes the gas grows hotter, reaching temperatures of several million Kelvin. Why the Sun's atmosphere is so hot remains a mystery, though astronomers believe that the Sun's magnetic field plays a role in somehow heating these low-density gases, as we will discuss in section 11.4.

The Sun's atmosphere consists of two main regions. Immediately above the photosphere lies the **chromosphere,** the Sun's lower atmosphere. It is usually invisible against the glare of the photosphere but can be seen at a total eclipse of the Sun as a thin red zone around the Sun (fig. 11.5). With a telescope, you can see that the chromosphere contains millions of thin columns called **spicules,** each a jet of hot gas thousands of kilometers long, as shown in figure 11.6.

The chromosphere's color* comes from the strong red emission line of hydrogen, Hα. We saw in chapter 3 that emission lines arise in hot, low-density gas. From these lines astronomers can infer the gas's temperature. Just above the photosphere, the temperature is about 4500 K, but 2000 kilometers higher, it reaches 50,000 K. Here, the chromosphere ends and the temperature shoots up to about 1 million degrees as we enter the **corona,** the Sun's outer atmosphere.

The corona's extremely hot gas has such low density that under most conditions we look right through it. But like the chromosphere, we can see it at a total solar eclipse when the Moon covers the Sun's brilliant disk. Then the pale glow of the corona can be seen to extend beyond the Sun's edge to several solar radii (fig. 11.7). Pictures of the Sun made at x-ray wavelengths show that the corona is not uniform but has streamers created by the Sun's magnetic field. In addition, the corona contains huge regions of cooler gas called **coronal holes** through which gas may escape from the Sun into space, as we will discuss in section 11.4.

Because the corona is so tenuous, it contains very little energy despite its high temperature. It is like the sparks from a Fourth of July sparkler: despite their high temperature, you hardly feel them if they land on your hand because they are so tiny and thus carry very little total heat.

---

*The prefix *chromo-* means "colored."

FIGURE 11.7
Photograph of the corona at a total eclipse of the Sun. (Courtesy Dennis di Cicco.)

11.2 HOW THE SUN WORKS

Our discussion so far has centered on the structure of the Sun. Our task now is to understand *why* it has that structure and how it works. For example, why is the Sun hot? What makes it shine? How does it generate energy?

Internal Balance (Hydrostatic Equilibrium)

The structure of the Sun depends on a balance between its internal forces. One force holds the Sun together. A second force prevents the Sun from collapsing. This balance is technically called **hydrostatic equilibrium.**

The Sun's inward force arises from its own gravity. The outward force arises from the rapid motion of its atoms, a motion that gives rise to a **pressure.** Thus, in the Sun, as in virtually all stars and planets, the balance of hydrostatic equilibrium requires that the *outward* force created by pressure exactly balance the *inward* force of the Sun's gravity (fig. 11.8).

The outward pressure force balances the inward gravitational force everywhere inside the Sun.

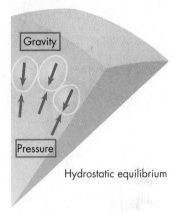

Hydrostatic equilibrium

FIGURE 11.8
Sketch illustrating the condition of hydrostatic equilibrium (balance of pressure and gravitational force) in the Sun.

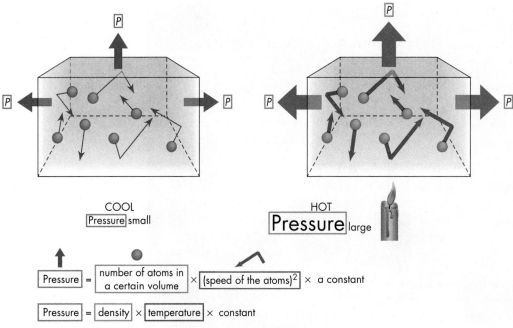

FIGURE 11.9
Sketch illustrating the perfect, or ideal, gas law. Gas atoms move faster at the higher temperature. When they collide, they therefore exert a greater force, creating a higher pressure. Thus, other things being equal, a hotter gas exerts a greater pressure.

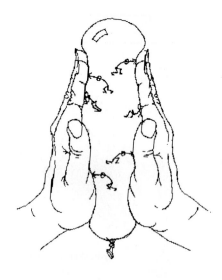

There are many examples of hydrostatic equilibrium in everyday life. Our atmosphere is in hydrostatic equilibrium, its gases pulled downward by the Earth's gravitational force but supported by collisions of the air molecules, creating pressure. Likewise, pressure in an automobile tire supports the weight of the car.

Without such a balance, the Sun would rapidly change. For example, if its pressure were too weak, the Sun's own gravity would rapidly crush it. Therefore, to understand the Sun, we need to discuss in more detail how its pressure arises.

Pressure in the Sun

Pressure in a gas comes from collisions among its atoms and molecules. If the gas is squeezed, atoms are pushed toward each other. As they collide, they rebound, resisting the compression, as you can feel by squeezing a balloon. The strength of the pressure depends on how often and how hard the collisions occur. Raising the density increases collisions by moving atoms closer together. Raising the temperature speeds atoms up, making them collide harder (fig. 11.9). Thus the strength of the pressure is proportional to the density times the temperature of the gas; that is:

$$\text{Pressure} = \text{Density} \times \text{Temperature} \times \text{a constant*}$$

a result known as the **perfect**, or **ideal, gas law.**

Although the gas law shows us that the pressure inside the Sun depends on temperature, it does not by itself show us how hot that temperature needs to be. We might reasonably guess that because the Sun has a huge mass (and therefore a huge gravity crushing it), it needs a huge temperature to offset that crushing force.

Figure 11.2 bears this out, showing that the Sun's core is very hot. To determine exactly how hot, astronomers solve a computer model of the Sun, and the answer turns out to be about 15 million K for the temperature and about 150 grams per cubic centimeter for the density, as we claimed earlier. But because heat flows out of the hot core, the Sun needs a mechanism to keep replacing its core heat.

---

*The value of the constant depends on the units used to measure the pressure, density, and temperature.

Powering the Sun

Energy that leaves the core eventually escapes into space as sunshine: heat and light. That heat loss must be replaced, or the Sun's internal pressure would drop, and the Sun would begin to shrink under the force of its own gravity. The Sun is therefore like an inflatable chair with tiny leaks through which the air escapes. If you sit in the chair, it will gradually collapse under your weight unless you pump air in to replace that which escapes. What acts as the energy pump for the Sun?

Early astronomers believed that the Sun might burn ordinary fuel such as coal. But even if the Sun were pure coal, it could burn only a few thousand years. At the end of the last century, the English physicist Lord Kelvin and the German scientist Hermann Helmholtz independently proposed that the Sun is not in hydrostatic equilibrium but that gravity slowly compresses it, making it shrink. In their theory, compression (increasing density) heats the gas and makes the Sun shine. However, gravity could power the Sun by this mechanism only for 10 million years, and we know the Sun has been shining for billions of years because we have fossils of life on Earth that old. Furthermore, if gravity were the power source, the Sun would be steadily shrinking, and it isn't. Therefore, something else must supply energy.

In 1899, T. C. Chamberlin suggested that subatomic energy—energy from the reactions of atomic nuclei—might power stars, but he could offer no explanation of how the energy was liberated. In 1905, Einstein proposed that energy might come from a body's mass. His formula,

$$E = mc^2$$

states that a mass, m, can become an amount of energy, E, equal to the mass multiplied by the square of the speed of light, c, where $c = 3 \times 10^8$ meters per second. The amount of energy available from reactions involving atomic nuclei is vastly larger than that from chemical processes such as burning. For example, if 1 gram of mass—less than the amount in an aspirin tablet—is converted to energy, it releases 9×10^{13} joules, the equivalent of about 20 kilotons of TNT* or a small nuclear weapon. If the Sun could convert even a tiny amount of its mass into energy, it would have an enormous source of power.

In 1919, the English astrophysicist A. S. Eddington, a pioneer in the study of the physics of stars, showed that the conversion of hydrogen into helium would release enough energy to power the Sun, but even his theory lacked the necessary details, which were supplied independently in the late 1930s by the physicists Hans A. Bethe and Carl F. von Weizsäcker. Their work showed that the Sun generates its energy by converting hydrogen into helium by a process called **nuclear fusion,** a process that bonds two or more nuclei into a single, heavier one.

Nuclear Fusion. Fusion is possible in the Sun because its interior is so hot. Under normal conditions, hydrogen nuclei repel each other, pushed apart by the electrical charge of the protons. However, at high temperatures, nuclei move so fast that when they collide, they are driven extremely close together. At such a collision, the nuclei may be brought sufficiently close so that the electrical repulsion between their protons is overwhelmed by the **nuclear** or **strong force,** the force that holds the protons together in a normal atom. Thus, the two separate nuclei can merge, or fuse, into a single new nucleus. Because this fusing process requires such a high temperature, the only place in the Sun hot enough for fusion to occur is its core. The core therefore is where the Sun makes its energy. However, before we can understand how fusion creates energy, we need to look at the structure of hydrogen and helium.

c = The speed of light: 3×10^8 m/s
m = An amount of mass
E = An amount of energy

---

*1 kiloton is about 4.2×10^{12} joules of energy.

Key

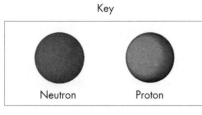

Neutron   Proton

^{1}H

^{2}H

^{3}He

^{4}He

FIGURE 11.10
Schematic diagrams of hydrogen and helium atoms and isotopes. (Electrons are omitted for clarity.)

Gamma rays are the highest energy form of electromagnetic radiation known. In some countries they are used to sterilize food, allowing such products as milk and meat to be stored, if covered, at room temperature without spoiling.

The Structure of Hydrogen and Helium. Hydrogen consists of one proton and an orbiting electron, and helium consists of two protons, two neutrons, and two orbiting electrons (fig. 11.10). In the Sun's hot interior, however, atoms collide so violently that the electrons are generally removed; that is, the atoms are completely ionized and form what scientists call a "plasma." Because the electrons are removed from most atoms and move independently in the gas, we can therefore ignore them in much of what follows.

Hydrogen and helium always have one and two protons respectively, but they have other forms (called "isotopes") with different numbers of neutrons. To identify the isotopes, we write their chemical symbol with a superscript that shows the total number of protons and neutrons. The usual form of hydrogen with one proton and no neutrons is ^{1}H, while the form of hydrogen containing a proton and a neutron is ^{2}H. Likewise, the normal form of helium with two protons and two neutrons is ^{4}He, while helium with two protons but only one neutron is ^{3}He. As we will see, these isotopes play a role in the energy supply of the Sun.

The Proton-Proton Chain. Hydrogen fusion in the Sun occurs in three steps called the **proton-proton chain**. In the first step, two ^{1}H nuclei collide and fuse to form the isotope of hydrogen, ^{2}H. In the collision, one proton becomes a neutron by ejecting two particles: a positron (denoted as e^+) and a neutrino (denoted by ν). The neutrinos so generated play no further role in the Sun's energy generation, but we will encounter them again because they can be detected when they leave the Sun and may help astronomers learn more about conditions in the Sun's interior.

This first step in converting the mass of an atom into the Sun's energy is depicted in figure 11.11A and can be written symbolically as

$$^1H + {}^1H \rightarrow {}^2H + e^+ + \nu + \text{Energy}$$

The terms to the left of the arrow are the normal hydrogen nuclei that start the process. The terms to the right are the heavy isotope of hydrogen that results (^{2}H), plus the positron and the neutrino. Energy is released by this reaction because the mass of ^{2}H is slightly less than the mass of the two ^{1}H's that were used to make it. The missing mass is converted to energy, as described in Einstein's formula $E = mc^2$, and ultimately that energy becomes the electromagnetic radiation of sunlight.

In the second step, the ^{2}H nucleus collides with a third ^{1}H to make the isotope of helium containing a single neutron, ^{3}He. This process releases a high-energy photon (gamma ray), which we denote by γ. Figure 11.11B shows this step, which can be written as:

$$^1H + {}^2H \rightarrow {}^3He + \gamma + \text{Energy}$$

Here again, the resulting particle, ^{3}He, has a smaller mass than the particles from which it was made, and again energy is released.

The third and final step is the collision and fusion of two ^{3}He nuclei. Here, the fusion results not in a single particle, but rather one ^{4}He and two ^{1}H nuclei. You can think about this reaction as the attempt to form a nucleus with 6 protons and 4 neutrons, except that 2 protons are ejected by their electric repulsion, as shown in figure 11.11C. Symbolically we can write it as:

$$^3He + {}^3He \rightarrow {}^4He + {}^1H + {}^1H + \text{Energy}$$

where again, the final mass is less than the initial mass.

We can find the quantity of energy released by comparing the initial and final masses and using the mass-energy relationship $E = mc^2$. Steps 1 and 2 use three ^{1}H, but the first two steps must occur twice to make the two ^{3}He's for the last step. Therefore, six ^{1}H's are used, but two are returned in step 3, and so a total of four ^{1}H's are needed to make each ^{4}He. If we now add up their masses, we find the following:

The mass of a single hydrogen nucleus is 1.673×10^{-27} kilograms. The mass of a helium nucleus is 6.645×10^{-27} kilograms. Thus,

$$
\begin{aligned}
4 \times 1 \text{ hydrogen} &= 6.693 \times 10^{-27} \text{ kg} \\
- 1 \text{ helium} &= 6.645 \times 10^{-27} \text{ kg} \\
\hline
\text{Mass lost} &= 0.048 \times 10^{-27} \text{ kg}
\end{aligned}
$$

Multiplying this by c^2 gives the energy yield per helium atom made, E. That is,

$$E = 0.048 \times 10^{-27} \times (3 \times 10^8)^2 = 0.048 \times 10^{-27} \times 9 \times 10^{16} = 4.3 \times 10^{-12} \text{ joules}$$

This tiny number by itself is insignificant, but when multiplied by the vast number of hydrogen atoms undergoing fusion in the Sun, the total energy released is equivalent to exploding 100 billion-megaton H-bombs per second! Our sunshine has a violent birth.

FIGURE 11.11
Diagram of the proton-proton chain.

11.3 PROBING THE SUN'S CORE

Although the immense bulk of the Sun hides its core from view, two experiments allow us an indirect look. Solar neutrinos escaping from the interior bear information of conditions there, and waves on the Sun's surface reveal the structure of the gas far below as waves in a stream reveal the presence of submerged rocks.

Solar Neutrinos

We saw in section 11.2 that the Sun makes neutrinos as it converts hydrogen into helium. The number of released neutrinos therefore tells us how rapidly hydrogen is being converted into helium, and from that we can deduce the temperature and density in the Sun's core.

But counting neutrinos is extremely difficult. **Neutrinos** have neither electric charge nor mass,* which gives them phenomenal penetrating power. They escape from the Sun's core through its outer 700,000 kilometers and into space like bullets through wet Kleenex. They pass straight through the Earth and anything on the Earth, such as you, and keep going. In fact, roughly a trillion neutrinos from the Sun pass harmlessly through your body every second.

The elusiveness that allows neutrinos to slip so easily through the Sun makes them slip with equal ease through detectors on Earth. Nevertheless, because so many neutrinos are produced, we need trap only a tiny fraction of them to get useful information. This trapping is done with neutrino detectors.

---

*Whether the mass is truly zero or just very small is still under study.

The penetrating power of neutrinos is the theme of a poem called "Cosmic Gall" by the novelist and poet, John Updike:

Neutrinos, they are very small.
They have no charge and have no mass
And do not interact at all.
The earth is just a silly ball
To them, through which they simply pass,
Like dustmaids down a drafty hall
Or photons through a pane of glass.
They snub the most exquisite gas,
Ignore the most substantial wall,
Cold-shoulder steel and sounding brass,
Insult the stallion in his stall,
And scorning barriers of class,
Infiltrate you and me! Like tall
And painless guillotines, they fall
Down through our heads into the grass.
At night they enter at Nepal
And pierce the lover and his lass
From underneath the bed—you call
It wonderful: I call it crass.

The first solar neutrino detector was a tank of cleaning fluid (perchloroethylene) the size of a swimming pool located a mile underground in an old gold mine in South Dakota. Currently the largest neutrino "telescopes" are the Super-Kamiokande detector (fig. 11.12) located deep in a zinc mine west of Tokyo and the Sudbury Neutrino Observatory more than a mile underground in a nickel mine in northern Ontario, Canada. The Super-Kamiokande detector contains 50,000 tons of water. The Sudbury detector contains about 1000 tons of heavy water, water consisting of molecules in which one of the ordinary hydrogen atoms (^{1}H) is replaced by "heavy hydrogen," ^{2}H.

The detectors are buried to shield them from the many other kinds of particles besides neutrinos that constantly bombard the Earth. For example, protons, electrons, and a variety of subatomic particles constantly shower our planet. These particles, traveling at nearly the speed of light, are called **cosmic rays** and are thought to be created in deep space when a massive star explodes.

Cosmic rays can penetrate only a short distance into the Earth; so if a detector is located deep underground, nearly all the cosmic rays are filtered out. Neutrinos, however, are unphased by a mere mile of solid ground: on the average, they could travel through a light year of lead!

Given their virtual unstoppability, how are neutrinos detected? In the water detectors, a neutrino streaking through the tank occasionally hits an electron. The impact imparts to the electron some of the neutrino's energy, causing the electron to recoil and emit a tiny burst of light. This light can be detected and used to deduce not only the number of neutrinos, but also the direction from which they come.

FIGURE 11.12
An inside view of the Super-Kamiokande neutrino detector before it was filled with water. The thousands of small round objects forming a grid are the detectors that record the light emitted when a neutrino strikes an electron in the water. The four bright white objects are simply lights so you can see what the interior of the detector looks like. (Courtesy Yoji Totsuka).

When astronomers examine the results of the chlorine and water detectors, they find that the number of neutrinos is about three times smaller than predicted—a large and worrisome discrepancy—if the Sun is making neutrinos as it burns hydrogen. Astronomers are therefore left in a quandary as to what to believe.

There are at least two explanations for the discrepancy seen in the chlorine and water detectors: (a) The model of the Sun's interior is wrong, and the Sun makes fewer neutrinos than astronomers think; or (b) neutrinos have properties that are not understood and so the experiments capture fewer than predicted.

In weighing the evidence at this time, astronomers increasingly believe that the discrepancy results from lack of knowledge of the properties of neutrinos. For example, scientists now believe that, in addition to the kind of neutrinos formed in the Sun, two other kinds exist. Moreover, they think that the solar neutrinos may change into either of these other neutrino types inside the Sun or on their way to Earth. Unfortunately, these other neutrino types are even harder to detect than ordinary neutrinos. Thus, the Sun may in fact be creating exactly the number of neutrinos expected, but some of them change into one or the other of the two less detectable forms before reaching Earth. It would indeed be ironic if studying the Sun led scientists to a better understanding of these tiny, mysterious particles. Yet that sort of unexpected outcome is one of the excitements of science.

Solar Seismology

Solar seismology is the study of the Sun's interior by analyzing waves in the Sun's atmosphere. We learned in chapter 4 that scientists can study the Earth's interior by analyzing earthquake waves. Astronomers can also learn about the Sun's interior by analyzing waves in its gases.

Waves similar to those of earthquakes travel through the Sun and make its surface heave like the ocean or a bubbling pan of hot oatmeal. The rising and falling surface gas makes a regular pattern (fig. 11.13), which can be detected as a Doppler shift of the moving material. Astronomers next use computer models of the Sun to predict how the observed surface waves are affected by conditions in the Sun's deep interior. With a relatively new technique, astronomers can now measure the density and temperature deep within the Sun from the pattern and speed of the waves in its atmosphere. The results they find agree well with those of the current models, indicating that our understanding of the Sun is correct.

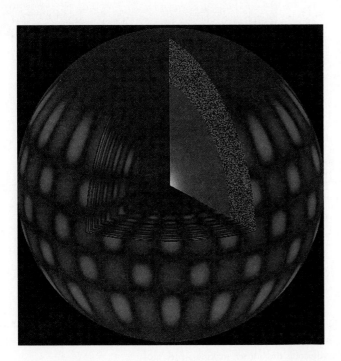

FIGURE 11.13
Computer diagram of solar surface waves. (Courtesy NOAO.)

11.4 SOLAR MAGNETIC ACTIVITY

The surface waves described above are but one of many disturbances in the Sun's outer regions. A wide class of dramatic and lovely phenomena on the Sun are caused by its magnetic field. This magnetic activity is also of interest because it affects the Earth, where it triggers auroral displays and appears to alter the climate.

Sunspots

Sunspots are the most common type of solar magnetic activity. They are large, dark-appearing regions (fig. 11.14) ranging in size from a few hundred to many thousand kilometers across. Spots last from a few days to over a month. They are darker than the surrounding gas because they are cooler* (4500 K as opposed to 6000 K of the normal photosphere), and they are cooler because they contain strong magnetic fields.

FIGURE 11.14
White-light photograph of photosphere with sunspots. (Courtesy NOAO.)

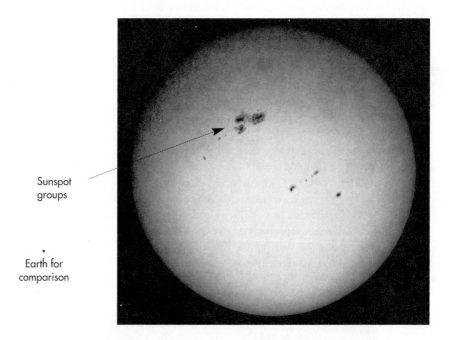

Sunspot groups

· Earth for comparison

FIGURE 11.15
Sketch of how magnetic fields cause sunspots.

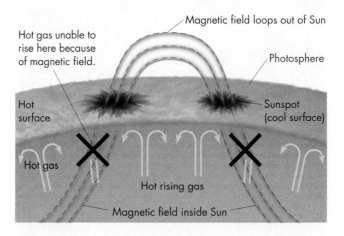

---

*Spattering a few drops of water onto a hot electric stove burner will show a similar effect. Each drop momentarily cools the burner, making a dark spot.

Solar Magnetic Fields

The magnetic field of sunspots is more than a thousand times stronger than the Earth's or the "normal" field of the Sun. In such intense fields, electrons and other charged particles spiral around the field, "frozen" to it, as shown in the sketch at left. They are thereby forced to follow the magnetic field. On the Earth, the field drives particles toward the poles and creates the auroras. In the Sun, the field slows the ascent of hot gas in the convection zone and, starved of heat from below, the surface cools and becomes darker, making a sunspot, as shown in figure 11.15.

Prominences and Flares

Prominences and flares are magnetic disturbances in the low-density, virtually transparent, hot gases of the Sun's atmosphere. **Prominences** are huge plumes of glowing gas that jut from the lower chromosphere into the corona. Figure 11.16 shows some of their variety. You can get some sense of their immensity from the white dot, which shows the size of the Earth.

Prominences form where the Sun's magnetic field reduces heat flow to a region. They are *cooler* than the gas around them, which means, according to the perfect gas law, that the pressure inside is less than outside. Thus the hot external gas "bottles up" the cooler gas of the prominence (fig. 11.17). Under favorable conditions, this cooling gas, trapped in its magnetic prison, may glow for weeks.

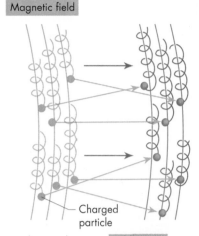

Magnetic field

Charged particle

Particles spiral around field lines

Field moves.
Because the particles must spiral around the field lines, they are dragged along as the field moves.

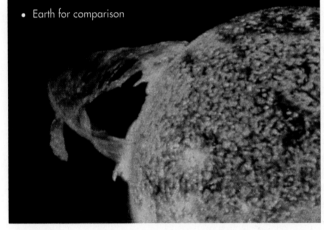

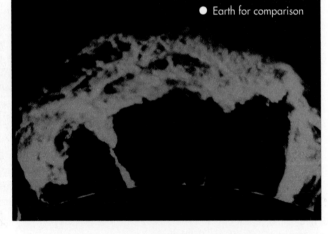

• Earth for comparison

● Earth for comparison

FIGURE 11.16
Photographs of some solar prominences. (Courtesy NOAO/NRL.)

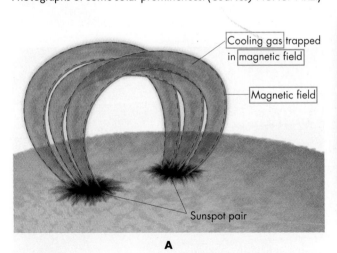

Cooling gas trapped in magnetic field

Magnetic field

Sunspot pair

A

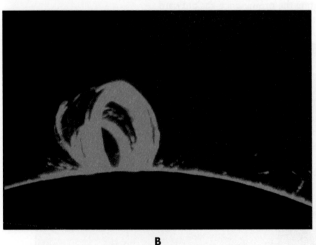

B

FIGURE 11.17
(A) Sketch illustrating how magnetic fields support a prominence. (B) Photograph of a prominence. (Courtesy NOAO.)

☀ Approximately how big is the prominence in (B)? Use the fact that the Sun's diameter is about 100 times the Earth's.

Time-lapse movies show that gas streams through prominences, sometimes rising into the corona, sometimes raining down onto the photosphere. The flow is channeled by and supported by the magnetic field, which often arcs between sunspots. Thus prominences also are related to sunspots.

Sunspots also give birth to **solar flares,** brief but bright eruptions of hot gas in the chromosphere (fig. 11.18). Over a few minutes or hours, gas near a sunspot may dramatically brighten. Such eruptions, though violent, are so localized they hardly affect the total light output of the Sun at all. Generally, you need a telescope to see their visible light, though they can increase the Sun's radio and x-ray emission by factors of a thousand in a few seconds.

Flares are poorly understood, but magnetic fields probably play an important role. One theory suggests that the field near a spot gets twisted by gas motions, a bit like winding up a rubber-band powered toy. But such twisting can go only so far before the rubber-band breaks. So, too, the magnetic field can be twisted only so far before it suddenly readjusts, whipping the gas in its vicinity into a new configuration. The sudden motion heats the gas, and it expands explosively. Some gas may even escape from the Sun and shoot across the inner Solar System to stream down on the Earth. Such a burst created the spectacular auroral displays seen in March 1989 and illustrated in figure 11.19.

Heating of the Chromosphere and Corona

Although the Sun's magnetic field cools sunspots and prominences, it heats the chromosphere and corona. Other stars also have chromospheres and coronas, and so it is additionally important to understand the Sun's. To begin, we need first to recall that the temperature of a gas is a measure of how fast its particles are moving. Anything that speeds atoms up increases their temperature.

An analogy may help you understand how magnetic waves can heat a gas. When you crack a whip, a slow motion of its handle travels as a wave along the whip. As the whip tapers, the wave's energy of motion is given to an ever smaller piece of material. With the same amount of energy and less mass to move, the tip accelerates and eventually breaks the sound barrier. The whip "crack" is a tiny sonic boom.

A similar speedup occurs in the Sun's atmosphere when magnetic waves formed in the photosphere move into the corona along the Sun's field lines (fig. 11.20). As the atmospheric gas thins, the wave energy is imparted to an ever smaller number of atoms, making them move faster. But "faster" in this case means hotter. Thus the upper atmosphere heats up as the waves travel into it.

FIGURE 11.18
(A) Photographs of solar flares. In (B), the flare has erupted below a prominence and is blasting the prominence gas outward. (Courtesy NOAO.)

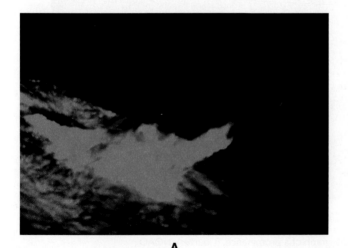

A   **B**

FIGURE 11.19
Photograph of the great aurora of March 1989. (Courtesy Eugene Lauria.)

Look up the origin of the word *aurora* in a dictionary. Does the appearance of the aurora indicate why that origin is appropriate?

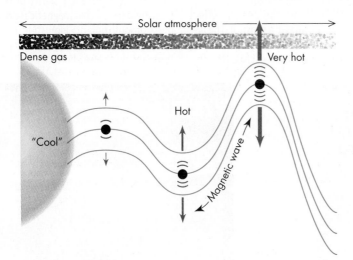

FIGURE 11.20
Diagram illustrating how magnetic waves heat the Sun's upper atmosphere. As the waves move outward through the Sun's atmosphere, they grow larger, imparting ever more energy to the gas through which they move, thereby heating it.

What generates the magnetic waves? They probably start in the convection zone where rising bubbles of gas shake the magnetic field and create magnetic waves just as shaking the loose end of a rope makes it wiggle. Thus the high temperature of the chromosphere and corona are another example of the importance of the Sun's magnetic field and its convection zone. This theory is supported by observations of stars other than the Sun with active convection zones, which also have active chromospheres and coronas.

The Solar Wind

In addition to the mass it loses in the outbursts of flares, the Sun undergoes a steady, less dramatic loss of mass. The corona's high temperature gives its atoms enough energy to escape the Sun's gravity. As these atoms stream into space, they form the **solar wind**, a tenuous flow of mainly hydrogen and helium that sweeps across the Solar System. The amount of material lost from the Sun is small: less than one ten-trillionth of its mass each year. Nevertheless, the solar wind creates comet tails, as we saw in chapter 10, and auroras, as we saw in chapter 4.

The solar wind arises because, unlike the rest of the Sun, the corona is *not* in hydrostatic equilibrium. Recall that the temperature in the Sun's atmosphere increases with altitude, making the corona much hotter than the photosphere. The corona's high temperature, according to the perfect gas law, creates a pressure within it larger than we might otherwise expect for its distance above the photosphere. The pressure is in fact sufficient to overcome the Sun's gravitational force on gas in its upper atmosphere. As a result it pushes that material outward into space. The expanding gas has a very low density (only a few hundred atoms in a cubic centimeter—the volume of a thimble). The gas atoms begin their outward motion slowly but accelerate with increasing distance as the Sun's gravitational attraction on them weakens. On the average, the wind speed is about 500 kilometers per second at the Earth's orbit, but it speeds up and slows down in response to changes in the Sun's magnetic field. From the Earth outward the wind coasts at a relatively steady speed that carries it at least to the orbit of Neptune. At some point, it impinges on the interstellar gas surrounding the Solar System, but where that happens has not yet been found.

EXTENDING OUR REACH

DETECTING MAGNETIC FIELDS: THE ZEEMAN EFFECT

Astronomers can detect magnetic fields in sunspots and other astronomical bodies by the **Zeeman effect,** a physical process in which the magnetic field splits some of the spectrum lines of the gas into two, three, or more components. The splitting occurs because the magnetic field alters the atom's electron orbits, which in turn alters the wavelength of its emitted light.

Figure 11.21A shows the Zeeman effect splitting spectrum lines in a sunspot. The line is single outside the spot but triple within, and by mapping the splitting across the Sun's face, astronomers can map the Sun's magnetic field, creating a magnetogram, as seen in figure 11.21B. The colors in a magnetogram show the strength and polarity of the magnetic field (recall that magnetic fields have a north or south polarity indicating their direction). Notice that the field is strong around spots and weak elsewhere.

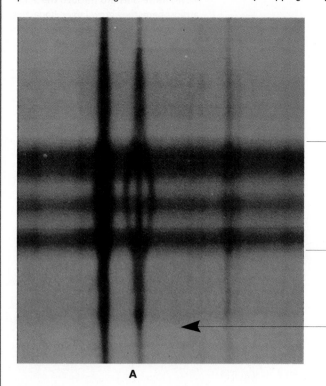

Region in sunspot

Region outside of sunspot

A

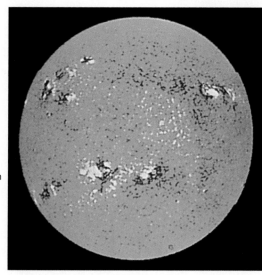

B

FIGURE 11.21

(A) Photograph of the Zeeman effect in a sunspot. Notice that the line is split over the spot where the magnetic field is strong but that the line is unsplit outside the spot where the field is weak or absent. (B) Magnetogram of the Sun. Yellow indicates regions with north polarity and dark blue indicates regions with south polarity. Notice that the polarity pattern of spot pairs is reversed between the top and bottom hemispheres of the Sun. (Courtesy NOAO.)

11.5 SOLAR CYCLE

Sunspot and flare activity change from year to year in what is called the **solar cycle**. This variability can be seen in figure 11.22, which shows the number of sunspots detected over the last 130 years. The numbers clearly rise and fall approximately every 11 years. For example, the cycle had peaks in 1947, 1958, 1969, 1980, and 1990.

Samuel Heinrich Schwabe, a pharmacist and amateur astronomer, discovered the solar cycle. His work prevented him from observing at night; so he studied the Sun during the day. (It is sometimes claimed jokingly that he invented and named the cotton schwabe.)

Flares and prominences also follow the solar cycle, and climate patterns on Earth may, too. For this reason, astronomers have sought to understand not only the cause of the solar cycle, but also how it influences terrestrial climate.

Cause of the Solar Cycle

As the Sun rotates, gas near its equator circles the Sun faster than gas near its poles; that is, it spins differentially, a property common in gaseous bodies (recall from chapter 9 that Jupiter and the other giant planets rotate differentially). The Sun's differential rotation is such that its equator rotates in about 25 days and its poles in 30. Thus a set of points arranged from pole to pole in a straight line would move over the course of time into a curve, as shown in figure 11.23.

FIGURE 11.22
Plot of sunspot numbers showing solar cycle. (Courtesy John A. Eddy.)

☀ Is the interval between the peaks always exactly 11 years? By how much does it change?

☀ Are we currently near a high or low of solar activity?

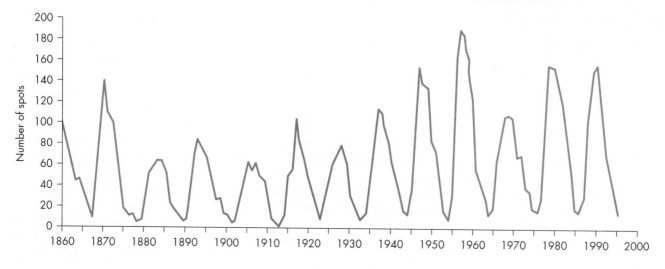

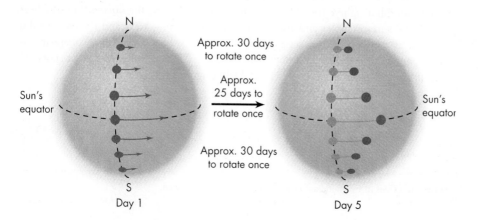

FIGURE 11.23
Sketch showing solar differential rotation. Points near the Sun's equator rotate faster than points near the poles.

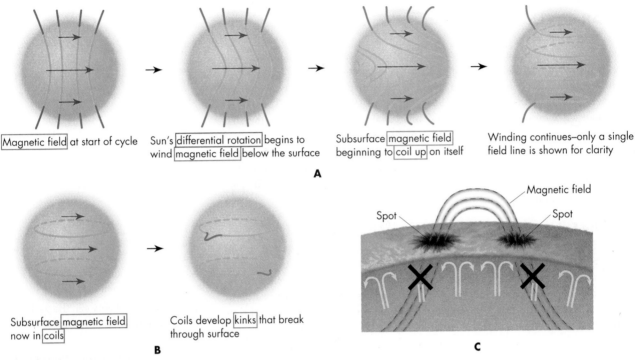

Magnetic field at start of cycle

Sun's differential rotation begins to wind magnetic field below the surface

Subsurface magnetic field beginning to coil up on itself

Winding continues—only a single field line is shown for clarity

A

Subsurface magnetic field now in coils

Coils develop kinks that break through surface

B

Spot   Magnetic field   Spot

C

FIGURE 11.24
Sketch showing, (A), the possible winding up of the subsurface magnetic field.
(B) Fields penetrating the Sun's surface.
(C) Formation of a spot pair.

Differential rotation should similarly distort the Sun's magnetic field, "winding up" the field below the Sun's surface.* Astronomers think such winding of the Sun's magnetic field may cause the solar cycle, though the exact mechanism is still not well understood. According to one hypothesis, the Sun's rotation wraps the solar magnetic field into coils below the surface, making the field stronger and increasing solar activity: spots, prominences, and flares. The wrapping occurs because the Sun's magnetic field is "frozen" into the gas, as discussed in section 11.4. Thus, if the gas moves, so does the field, and vice versa.

Because the field and gas are tightly connected, differential rotation causes gas at the equator, which is moving faster than the gas at the poles, to drag the magnetic field with it so that a field, initially straight north to south, is wound into two subsurface loops, as shown in figure 11.24A. As the loops are wound tighter, they develop kinks, as when you twist a rubber band too tight. The cycle ends when the field twists too "tightly" and collapses, and the process repeats.

Sunspots form when kinks in the magnetic field rise to the Sun's surface and break through the photosphere (fig. 11.24B). Here, the field slows the outward flow of heat, making the surface cooler and darker than in surrounding areas and thereby creating sunspots, as we discussed in section 11.4. Each kink breaks the photosphere in *two* places—one where it leaves and one where it enters. We therefore expect that spots will occur in pairs or paired groupings, as sketched in figure 11.24C.

---

*Spots and cycles of magnetic activity occur on other stars as well.

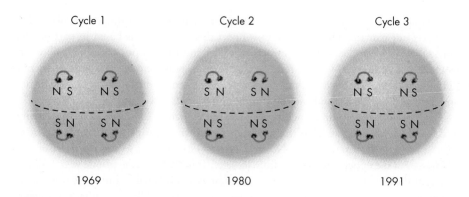

FIGURE 11.25
Diagram illustrating the approximate 22-year periodicity of solar magnetic activity.

In such pairs, one grouping has a north polarity and the other a south polarity.* That is, in one the field emerges from the surface, while in the other it descends so that the field direction is reversed. The false-color map of the Sun's field that we saw in figure 11.21B shows this effect clearly. Yellow areas have a north polarity and dark blue areas a south polarity. Notice that the pattern of the polarity differs between the two hemispheres of the Sun. In one, the yellow areas are on the right, while in the other they are on the left. This reversal arises because the subsurface field is coiled in opposite directions in the two hemispheres, additional evidence that sunspots and the solar cycle are caused by winding of the Sun's magnetic field.

Changes in the Solar Cycle

The solar cycle is not always 11 years: it may be as short as 7 or as long as 16 years. Moreover, if we consider the polarity of the spot groups, the cycle averages 22 years, rather than 11, because the polarity of the Sun's field reverses at the end of each 11-year cycle. It therefore takes *two* 11-year cycles for the field to return to its original configuration.

Figure 11.25 illustrates this effect. In the first frame we see spot pairs as the cycle begins. A right-hand spot in the top hemisphere (technically the "leading spot" because it leads in the direction the Sun rotates) has a south polarity, while a left-hand spot (technically called a "trailing spot") has a north polarity.

You can also see an additional feature of the cycle: all leading spots in one hemisphere have the same polarity. In the other hemisphere, the leading spots have the opposite polarity. Eleven years later, the polarity of the pairs will be reversed, as you can see by looking at the second frame in figure 11.25. Only in the next cycle, another 11 years later, will the spot fields have returned to their original directions. Thus the full cycle of magnetic activity takes 22 years on the average.

Links between the Solar Cycle and Terrestrial Climate

Climatologists find this 22-year period interesting because of a cycle in which droughts recur approximately every 22 years in the midwestern United States and Canada. Does the Sun's cycle affect the Earth's climate cycle? If so, how? One possibility is as follows.

The Sun's magnetic field heats the corona. The corona drives the solar wind. The solar wind alters the Earth's upper atmosphere; in particular it changes the way the temperature varies with altitude. This in turn alters the atmosphere's circulation and may shift the jet stream to a new location. The jet stream steers storms and hence rainfall.

---

*Polarity of magnetic fields was described briefly in chapter 4 when we discussed the Earth's magnetic field.

Although this hypothesis cannot be verified yet, many scientists think that solar activity affects our climate. The evidence to support this hypothesis is based in part on the work of E. W. Maunder, a British astronomer who studied sunspots. Maunder noted in 1893 that, according to old solar records, very few sunspots were seen between 1645 and 1715 (fig. 11.26). He concluded that the solar cycle turned off during that period. The period is now called the **Maunder minimum** in honor of his discovery.

The Maunder minimum coincides with an approximately 70-year spell of abnormally cold winters in Europe and North America. Glaciers in the Alps advanced; rivers froze early and remained frozen late; the North Sea froze. The cold was so abnormal that meteorologists call the epoch part of the "little ice age." If only one such episode were known, we might dismiss the sunspot-climate connection as a coincidence, but three other cold periods have also occurred during times of low solar activity. This strengthens our belief that somehow the Sun's magnetic activity affects our climate.

Although scientists are still unsure about what creates the link between solar magnetic activity and our climate, they no longer doubt that such a link exists. Figure 11.27 shows how the ocean temperature (expressed as deviations from the normal average) has changed from 1860 to 1980. The figure also shows how the number of sunspots has changed over the same time span. Notice that when the number of spots is high, the ocean is warmer than average and when the number of spots is low, the ocean is colder than average.

FIGURE 11.26
Plot illustrating that the number of sunspots changes with time, showing the Maunder minimum and the solar cycle. (Courtesy John A. Eddy.)

Can you see a second, slower cycle in the sunspot numbers? (Compare the peak numbers near 1810 and 1900 with the numbers near 1850 and 1950.)

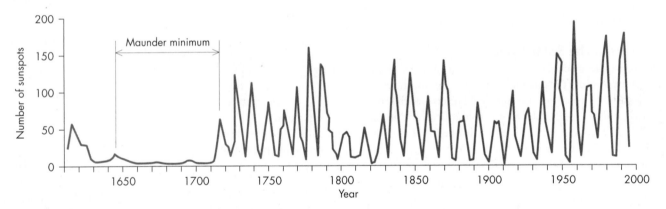

FIGURE 11.27
Curves showing the change in ocean temperatures on Earth and the change in sunspot numbers over several decades. Notice that the curves change approximately in step. Astronomers deduce from these curves that solar magnetic activity alters our climate. (The spot numbers are averaged over 11 year intervals).

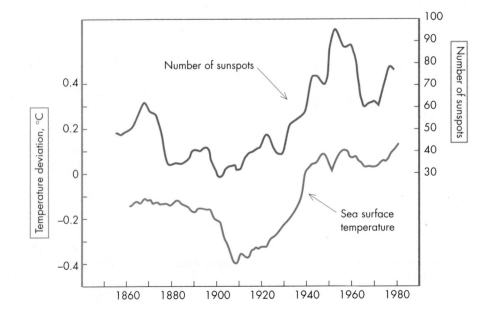

SUMMARY

The Sun is composed mostly of hydrogen and helium atoms. It is held together by gravity and supported against gravitational collapse by the pressure created by its high internal temperature. This temperature is high enough in the core that hydrogen can be converted into helium by nuclear fusion, thereby supplying the Sun with energy. That energy then flows to the surface by radiation and convection.

Above the Sun's visible surface (the photosphere), lie the chromosphere and corona—low-density, hot atmospheric layers. They are heated by magnetic waves traveling outward from the photosphere. In the corona, the temperature reaches a million degrees and drives the corona gases into space, creating the solar wind.

The Sun's magnetic field varies in structure and strength. As the subsurface field strengthens, sunspots form where the field breaks through the surface. Near spots, prominences and flares may occur. Spot numbers rise and fall with a roughly 11-year period. For reasons that are not yet well understood, this variation in solar magnetic activity also creates climatic variation on the Earth.

QUESTIONS FOR REVIEW

1. How big is the Sun compared to the Earth?
2. How can we measure the Sun's size, mass, and temperature?
3. What is the Sun made of? How do we know?
4. What holds the Sun together?
5. Why doesn't the Sun collapse?
6. Why must the interior of the Sun be so hot?
7. How is solar energy generated?
8. How does energy get to the Sun's surface from its core?
9. What visible evidence do we have that the Sun has a convection zone?
10. What are the photosphere, chromosphere, and corona?
11. Which of these layers is hottest? How do we know?
12. What is meant by solar activity?
13. Why do sunspots appear dark?
14. How do a prominence and a flare differ?
15. What role does magnetic activity play in solar activity?
16. How do we know there are magnetic fields in the Sun?
17. What is meant by solar neutrinos? Why do astronomers say there is a discrepancy? Why is the discrepancy of interest?
18. What is solar seismology? What does it tell us about the Sun?
19. What is the solar cycle?
20. What is the period between maximum sunspot numbers? How does this differ from the period of the full solar cycle?
21. What is the Maunder minimum? Why is it of interest?

THOUGHT QUESTIONS

1. Why can the Sun not be powered by a chemical process such as the combustion of hydrogen and oxygen to form water?
2. Why is hydroelectric power generation an indirect form of solar power?
3. The Sun's corona has a temperature of millions of degrees K. Why does it not incinerate us?
4. Why is sailing an indirect form of solar power?

PROBLEMS

1. Given that the angular diameter of the Sun is 1/2 degree and that its distance is 1.5×10^8 kilometers, go through the math to determine the Sun's diameter.
2. Suppose you were an astronomy student on Jupiter. Use the orbital data for Jupiter (distance from Sun = 5.2 AU; period = 11.8 years) to measure the Sun's mass using the modified form of Kepler's third law.
3. In this problem you will calculate approximately the temperature of the Sun's core. You can do it either step by step, or by writing out all the algebra to obtain a final result. You will need the following ideas: The pressure in a gas is given by the perfect gas law ($P = 8300\,\rho T$); the density of a body, ρ, is its mass per volume, which for a sphere is $M/(4\pi R^3/3)$; the pressure force from the interior is $P \times A$, where A is the area over which

the pressure acts. You can take that as the Sun's cross-section, πR^2. Thus the pressure force is $\pi R^2 P$. Finally, you need to invoke hydrostatic equilibrium: pressure forces must balance gravitational forces. Approximate the gravitational force holding the Sun together by assuming it is split into two equal halves and apply Newton's law of gravity to calculate the force between the halves. Assume they are separated by 1 solar radius.

4. Calculate the escape velocity from the Sun.

TEST YOURSELF

1. The diameter of the Sun is about how large compared with the Earth's?
   (a) Twice as big.
   (b) One half as big.
   (c) 10 times as big.
   (d) 100 times as big.
   (e) 10,000 times as big.

2. About how many years elapse between times of maximum solar activity?
   (a) 3.
   (b) 5.
   (c) 11.
   (d) 33.
   (e) 105.

3. Sunspots are dark because
   (a) they are cool relative to the gas around them.
   (b) they contain 10 times as much iron as surrounding regions.
   (c) nuclear reactions occur in them more slowly than in the surrounding gas.
   (d) clouds in the cool corona block our view of the hot photosphere.
   (e) the gas within them is too hot to emit any light.

4. The Sun is supported against the crushing force of its own gravity by
   (a) magnetic forces.
   (b) its rapid rotation.
   (c) the force exerted by escaping neutrinos.
   (d) gas pressure.
   (e) the antigravity of its positrons.

5. The Sun produces its energy from
   (a) fusion of neutrinos into helium.
   (b) fusion of positrons into hydrogen.
   (c) disintegration of helium into hydrogen.
   (d) fusion of hydrogen into helium.
   (e) electric currents generated in its core.

FURTHER EXPLORATIONS

Akasofu, Syun-Ichi. "The Shape of the Solar Corona." *Sky and Telescope* 88 (November 1994): 24.

Edberg, Stephen J. "Discover the Daytime Star." *Astronomy* 23 (February 1995): 66.

Foukal, Peter V. "The Variable Sun." *Scientific American* 262 (February 1990): 34.

Hathaway, David H. "Journey to the Heart of the Sun." *Astronomy* 23 (January 1995): 38.

Kennedy, James R. "GONG: Probing the Sun's Hidden Heart." *Sky and Telescope* 92 (October 1996): 20.

Kippenhahn, Rudolph. *Discovering the Secrets of the Sun.* New York: Wiley, 1994.

Lang, Kenneth R. *Sun, Earth, and Sky.* Berlin, New York: Springer-Verlag, 1995.

Lang, Kenneth R. "Unsolved Mysteries of the Sun: [Parts I and II]." *Sky and Telescope* 92 (August 1996): 38, no. 92 (September 1996): 24.

Marsden, Richard G., and Edward L. Smith. "Ulysses: Solar Sojourner." *Sky and Telescope* 91 (March 1996): 24.

Mechler, Gary. *The Sun and the Moon.* New York: Knopf, 1995.

Mims, Forrest M. "Sunspots and How to Observe Them Safely." *Scientific American* 262 (June 1990): 130.

Nesme-Ribes, Elizabeth, Sallie L. Baliunas, and Dmitry Sokoloff. "The Stellar Dynamo." *Scientific American* 275 (August 1996): 45.

Noyes, Robert W. *The Sun, Our Star.* Cambridge, Mass.: Harvard University Press, 1982.

Web Sites

The High Altitude Observatory has an attractive set of images of the Sun showing spots, prominences, and so forth at
www.hao.ucar.edu/public/slides/slides.html

The sites below all have some Sun information.
bang.lanl.gov/solarsys/sun.htm
www.seds.org/billa/tnp/sol.html

KEY TERMS

chromosphere, 330

convection zone, 329

corona, 330

coronal holes, 330

cosmic rays, 336

granulation, 329

hydrostatic equilibrium, 331

Maunder minimum, 346

neutrinos, 335

nuclear force, or strong force, 333

nuclear fusion, 333

perfect gas law, or ideal gas law, 332

photosphere, 328

pressure, 331

prominences, 339

proton-proton chain, 334

radiative zone, 329

solar cycle, 343

solar flares, 340

solar wind, 341

spicules, 330

sunspots, 338

Zeeman effect, 342

 ## PROJECTS

1. Measure the diameter of the Sun. Take a piece of thin, dark, cardboard and put a small hole in it. Hold it in sunlight about one meter (about 3 feet) from a piece of white paper so that a small image of the Sun appears on the paper (see fig. 11.28A). Carefully measure the distance (d) between the cardboard and the piece of paper and the size of the Sun's image (s) on the paper. On a separate piece of paper draw two straight lines that cross with a small angle between them (see fig. 11.28B). Draw two small circles between the lines as shown in figure 11.28B. Convince yourself that if D is the distance to the Sun (1 AU) and S is its diameter, then $S/D = s/d$, where s is the size of Sun's image on the paper and d is the distance from the paper to the cardboard. Look up the value of D in table 11.1. Then solve for S, the Sun's diameter. Does your value agree with the value you calculate from the value of the Sun's radius in table 11.1?

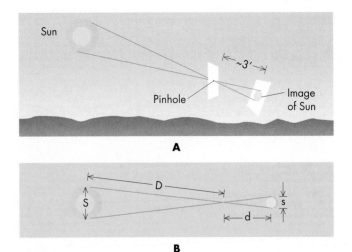

FIGURE 11.28
A simple way to measure the Sun's diameter.

Continued

PROJECTS—*Continued*

2. Observing sunspots. Mount a pair of binoculars on a tripod or some other fixed object. You can use duct tape if you cannot find a binocular clamp and tripod. Hold a piece of white paper about 60 centimeters (roughly two feet) from the eyepieces of the binoculars and adjust them so that the Sun's image falls on the paper (fig. 11.29). You may need to adjust the focus wheel of the binoculars to make the Sun's image sharp and clear on the paper. DO NOT LOOK AT THE SUN THROUGH THE BINOCULARS. THE SUN'S LIGHT CAN QUICKLY AND PERMANENTLY BLIND YOU IF YOU DO. Can you see any sunspots on the Sun's image on the paper? (They will appear as small dark blotches). Draw a circle on the paper to match the size of the Sun's image. Mark the date on the paper and sketch the location of the spots that you see on the circle. Repeat your observations a few days later. Are the spots in the same location? If they have shifted their position, can you think of a reason why?

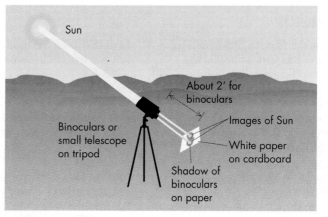

FIGURE 11.29
Observing sunspots with a pair of binoculars.

CHAPTER TWELVE

MEASURING THE PROPERTIES OF STARS

A region where stars are forming from interstellar dust and gas clouds (Courtesy of HST)

It is hard to believe that the stars we see in the night sky as tiny glints of light are in reality huge, dazzling balls of gas, many vastly larger and brighter than our Sun. Stars (except for the Sun) look dim to us only because they are so far away—several light years (trillions of miles) even to the nearest. Such remoteness creates tremendous difficulties for astronomers trying to understand the nature of stars; direct probing is clearly impossible. In this chapter we will see how astronomers overcome the distance barrier that separates us from stars and how they learn the physical properties of these distant objects. How far away are they? How big? What are they made of?

The answers to these questions show us that most stars are remarkably like the Sun. For example, like the Sun, they are composed mostly of hydrogen and helium and have about the same mass. A small percentage, however, are as massive as 30 times the Sun's mass (30 $M_\odot$) and are much hotter than the Sun and blue in color. Others are much less massive than the Sun, only one tenth its mass, and are cool, red, and dim. Even in stars similar to the Sun in composition and mass, their radii and therefore density may differ enormously. Some giant stars have a radius hundreds of times larger than the Sun's—so big that were the Sun their size it would extend beyond the Earth's orbit. On the other hand, some stars are white dwarfs, with as much material as the Sun packed into a volume the size of the Earth.

Astronomers can learn all these properties of stars by using physical laws and theories to interpret measurements made from the Earth. For example, theories of light yield the surface temperature, distance, and motion of a star; theories of atoms yield the composition of a star; and a modified form of Kepler's law yields the mass of a star. In using such laws, astronomers may sometimes employ more than one method to determine a desired property of a star. For example, a star's temperature may be measured from either its color or its spectrum. Such alternative methods serve as checks on the correctness of the procedures astronomers use to determine the properties of stars.

Using parallax to find distance may seem unfamiliar, but parallax creates our stereovision, the ability to see things three dimensionally. When we look at something, each eye sends a slightly different image to the brain, which then processes the pictures to determine the object's distance. You can demonstrate the importance of the two images by covering one eye and trying to pick up a pencil quickly. You will probably not succeed in grasping it on the first attempt.

12.1 MEASURING A STAR'S DISTANCE

One of the most difficult problems astronomers face is measuring the distance to stars and galaxies. Yet knowing the distances to such bodies is vital if we are to understand their size and structure. For example, without knowledge of a star's distance, astronomers find it difficult to learn many of the star's other properties, such as its mass, radius, or energy output.

Measuring Distance by Triangulation and Parallax

Astronomers have several methods for measuring a star's distance, but for nearby stars the fundamental technique is **triangulation**, the same method we described for finding the distance to the Moon. In triangulation, we construct a triangle in which one side is the distance we seek but cannot measure directly and another side is a distance we can measure—a baseline, as shown in figure 12.1A. For example, to measure the distance across a deep gorge, we construct an imaginary triangle with one side spanning the gorge and another side at right angles to it and running along the edge we are on, as shown in figure 12.1B. By measuring the length of the side along the gorge edge and the angle A, we can determine the distance across the gorge either by a trigonometric calculation or from a scale drawing of the triangle.

Astronomers use a method of triangulation called **parallax** to measure the distance to stars. Parallax is a change in an object's apparent position caused by a change in the observer's position. A familiar example is the apparent motion of telephone poles as you drive past them. Another example—easy to demonstrate in your room—is to hold your hand motionless at arm's length and shift your head from side to side. Your hand seems to move against the background even though in reality it is your head that has changed position, not your hand.

This simple demonstration illustrates how parallax gives a clue to an object's distance. If you hold your hand at different distances from your face, you will notice that the apparent shift in your hand's position—its parallax—is larger if it is close to your face than if it is at arm's length. That is, *nearby objects exhibit more parallax than more remote ones*, for a given motion of the observer, a result true for your hand and for stars.

To observe stellar parallax, astronomers take advantage of the Earth's motion around the Sun, as shown in figure 12.2A. They observe a star and carefully measure its position against background stars. They then wait 6 months until the Earth has moved to the other side of its orbit, a known distance of 2 AU (about 300 million kilometers), and make a second measurement. As figure 12.2B shows, the star will have a slightly different position

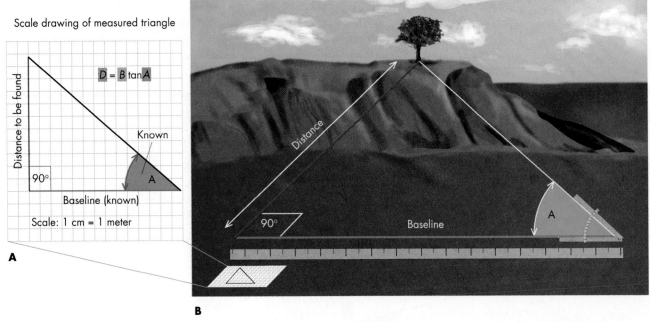

FIGURE 12.1
Sketch illustrating the principle of triangulation.

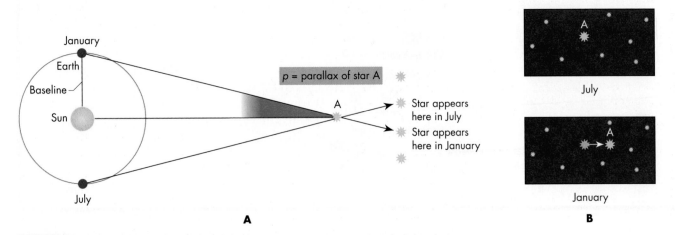

FIGURE 12.2
(A) Triangulation to measure a star's distance. The radius of the Earth's orbit is the baseline. (B) As the Earth moves around the Sun, the star's position changes as seen against background stars. Parallax is defined as one half the angle by which the star's position shifts. Size of bodies and their separation are exaggerated for clarity.

compared to the background of stars as seen from the two points. The amount by which the star's apparent position changes depends on its distance from us. The change is larger for nearby stars than for remote stars, but for all stars it is extremely small—so small that it is measured not in degrees but in fractions of a degree called "arc seconds."*

For reasons explained below, astronomers define a star's parallax, p, not by the angle its position appears to shift, but by half that angle (see fig. 12.2B). With that definition for parallax, the star's distance, d, is simply $1/p$ if we measure p in arc seconds, and d, not in kilometers or light years, but in a new unit called "parsecs" (abbreviated pc). That is,

$$d_{pc} = \frac{1}{p_{arc\ seconds}}$$

With this choice of units, one **parsec** equals 3.26 light years (3.09×10^{13} kilometers). The word *parsec* comes from a combination of *par*allax and "arc *sec*ond."

To determine a star's distance, we measure its parallax, p, and use $d_{pc} = 1/p_{arc\ seconds}$. For example, suppose from the shift in position of a nearby star we find its parallax is 0.25 arc seconds. Its distance is then $d = \frac{1}{0.25} = 4$ parsecs. Similarly, a star whose parallax is 0.1 arc second is $\frac{1}{0.1} = 10$ parsecs from the Sun. From this technique astronomers have discovered that at present the nearest star is Proxima Centauri, which lies 1.3 parsecs (4.3 light years) from the Sun. Its companion, Alpha Centauri, a bright star visible in the southern sky, is only slightly farther. This spacing (about 1 parsec) is typical for stars near the Sun.

EXTENDING OUR REACH

MEASURING THE DISTANCE TO SIRIUS

We can find a star's distance by a geometric construction similar to the one we used in chapter 1 to relate a body's true diameter to its angular diameter and distance. We begin by measuring the angle, A, that the star (Sirius, for example) appears to shift in the sky. We next construct the triangle shown in figure 12.3. The line AC is the radius of the Earth's orbit, with point A chosen so that an imaginary line from Sirius (S) to the Sun, C, will be perpendicular to AC. This makes ACS a right triangle, with angle C equal to 90 degrees and the side AC—the distance from the Earth to the Sun—1 astronomical unit. Now let p be half the measured shift in Sirius's position, A. We can now calculate Sirius's distance from the Sun, d, in a manner similar to the one we used to measure an object's true

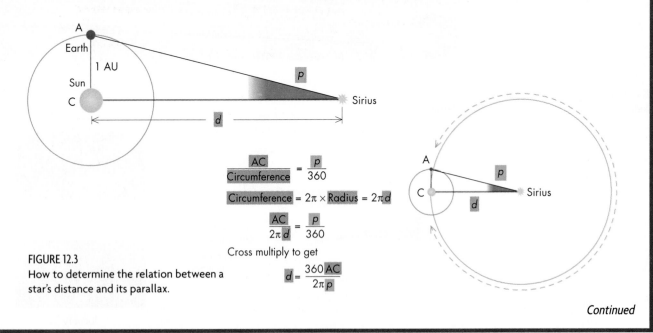

FIGURE 12.3
How to determine the relation between a star's distance and its parallax.

$$\frac{AC}{Circumference} = \frac{p}{360}$$

$$Circumference = 2\pi \times Radius = 2\pi d$$

$$\frac{AC}{2\pi d} = \frac{p}{360}$$

Cross multiply to get

$$d = \frac{360\ AC}{2\pi p}$$

Continued

---

*One arc second is $\frac{1}{3600}$ of a degree because an arc second is $\frac{1}{60}$ of an arc minute, which is $\frac{1}{60}$ of a degree.

EXTENDING OUR REACH   *Continued*

diameter from its angular diameter, as discussed in the box in chapter 1 (fig. 1.23).

We draw a circle centered on Sirius and passing through the Earth and the Sun. We call the radius of the circle *d*, to stand for Sirius's distance from the Sun. We next form the following proportion: *AC*, the radius of the Earth's orbit, is to the circumference of the circle as *p* is to the total number of degrees around the circle, which we know is 360. That is,

$$\frac{AC}{2\pi d} = \frac{p}{360}$$

If we measure *d* and *AC* in astronomical units, we can set *AC* = 1. Then, solving for *d* we obtain

$$d = \frac{360}{2\pi p} AU$$

This equation takes on a much simpler form if we express *p* in arc seconds rather than degrees, and *d* in parsecs (as defined above) rather than in AU. With these new units, the factors 360 and 2π disappear, leaving *d* = 1/*p*, as we claimed earlier. (In

the above we assume that Sirius itself has not changed its location significantly compared to its distance in so brief a time as 6 months—a very good assumption.)

Measurements show that for Sirius the angle *p* is 0.377 arc seconds. Thus its distance is 1/0.377 = 2.65 parsecs. To express this in light years, we multiply by 3.26 (the number of years in a parsec) to get Sirius's distance as about 8.6 light years.

Although the parallax-distance relation is mathematically a very simple formula, obtaining a star's parallax to use in the formula is very difficult because the angle by which the star shifts is extremely small. It was not until the 1830s that the first parallax was measured by the German astronomer Friedrich Bessel at Königsberg Observatory (now in Kaliningrad). Even now, the method fails for most stars further away than about 100 parsecs because the Earth's atmosphere blurs the tiny angle of their shift, making it unmeasurable.* Astronomers can avoid such blurring effects by observing from above the atmosphere, and an orbiting satellite, *Hipparcos,* is now making parallax measurements from space. With its data, astronomers should be able to accurately measure distances to stars as far away as 250 parsecs. For more remote stars, they must use a different method, one based on how bright objects look to us.

Measuring Distance by the Standard-Candle Method

If you look at an object of known brightness, you can estimate its distance from how bright it appears. For example, if you look at two 50-watt light bulbs, one close and one far away, you can tell fairly accurately how much farther away the dim one is. In fact, if you drive at night, your life depends on making such distance estimates when you see traffic lights or oncoming cars. Astronomers call such distance measurements the method of standard candles and use a similar but more refined version of it to find the distance to stars and galaxies. But to use this method to measure an object's distance, astronomers must first determine its true brightness. So until we learn ways to make such brightness determinations, we must set this method aside, but we will return to it later in the chapter.

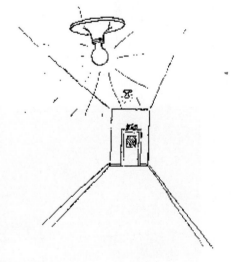

12.2 MEASURING THE PROPERTIES OF STARS FROM THEIR LIGHT

If we were studying flowers or butterflies, we would want to know something about their appearance, size, shape, colors, and structure. So, too, astronomers want to know the sizes, colors, and structure of stars. Such knowledge not only helps us better understand the nature of stars, but also is vital in unraveling their life story and finding patterns of similarity

---

*With electronic techniques, astronomers can measure, from the ground, distances to some bright stars as far away as 250 parsecs.

that could apply to our own star, the Sun. Obtaining knowledge of a star's physical properties is not easy, however, because we cannot directly probe it. But by analyzing a star's light, astronomers can deduce many of its properties, such as its temperature, composition, radius, mass, and motions.

Temperature

Stars are extremely hot by earthly standards. The surface temperature of even cool stars is far above the temperatures at which most substances vaporize, and so using a physical probe to take a star's temperature would not succeed, even if we had the means to send the probe to the star. If astronomers want to know how hot a star is, they must once more rely on indirect methods. Yet the method used is familiar. You use it yourself in judging the temperature of an electric stove burner.

A body's temperature can often be deduced from the color of its emitted light. As we saw in chapter 3, hotter bodies emit more blue light than red. Thus hot bodies tend to glow blue, and cooler ones red. You can see such color differences if you look carefully at stars in the night sky. Some, such as Rigel in Orion, have a blue tint. Others, such as Antares in Scorpius, are obviously reddish. Thus, even our naked eye can tell us that stars differ in temperature.

We can use color in a more precise way to measure a star's temperature with Wien's law, which we discussed in chapter 3. It states that a body's temperature, T, in Kelvin is given by the following:

T = Temperature of star
λ_m = Strongest emitted wavelength of starlight

$$T = \frac{3 \times 10^6}{\lambda_m}$$

where λ_m is the wavelength in nanometers (nm) at which it radiates most strongly.* Thus the longer the wavelength of the maximum emitted energy, the lower is the temperature of the radiating body.

For an illustration of how we can use Wien's law to measure a star's temperature, suppose we pick Sirius, a bright nearby star. To determine the wavelength at which it radiates most strongly, we need to measure its energy at many different wavelengths, as illustrated in figure 12.4. We find that the strongest emission is at 300 nanometers, in the ultraviolet part of the spectrum. That is, λ_m = 300 nanometers. Inserting this value in

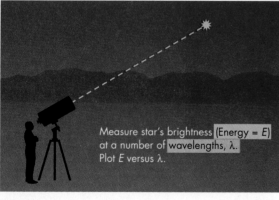

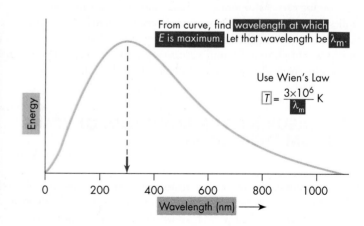

FIGURE 12.4
Measuring a star's surface temperature using Wien's law.

*The subscript *m* on λ is to remind us that it is the wavelength at which the star's emitted energy is at a *maximum*.

Wien's law, we find

$$T = \frac{3 \times 10^6}{300} = 10,000 \text{ K}$$

For another example, we pick the red star Betelgeuse, which radiates most strongly at about 1000 nanometers. Its temperature is therefore about

$$\frac{3 \times 10^6}{1000} = \frac{3 \times 10^6}{10^3} = 3 \times 10^{6-3} = 3 \times 10^3 = 3000 \text{ K}$$

A star's temperature is only one of several properties we can deduce from its light. For example, in some cases we can also measure the star's radius, but before we discuss how, we need to discuss briefly another general property of stars and other hot objects—the amount of energy they radiate.

Luminosity

Astronomers call the amount of energy a body radiates each second its **luminosity** (abbreviated as L). An everyday example of luminosity is the wattage of a light bulb: a typical table lamp has a luminosity of 50 watts, whereas a bulb for an outdoor parking lot may have a luminosity of 1500 watts. Stars, of course, are enormously more luminous. For example, the Sun has a luminosity of about 4×10^{26} watts, which it obtains by "burning" its hydrogen into helium. Thus a star's luminosity measures its rate of fuel consumption, a vital quantity for determining its lifetime. Knowing a star's luminosity is also important because from it astronomers can measure a star's radius and distance. But to understand how such measurements are made, we must first discuss a relation between how bright an object appears, its luminosity, and its distance—a relation known as the inverse-square law.

The Inverse-Square Law and Measuring a Star's Luminosity

The **inverse-square law** relates a body's luminosity to its distance and its apparent brightness, that is, how bright it looks to us. We all know that a light looks brighter when we are close to it than when we are farther from it. The inverse-square law puts that everyday experience into a mathematical form, describing how light energy spreads out from a source such as a star. As the light travels outward, it moves in straight lines, spreading its energy uniformly in all directions, as shown in figure 12.5A. Near the source, the light will have spread only a little, and so more strikes an observer, making it look brighter than if the observer were far away.

FIGURE 12.5
The inverse-square law. (A) Light spreads out from a point source in all directions. (B) As photons move out from a source, they are spread over a progressively larger area as the distance from the source increases. Thus a *given* area intercepts fewer photons the farther it is from the source. (C) The area over which light a distance d from the source is spread is $4\pi d^2$.

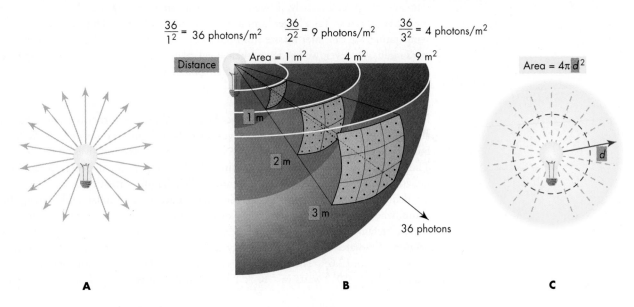

$$\frac{36}{1^2} = 36 \text{ photons/m}^2 \qquad \frac{36}{2^2} = 9 \text{ photons/m}^2 \qquad \frac{36}{3^2} = 4 \text{ photons/m}^2$$

Distance   Area = 1 m²   4 m²   9 m²

1 m

2 m

3 m

36 photons

Area = $4\pi d^2$

d

A   B   C

This decrease in brightness with increasing distance can also be understood if you think of light as photons. Photons leaving a light source such as a star spread evenly in all directions. If you imagine a series of progressively larger spheres drawn around the source, the same number of photons pass through each sphere, but because more distant spheres are larger, the number of photons passing through a given *area* grows smaller as the spheres become more distant, as shown in figure 12.5B. Similarly, as an observer moves away from a light source, photons reaching the observer are spread more widely, and so fewer enter the observer's eye, making the source appear dimmer. Thus a more distant source is dimmer because most of its light has spread along lines that never reach the observer.

At a distance d from a light source, its luminosity, L, has spread over a sphere whose radius is d, as shown in figure 12.5C. We use d to stand for radius here to emphasize that we are finding a distance. Any sphere has a surface area given by $4\pi d^2$, and so the brightness we observe, B, is just

$$B = \frac{L}{4\pi d^2}$$

This relationship is called the inverse-square law because distance appears in the denominator as a square.

The inverse-square law is one of the most useful mathematical tools available to astronomers, not only for measuring a star's luminosity, but also its distance. To find a star's luminosity, astronomers measure its distance, d, by parallax, as described in section 12.1. Next, with a photometer, a device similar to the electric exposure meter in a camera, they measure how bright the star appears from Earth, B. Finally, with B and d known, they calculate the star's luminosity, L, using the inverse-square law. Such measurements show that although most stars have a luminosity similar to the Sun's, their range in luminosity is enormous. For example, some stars are thousands of times less luminous than the Sun while others are millions of times more luminous. Much of this range in luminosity comes from the great range in stellar radii. That is, some stars are vastly larger than others, as we will now show.

Radius

Common sense tells us that if we have two objects of the same temperature but of different sizes the larger one will emit more energy than the smaller one. For example, three glowing charcoal briquettes in a barbecue emit more energy than just one briquette at the same temperature. Similarly, if two stars have the same temperature, the larger star emits more energy and so has a larger luminosity than the smaller star. Thus, if we know a star's temperature, we can infer its size from the amount of energy it radiates. To calculate the star's radius, however, we need a mathematical relation between luminosity, temperature, and radius—a relation known as the "Stefan-Boltzmann law."

The Stefan-Boltzmann Law

Imagine watching an electric stove burner heat up. When the burner is on low and is relatively cool, it glows dimly red and gives off only a slight amount of heat. When the burner is on high and is very hot, it glows bright yellow-orange and gives off far more heat. Thus you can both see and feel that raising a body's temperature increases the amount of radiation it emits per second—its luminosity. This familiar situation is an example of a law deduced in the late 1800s by two German scientists, Josef Stefan and Ludwig Boltzmann, who showed that the luminosity of a hot object depends on its temperature.

The Stefan-Boltzmann law, as their discovery is now called, states that a body of temperature T radiates an amount of energy each second equal to σT^4 per square meter, as shown in figure 12.6A. The quantity σ is called the "Stefan-Boltzmann constant," and its value is 5.67×10^{-8} watts m^{-2} K^{-4}. The Stefan-Boltzmann law affords a mathematical

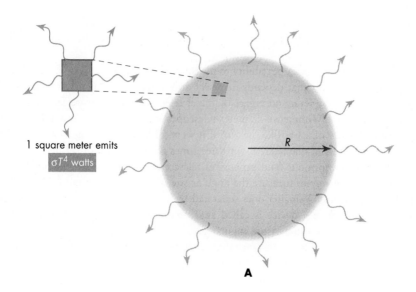

Total energy radiated per second by the star is its

Luminosity = L

L = Energy emitted by one square meter
× Number of square meters of its surface
= σT^4 × Star's surface area

For a spherical star of radius R, the surface area is $4\pi R^2$

Thus, $L = \sigma T^4 \times 4\pi R^2$

or

$L = 4\pi R^2 \sigma T^4$

B

A

FIGURE 12.6
The Stefan-Boltzmann law can be used to find a star's radius. (A) Each part of its surface radiates σT^4. (B) Multiplying σT^4 by the star's surface area ($4\pi R^2$) shows that its total power output—its luminosity, L—is $4\pi R^2\,\sigma T^4$. To find the star's radius, we solve this equation to get $R = \sqrt{L/(4\pi\sigma T^4)}$. Finally, we use this equation and measured values of the star's L and T to evaluate R.

1 square meter emits σT^4 watts

explanation of what we noticed for the electric stove burner, because a hotter burner has a larger T, and therefore, according to the law, the burner radiates more strongly. Similarly, the Stefan-Boltzmann law is the basis for dimmer switches on lights: changing the amount of electricity going to the lamp changes its temperature and thereby its brightness. However, the Stefan-Boltzmann law does not apply to all hot objects. For example, it does not accurately describe the radiation from hot, low-density gas, such as that in fluorescent light bulbs or interstellar clouds.

We can apply the Stefan-Boltzmann law to determine a star's luminosity, as follows. According to the law, if a star has a temperature T, each square meter of its surface radiates an amount of energy per second given by σT^4. We can find the total energy the star radiates per second—its luminosity, L—by multiplying the energy radiated from one square meter (σT^4) by the number of square meters of its surface area (fig. 12.6B). If we assume that the star is a sphere, its area is $4\pi R^2$, where R is its radius. Its luminosity, L, is therefore

$$L = 4\pi R^2\,\sigma T^4$$

That is, a star's luminosity equals its surface area times σT^4. This relation between L, R, and T, may at first appear complex, but its content is very simple: increasing either the temperature or the radius of a star makes it more luminous. Making T larger makes each square meter of the star brighter. Making R larger increases the number of square meters.

In the above expression, L is measured in watts, but the "wattage" of stars is so enormous that it is more convenient to use the Sun's luminosity as a standard unit. Likewise, it is easier to use the Sun's radius as a standard size unit rather than meters or even kilometers. If we need to convert to watts or meters, we simply remember that one solar luminosity, $L_\odot$, is 4×10^{26} watts and one solar radius, $R_\odot$, is 7×10^5 kilometers.

Because a star's luminosity depends on its radius and temperature—as given by the Stefan-Boltzmann law—if we know its luminosity and temperature, we can find its radius, R. The method works because if we have a mathematical relation between three quantities (in this case L, T, and R) we can find any one of these given the other two. Thus, if we know a star's luminosity and its temperature, we can use the Stefan-Boltzmann law to solve for its radius. The following box explains how.

Astronomers can also measure a star's radius from its angular size, the technique we used in chapter 11 to measure the radius of the Sun. Unfortunately, the angular size of all stars except the Sun is extremely tiny because they are so far from the Earth, and so even in powerful telescopes stars generally look like a smeary spot of light. That smearing is made worse by the blurring effects of our atmosphere and by a physical limitation of telescopes called "diffraction," which we discussed briefly in chapter 5, "Telescopes."

Diffraction limits a telescope's ability to measure tiny angular sizes, hopelessly blurring the light by an amount that depends inversely on the telescope's diameter. That is, diffraction effects are less severe on bigger-diameter telescopes. To measure the angular size of stars, however, truly immense-diameter telescopes are required. For example, to measure the angular size of a star like the Sun if it were 50 light-years away would require a telescope 300 meters in diameter, about three times the size of a football field. To avoid the need for such enormous (and expensive) telescopes, astronomers have therefore devised an alternative way to measure the angular size of stars by using not one huge telescope but two (or more) smaller ones separated by a large distance (for example, several hundred meters). Such a device is called an "interferometer" (discussed more fully in chapter 5), and its ability to measure angular sizes is equivalent to that of a single telescope whose diameter is equal to the distance that separates the two smaller ones. (Of course, a single enormous telescope would gather more light and enable astronomers to measure the sizes of very faint stars, but it may never be possible to build such a huge instrument.) A computer then combines the information from the two telescopes to give a crude picture of the star. Such interferometric observations are still hampered by blurring caused by

EXTENDING OUR REACH

MEASURING THE RADIUS OF THE STAR SIRIUS

From its color (the wavelength at which it radiates most strongly), we found that the temperature of Sirius is about 10,000 K. From the amount of energy we receive from Sirius and its distance, which can be found by parallax, we can find that its luminosity, L, is about 25 $L_\odot$. We now insert these values into the Stefan-Boltzmann law and solve for R. It is much easier, however, to approach the problem in stages, first writing down $L = 4\pi R^2 \sigma T^4$ separately for both Sirius (L_s) and the Sun ($L_\odot$) as follows:

$$L_s = 4\pi R_s^2 \sigma T_s^4$$

$$L_\odot = 4\pi R_\odot^2 \sigma T_\odot^4$$

L = Luminosity of star and Sun
R = Radius of star and Sun
σT^4 = Energy emitted by 1 sq. m. of star and Sun
T = Temperature of star and Sun

Next we divide the bottom expression into the top one to obtain

$$\frac{L_s}{L_\odot} = \frac{4\pi R_s^2\, \sigma T_s^4}{4\pi R_\odot^2\, \sigma T_\odot^4}$$

We can simplify this expression by canceling the identical 4π and σ factors to get

$$\frac{L_s}{L_\odot} = \frac{R_s^2\, T_s^4}{R_\odot^2\, T_\odot^4}$$

We next collect the R's and T's as separate factors, giving us

$$L_s/L_\odot = (R_s/R_\odot)^2\, (T_s/T_\odot)^4$$

We now solve this expression for $(R_s/R_\odot)^2$ to get

$$(R_s/R_\odot)^2 = (L_s/L_\odot)\, (T_\odot/T_s)^4$$

Finally, we take the square root of both sides to get

$$R_s/R_\odot = (L_s/L_\odot)^{1/2}\, (T_\odot/T_s)^2$$

We can now evaluate $R_s/R_\odot$ by inserting the values for the luminosity and temperature of Sirius and the Sun. Notice that $T_\odot = 6000$ K, and by definition of our units for luminosity, $L_\odot = 1$. This gives us

$$R_s/R_\odot = (25)^{1/2}\, (6000/10,000)^2 = 1.8$$

That is, the radius of Sirius* is 1.8 $R_\odot$, a little less than twice the Sun's radius.

---

*In Classical times, the Persians, Greeks, and Romans sometimes called Sirius the Dog Star. It is the brightest star in the constellation Canis Major, the Big Dog, and was associated by many early people with misfortune and fevers. The ancient Greeks and Romans also blamed Sirius for the extreme heat of July and August because at that time of year it rose at about the same time as the Sun. They therefore believed it added its brilliance to the Sun's, making the season extra warm. The heat we receive from Sirius is of course negligible, but, like the ancient Greeks and Romans, we still refer to the hot days of late summer when the Dog Star shines with the Sun as the "Dog Days."

our atmosphere. Those effects can be partially offset by a technique called "speckle inter-ferometry" in which a high-speed camera takes many very short exposures of the star's light. A computer then combines the separate short exposures into a "deblurred" image, such as that of Betelgeuse, illustrated in figure 12.7. With interferometers, astronomers can measure the radius of several dozen nearby stars and a few giant stars.

The above methods (the Stefan-Boltzmann law and interferometer observations) show that stars differ enormously in radius. Although most stars have approximately the same radius as the Sun has, some are hundreds of times larger, and so astronomers call them **giants.** Stars much smaller than our Sun are called **dwarfs.**

In using the Stefan-Boltzmann law to measure a star's radius, we have seen that L can be measured either in watts or in solar units for L. Astronomers sometimes use a different set of units, however, to measure stellar brightness.

The Magnitude System

About 150 B.C., the ancient Greek astronomer Hipparchus measured the apparent brightness of stars in the night sky using units he called **magnitudes.** He designated the stars that looked brightest magnitude 1, and the dimmest ones he could just barely see magnitude 6. For example, Betelgeuse, a bright red star in the constellation Orion, is magnitude 1, while the somewhat dimmer stars in the Big Dipper's handle are approximately magnitude 2. Astronomers still use this scheme to measure the brightness of astronomical objects, but they now use the term *apparent magnitude* to emphasize that they are measuring how bright a star *looks* to an observer. A star's apparent magnitude depends on its luminosity and its distance; it may look dim either because it has a small luminosity—it does not emit much energy—or because it is very far away.

Astronomers use the magnitude system for many purposes (for example, to indicate the brightness of stars on star charts), but it has several confusing properties. First, the scale is "backwards" in the sense that bright stars have small magnitudes, while dim stars have large magnitudes. Moreover, modern measurements show that Hipparchus underestimated the magnitudes of the brightest stars, and so the magnitudes now assigned them are negative numbers.

Second, magnitude *differences* correspond to brightness *ratios.* That is, if we measure the brightness of a first-magnitude star and a sixth-magnitude star, the former is 100 times brighter than the latter. Thus a *difference* of 5 magnitudes corresponds to a brightness *ratio* of 100; so, when we say a star is 5 magnitudes brighter than another, we mean it is a *factor* of 100 brighter. Each magnitude difference corresponds to a factor of 2.512 . . . —the fifth root of 100—in brightness. Thus a first-magnitude star is 2.512 times brighter than a second-magnitude star and is 2.512 × 2.512, or 6.310, times brighter than a third-magnitude star. Table 12.1 lists the ratios that correspond to various

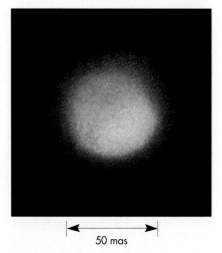

FIGURE 12.7
Image of the star Betelgeuse made by interferometry. The lack of detail in this image reflects the current difficulty in observing the disk of any star but our Sun. One mas is approximately 2.8×10^{-7} degrees. (Courtesy Mullard Radio Obs).

☀ Use this image of Betelgeuse to measure its diameter. You will need the formula developed in the box in chapter 1, section 1.2, describing how to determine the diameter of a body from its angular size and distance. Betelgeuse is about 6×10^{15} kilometers away (200 parsecs). Its angular size is measured in milli-arc seconds (mas).

TABLE 12.1	Relating Magnitudes to Brightness Ratios
Magnitude difference	**Ratio of brightness**
1	2.512:1
2	$2.512^2 = 6.31:1$
3	$2.512^3 = 15.85:1$
.	. .
5	$2.512^5 = 100:1$
10	$2.512^{10} = 10^4:1$
20	$2.512^{20} = 10^8:1$

TABLE 12.2   Relating Absolute Magnitude to Luminosity

Absolute magnitude	Approximate luminosity in solar units
−5	10,000
0	100
5	1
10	0.01

differences in magnitude. For example, let us compare the apparent brightness of the planet Venus with the star Aldebaran. At its brightest, Venus has an apparent magnitude of −4.2; Aldebaran's apparent magnitude is 0.8, and so the difference in their magnitudes is 0.8 − (−4.2) = 5.0. Therefore we see from table 12.1 that Venus is 100 times brighter to our eye than Aldebaran.

Third, astronomers often use a quantity called "absolute magnitude" to measure a star's luminosity. Recall that apparent magnitude is how bright a star looks to an observer. If seen from nearby, a given star will look bright, but if seen from a great distance, the same star will look dim. That is, a given star's apparent magnitude depends on its distance from the observer. Absolute magnitude, on the other hand, is how bright the star would appear—its apparent magnitude—if the star were at a standard distance, chosen to be 10 parsecs. Table 12.2 illustrates how absolute magnitude is related to a star's luminosity.

12.3 SPECTRA OF STARS

A star's spectrum is a record of the amount of energy it emits at each wavelength and is perhaps the single most important thing we can know about the star. The spectrum tells us the star's composition, temperature, luminosity, velocity in space, rotation speed, and other properties. Under some circumstances, it may also reveal the star's mass and radius.

Figure 12.8 shows the spectra of several different stars. You can easily see differences between them. Spectrum B is from a star similar to the Sun; spectrum A is from a star hotter than the Sun; and spectrum C is from a star cooler than the Sun. Spectrum A has only a few lines, but they are very strong and their spacing follows a regular pattern. Spectrum C, however, shows a welter of lines with no apparent regularity. Understanding such differences and what they tell us about a star is one of the goals of studying stellar spectra.

FIGURE 12.8
Computer generated spectra of three stars. A is hotter than the Sun. B is similar to the Sun. C is cooler than the Sun. (Courtesy Roger Bell and Michael Briley.)

Measuring a Star's Composition

As light moves from a star's core through the gas in its surface layers, atoms there absorb the radiation at some wavelengths, creating dark absorption lines in the star's spectrum, as shown in figure 12.9. Each type of atom—hydrogen, helium, calcium, and so on—absorbs at a unique set of wavelengths. For example, hydrogen absorbs at 656, 486, and 434 nanometers, in the red, blue, and violet part of the spectrum respectively. Gaseous calcium, on the other hand, absorbs strongly at 393.3 and 396.8 nanometers, producing a strong double line in the violet. Because each atom absorbs a unique combination of wavelengths of light, each has a unique set of absorption lines. From such absorption lines, we can determine what a star is made of.

To measure a star's composition we make a spectrum of its light and then compare the absorption lines we see with tables that list the lines made by each atom. When we find matching absorption lines, we infer that the element exists in the star. To find the quantity of each atom in the star—each element's abundance—we use the darkness of the absorption line, as discussed in chapter 3. Such determinations of the identity and quantity of elements present are difficult, however, because even though an element may be present in the star, the element's atoms may, because of temperature effects, be unable to absorb light and make a spectral line.

How Temperature Affects a Star's Spectrum

To see why temperature affects a star's spectrum, recall that light is absorbed when its energy matches the energy difference between two electron orbits. For an atom to absorb light, its electrons must therefore be in the proper orbit, or, more technically, energy level. An atom may be abundant in a star's atmosphere and create only very weak lines at a particular wavelength simply because the gas is so hot or so cold that its electrons are in the "wrong" level to absorb light at that wavelength. Hydrogen illustrates this situation dramatically.

The absorption lines of hydrogen that we see at visible wavelengths are made by electrons orbiting in its second level. These lines are sometimes called the "Balmer lines," for the spectroscopist who first studied their pattern, to distinguish them from other hydrogen lines in the ultraviolet and infrared wavelengths. Balmer lines occur at wavelengths of light that are absorbed because they have exactly the amount of energy to lift an electron from hydrogen's second energy level to the third or higher. These lines are especially important because their wavelengths are in the visible spectrum and are therefore easily observable. If the hydrogen atoms in a star have no electrons orbiting in level 2, no Balmer absorption lines will appear even though hydrogen may be the most abundant element in the star.

FIGURE 12.9
Formation of stellar absorption lines. Atoms in the cooler atmospheric gas absorb radiation at wavelengths corresponding to jumps between electron orbits.

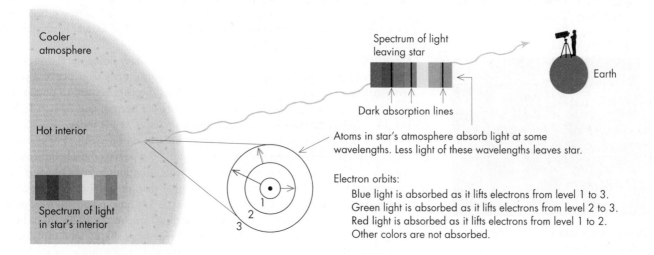

Cooler atmosphere

Spectrum of light leaving star

Earth

Dark absorption lines

Hot interior

Spectrum of light in star's interior

Atoms in star's atmosphere absorb light at some wavelengths. Less light of these wavelengths leaves star.

Electron orbits:
Blue light is absorbed as it lifts electrons from level 1 to 3.
Green light is absorbed as it lifts electrons from level 2 to 3.
Red light is absorbed as it lifts electrons from level 1 to 2.
Other colors are not absorbed.

Hydrogen Balmer lines will appear weakly or not at all if a star is either very cold or very hot. In a cool star, most hydrogen atoms have their electrons in level 1. In a hot star the atoms move faster, and when they collide, electrons may be excited ("knocked") into higher orbits: the hotter the star, the higher the orbit. In fact, in very hot stars, electrons may be knocked out of the atom, in which case the atom is said to be ionized. As a result of this excitation, electrons in a very hot star will generally be in level 3 or higher. Only if the hydrogen has a temperature between about 8000 K and 15,000 K will enough electrons be in level 2 to make strong hydrogen Balmer lines. If we are therefore to deduce correctly the abundance of elements in a star, we must correct for such temperature effects. With these corrections, we discover that virtually all stars are composed mainly of hydrogen: their composition is similar to that of our Sun—about 71% H, 27% He, and a 2% mix of the remaining elements. But despite their uniform composition, stars exhibit a wide range in the appearance of their spectra, as astronomers noted when they began to study stellar spectra.

Classification of Stellar Spectra

Stellar spectroscopy, the study and classification of spectra, was born early in the last century when the German scientist Joseph Fraunhofer discovered dark lines in the spectrum of the Sun. He later observed similar lines in the spectra of stars and noted that different stars had different patterns of lines. In 1866, Father Pietro Angelo Secchi, an Italian priest and scientist, noticed that the line patterns depended on the star's color. He assigned stars to four color classes—white, yellow, red, and deep red—and considered these classes as evidence for stellar evolution—hot stars cooling from white heat to deep red as they aged.

Secchi's four classes form the basis of the spectral classification system we use today. He noted that white stars showed a spectrum with mostly hydrogen absorption lines; yellow stars had spectra like the Sun with many lines, the darkest created by calcium and hydrogen atoms; and red stars had spectra with numerous dark lines that formed bands. Secchi made his studies visually, sitting at a telescope and drawing pictures of what he saw through the eyepiece. Such visual observations were soon replaced by photography.

About 1840, John William Draper, an American professor of chemistry and physiology (and a historian!), introduced photography to astronomy by making the first known photograph of an astronomical object—the Moon. In 1872, his son, Henry Draper, a physician and amateur astronomer, applied the new field of photography to stellar spectroscopy. On the death of the younger Draper, his widow endowed a project at Harvard to create a compilation of stellar spectra, known as the Henry Draper Catalog.

The use of photography in compiling the Draper Catalog made it possible to obtain many more spectra and with better details than could be observed visually. In fact, details were so good that E. C. Pickering, the astronomer in charge of the project, began to use letters to subdivide the spectral classes, assigning the letters A to D to Secchi's white class, E to L to the yellow class, and M and N to the red classes.

About 1901, Annie Jump Cannon, the astronomer actually doing the classification for the Draper catalog, discovered that the classes fell in more orderly sequence of appearance if rearranged by temperature. She therefore reordered Secchi's classes, to obtain the sequence O, B, A, F, G, K, and M, with O stars being the hottest, and M stars being the coolest. Her work is the basis for the stellar **spectral classes** we use today.

Stellar spectral classes are based on the appearance of the spectrum. For example, A-type stars show extremely strong hydrogen lines, while B-type stars show helium and weak hydrogen lines. This scheme, however, despite its wide adoption, lacked a physical basis in that it employed the appearance of the spectra rather than the physical properties of the stars that produced them. For example, astronomers at that time did not know what made the spectra of A and F stars differ. That understanding came in the 1920s from the work of the American astronomer Cecilia Payne (later Payne-Gaposhkin), who explained why the strength of the hydrogen lines depended on the star's temperature, as we described above. Her discovery was based on work by the Indian astronomer M. Saha, who showed

Annie Jump Cannon. (Courtesy Harvard College Observatory.)

how to calculate the level in which an atom's electrons were most likely to be found. Saha showed that the levels occupied by electrons—a quantity crucial for interpreting the strength of spectral lines—depend on the star's temperature and density. With Saha's equation, Payne then showed mathematically the correctness and reasonableness of Cannon's order for the spectral classes. She thereby demonstrated that a star's spectral class is determined mainly by its temperature.*

Payne's theory unfortunately left astronomers with Cannon's odd nonalphabetical progression for the spectral classes, which—from hot to cold—ran O, B, A, F, G, K, and M. So much effort, however, had been invested in classifying stars using this system that it was easier to keep the classes as assigned with their odd order than to reclassify them (Cannon, in her life, classified some quarter million stars). As a help to remember the peculiar order, astronomers have devised mnemonics. One of the first is "Oh, be a fine girl/guy. Kiss me." A slightly less sexist but more violent version is, "Oh, big and furry gorilla, kill my roommate."† A third version is, "Only brilliant, artistic females generate killer mnemonics." Choose whichever appeals most, or make up your own version, to help you learn this important sequence.

Definition of the Spectral Classes

A star's spectral class is determined by the lines in its spectrum. Figure 12.10 illustrates the seven main types—O through M—and the differences in the line patterns are easy to see even without knowing the identity of the lines. However, knowing which element makes which line can increase our understanding of the different line patterns. As mentioned earlier, almost all stars are made of the same mix of elements, but differing conditions result in differing spectra. For example, O stars have weak absorption lines of hydrogen but strong absorption lines of helium, the second most abundant element, because they are so hot. At their high temperature, an O star's hydrogen atoms collide so violently and are excited so much by the star's intense radiation that the electrons are stripped from most of the hydrogen, ionizing it. With its electron missing, a hydrogen atom cannot absorb light. Because most of the O star's hydrogen is ionized, such stars have extremely weak hydrogen absorption lines. Helium atoms are more tightly bound, however, and most retain at least one of their electrons, allowing them to absorb light. Thus O stars have absorption lines of helium but only weak lines of hydrogen.

Stars of spectral class A have very strong hydrogen lines. Their temperature is just right to put lots of electrons into orbit 2 of hydrogen, which makes for strong Balmer lines. Balmer lines also appear in F stars but are weaker. F stars are distinguished by the multitude of lines from metals such as calcium and iron, elements that also appear strongly in

Cecilia Payne. (Courtesy Katherine Haramundanis.)

FIGURE 12.10
The Stellar spectral classes. (A computer generated depiction, courtesy Roger Bell and Michael Briley.)

---

*Payne also made clear that stars are composed mainly of hydrogen, as mentioned earlier.
†Devised by an astronomy class at Harvard, the R for "roommate" is one of four rare spectral classes: R, N, S, and W.

G and *K* stars. Such elements are present in hotter stars but are ionized and generally create only very weak spectral lines under those conditions. In cooler stars, however, metal lines, particularly ionized calcium, are moderately strong, and hydrogen lines become weak.

In the very cool *K* and *M* stars, hydrogen is almost invisible because its electrons are mostly in level 1 and therefore cannot make Balmer lines. These stars have such cool atmospheres that molecules form, and as we saw in chapter 3, they produce very complex spectra. As a result, *K* and *M* stars have numerous lines from such substances as the carbon compound cyanogen (CN) and the carbon radical methylidyne (CH), as well as gaseous titanium oxide (TiO). These features and those found in the hotter stars are summarized in table 12.3.

We have seen above that *O* stars are hot and *M* stars are cool, but what in fact are their temperatures? Application of Wien's law and theoretical calculations based on Saha's law of how electrons are distributed in atomic orbits show that temperatures range from more than 25,000 K for *O* stars to less than 3500 K for *M* stars, with *A* stars being about 10,000 K and *G* stars, like our Sun, being about 6000 K.

Because a star's spectral class is set by its temperature, its class also indicates its color. We know that hot objects are blue and cool objects are red, and so we find that *O* and *B* stars (hot classes) are blue, while *K* and *M* stars (cool classes) are red.

To distinguish finer gradations in temperature, astronomers subdivide each class by adding a numerical suffix—for example, B1, B2, . . . B9—with the smaller numbers indicating higher temperatures. With this system, the temperature of a B1 star is about 20,000 K, whereas that of a B5 star is about 13,500 K. Similarly, our Sun, rather than being just a G star, is a G2 star.

Measuring a Star's Motion

All stars move through space and probably, like our Sun, also spin on their rotation axes. Such motions turn out to be important clues to a star's history and are vital in determining star masses and the structure of galaxies.

We find a star's motion from its spectrum using the Doppler shift which, as we saw in chapter 3, is the alteration of the wavelength of light from a moving source. If a source moves toward us, its wavelengths are shortened, while if it moves away, they are lengthened, as illustrated in figure 12.11A. The amount of the wavelength shift depends on the source's speed along the line of sight, what astronomers call **radial velocity**. We find the shift by subtracting the wavelength the star would absorb (or emit) if it were not moving relative to the Earth, λ_z, from the new wavelength, λ. We will call that difference $\Delta\lambda$. With these definitions, the Doppler shift law states that

$$\frac{\Delta\lambda}{\lambda_z} = \frac{v}{c}$$

where v is the source's radial velocity, and c is the speed of light. That is, the change in wavelength created by motion of the star (or observer) divided by the wavelength in the absence of motion is just the radial velocity of the star divided by the speed of light. Thus, if we can measure $\Delta\lambda$, we can solve for v. The quantity $\Delta\lambda$ may turn out to be either positive or negative depending on whether the wavelengths of the source are increased or decreased. If the wavelengths are increased, the distance between the source and observer is increasing, meaning that from the observer's point of view the source is moving away. If the wavelengths are decreased, the distance between the source and observer is decreasing, and so from the observer's point of view the source is approaching. Thus, from the observer's point of view, a positive $\Delta\lambda$ implies the source is receding, while a negative $\Delta\lambda$ implies it is approaching.

Astronomers measure the Doppler shift of a star by recording its spectrum directly beside the spectrum of a standard, nonmoving source attached to the telescope, as shown in figure 12.11B. They then measure the wavelength shift by comparing the spectrum of the star to the standard. From that shift, they calculate the star's radial velocity with the Doppler shift formula. Notice that a star will have a radial velocity only if its motion has some component toward or away from us. Stars moving across our line of sight, maintaining a constant distance from Earth, have no Doppler shift.

Spectral class	Temperature range (K)	Features
O	Hotter than 25,000	Ionized helium, weak hydrogen
B	11,000–25,000	Neutral helium, hydrogen stronger
A	7500–11,000	Hydrogen very strong
F	6000–7500	Hydrogen weaker, metals— especially ionized Ca—moderate
G	5000–6000	Ionized Ca strong, hydrogen weak
K	3500–5000	Metals strong, CH and CN molecules appearing
M	Cooler than 3500	Molecules strong, especially TiO

TABLE 12.3   Summary of Spectral Classes

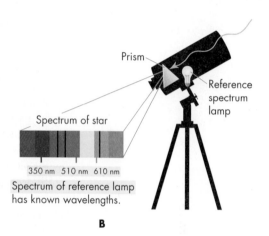

FIGURE 12.11

Measuring a star's radial velocity from its Doppler shift. (A) Spectrum lines from a star moving away from Earth are shifted to longer wavelengths (a red shift). Spectrum lines from a star approaching Earth are shifted to shorter wavelengths (a blue shift). (B) A standard lamp attached inside the telescope serves as a comparison spectrum to allow the wavelength shift of the star's light to be found.

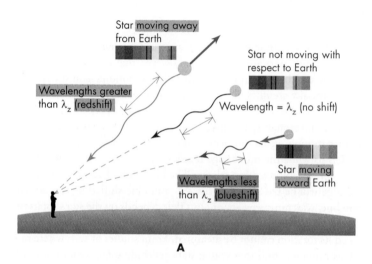

We illustrate how to measure a star's motion with the following example. Suppose we observe a star and see a strong absorption line in the red part of the spectrum with a wavelength of 656.38 nanometers. From laboratory measurements on nonmoving sources, we know that hydrogen has a strong red spectral line at a wavelength of 656.28 nanometers. Assuming these are the same lines, we infer that the wavelength is shifted by a $\Delta\lambda$ of 0.10 nanometers. The Doppler shift law tells us that

$$\frac{\Delta\lambda}{\lambda_z} = \frac{v}{c}$$

Therefore,

$$\frac{v}{c} = \frac{0.1}{656.28}$$

To find the radial velocity, v, we multiply both sides by c and obtain

$$v = \frac{0.1c}{656.28} = 45.7 \text{ kilometers per second}$$

Because $\Delta\lambda$ is positive, the star is moving *away* from us.

Such measurements reveal that all stars are moving and that those near the Sun share approximately its direction and speed of revolution (about 200 kilometers per second) around the center of our galaxy, the Milky Way. Superimposed on this orbital motion, however, are small random motions of about 20 kilometers per second. Stars therefore

$\Delta\lambda$ = Change in wavelength due to motion of light source
λ_z = Wavelength of unmoving light source
v = Radial velocity of light source
c = Speed of light (300,000 km/s)

FIGURE 12.12
Measuring a star's rotation velocity. As a star spins, one edge approaches, making a blue shift, and the other edge recedes, making a red shift. The result is that spectral lines are widened. The line's width can give the star's rotation velocity.

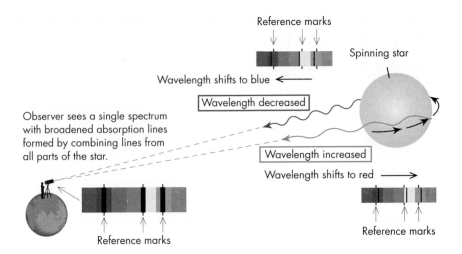

FIGURE 12.12
Measuring a star's rotation velocity. As a star spins, one edge approaches, making a blue shift, and the other edge recedes, making a red shift. The result is that spectral lines are widened. The line's width can give the star's rotation velocity.

move rather like cars on a freeway—some move a little faster and overtake slower-moving ones, which gradually fall behind. Stars are so far away, however, that their motion is imperceptible to the eye, much as a distant airplane seems to hang motionless in the sky. In fact, stellar motions across the sky are difficult to detect even in pictures of the sky taken years apart.

In addition to their motion through space, most stars spin on their rotation axes, just as the Earth does, another motion we can measure with the Doppler shift. As a star rotates, one edge approaches us while the other edge recedes, as illustrated in figure 12.12. Thus light emitted by gas on the approaching edge will be blue-shifted and light emitted by gas on the receding edge will be red-shifted. This red- and blue-shifted light is added to the original spectrum line, widening it by an amount that depends on the star's rotation speed. If the star happens to be spinning so that its pole points toward the Earth, no Doppler shift occurs, and its rotation cannot be measured. From studies of stars where rotation can be detected, astronomers find that young stars generally rotate faster than old stars like the Sun. For example, some extremely young *B* stars spin at their equator at nearly 300 kilometers per second, whereas the Sun rotates at about 2 kilometers per second.

12.4 BINARY STARS

Many stars have a special motion: they orbit around each other, held together as stellar companions by their mutual gravitational attraction, as illustrated in figure 12.13. Astronomers call such stellar pairs **binary stars** and value them greatly because they offer one of the few ways to measure stellar masses. To understand how, recall that the gravitational force between two bodies depends on their masses. The gravitational force, in turn, determines the stars' orbital motion. If that motion can be measured, we can work backwards to find the mass.

A star's mass, as we will discover in the next chapter, critically controls its existence, determining both how long it lasts and how its structure changes as it ages. Thus, if we are to understand the nature of stars, we must know their mass. Fortunately, this is not difficult to determine for stars belonging to a binary system, and such stars are very common. At least 40% of all stars known have orbiting companions, and the fraction may be much larger. For example, although the Sun has no true stellar companion, it is orbited by a family of planets. Although we have described binaries as star pairs, in many cases more than two stars are involved. Some stars are triples, others quadruples, and in at least one case a six-member system is known.

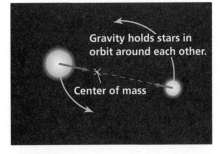

FIGURE 12.13
Two stars orbiting in a binary system, held together as a pair by their mutual gravitational attraction.

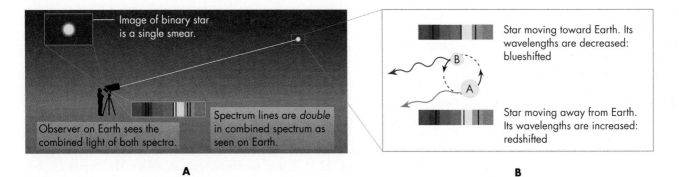

A

B

FIGURE 12.14
Spectroscopic binary star. (A) The two stars are generally too close to be separated by even powerful telescopes. (B) Their orbital motion creates a different Doppler shift for the light from each star. Thus the spectrum of the stellar pair contains two sets of lines, one from each star.

Most binary stars are only a few astronomical units apart. A few, however, are so close that they "touch," orbiting in contact with each other in a common envelope of gas. These contact binaries and other close pairs probably formed at the same time from a common parent gas cloud; that is, rather than becoming a single star surrounded by planets like our Sun did, the cloud formed a close star pair. Widely spaced binaries, on the other hand, perhaps formed when one star captured another that happened to pass nearby, but astronomers still have many unanswered questions about how binary stars form.

A few binary stars are easy to see with even a small telescope. Mizar, the middle star in the handle of the Big Dipper, is a good example. When you look at this star with the naked eye, you will first see a dim star, Alcor, which is not a true binary companion, but simply a star lying in the same direction as Mizar. With a telescope, however, Mizar can be seen to be two very close stars: a true binary.

Visual and Spectroscopic Binaries

For some stars (like Mizar) we can directly see orbital motion of one star about the other* by comparing photographs taken several years apart. Such star pairs are called **visual binaries** because we can see two separate stars and their individual motion.

Some binary stars may be so close together, however, that their light blends into a single blob that defies separation with even the most powerful telescopes. In such cases, the orbital motion cannot be seen directly, but it may nevertheless be inferred from their combined spectra. Figure 12.14A shows a typical **spectroscopic binary**, as such stars are called. As the stars orbit each other about their shared center of mass, each star alternately moves toward and then away from Earth. This motion creates a Doppler shift, and so the spectrum of the star pair shows two sets of lines, one set from each star—as you can see in figure 12.14B—and the spectrum of each shifts first to the red as the star swings away from us and then toward the blue as it approaches. From a series of spectra, astronomers can measure the orbital speed from the Doppler shift of each star, and by observing a full cycle of the motion they can find the orbital period. From the orbital period and speed they can find the size of the orbit, and with the size of the orbit and the period they can find the stars' masses using a modified form of Kepler's third law, as described below.

Measuring Stellar Masses with Binary Stars

In the early seventeenth century, Kepler showed that the time required for a planet to orbit the Sun is related to its distance from the Sun. If P is the orbital period and a is the semimajor axis (half the long dimension) of the planet's orbit, then $P^2 = a^3$, a relation called Kepler's third law.

---

*More technically, the two stars orbit a common center of mass located on the imaginary line joining them. The exact location of the center of mass depends on the star masses—if the masses are the same, the center of mass lies exactly halfway between them.

Newton discovered that Kepler's third law in a generalized form applies to any two bodies in orbit around each other. If their masses are m and M and they follow an elliptical path of semimajor axis, a, relative to each other with an orbital period, P, then

$$(m + M)P^2 = a^3$$

where P is expressed in years, a in astronomical units, and m and M in solar masses. This relation is our basic tool for measuring stellar masses, as we will now show.

To find the mass of the stars in a visual binary, astronomers first plot their orbital motion, as depicted in figure 12.15. It may take many years to observe the entire orbit, but eventually the time required for the stars to complete an orbit, P, can be determined. From the plot of the orbit and with knowledge of the star's distance from the Sun, astronomers next measure the semimajor axis, a, of the orbit of one star about the other. Suppose P and a turn out to be 10 years and 6 AU respectively. We can then find their combined mass, $m + M$, by solving the modified form of Kepler's law, and we get (after dividing both sides by P^2)

$$m + M = \frac{a^3}{P^2}$$

Inserting the measured values for P and a, we see that

$$m + M = \frac{6^3}{10^2}$$
$$= \frac{216}{100}$$
$$= 2.16 \; M_\odot$$

That is, the combined mass of the stars is 2.16 times the Sun's mass.

Additional observations of the stars' orbits allow us to find their individual masses, but for simplicity we will omit the details here. From analyzing many such star pairs, astronomers have discovered that star masses fall within a fairly narrow range from about 30 to 0.1 $M_\odot$. We will discover in the next chapter why the range is so narrow, but once again we find our Sun is about midway between the extremes, with some stars being a few tens of times more massive while others are a few tens of times less.

In the above discussion we assumed that we can actually see the two stars as separate objects, that is, that they are a visual binary. If the stars are spectroscopic binaries and cannot be distinguished as separate stars, their period and separation can be determined from a series of spectra.

FIGURE 12.15
Measuring the combined mass of two stars in a binary system using the modified form of Kepler's third law.

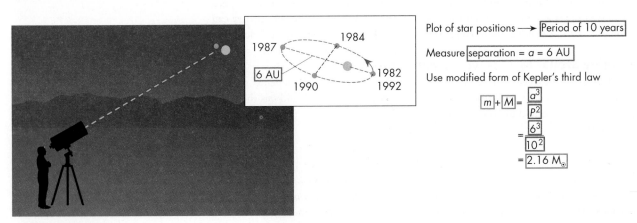

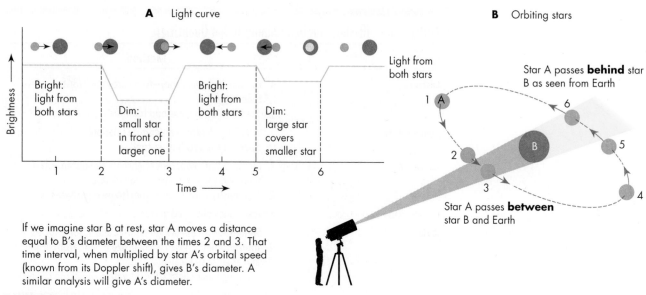

If we imagine star B at rest, star A moves a distance equal to B's diameter between the times 2 and 3. That time interval, when multiplied by star A's orbital speed (known from its Doppler shift), gives B's diameter. A similar analysis will give A's diameter.

FIGURE 12.16
An eclipsing binary and its light curve. From the duration of the eclipse the stars' radii may be found.

Eclipsing Binary Stars

On rare occasions, the orbit of a binary star will be almost exactly edge on as seen from the Earth. Then, as the stars orbit, one will eclipse the other as it passes between its companion and the Earth. Such systems are called **eclipsing binary** stars, and if we watch such a system, its light will periodically dim. During most of the orbit, we see the combined light of both stars, but at the times of eclipse the brightness of the system decreases as one star covers the other, producing a cycle of variation in light intensity called a "light curve." Figure 12.16A is a graph of such change in brightness over time.

Eclipsing binary stars are useful to astronomers because the duration of the eclipses depends on the stars' radii. Figure 12.16B shows why. The eclipse begins as the edge of one star first lines up with the edge of the other, as shown in the figure. The eclipse ends when the opposite edges of each star line up. During the eclipse, the covering star has moved relative to the other a distance equal to its diameter, plus that of the star it has covered. From such observations and knowledge of the orbital velocities of the eclipsing stars, astronomers can calculate the diameter of each star.

Light curves of eclipsing stars can also give information about the stars' shape and the distribution of brightness across their disks. For example, "star spots"—analogous to sunspots—can be detected on some stars by this method. If, in addition, astronomers can obtain spectra of the eclipsing system, the stars' masses may be found by use of the method we described above for spectroscopic binaries. Thus astronomers can obtain especially detailed information about eclipsing binary stars.

12.5 SUMMARY OF STELLAR PROPERTIES

Measuring the properties of stars is not easy and is possible for only a tiny fraction of the stars known. Nevertheless, using the methods we have described above, astronomers can determine a star's distance, temperature, composition, radius, mass, and radial velocity. Some of the methods used are summarized in table 12.4. We find from such measurements

TABLE 12.4 **Methods for Determining Stellar Quantities**

Quantity	Method
Distance	1. Parallax (triangulation)—for nearby stars (distance less than 250 pc) 2. Standard-candle method for more distant stars
Temperature	1. Wien's law (color-temperature relation) 2. Spectral class (*O* hot; *M* cool)
Luminosity	1. Measure star's apparent brightness and distance and then calculate with inverse square law. 2. Luminosity class of spectrum (to be discussed later)
Composition	Spectral lines observed in star
Radius	1. Stefan-Boltzmann law (measure *L* and *T*, solve for *R*) 2. Interferometer (gives angular size of star; from distance and angular size, calculate radius.) 3. Eclipsing binary light curve (duration of eclipse phases)
Mass	Modified form of Kepler's third law applied to binary stars
Radial velocity	Doppler shift of spectrum lines

that all stars have nearly the same composition of about 71% hydrogen and 27% helium, with a trace of the heavier elements. Most have surface temperatures between about 30,000 and 3000 K and masses between about 30 and 0.1 M⊙. We will see in the next chapter that the need to balance their internal gravity makes stars very uniform in their properties, but before we do, we will discuss how astronomers illustrate diagrammatically many of the features we have described above.

12.6 THE H-R DIAGRAM

By the early 1900s, astronomers had learned to *measure* stellar temperatures, masses, radii, and motions but had *understood* little of how stars worked. What made an *A* star different from a *K* star? Why were some *M* stars more luminous and others much less luminous? What supplied the power to make stars shine? We have now reached a similar point in our study of stars. We know that there are many varieties of stars. But what creates that variety and what does it mean? A fundamental step in the discovery of the nature of stars remains: the Hertzsprung-Russell diagram.

In 1912, the Danish astronomer Ejnar Hertzsprung and the American astronomer Henry Norris Russell working independently found that if stars are plotted on a diagram according to their luminosity and their temperature (or, equivalently, spectral class) then most of them lie along a smooth curve. The nearly simultaneous discovery of this relation by astronomers working on opposite sides of the Atlantic Ocean is an interesting example of how conditions can be "ripe" for scientific advances.

A typical Hertzsprung-Russell diagram, now generally called an **H-R diagram** for short, is shown in figure 12.17. Stars fall in the diagram along a diagonal line, with hot luminous stars at the upper left and cool dim stars at the lower right. The Sun lies almost in the middle. The H-R diagram does *not* depict the position of stars at some location in space. It shows merely a correlation between stellar properties much as a height-weight table does for people.

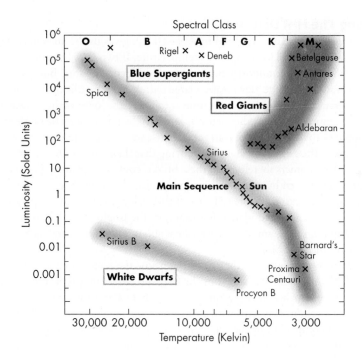

FIGURE 12.17
The H-R diagram. The long diagonal line from upper left to lower right is the main sequence. Giant stars lie above the main sequence. White dwarf stars lie below it. Notice that bright stars are at the top of the diagram and dim stars at the bottom. Also notice that hot (blue) stars are on the left and cool (red) stars are on the right.

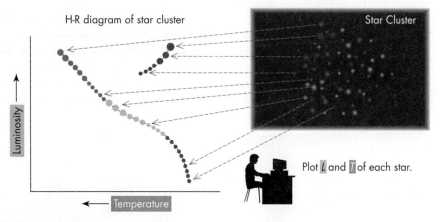

Measure luminosity (*L*) and spectrum of each star in the cluster. From spectrum, deduce the temperature, *T*.

FIGURE 12.18
Constructing an H-R diagram from the spectra and luminosities of stars in a cluster.

Constructing the H-R Diagram

To construct an H-R diagram of a group of stars, astronomers plot each according to its temperature (or spectral class) and luminosity, as shown in figure 12.18. By tradition, they put bright stars at the top and dim stars at the bottom—hot stars on the left and cool stars on the right. Equivalently, blue stars lie to the left and red stars to the right. Notice that temperature therefore increases to the left, rather than to the right, as is conventional in graphs. The approximately straight line along which the majority of stars lies is called the **main sequence**.

When a group of stars is plotted on an H-R diagram, generally about 90% will lie along the main sequence, but some will lie off it. Of these, a few will be in the upper right where stars are cool but very luminous, and some will be below the main sequence to the lower center, where stars are hot but dim. What makes these stars different?

Analyzing The H-R Diagram

The Stefan-Boltzmann law, described in section 12.2, gives us the answer. A star's luminosity depends on its surface area and its temperature. If two stars have the same temperature, any difference in luminosity must reflect a difference in area. But a star's area depends on its radius—a large area means a large radius. Stars that lie in the upper right of the H-R diagram have the same temperature as those on the lower main sequence, but because they are more luminous, they must be larger. In fact, they must be immensely larger because some of these bright cool stars emit thousands of times more energy than main-sequence stars of the same temperature, implying that their surface areas are thousands of times larger. Astronomers therefore call these bright, cool stars "giants." The region in the H-R diagram populated by giants is sometimes called the "giant branch." Because many of the stars are cool and therefore red (recall that color is related to temperature), these huge cool stars are called **red giants.** For example, the bright star Aldebaran in the constellation Taurus is a red giant. Its temperature is about 4000 Kelvin and its radius is about 30 times larger than the Sun's.

A similar analysis shows that the stars lying below the main sequence must have very tiny radii if they are both hot and dim. In fact, the radius of a typical hot, dim star is roughly 100 times smaller than the Sun's, making them about the same size as the Earth. Because they are so hot, they glow with a white heat and are therefore called **white dwarfs.** Sirius B, a dim companion of the bright star Sirius, is a white dwarf. Its radius is about 0.008 the Sun's—a little smaller than the Earth's radius—and its surface temperature is about 27,000 Kelvin.

We can illustrate such extremes in diameter by drawing lines in the H-R diagram that represent a star's radius. That is, we use the Stefan-Boltzmann law, which relates the star's luminosity L, radius R, and temperature T. From this law, $L = 4\pi R^2 \sigma T^4$, we plot the values of L that result from a fixed choice of R as we vary T. Any stars lying along the plotted line must have equal radii. Figure 12.19 shows that such lines run diagonally from upper left to lower right. Notice that as we move to the upper right, we cross lines of progressively larger radius, implying that stars there are very large. That is, not only are they giants in luminosity, as we mentioned earlier, but they are also giants in radius. Comparing now the radii of stars across the H-R diagram, we see three main types: main-sequence stars, red giants, and white dwarfs.

FIGURE 12.19
Lines showing where in the H-R diagram stars of a given radius will lie.

☀ On the basis of these lines, what is the approximate radius of Betelgeuse? Does that agree with the number you found in figure 12.7?

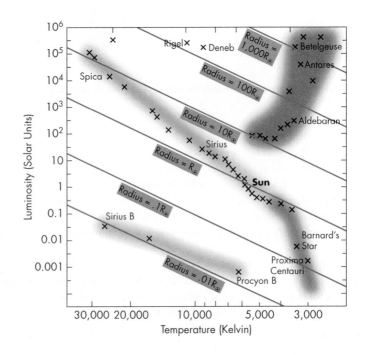

Giants and Dwarfs

Giants, dwarfs, and main-sequence stars differ in more than just diameter. They also differ dramatically in density.* Recall that density is a body's mass divided by its volume. For a given mass, a larger body will therefore have a lower density, and so a giant star is much less dense than a main-sequence star if their masses are similar. For example, the density of the Sun, a main-sequence star, is about 1 gram per cubic centimeter, while the density of a typical giant star is about 10^{-6} gram per cubic centimeter—1 million times less.

Why do stars have such disparate densities? Russell suggested about 1918 that as a star evolves, its density changes. He believed that stars began their life as diffuse, low-density red giants. Gravity then drew the stars' atoms toward one another, compressing their matter and making stars smaller, denser, and hotter so that they turned into hot, blue, main-sequence stars. Russell, in fact, believed that gravity compressed these gaseous bodies so much that by the time they were on the main sequence they had become liquid. Because liquids are difficult to compress, he believed that once a star reached the main sequence it could no longer shrink. Unable to be compressed further, the star would have no source of heat and would therefore simply cool off and gradually evolve down the main sequence, growing dimmer and dimmer.

As we will discuss in the next chapter, today we know that stars evolve in the opposite direction. They begin life as main-sequence stars and turn into giants as they age. But although Russell was wrong about how stars evolve, his ideas led to many other important discoveries about stars, such as the mass-luminosity relation.

The Mass-Luminosity Relation

In 1924 the English astrophysicist A. S. Eddington discovered that the luminosity of a main-sequence star is determined by its mass. That is, main-sequence stars obey a **mass-luminosity relation** such that the larger a star's mass, the larger its luminosity is, as illustrated in figure 12.20. A consequence of this relation is that stars near the top of the main sequence, brighter stars, are more massive than stars lower down.

FIGURE 12.20

(A) The mass-luminosity relation shows that more massive stars are more luminous. (B) On the H-R diagram, high-mass stars lie higher on the main sequence than low-mass stars.

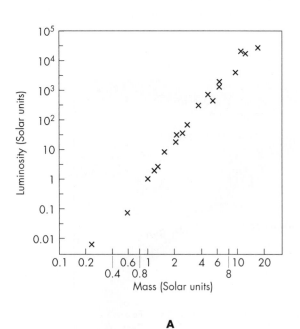

A

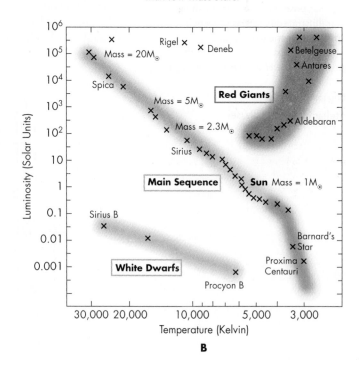

B

---

*Technically we should use the term *average density* here and in the following.

$\mathcal{L}$ = Luminosity of stars in solar units
$\mathcal{M}$ = Mass of stars in solar units

Eddington made his discovery by measuring the masses of stars in binary systems and determining their luminosity from their distance and apparent brightness. Moreover, he discovered that the mass-luminosity relation could be expressed by a simple formula. If $\mathcal{M}$ and $\mathcal{L}$ are given in solar units, then for many stars

$$\mathcal{L} = \mathcal{M}^3$$

approximately. If we apply the relation to the Sun, its mass is 1 and its luminosity is 1 in these units, and the relation is clearly obeyed because $1 = 1^3$. If we consider instead a more massive star, for example, a 2 $M_\odot$ one, $\mathcal{L}$ (according to the relation) should be 2^3 (= 8) $L_\odot$. Eddington went on to show from the physics of stellar structure that just such a mass-luminosity relationship is required if a star is to avoid collapsing under its own gravitational forces and if its energy flows through its interior by radiation. In the next chapter we will see how this relation arises and its importance in determining how long a star can live.

Luminosity Classes

Because a star's luminosity is important in determining its distance, radius, and life-span, astronomers have sought other ways to measure it. In the late 1800s Antonia Maury, an early stellar spectroscopist at Harvard, noticed that absorption lines were extremely narrow in some stars compared with other stars of the same temperature. In the early 1900s Hertzsprung, too, noticed that narrowness and, even before developing the H-R diagram, recognized that luminous stars had narrower lines than less luminous stars, as illustrated in figure 12.21.

The width of the absorption lines in a star's spectrum turns out to depend on the star's density: the lines are wide in high-density stars and narrow in low-density ones. The density of a star's gas is related in turn to its luminosity because a large-diameter star has—other things being equal—a large surface area, causing it to emit more light, and a large volume, giving it a lower density. A small-diameter star, on the other hand, generally has a high density because its gas is compressed into a small volume. Such a star also tends to be less luminous because its surface area is small and thus emits less light.

Using this relationship between spectral line width and luminosity, astronomers divide stars into five luminosity classes (fig. 12.22), denoted by the Roman numerals I to V. Class

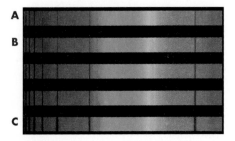

FIGURE 12.21
Spectral lines are narrow in giant (luminous) stars (A and B) and wide in main-sequence (dim) stars (C). (Based on a figure courtesy Roger Bell and Michael Briley.)

☀ Suppose the spectrum of the giant star illustrated here was of Deneb and the spectrum of the main-sequence star was of Sirius. Is that consistent with their position in the H-R diagram (see Fig. 12.17)?

FIGURE 12.22
Stellar luminosity classes.

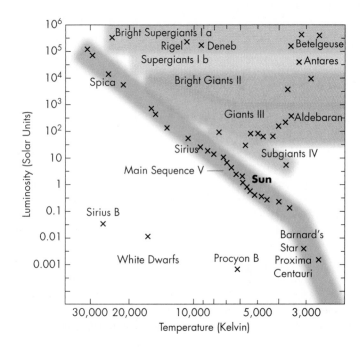

TABLE 12.5	Stellar Luminosity Classes	
Class	**Description**	**Example**
Ia	Supergiants	Betelgeuse, Rigel
Ib	Dimmer supergiants	Polaris (the North Star)
II	Bright giants	Mintaka (a star in Orion's belt)
III	Ordinary giants	Arcturus
IV	Subgiants	Achernar (a bright southern star)
V	Main sequence	The Sun, Sirius

V stars are the dimmest, and class I stars are the brightest. In fact, class I stars are split into two classes, Ia and Ib. Table 12.5 shows the correspondence between class and luminosity, and although the scheme is not very precise, it allows astronomers to get an indication of a star's luminosity from its spectrum.

A star's luminosity class is often added to its spectral class to give a more complete description of its light. For example, our Sun is a G2V star, while the blue giant Rigel is a B8Ia star. The luminosity class is especially useful for indicating the difference in luminosity between main-sequence and giant stars of the same spectral class, such as the low-luminosity, nearby, main-sequence star 40 Eridani (K1V) and the high-luminosity giant star Arcturus (K1 III).

Summary of the H-R Diagram

The H-R diagram offers a simple, pictorial way to summarize stellar properties. Most stars lie along the main sequence, with hotter stars being more luminous. Of these, the hottest are blue, and the coolest are red. Yellow stars, like the Sun, have an intermediate temperature.

Along the main sequence, a star's mass determines its position, as described by the mass-luminosity relation. Massive stars lie near the top of the main sequence; they are hotter and more luminous than low-mass stars. Thus, from top to bottom along the main sequence, the star masses decrease but nevertheless lie in a relatively narrow range, from about 30 to 0.1 $M_\odot$.

Finally, the H-R diagram reveals that most stars fall into three classes—main-sequence stars, giants, and white dwarfs. There are some peculiar stars that fall into none of the three main classes, for example, variable stars.

12.7 VARIABLE STARS

Not all stars have a constant luminosity. Some change in brightness and are called **variable stars.** Probably all stars vary slightly in brightness—even the Sun does in the course of its magnetic activity cycle—but relatively few change enough to be noticeable to the naked eye. Among those that do vary, however, are several varieties of stars that are very valuable to astronomers as distance indicators, as we will discover in the next chapter. Especially valuable are the pulsating variables—stars whose radius changes, rhythmically swelling and shrinking. As the radius changes (sometimes by 30%), the star's luminosity must also change according to the Stefan-Boltzmann law. The luminosity change, however, is complicated because, as the star pulsates, its temperature also changes, sometimes by as much as a factor of about 2. The result of both changes is that, as the star pulsates, its luminosity changes by a factor of 2 or 3. Although such large variations in luminosity are typical for many pulsating variables, ordinary stars also pulsate but much less dramatically. For them, the pulsations are more like the vibrations of a ringing bell. For example, our Sun's surface vibrates up and down in a complex pattern, as shown in figure 11.13.

The variation in luminosity and temperature are periodic for the majority of pulsating stars; that is, they repeat over some definite time interval called the **period**, as shown in figure 12.23. Astronomers call such a depiction of a star's brightness change its "light curve" and they use the shape of the curve and the time for the star to complete a cycle of brightness—the period—to classify variable stars. On the basis of such light curves, astronomers identify more than a dozen types of regularly pulsating variable stars. For example, some stars pulsate with a period of about half a day, while others take more than a year to complete a pulsation. If variable stars are plotted on the H-R diagram according to their average brightness, most lie in a narrow region called the "instability strip," illustrated in figure 12.24. In the next chapter we will learn why pulsating stars are confined to this region.

Although many variable stars pulsate regularly, some variable stars change their brightness erratically. Such stars are called "irregular variables" and many very young and very old stars are of this category.

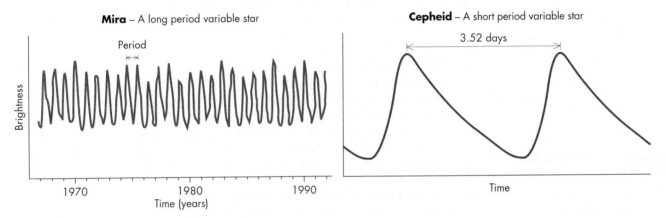

FIGURE 12.23
Some variable star light curves. (Courtesy AAVSO International Data Base and Douglas L. Welch, McMaster University).

☀ Approximately how long is Mira's pulsation period?

FIGURE 12.24
The instability strip in the H-R diagram. Stars that fall in this band generally pulsate.

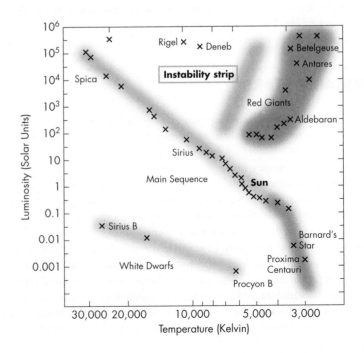

12.8 FINDING A STAR'S DISTANCE BY THE METHOD OF STANDARD CANDLES

In our discussion at the start of this chapter we claimed that astronomers can determine a star's distance from knowledge of its luminosity. As an example of the method, we mentioned using the brightness of an oncoming car's headlights to estimate its distance. This method works, however, only if we know the luminosity of the light source with some assurance; that is, we know that the luminosity of the light source is some "standard" value. For that reason, astronomers sometimes call this distance-finding scheme the **method of standard candles,** and it is perhaps their single most powerful tool for finding distances.

To use the method of standard candles, we compare mathematically how bright an object looks—its apparent brightness, B, with how bright it truly is—its luminosity, L. If we know the apparent brightness and luminosity of a source, we can find its distance using the inverse-square law. To measure a star's distance, an astronomer first measures its apparent brightness, B, with a photometer, as mentioned in section 12.1. Next, the star's luminosity, L, is determined from its spectrum, its position on the H-R diagram, or, in the case of a pulsating variable star, from its period, as we will discuss in the next chapter. With B and L known, the astronomer then calculates the distance to the star, d, using the inverse-square law.

In using the inverse-square law, however, it is generally much easier to find a star's distance by comparing it to the distance of a known star. For example, suppose we have two stars whose luminosity we know to be the same because they have identical spectra. Suppose one star, the nearer one, is close enough that we can find its distance by parallax to be 10 parsecs. Furthermore, suppose the star whose distance we seek looks 25 times dimmer. We find its distance by writing out the inverse-square law for both stars:

$$B_{\text{near}} = \frac{L_{\text{near}}}{4\pi d_{\text{near}}^2}$$

$$B_{\text{far}} = \frac{L_{\text{far}}}{4\pi d_{\text{far}}^2}$$

L = Luminosity of star
B = Brightness of star
d = Distance to star

Next, we divide the bottom expression into the top one to obtain after cancellation

$$\frac{B_{\text{near}}}{B_{\text{far}}} = \frac{4\pi L_{\text{near}}\, d_{\text{far}}^2}{4\pi L_{\text{far}}\, d_{\text{near}}^2} = \frac{L_{\text{near}}\, d_{\text{far}}^2}{L_{\text{far}}\, d_{\text{near}}^2}$$

If the two stars have the same luminosity, we can also cancel the L's, leaving us with

$$\frac{B_{\text{near}}}{B_{\text{far}}} = \left(\frac{d_{\text{far}}}{d_{\text{near}}}\right)^2$$

That is, for two stars of equal luminosity, the brightness ratio is the inverse of the distance ratio squared. We can therefore find the distance ratio by taking the square root of both sides to get $d_{\text{far}}/d_{\text{near}} = \sqrt{B_{\text{near}}/B_{\text{far}}}$. For the problem we are solving, the brightness ratio is 25. Its square root is 5. Therefore the further star is 5 times more distant than the nearer one. Because we know that the nearer star is 10 parsecs away, the further star is 50 parsecs from us.

This method may seem to be rather roundabout, but astronomers have few other options for finding star distances. Moreover, with a little practice, such calculations can be done quickly. **To find a star's distance compared to another star of the same luminosity, find the apparent brightness ratio and take its square root.** This is then the distance ratio. If the stars have different luminosities, the calculation is slightly more difficult. In that case, we must use the full equation above.

SUMMARY

Astronomers use many techniques to measure the properties of a star (fig. 12.25). Parallax—triangulation of the star from opposite sides of the Earth's orbit—gives a distance for nearby stars, with the nearest being about 1.3 parsecs (about 4 light years) away from the Sun. The parallax method fails beyond about 250 parsecs because the angle becomes too small to measure. Astronomers must therefore use the inverse-square law, comparing apparent brightness and luminosity to find the distance of more remote stars.

Stars have a wide range of surface temperature as measured by their color. According to Wien's law, hot stars will be blue and cool stars red. Once the temperature and luminosity of a star are known we can calculate its radius with the Stefan-Boltzmann law.

We can find a star's composition by looking at the lines in its spectrum. The spectrum can also be used to assign stars to the spectral classes: *O, B, A, F, G, K, M*, where *O* stars are the hottest, and *M* stars are the coolest.

We can find a star's mass if it is in a binary system—two or more stars bound together gravitationally. For such stars, the orbital period and size when combined in the modified form of Kepler's third law give the masses of the orbiting stars.

Once a star's luminosity and temperature are known, it may be plotted on an H-R diagram. In this plot, most stars lie along a diagonal line called the main sequence, which runs from hot, blue, luminous stars to cool, red, dim ones. Main-sequence stars obey a mass-luminosity relation such that high-mass stars are more luminous than low-mass ones.

A small percentage of stars fall above the main sequence, being very luminous but cool. Their high luminosity implies they must have a large diameter, and so they are called red giants. Similarly, a few stars fall below the main sequence and are hot but dim. Their low luminosity implies a very small diameter, comparable to that of the Earth. These hot, small stars are called white dwarfs.

Some stars vary in brightness and are called variable stars. Many of these variables lie in a narrow region of the H-R diagram called the "instability strip."

FIGURE 12.25
Summary of how astronomers measure the distance, temperature, mass, composition and radius of stars.

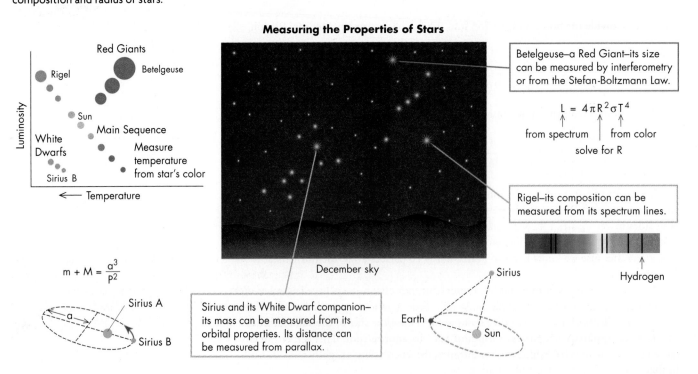

Measuring the Properties of Stars

Betelgeuse—a Red Giant—its size can be measured by interferometry or from the Stefan-Boltzmann Law.

$$L = 4\pi R^2 \sigma T^4$$

from spectrum | from color

solve for R

Rigel—its composition can be measured from its spectrum lines.

Hydrogen

December sky

Sirius and its White Dwarf companion—its mass can be measured from its orbital properties. Its distance can be measured from parallax.

$$m + M = \frac{a^3}{p^2}$$

Sirius A

Sirius B

Earth   Sirius   Sun

Red Giants

Rigel   Betelgeuse

Sun   Main Sequence

White Dwarfs   Measure temperature from star's color

Sirius B

Luminosity

⟵ Temperature

QUESTIONS FOR REVIEW

1. Describe the method of distance measurement by triangulation.
2. How do astronomers triangulate a star's distance?
3. How is the parsec defined? How big is a parsec compared with a light-year?
4. How do astronomers measure a star's temperature?
5. Why do stars have dark lines in their spectra?
6. What are the stellar spectral classes? Which are hot and which cool?
7. What distinguishes the spectral classes of stars?
8. What is a binary star?
9. How do visual and spectroscopic binaries differ?
10. Why are binary stars useful to astronomers?
11. What is an eclipsing binary? What can be learned from eclipsing binaries?
12. What is the H-R diagram? What are its axes?
13. What is the main sequence?
14. How do we know that giant stars are big and dwarfs small?
15. How does mass vary along the main sequence?
16. What is the mass-luminosity relation?
17. What is a variable star?
18. What is meant by the period of a variable star?
19. Where in the H-R diagram are variable stars found?

THOUGHT QUESTIONS

1. Measure approximately the distance from where you live to the neighboring building by triangulation. Make a scale drawing, perhaps letting 1/4 inch be 1 foot, and choose a base line. Construct a scale triangle on your drawing and find how far apart the buildings are.
2. Would it be easier to measure a star's parallax from Pluto? Why?
3. Suppose a binary star's orbit is in a plane perpendicular to our line of sight. Can we measure its mass using the methods described in this chapter? Why?
4. Use the data in table 12.6 below to plot an H-R diagram. Which star is a red giant? Which is a white dwarf? Note: Plotting will be much easier if you plot the logarithm of the luminosity; that is, express it in powers of 10 and use the power. For example if the luminosity is 100, plot it as 2 for 10^2. Alternatively, use a pocket calculator as follows. Enter the luminosity in solar units and hit the "log" key. If the luminosity is 300,000, the answer you get should be 5.477. . . .

TABLE 12.6   Data for Plotting an H-R Diagram*

Star name	Temperature	Spectrum	Luminosity (solar units)
Sun	6000	G2	1
Sirius	10,000	A1	25
Rigel	12,000	B8	350,000[†]
Betelgeuse	4000	M2	150,000[†]
40 Eridani	15,000	dA[‡]	0.01
Barnard's Star	3000	M5	0.001
Spica	20,000	B1	3000

*These data have been rounded off to make plotting easier.
[†]The luminosity of Rigel and Betelgeuse are not accurately known because the distance to them is extremely uncertain.
[‡]Given the low luminosity of this star, what do you think the *d* in its spectral class stands for?

PROBLEMS

1. The star Rigel radiates most strongly at about 200 nanometers. How hot is it?
2. The bright southern star α Centauri radiates most strongly at about 500 nanometers. What is its temperature? How does this compare to the Sun's.
3. Arcturus is about half as hot as the Sun but is about 100 times more luminous. What is its radius compared to the Sun?
4. A stellar companion of Sirius has a temperature of about 27,000 K and a luminosity of about 10^{-2} L$_\odot$. What is its radius compared to the Sun? What is its radius compared to the Earth?
5. Sirius has a parallax of 0.377 arc seconds. How far away is it?
6. The parallax of Proxima Centauri is about 0.763 arc seconds. How far away is it?

7. The parallax of the red giant Betelgeuse is just barely measurable and has a value of about 0.005 arc seconds. What is its distance? Suppose the measurement is in error by + or −0.003 arc seconds. What limits can you set on its distance?

8. Two stars in a binary system have an orbital period, *P*, of 5 years and an orbital separation, *a*, of 10 AU. What is their combined mass?

9. Two stars in a binary system have an orbital period, *P*, of 2 years and an orbital separation, *a*, of 4 AU. What is their combined mass?

10. Two stars in a binary system are determined from their position on the H-R diagram and the mass-luminosity relation to have a combined mass of 8 M☉. Their orbital period, *P*, is 1 year. What is their orbital separation, *a*?

11. A line in a star's spectrum lies at 400.0 nanometers. In the laboratory, that same line lies at 400.2 nanometers. How fast is the star moving along the line of sight; that is, what is its radial velocity? Is it moving toward or away from us?

TEST YOURSELF

1. A star has a parallax of 0.04 arc seconds. What is its distance?
   (a) 4 light years.
   (b) 4 parsecs.
   (c) 40 parsecs.
   (d) 25 parsecs.
   (e) 250 parsecs.

2. A star radiates most strongly at 400 nanometers. What is its surface temperature?
   (a) 400.
   (b) 4000.
   (c) 40,000.
   (d) 75,000.
   (e) 7500.

3. Which of the following stars is hottest?
   (a) An *M* star.
   (b) An *F* star.
   (c) A *G* star.
   (d) A *B* star.
   (e) An *O* star.

4. A star that is cool and very luminous must have
   (a) a very large radius.
   (b) a very small radius.
   (c) a very small mass.
   (d) a very great distance.
   (e) a very low velocity.

5. In what part of the H-R diagram do white dwarfs lie?
   (a) Upper left.
   (b) Lower center.
   (c) Upper right.
   (d) Lower right.
   (e) Just above the Sun on the main sequence.

FURTHER EXPLORATIONS

Cannizzo, John K., and Ronald H. Kaitchuck. "Accretion Disks in Interacting Binary Stars." *Scientific American* 266 (January 1992): 92.

Davis, John. "Measuring the Stars." *Sky and Telescope* 82 (October 1991): 361.

Hearnshaw, John B. "Origin of the Stellar Magnitude Scale." *Sky and Telescope* 84 (November 1992): 494.

Kaler, James B. "The Brightest Stars in the Universe." *Astronomy* 19 (May 1991): 30.

———. "The Faintest Stars." *Astronomy* 19 (August 1991): 26.

———. "The Smallest Stars in the Universe." *Astronomy* 19 (November 1991): 50.

———. *Stars and Their Spectra: an Introduction to the Stellar Spectral Sequence.* Cambridge, New York: Cambridge University Press, 1989. (Much of this material is also covered in a series of articles that appeared in *Sky and Telescope* 71 (February 1986): 129 and continued at 3-month intervals, culminating with

"Journeys on the H-R Diagram." *Sky and Telescope* 75 (May 1988): 482.)

Kidwell, P. A. "Three Women of American Astronomy." *American Scientist* 78 (May/June 1990): 244.

MacRobert, Alan M. "The Spectral Types of Stars." *Sky and Telescope* 92 (October 1996): 48.

———. "The Stellar Magnitude System." *Sky and Telescope* 91 (January 1996): 42.

Steffey, Philip C. "The Truth about Star Colors." *Sky and Telescope* 84 (September 1992): 266.

Terrell, Dirk. "Demon Variables (the Algol Family)." *Astronomy* 20 (October 1992): 34.

Trimble, Virginia. "White Dwarfs: The Once and Future Suns." *Sky and Telescope* 72 (October 1986): 348.

KEY TERMS

binary stars, 368

dwarfs, 361

eclipsing binary, 371

giants, 361

H-R diagram, 372

inverse-square law, 357

luminosity, 357

magnitudes, 361

main sequence, 373

mass-luminosity relation, 375

method of standard candles, 379

parallax, 352

parsec, 354

period, 378

radial velocity, 366

red giants, 374

spectral classes, 364

spectroscopic binary, 369

triangulation, 352

variable stars, 377

visual binaries, 369

white dwarfs, 374

 PROJECTS

With the help of the star charts, find five of the following stars: Aldebaran, Altair, Antares, Betelgeuse, Capella, Castor, Deneb, Pollux, Procyon, Regulus, Rigel, Spica, Vega. What color does each star look to you? (Try not to let its twinkling affect the color you decide on). Do your colors match those that you infer from the star's spectrum class as listed in appendix table 8? Which of these stars are giants? How many times bigger radius does each have compared to the Sun? Using the information in the appendix, about how long ago did the light from each of the stars start on its journey to Earth? Where were you when the light started its trip? Were you even born?

CHAPTER THIRTEEN
STELLAR EVOLUTION

Gas ejected from a dying low mass star (Anglo-Australian Observatory, David Malin)

A supernova remnant—the debris blown into space by a dying high mass star (R. Wainscoat)

Young stars forming from interstellar gas (Anglo-Australian Observatory, David Malin)

To us stars appear permanent and unchanging, but like many other things, they are born, grow old, and die. Stars are not alive, of course. They are merely immense balls of gas, but astronomers nevertheless refer to the changes that stars undergo as stellar evolution. Those changes are many, and they are driven by the physical laws that govern a star's structure. For example, gravity holds a star together, and the pressure of its gas supports it against gravity's inward pull. A star generates its supporting pressure from heat energy created by gravity compressing the material in its core. That heat energy steadily escapes from the star as the starlight we see, and it must be replenished, or the star will collapse. A star replenishes its energy by nuclear fusion in its core. In the early stages of its evolution, a star fuses hydrogen into helium. It eventually consumes the hydrogen in its core, however, and changes its structure, swelling into a red giant. Stars like the Sun consume their core's hydrogen in about 10 billion years, but more massive stars use up their hydrogen much faster. They therefore evolve much faster than the Sun and burn out in only a few million years. To understand why massive stars evolve faster, recall that according to the mass-luminosity relation, massive stars are far more luminous than stars like the Sun. Thus, to supply their greater luminosity, massive stars must burn their fuel faster than less-massive stars. Consequently they run out of fuel faster.

When a star runs out of fuel, its evolution is nearly finished. Stars like the Sun die slowly but quietly and become white dwarfs. More massive stars explode and become neutron stars or black holes. Such stellar explosions, though marking the death of one star, may trigger the birth of another. Moreover, the explosion of a dying star blasts into space elements vital to human life such as carbon and iron, elements that eventually become incorporated into new stars, planets, and ultimately people. We therefore owe our very existence to stellar evolution. The goal of this chapter is to illustrate how such evolution occurs.

13.1 THE EVOLUTION OF A STAR

The evolutionary changes that affect all stars happen incredibly slowly by human standards and require millions to billions of years. However, we need not wait such a vast span of time to see a star aging. We can "see" a star's aging with computer calculations that solve the equations that govern the star's physics. Stellar models, as such calculations are called, allow us to trace a star's life, revealing on a single page a billion years of a star's history. Figure 13.1 summarizes a few of the most important features of such calculations and may serve as a road map in our subsequent discussion.

Stellar models are only one tool in the study of stellar evolution. Astronomers can also deduce the basic features of how a star evolves from observations. For example, the existence of main-sequence stars, red giants, and white dwarfs suggests to astronomers a picture of how stars age, much the way a snapshot of a baby with its parents and grandparents suggests to a viewer a picture of human aging.

With these ideas in mind, let us follow the life story of a star from its birth to its death. In telling that story, we need to take account of the great difference in how stars end their lives. Those differences depend on the star's mass, and so we will divide stars into two groups—low-mass stars, such as the Sun, and high-mass stars. For convenience, we will consider high-mass stars to be those whose mass is about five or more times larger than the

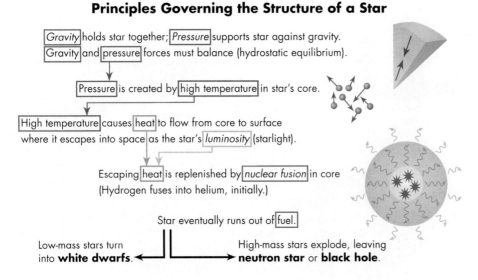

Principles Governing the Structure of a Star

Gravity holds star together; Pressure supports star against gravity.
Gravity and pressure forces must balance (hydrostatic equilibrium).

Pressure is created by high temperature in star's core.

High temperature causes heat to flow from core to surface
where it escapes into space as the star's luminosity (starlight).

Escaping heat is replenished by nuclear fusion in core
(Hydrogen fuses into helium, initially.)

Star eventually runs out of fuel.

Low-mass stars turn into **white dwarfs**.   High-mass stars explode, leaving **neutron star** or **black hole**.

Note: High-mass stars require more pressure to support their greater mass.
       Greater pressure is produced by higher temperature.
       Higher temperature produces higher luminosity.
       Higher luminosity leads to faster fuel usage.
       Faster fuel usage means a high-mass star burns out sooner than a low-mass star.

FIGURE 13.1
Outline of the processes that govern the structure of stars.

Sun's. Those whose mass is less than five times larger than the Sun's we will consider low-mass stars. We will begin with the life history of our Sun, a typical low-mass star.

The Sun's Life Story

The Sun began its life as an **interstellar cloud,** a tenuous, cold, dark mass of gas drifting around the Milky Way galaxy. The cloud—perhaps because of a collision with a neighboring cloud—began to collapse. As it shrank under the influence of its increasing gravity, it grew smaller and hotter. Within a few million years—an instant, to the Sun—hydrogen began to fuse into helium in its core. The energy released by the fusion raised the pressure inside the Sun, stopping its collapse and leaving it looking as it does today: a small, yellow star. If we plotted it on the H-R diagram, it would lie on the main sequence.

The Sun will remain a main-sequence star until it consumes about 90% of the hydrogen in its core, a process that will take a total of 10 billion years, about half of which time the Sun has already used up. When the hydrogen in the Sun's core is nearly spent, its core will shrink and grow hotter. The rising temperature in the core will make the remaining hydrogen there burn faster, generating more energy. As that energy flows outward through the Sun's outer layers, it will lift them and, paradoxically, cool them as they expand farther from the source of heat. As the surface of this bloated Sun cools, it will turn red, and its light will reveal to our distant descendants a Sun swollen into a red giant. Their view will be brief, however, because the expanding Sun will nearly engulf them and the Earth.

The Sun will shine as a red giant for perhaps another billion years but will then shrink and grow hotter, transforming—when its core finally becomes hot enough to fuse helium—into a yellow giant. During this stage, it will pulsate as if taking slow, deep breaths, but these are dying gasps. Eventually the Sun will consume most of the helium in its core and will once again change into a red giant, but a giant even larger and more luminous than before. That high luminosity is its death warrant: radiation streaming outward through its atmosphere will drive its gas into space, stripping the Sun to its bare core. The ejected gas will form a shell that will gradually disperse. The tiny core, fiercely hot but

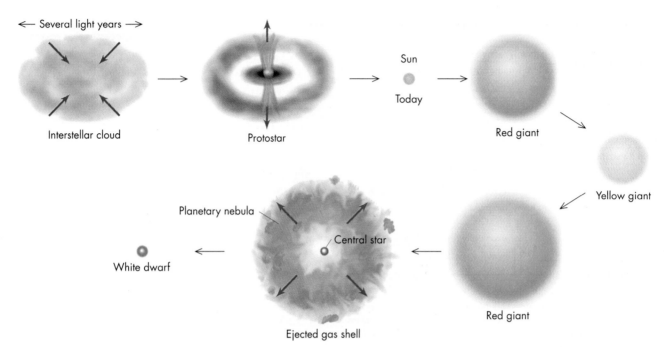

FIGURE 13.2
Evolution of a 1 solar-mass star like the Sun, based on computer calculations.

with no further energy supply, will cool and dwindle to a white dwarf. These changes, summarized in figure 13.2, are the fate of the Sun.

The Life Story of a High-Mass Star

A very different fate awaits stars significantly more massive than the Sun: they explode when they run out of fuel. Astronomers are uncertain precisely what mass is needed to make a star explode but believe it is about 5 $M_\odot$. For that reason we take 5 $M_\odot$ as the dividing line between high-mass and low-mass stars.

We chart the history of a high-mass star in figure 13.3, where we see that its early life is similar to the Sun's: it originates from the collapse of an interstellar cloud but from a larger and more massive clump than the one that formed the Sun. The clump's greater mass compresses it more, thereby heating it more than the young Sun was heated. Thus, when the clump becomes a main-sequence star, it is much hotter, bluer, and more luminous than the Sun. Its greater luminosity implies that it burns its fuel faster, and so rather than taking 10 billion years to expend its hydrogen, it squanders its fuel in 100 million years or less, aging far more rapidly than the Sun. As it runs out of hydrogen, a massive star behaves much like a low-mass star, swelling steadily in size and growing cooler. However, a massive star passes through a pulsating yellow-giant stage before it turns into a red giant.

The greater mass of such a star now becomes a primary influence on its evolution, creating an intense gravitational compression of its core. As the core is compressed, its temperature rises, and it burns fuel ever more furiously to maintain the pressure that supports the star. The higher temperature also permits the star to fuse progressively heavier elements to supply its energy—helium is fused into carbon, carbon into oxygen, and ultimately, silicon into iron. Iron, however, does *not* release energy when it is fused, and thus a star with a massive iron core cannot support itself against gravity's inward pull. Lacking support, its core collapses in less than a second, triggering a cataclysmic explosion that showers the heavy elements it has made into space. Such a star does not leave a white dwarf, rather it becomes either a tiny, incredibly compressed ball of neutrons (a neutron star) or an even denser body—a black hole, gravity's ultimate triumph.

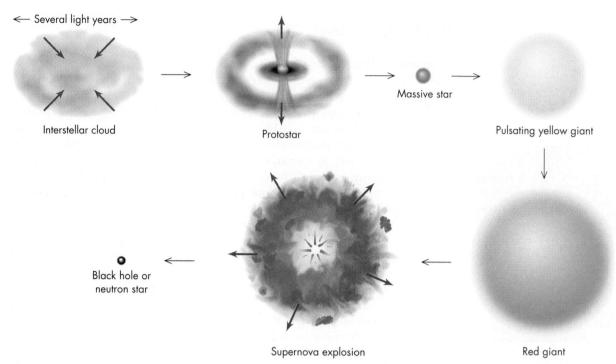

← Several light years →

Interstellar cloud

Protostar

Massive star

Pulsating yellow giant

Black hole or
neutron star

Supernova explosion

Red giant

FIGURE 13.3
Evolution of a high-mass star, based on computer calculations.

The Importance of Gravity

Gravity drives stellar evolution. From the moment of its birth, when an interstellar gas cloud collapses to become a star, to the moment of its death, gravity constantly works to crush the star. That crushing force makes a star evolve, because physical laws dictate that the star's core must be hot enough to create a pressure sufficient to support the star against gravity's inexorable pull.

To keep its core hot, a star must burn* its fuel to generate new energy to offset the energy it constantly loses into space as it shines. Recall the analogy we used in chapter 11: the Sun behaves like a leaky inflatable chair. To keep the chair from collapsing, air must constantly be pumped in. Similarly, energy must constantly be produced in a star. Any interruption or change in its energy supply will alter its structure. For example, before its main-sequence life, the Sun was heated by gravity's compression of its gas. Now, as a main-sequence star, hydrogen fusion powers its core. As a red giant, hydrogen will burn in a shell around its core. As a yellow giant, helium fusion will power the Sun. When its helium is consumed, the Sun will become a red giant for the last time and blow off its outer layers. It will then have no energy source and thus will become a white dwarf.

All stars need fuel to keep their cores hot, and the radically different evolution of high- and low-mass stars results from the former's ability to burn nuclear fuels heavier than helium. Thus a major emphasis in this chapter will be on how stars supply their energy needs, but we must first understand how they are born.

---

*Astronomers often refer to nuclear fusion as "burning," even though fusion is *not* burning in the same sense as a candle or gasoline burns. Such burning involves chemical reactions in which oxygen combines with some fuel.

13.2 STAR FORMATION

Interstellar Gas Clouds

Stars form from interstellar matter—huge gas and dust clouds that orbit with the stars inside our galaxy. The gas is mostly hydrogen (71%) and helium (27%); the dust is composed of solid, microscopic particles of silicates, carbon, and iron compounds. Interstellar gas is generally cold, perhaps only 10 K (hundreds of degrees below zero on the Fahrenheit scale). At such low temperatures, atoms and molecules in a gas move too slowly to generate much pressure. As a result, cold gas may not have adequate pressure to support itself against gravity, and it may form vast clumps—gas clouds—that collapse. The collapse may be triggered by collision with a neighboring cloud, the explosion of a nearby star, or other processes.

Maps made with radio telescopes show that gas clouds are not uniform: they contain smaller clumps of gas where the density is higher than the average. When a cloud with such clumps collapses, each clump is also compressed, growing denser. Thus a single cloud may break into many smaller pieces, each of which may form a star. The result is that stars generally form in groups, not in isolation, and all the stars within a group form at approximately the same time.

Astronomers envisage the transformation from gas cloud to star proceeding in several distinct stages. In the first stage, a dense clump within a cloud begins to collapse, its gas drawn inward by gravity, which compresses and heats it. In the second stage, any rotation of the gas clump makes it flatten into a disk, as we described in chapter 7 on the origin of the Solar System. In about a million years, the disk forms a small, hot, dense core at its center called a **protostar**, which marks the third stage. These stages are illustrated in figure 13.4.

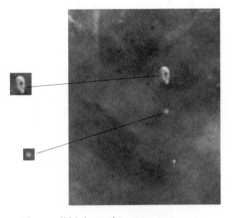

The small blobs in this picture are protostars in the Orion nebula, a huge gas cloud about 1500 light years from Earth. (Courtesy Space Telescope Science Institute.)

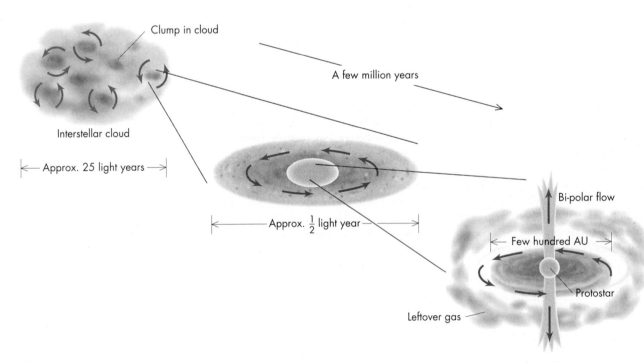

Clump in cloud

A few million years

Interstellar cloud

Approx. 25 light years

Approx. $\frac{1}{2}$ light year

Bi-polar flow

Few hundred AU

Protostar

Leftover gas

FIGURE 13.4
Artist's depiction of the birth of a star.

Protostars

A protostar is hotter than the gas from which it condensed—perhaps 1500 K—but it is still much cooler than ordinary stars. Its relatively low temperature makes a protostar "shine" mainly at infrared and radio wavelengths rather than at visible ones. Thus to observe a protostar, astronomers use telescopes that detect infrared or radio wavelengths. With such telescopes, astronomers can make infrared and radio maps (like those in fig. 13.5) of regions in which stars are forming. The small, bright blobs are protostars. These objects are virtually undetectable at visible wavelengths because they are cool and because the dust around them absorbs what little visible light they emit. But protostars do not remain cool for long. As their gravity continues to draw surrounding material inward, they grow hotter from the gravitational energy released by the infalling material.

Gravitational energy is released whenever something falls. For example, when a cinderblock falls onto a box of tennis balls, the impact scatters the balls in all directions, giving them kinetic energy—energy of motion. In a gas, energy of motion appears as heat, raising the temperature; as material falls onto a protostar, its gas atoms speed up and it grows hotter.

Eventually the protostar becomes hot enough for nuclear reactions to begin in its core. We saw in chapter 11 that hydrogen can fuse into helium at about 7 million K. This new energy supply raises the temperature and pressure in the central regions of the protostar, and eventually the outward pressure grows strong enough to stop further collapse in the core. Surrounding material, especially from the disk, continues to fall onto the star for millions of years, however. The infalling matter releases gravitational energy, which supplies additional heat to the protostar. In fact, for many young stars, the gravitational energy of infalling matter greatly exceeds that released by nuclear reactions in the core.

A protostar passes through these stages in a time that is brief by astronomical standards: a few million years for a star like the Sun and even less time for a more massive star. But for all types of protostars, infall creates violent changes in brightness and, more strangely, ultimately creates a strong *outflow* of gas. The reason for the outflow is not yet understood, but it is probably caused by energy that is released as matter falls from the inner parts of the disk onto the star. Regardless of how the outflow is generated, its effects are striking, as you can see in figure 13.6.

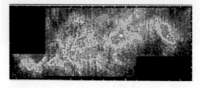

FIGURE 13.5
False-color infrared and radio maps of protostars. (Courtesy Karen Strom, Mark Heyer, Ron Snell, FCAD and FCRAO.)

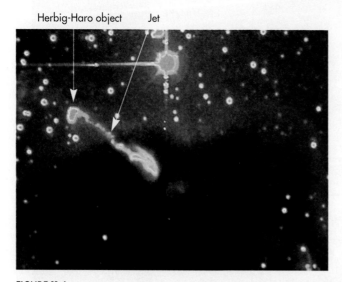

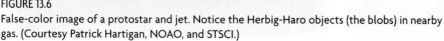

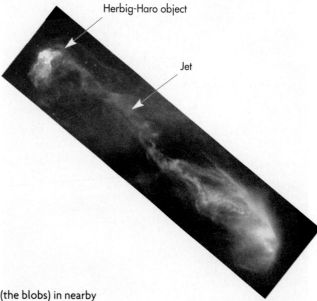

FIGURE 13.6
False-color image of a protostar and jet. Notice the Herbig-Haro objects (the blobs) in nearby gas. (Courtesy Patrick Hartigan, NOAO, and STSCI.)

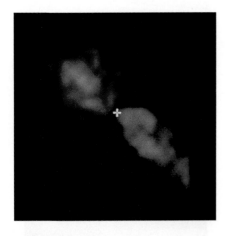

FIGURE 13.7
Bipolar flow of gas ejected from a young star. This is a radio picture in which the gas receding from us is shown in red, whereas the gas approaching us is shown in blue. (Courtesy Ronald Snell, FCRAO.)

Bipolar Flows from Young Stars

Figure 13.6 shows photographs of a young star still surrounded by gas left over from its birth. A long, thin jet of gas squirts out from it and has carved a cavity in the gas that remains around the young star. Farther out along the jet's direction lie several small, bright blobs called "Herbig-Haro objects." Herbig-Haro objects, named for the astronomers who first discovered them, appear to be splashes similar to those made when water from a hose hits a window and spreads out into blobs. The jet of gas from the star spreads when it hits surrounding gas, and the impact heats both the jet and the gas around it, making them glow, as shown in the electronically made pictures. The jet also pushes gas outward from around the protostar to create **bipolar flows**. Bipolar flows can be seen in figure 13.7, which shows a radio map of the area around a protostar. Here, a computer has been used to color red the gas moving away from us (that is, red-shifted material) and to color blue the gas moving toward us.

FIGURE 13.8
Photograph of two young star clusters still surrounded by the interstellar gas from which they formed. NGC 6559/IC 1274-75 is on the left, and NGC 2264 is on the right. (Courtesy Anglo-Australian Observatory; photographs by David Malin.)

☀ What creates the red glow of the gas near these stars? What might account for the dark cone in the bottom part of the cluster on the right? What do you think the squiggly dark blobs silhouetted against the pink glow in the picture on the left might be?

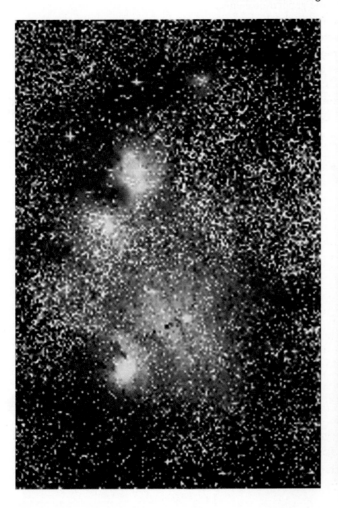

Bipolar flows are important because they clear away gas and dust from around protostars and thereby allow astronomers to see them directly with visible light. However, even powerful bipolar flows leave some of the surrounding material behind. Thus many young stars are partially immersed in interstellar matter, as illustrated in figure 13.8. Such stars often vary erratically in brightness and have gas streaming away from their surfaces, as shown by the Doppler shift of their spectral lines. This flow occurs in the star's upper atmosphere and is distinct from the bipolar flow we described above. Rather, it is similar to the solar wind but far more intense. Stars that exhibit these features (variable light and outflowing gas) are called **T Tauri stars,** and some of their variability is probably caused by magnetic activity much like that observed on our Sun. For example, T Tauri stars show evidence of giant "starspots" and magnetic flares thousands of times more intense than the activity we observe on our Sun. Astronomers cannot see the spots directly with optical telescopes, but by combining data on variations of the star's brightness obtained with x-ray, radio, and infrared telescopes, they can deduce the presence of spots on these peculiar young stars.

If T Tauri stars are plotted on an H-R diagram, they lie a little above the lower main sequence because they are still shrinking and are thus larger and brighter than main-sequence stars. Figure 13.9 shows the H-R diagram for a very young star group and a photograph of the cluster itself. Such H-R diagrams help verify our ideas about star formation because older clusters never contain T Tauri stars.

You can see that surrounding the cluster is gas left over from its formation, and if you look closely in the central regions, you can see small, dark blobs that may be protostars that have not yet become hot enough to emit visible light. These tiny, dark globs are called Bok globules in honor of Bart Bok, the astronomer who first studied them in detail. Globules exhibit a diversity of sizes, with diameters ranging from about 0.1 to 2 light years and with masses ranging from a few up to about 200 times the Sun's mass.

Stellar Mass Limits

We saw in the previous chapter that the mass of most stars is between about 0.1 and 30 $M_\odot$. These limits are not absolute—higher- and lower-mass stars may exist—but the theory of star formation described above can help us understand why very high- or very low-mass stars are uncommon.

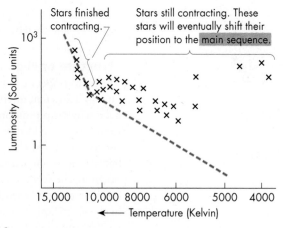

FIGURE 13.9
The young star cluster NGC 2264 and its H-R diagram. Notice the large amounts of leftover gas heated by the brilliant young stars. (Courtesy Anglo-Australian Observatory/photograph by David Malin; H-R diagram adapted from work of M. Walker.)

Stars much smaller than about 0.1 $M_\odot$ are rarely seen because their mass is too small to compress them enough to begin nuclear burning. Although very low-mass objects do form, they remain so cool and dim that they are extremely difficult to see. Astronomers have searched diligently for these very low-mass stars—often called "brown dwarfs" because of their dimness—but they have so far found only a few.

Very high-mass stars are rare for a different reason. As their great gravity compresses them, they rapidly become extremely hot and luminous, emitting such intense radiation that they heat the gas around them, raising its pressure and preventing additional material from falling onto them. Thus their high temperature limits the amount of material they can accumulate. Furthermore, once formed, high-mass stars are so luminous (recall the mass-luminosity relation) that their radiation strips gas from their outer layers, driving it into space. Thus a 100 $M_\odot$ star would quickly "shine" itself apart.

We have followed the protostar from its birth to the shedding of its natal dust and gas. Now we will proceed to the next stage of its development.

13.3 MAIN-SEQUENCE STARS

A star stops being a protostar and becomes a main-sequence star when nuclear fuel in its core can supply energy to raise its pressure and stop its collapse. At this stage of its life, the star's interior structure is like the Sun's: it contains a core, where hydrogen is burnt into helium, and an envelope of gas around the core that transports energy to the star's surface. The properties of the core and envelope, however, depend greatly on the star's mass.

Why a Star's Mass Determines Its Core Temperature

We saw in the last chapter that high-mass stars are both hotter and more luminous than low-mass stars. This relation between mass, temperature, and luminosity results from the condition that the gravity and pressure inside a star must be in balance, a condition we called "hydrostatic equilibrium." To see how this relation arises, recall that the gravitational attraction of a large mass is greater than that of a small mass, other things being equal. That greater gravitational attraction therefore requires a greater pressure to offset it. Thus high-mass stars must have a higher internal pressure than low-mass stars have. That higher pressure can be achieved—according to the perfect gas law—by a higher temperature. Thus a high-mass star must have a hotter core than a low-mass star has. The hotter core then leads to a higher luminosity for massive stars. In summary, high-mass stars must be hotter (and thereby more luminous) to offset their larger gravity.

Structure of High-Mass and Low-Mass Stars

The high temperature at the core of a massive star allows it to fuse its hydrogen into helium by a different mechanism from that of the Sun. In the Sun, hydrogen is converted to helium by the proton-proton chain: two protons fuse to form heavy hydrogen, to which a third proton is added to make ^{3}He. The ^{3}He then fuses to form ^{4}He. In a massive star, the conversion of hydrogen to helium takes place by means of the CNO cycle, whereby carbon, nitrogen, and oxygen atoms already present in the star's core act as catalysts to aid the reaction. The CNO cycle, however, requires a very high temperature to operate and thus can only generate energy at the very center of the star. Photons are unable to carry this huge energy flow from so small a volume, and so the gas in the core begins to rise in irregular clumps and carries the heat outward by convection. Similar convective motions develop in the Sun below its surface, where the radiant energy flow is impeded by the strong absorption of the cool subsurface gas. In a high-mass star, however, the convection currents are confined to the core. They do not extend close enough to the star's surface to allow hydrogen-rich gas there to mix into its core and replenish the depleted fuel there. Neither high-mass stars nor low-mass stars like the Sun can use the fuel in their outer layers. Only very low-mass stars—less than about 0.4 M$_\odot$—mix completely. Thus a star can be starved for fuel even though its outer layers are fuel-rich. This strongly affects a star's lifetime.

Main-Sequence Lifetime of a Star

The time a star stays on the main sequence is called its **main-sequence lifetime**. A star's main-sequence lifetime depends on its mass and luminosity. To see why, we can use a simple analogy. Suppose we want to know how long a car can run on a tank of gasoline. That time clearly depends on how much fuel it has and how rapidly the fuel is consumed. For example, if the car has a 16-gallon tank of gasoline and burns fuel at the rate of 2 gallons per hour, it can cruise for 8 hours—$^{16}/_2$—before running out of gas. When the same formula is applied to a star, its mass determines how much fuel it has, and its luminosity determines how rapidly the fuel is burned. Thus to calculate the star's main-sequence lifetime, we divide the amount of fuel by the rate at which it is consumed.

The rate at which a star burns its fuel depends on its luminosity: Brighter stars burn their fuel faster. We can show, using the information in chapter 11, that the Sun consumes its fuel at the rate of about 2×10^{19} kilograms of hydrogen per year. For a star whose luminosity is $\mathscr{L}$ solar units, the rate of fuel consumption in kilograms per year is therefore $2 \times 10^{19} \mathscr{L}$. Using the car analogy, the lifetime of the star, t, is then its mass divided by $2 \times 10^{19} \mathscr{L}$, or

$$t = \frac{\text{Mass (kg)}}{2 \times 10^{19} \mathscr{L}} \text{ years}$$

A star cannot use all its hydrogen for fuel, however. Only its core is hot enough for nuclear reactions to occur, and the core is generally only about one-tenth of its total mass. Thus the value of the mass we use above should be about 0.1 times the star's mass, and so for a star whose mass is M, the mass of available fuel is $0.1 M$ kg. For the Sun, whose mass is 2×10^{30} kilograms, the mass of available fuel is $2 \times 10^{30} \times 0.1 = 2 \times 10^{29}$ kilograms. For a star whose mass is $\mathscr{M}$ times the Sun's, the mass of available fuel is therefore 2×10^{29} $\mathscr{M}$ kilograms. Using this expression for the amount of fuel available in our expression for a star's lifetime, t, we obtain

$$t = 2 \times \frac{10^{29} \mathscr{M}}{2 \times 10^{19} \mathscr{L}} \text{ Years}$$

$$= 10^{10} \frac{\mathscr{M}}{\mathscr{L}} \text{ Years}$$

where $\mathscr{M}$ and $\mathscr{L}$ are in solar units.

For the Sun ($\mathscr{M} = 1$ and $\mathscr{L} = 1$), this is 10^{10} (or 10 billion) years. For a star like Sirius, whose mass is about 2 M$_\odot$ and whose luminosity is about 20 L$_\odot$, its main sequence lifetime is 10^9 years. For a star whose mass is 10 M$_\odot$ and whose luminosity is 10^5 L$_\odot$, the main-sequence lifetime is a mere 10^6 years. Notice that *massive stars have much shorter lifetimes* than the Sun, despite having more fuel, because they burn it so much faster to supply their greater luminosity. They are like gas-guzzler cars which, despite having 20-gallon fuel tanks, run out of fuel sooner than economy-size cars with only 10-gallon tanks.

The shortness of massive stars' lives implies that those we see must be relatively young. Because massive stars are blue, we can conclude that, in general, *blue stars have formed recently.* Their youth is also indicated by the presence of gas, left over from their formation, still surrounding them, as shown in figures 13.8 and 13.9.

13.4 GIANT STARS

Leaving the Main Sequence

When a main-sequence star exhausts the fuel in its core, the pressure there begins to drop. Gravity, now less strongly opposed, compresses the core and heats it, as shown in figure 13.10A. Because the core has exhausted its hydrogen, as its temperature rises, hydrogen begins burning *outside* the core in what is called a **shell source**, as illustrated in figure 13.10B.

Heat from the shell source and contracting core raises the pressure around the core. That stronger pressure pushes the surrounding gas outward, making the star expand and its radius grow by a factor of anywhere from five to several hundred, depending on the star's mass. But this expansion cools the outer layers, and so the star grows red (recall that cooler bodies radiate more strongly at long [red] wavelengths). Because of its size and color, we call such a star a red giant. Thus a main-sequence star becomes a red giant as it uses up the hydrogen in its core.

A giant star has a very different structure from that of a main-sequence star. As its name implies, its radius is much larger: hundreds of times larger than the Sun's, in some cases. But this enormous size is deceptive. Although most of the star's *volume* is taken up by its very low-density, tenuous atmosphere, most of its *mass* lies in a tiny, hot, compressed core not much larger than the Earth. The core, or its surrounding shell source, supplies the star's energy. The huge envelope is relatively cool and opaque to photons, and so convection rather than radiation carries energy to the star's surface. Some of that energy continues to come from burning hydrogen, but eventually the star must begin to burn helium.

Nuclear Fuels Heavier Than Hydrogen

Helium burning begins very differently in high-mass and low-mass stars and creates significant differences in their structures. To see why, we need to look in more detail at how nuclear fusion occurs and why it requires a very high temperature.

Two nuclei fuse when they are brought close enough together for the nuclear force to bind them to one another. That force, which is also what holds the protons and neutrons together in a nucleus, operates only over very short distances. If nuclei are more than a few diameters apart, the nuclear force is too weak to bond them. In fact, they will be repelled by the electrical force between the similarly charged protons, as illustrated in figure 13.11. That electric repulsion grows as the number of protons increases. It is weak for hydrogen, with its single proton, but is stronger for heavier elements, with more protons. It is enor-

FIGURE 13.10
(A) The core of a star begins to shrink as a star uses up the hydrogen in its core. This compresses and heats the core. (B) The heated core ignites the surrounding gas to make a shell source, and the outer layers of the star expand, turning it into a red giant.

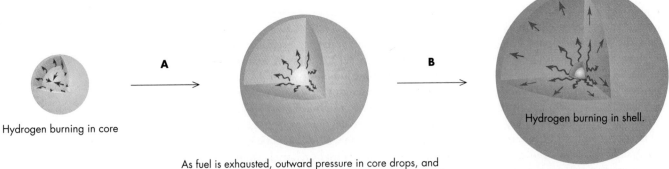

Hydrogen burning in core

A

As fuel is exhausted, outward pressure in core drops, and gravity compresses it. Core contracts. Outer layers expand. Rising heat in contracting core creates pressure that causes outer layers to expand. Shell of H outside core ignites.

B

Hydrogen burning in shell.

Outer layers expand and cool. A red giant forms.

mously more difficult, therefore, for two carbons, with their six protons, to fuse than for two hydrogen atoms to do so.

Nuclei can overcome this electrical repulsion and fuse if they collide at enormously high speed. That speed is achieved in hot gases: the hotter a gas, the faster its nuclei move, and so the more easily it can fuse heavy nuclei. Hydrogen can fuse at about 7 million Kelvin, but helium, with its two protons, must be heated to about 100 million Kelvin. Carbon, with its six protons, must be hotter still to fuse.

If a star's core is as hot as 100 million Kelvin, helium nuclei there will fuse into carbon. The nuclear reaction combines 3 ^{4}He into a ^{12}C. Helium nuclei are sometimes called "alpha particles," and so this reaction is sometimes called the **triple alpha process**, or simply "helium burning." The reaction releases energy but only, it turns out, about one-tenth the amount per kilogram of fuel that hydrogen burning yields. Nevertheless, once a star has consumed all its core hydrogen, it must use its helium or collapse. How does a star make the switch from hydrogen to helium fusion? It must heat its core to 100 million K, and any star more massive than about 0.5 $M_\odot$ can do this by compressing its core.

A high-mass star needs to compress its core only a little before helium begins to fuse because its core is already extremely hot. Thus helium burning begins easily for high-mass stars. A low-mass star like the Sun, however, must compress its core enormously to make it hot enough for helium fusion. The compression of such a star packs its nuclei so closely that they no longer behave like the nuclei of an ordinary gas—in technical terms, the gas becomes **degenerate**.

Degeneracy in Low-Mass Stars

Atomic nuclei in a degenerate gas are packed so tightly that they nearly touch their neighbors. Such a gas behaves almost like a solid: heating does not raise its pressure or make it expand any more than heating a brick makes it swell. In the next chapter, we will discuss degeneracy in more detail. Here we will consider only how it affects nuclear burning.

When a nuclear fuel begins to fuse in a normal gas, the energy released heats the gas, raising its pressure so that the gas expands. The expansion cools the gas and reduces the rate of nuclear burning, acting like a thermostat that switches off the furnace when the house gets too hot. The reason cooling slows nuclear burning is that only at high temperatures do nuclei collide violently enough to fuse. Thus, as a gas cools, the rate at which its atoms fuse decreases, so that less energy is released. In normal stars, this mechanism keeps them from collapsing or exploding. In degenerate matter, this thermostating mechanism fails.

For a low-mass star with a degenerate core, when a fuel ignites, the energy released does not raise the pressure because the pressure does not depend on the temperature. Thus the gas does not expand and cool. Instead, it simply gets hotter and releases more energy, which makes it hotter still. Under these circumstances, the release of energy accelerates explosively in what is called the **helium flash**. In only a few minutes, the star's energy production increases several thousand times, yet this outburst is totally hidden from our view by the star's outer layers, just as a firecracker set off under a mattress creates little visible disturbance. However, the energy released heats the core enough to change it back to a normal gas, destroying its degeneracy. With its degeneracy gone, the gas can now expand. That expansion in turn makes the star's outer layers adjust—shrinking, compressing, and heating. The star's reheated surface changes color from red to yellow, and the star becomes a yellow giant. The helium flash therefore ends the red-giant stage of low-mass stars.

High-mass stars do not have a helium flash. As we said earlier, they are so hot that their cores ignite helium with only a little compression and so they do not become degenerate at this stage of their lives. They continue to expand and change gradually from yellow to red giants, but for both high-mass and low-mass stars, the onset of helium burning may trigger instabilities in the star and make it pulsate.

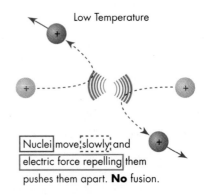

Low Temperature

Nuclei move slowly and electric force repelling them pushes them apart. **No** fusion.

High Temperature

Nuclei move faster and electrical force repelling them is overwhelmed. They collide and **fuse**.

FIGURE 13.11
How a high temperature acts to overcome the electrical repulsion of the nuclei and brings them close enough for the nuclear force to fuse them.

13.5 YELLOW GIANTS AND PULSATING STARS

When stars are plotted on the H-R diagram, some lie between the hot, luminous, blue, main-sequence stars and the red giants. Such stars are called "yellow giants." The most luminous yellow giants are old high-mass stars. Less luminous yellow giants are old low-mass stars that have completed their first red-giant stage and are burning helium in their cores. Regardless of their mass, many yellow giants have the unusual property of swelling and shrinking rhythmically: they pulsate. As they change size, their luminosity changes, and so astronomers refer to them as "variables," or "pulsating stars."

Two particularly important types of pulsating stars are the RR Lyrae (pronounced /lie-ree/) and Cepheid (pronounced /sef-ee-id/) variables. RR Lyrae stars have a mass comparable to the Sun's and are yellow-white giants with about 100 times the Sun's luminosity. The time it takes them to complete a pulsation cycle from bright to dim and back to bright is called their "period," as illustrated in figure 13.12. RR Lyrae variables have periods of about 1 day or a little less and are named for RR Lyrae, a star in the constellation Lyra, the harp, which was the first star of this type to be identified.

Cepheid variables are yellow supergiants that are more massive than the Sun and about 20,000 times its luminosity. They are named for the star Delta Cephei, and their periods can be as short as about 1 day or as long as about 70 days.

Why Do Stars Pulsate?

Giant stars pulsate because their atmospheres trap some of their radiated energy. This heats their outer layers, raising the pressure and making the layers expand. The expanded gas cools, and the pressure drops, so gravity pulls the layers downward and recompresses them. The recompressed gas begins once more to absorb energy, leading to a new expansion. These stars continue alternately to trap and release the energy, and so they continue to swell and shrink, as shown in figure 13.13.

A covered pan of water boiling on a stove behaves similarly. The lid will trap the steam so that pressure inside rises. Eventually, the pressure becomes strong enough to tip the lid, and steam escapes. The pressure decreases, and the lid falls back. It again traps the steam, the pressure again builds up, and the cycle is repeated.

FIGURE 13.12
Schematic light curves of two pulsating variable stars: (A) an RR Lyrae star. (B) a Cepheid variable.

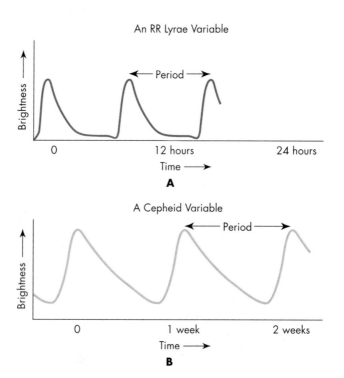

A similar process occurs in pulsating stars, with the role of steam played by the star's radiation and the role of the lid played by the star's atmosphere. For a star to trap radiation this way, its atmosphere must have special absorbing properties—technically called "opacity"—that occur only if its surface temperature and radius fall in a narrow range. That range, called the **instability strip,** is shown in the H-R diagram illustrated in figure 13.14. A star that assumes these characteristics as it evolves will begin to pulsate and will continue to pulsate until its temperature or radius changes enough to remove it from the instability strip. For example, in the distant future, the Sun will pulsate as it alters its structure from red to yellow giant and enters the instability strip. Because of its low mass, the Sun will at that time become an RR Lyrae star. As stars more massive than the Sun evolve, their position on the H-R diagram lies above the position of RR Lyrae variables, and these more massive stars become the highly luminous Cepheid variables. The amount of time a given star spends in the instability strip depends on its mass. Massive stars such as Cepheids evolve across the strip in less than 1 million years, but they cross the region several times as their interior structures alter. Low-mass stars such as the RR Lyrae variables spend more time in the strip, perhaps a few million years, but cross it less often. In either case, stars pulsate for only a brief period of their lives.

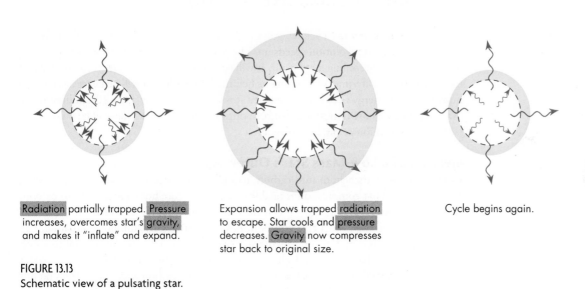

Radiation partially trapped. Pressure increases, overcomes star's gravity, and makes it "inflate" and expand.

Expansion allows trapped radiation to escape. Star cools and pressure decreases. Gravity now compresses star back to original size.

Cycle begins again.

FIGURE 13.13
Schematic view of a pulsating star.

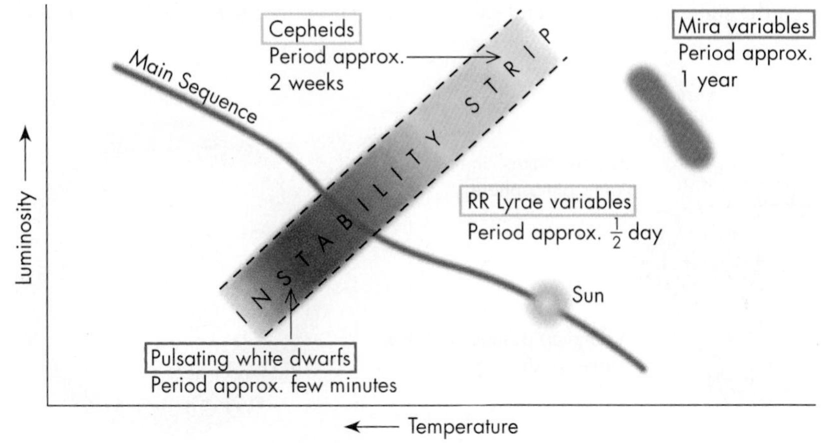

FIGURE 13.14
The instability strip in the H-R diagram. Stars in this narrow region pulsate. The locations of a few other types of variable stars are also indicated.

Main Sequence

Cepheids
Period approx. 2 weeks

INSTABILITY STRIP

Mira variables
Period approx. 1 year

RR Lyrae variables
Period approx. $\frac{1}{2}$ day

Sun

Pulsating white dwarfs
Period approx. few minutes

Luminosity →

← Temperature

Astronomers have identified many other types of pulsating variables besides Cepheids and RR Lyrae stars. For example, Mira variables, which have pulsation periods of about 1 year, lie in the upper right of the H-R diagram. Many other kinds of variable stars, such as ZZ Ceti stars, a kind of pulsating white dwarf with periods as short as a few minutes, lie along a rough extension of the instability strip across the main sequence and downward toward the lower left of the H-R diagram.

The Period-Luminosity Relation

Many varieties of pulsating stars obey the **period-luminosity relation,** which shows that more luminous stars have longer pulsation periods. To see why, we need to recall an important property of stars: their luminosity depends on their temperature and surface area. Stars with larger surface areas tend to be more luminous. The larger surface area implies a larger volume, and when the star's matter is spread throughout that larger volume, the gravitational attraction among its atoms is weakened. Thus the weak gravity of a large-diameter star pulls its atmosphere inward slowly compared with a more compact star. The slowness of their atmospheric motion thus makes big, bright stars pulsate more slowly than small, dim stars.

13.6 DEATH OF STARS LIKE THE SUN

Once the Sun has become a yellow, pulsating giant, it is nearing the end of its life, and as death approaches, its evolution speeds up. The Sun will spend 10 billion years consuming its hydrogen and becoming a red giant. It will spend only 1 billion years (roughly 10% of its main-sequence lifetime) consuming its helium. The evolution will be faster because helium yields less energy than hydrogen when it is burned. In only a few 100 million years, the helium in the core is mostly consumed, during which time the star will lie on the horizontal branch region of the H-R diagram.

Ejection of a Low-Mass Star's Outer Layers

As the core burns its helium into carbon, its radius will once again shrink, compressing and heating it. That compression is unable to make the core hot enough to burn carbon, but it does make it hot enough to increase significantly the rate at which the star burns its fuel. The more rapid fuel consumption makes the star more luminous, inflating it to an even larger radius than before. As the star swells in size, its outer layers cool to about 2500 K—so cool that carbon and silicon atoms condense and form "flakes," just as water in our atmosphere forms snowflakes when it cools. These carbon and "rock" flakes do not fall into the star, however. Rather, they are pushed *outward* by the flood of photons pouring from the star's luminous core, just as dust in a comet is blown into a tail by the Sun's radiation. The rising flakes drag gas with them and drive it into space, forming a huge shell around the star (fig. 13.15). The carbon flakes (as well as other solid particles that condense in the star's cool atmosphere) survive their trip out of the star and drift through

FIGURE 13.15
Artist's conception of how radiation pressure in a luminous red giant pushes on carbon flakes in its atmosphere and drives off the gas. The ejected gas forms a shell around the star—a planetary nebula.

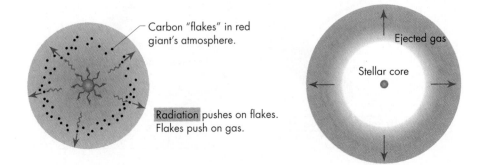

space. Some have been found in meteoritic material within our solar system, "fossils" from a bygone star.

The gas shell expands, thins, and becomes transparent, allowing us to see through to the star's furiously hot core. Because it is so hot, the core's radiation is rich in ultraviolet light, which heats and ionizes the shell around it, making it glow. Astronomers have observed many such glowing shells around dying stars (fig. 13.16) and call them **planetary nebulas**. This term is unfortunate because planetary nebulas have nothing to do with planets. The usage survives from times when observers had only poor telescopes through which planetary nebulas looked like small disks, similar to planets.

Planetary Nebulas

A planetary nebula shell contains about one-fourth solar mass of glowing gas but may have as much as several solar masses of cooler, nonluminous gas around it. Because of this uncertainty about how much matter is expelled, astronomers are not sure whether the Sun will form such a luminous shell or not. Shells are typically about one-fourth of a light year in diameter and expand at about 20 kilometers per second. We can see this expansion by

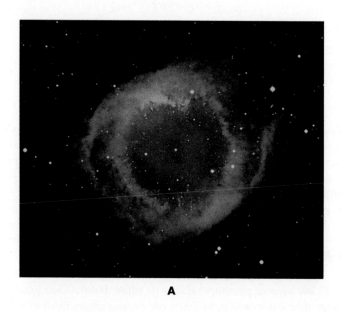

A

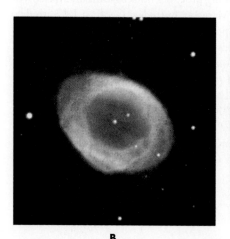

B

C

FIGURE 13.16
Photographs of several planetary nebulas. (A) The Helix nebula. (Courtesy Anglo-Australian Observatory/photograph by David Malin.) (B) The Ring nebula and (C) NGC 6781. (Courtesy Cal Tech and Carnegie Institute of Washington.) Notice the central star in each. Other stars that look as if they are inside the shell are foreground or background stars.

☼ What will eventually happen to the central star? What will eventually happen to the nebula itself?

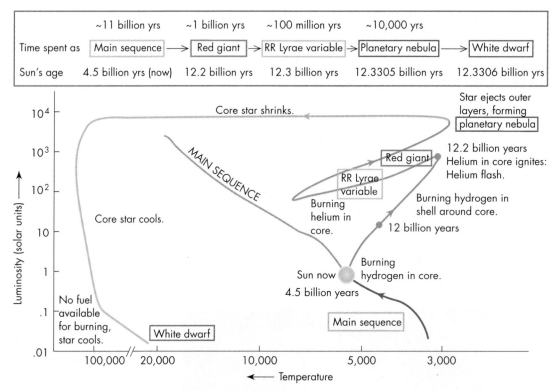

Time spent as	Main sequence	→	Red giant	→	RR Lyrae variable	→	Planetary nebula	→	White dwarf
	~11 billion yrs		~1 billion yrs		~100 million yrs		~10,000 yrs		
Sun's age	4.5 billion yrs (now)		12.2 billion yrs		12.3 billion yrs		12.3305 billion yrs		12.3306 billion yrs

FIGURE 13.17

The evolution of a low-mass star like the Sun plotted in an H-R diagram.

comparing photographs taken several decades apart. The shells eventually grow so big and diffuse that they mingle with the interstellar medium. Thus the cycle of a star's life is complete: gas to star and back to gas. But the core of the star remains behind as a tiny, glowing ball that astronomers call a "white dwarf." Figure 13.17 shows these changes in an H-R diagram, affording a shorthand view of the life of a star like the Sun. A very different fate, however, awaits a more massive star.

13.7 OLD AGE OF MASSIVE STARS

Massive stars do not become planetary nebulas or white dwarfs. Their great mass compresses and heats their cores enough to ignite carbon and allows them to keep burning when their helium is gone. A variety of other nuclear reactions may also occur that create heavy elements and at the same time supply energy for the star. Astronomers call the formation of heavy elements by such nuclear burning processes **nucleosynthesis.** Nucleosynthesis in stars was discussed in detail by E. Margaret Burbidge, Geoffrey R. Burbidge, William A. Fowler, and Fred Hoyle, who proposed that *all* the chemical elements in our Universe heavier than helium were made in this way. Their idea is one of the triumphs of stellar evolution theory, and an enormous amount of theory and observation support it.

Formation of Heavy Elements: Nucleosynthesis

Heavy elements form when two or more light nuclei combine and fuse into a single heavier one. The new nucleus must have a number of protons equal to the sum of the numbers found in its parent nuclei.* Likewise, it must contain as many neutrons as they possessed. For example, suppose ^{12}C and ^{16}O fuse. Carbon has 6 protons and 6 neutrons, and

---

*In some reactions, such as the p-p chain, a proton may turn into a neutron and a positron.

TABLE 13.1   Nuclear Reactions That Form Heavy Elements

$4\ ^{1}H$	$\rightarrow$	^{4}He	$2\ ^{12}C$	$\rightarrow$	$^{24}Mg + \gamma$
$3\ ^{4}He$	$\rightarrow$	^{12}C	$2\ ^{12}C$	$\rightarrow$	$^{16}O + 2\ ^{4}He$
$2\ ^{12}C$	$\rightarrow$	$^{20}Ne + ^{4}He$	$2\ ^{16}O$	$\rightarrow$	$^{32}S + \gamma$
$^{20}Ne + ^{4}He$	$\rightarrow$	$^{24}Mg + \gamma$	$2\ ^{16}O$	$\rightarrow$	$^{28}Si + ^{4}He + \gamma$

KEY: H, hydrogen; He, helium; C, carbon; O, oxygen; Ne, neon; Mg, magnesium; S, sulfur; Si, silicon; γ, a gamma ray (high-energy photon).

NOTE: Helium is not considered a heavy element, but its production is shown in the table because it is the first step of nucleosynthesis in stars.

oxygen has 8 protons and 8 neutrons. Their fusion makes a nucleus with $6 + 8 = 14$ protons and $6 + 8 = 14$ neutrons (that is, an atom with 14 protons and 28 particles total in the nucleus). Reference to a table of the elements shows that such an atom is silicon, ^{28}Si. Table 13.1 shows some of the reactions that make heavy elements in high-mass stars. The symbol γ is used for a high-energy photon (a gamma ray), which is produced in many of the reactions.

In massive stars, these reactions provide enough energy to support the stellar mass against gravity's inexorable pull. As each fuel—hydrogen, helium, carbon—is exhausted, the star's core contracts and heats by compression. The higher temperature then allows the star to burn still heavier elements, the "ashes" of one set of nuclear reactions becoming the fuel for the next set. Oxygen, neon, magnesium, and eventually silicon are formed, but because progressively higher temperatures are needed for each new burning process, each new fuel is confined to a smaller and hotter region around the star's core. Thus the star develops a layered structure, as illustrated in figure 13.18. The star's surface, where the temperature is too low for nuclear reactions to occur, remains hydrogen, but beneath the surface lies a series of nested shells, each made of a heavier element than the one surrounding it. Helium surrounds a shell of carbon, which in turn surrounds a shell of oxygen, and so on to the core. By the time the star is burning silicon into iron, its core has shrunk to a diameter smaller than the Earth's, and the core's temperature is about 2 billion K. A massive star (10 $M_\odot$ or more) may take less than 10^7 years to form an iron core. This brief lifetime results because (1) the star burns its fuel rapidly to offset the energy lost because of its high luminosity, and (2) the structure of atomic nuclei is such that fusion of elements heavier than hydrogen yields much less energy than the fusion of a similar amount of hydrogen, and so more fuel must be burned to supply the same amount of energy.

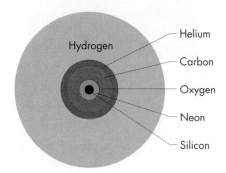

FIGURE 13.18
The layered structure of a massive star as it burns progressively heavier fuels in its core.

Core Collapse of Massive Stars

The formation of an iron core signals the end of a massive star's life. Iron cannot burn and release energy: an iron atom is so tightly bound by the nuclear force that adding additional protons or neutrons weakens it and makes it break up. Thus nuclear fusion stops with iron, and a star with an iron core is out of fuel.

Fuel exhaustion causes a star's core to shrink and heat. But in high-mass stars the shrinkage presses the iron nuclei so tightly together that a new reaction can occur: protons and electrons may themselves "merge," neutralizing their charge and becoming neutrons. The shrinking core is thus transformed from a sphere of iron into a sphere of neutrons, with catastrophic results for the star. Most of the pressure that supported the core was supplied by electrons, but they have been absorbed by protons. Thus the star's core pressure suddenly drops. Nothing remains to support the star, and so its interior begins to collapse. However, because the matter is so dense, its force of gravity is immense, and it crushes the core. In less than a second, the core shrivels from an iron ball the size of the Earth to a ball of neutrons about 10 kilometers (about 6 miles) in radius. The star's outer layers, with nothing to support them, plummet inward like a tall building whose first floor is suddenly blown away. The plunging outer layers of the star strike the neutron core, crushing it still more, while the impact heats the infalling gas to billions of degrees. The pressure surges and lifts the outer layers away from the star in a titanic explosion—a **supernova.**

Supernova Explosions

The explosion of a massive star mixes the elements synthesized by nuclear burning during its evolution with the star's outer layers and blasts them into space. This incandescent spray expands away from the star's collapsed core at more than 10,000 kilometers per second. Depending on the star's mass, 10 or so solar masses of matter may be flung outward, ultimately to mix with surrounding material and, in time, to form new generations of stars. Interstellar gas is thereby enriched in heavy elements, the atoms needed to build the rock of planets and the bones of living creatures. Indeed, the supernova outburst itself generates even more heavy elements. The explosion creates free neutrons that rapidly combine with atoms in the star to build up heavy and rare elements such as gold, platinum, and uranium.

For the star, the supernova explosion is a quick and glorious death. In a few minutes, it releases more energy than it has generated by nuclear burning during its entire existence. In a matter of hours, the dying star brightens to several billion times the luminosity of the Sun, radiating energy in its death throes at a greater rate than all the other stars in the galaxy combined. Figure 13.19 is a photograph of a galaxy with a supernova: the exploding star gleams like a beacon.

A supernova emits more than just visible light: most of the energy of its blast is carried by a burst of neutrinos, the same tiny, highly penetrating particles generated in the Sun as it converts hydrogen into helium. Just as the Sun's neutrinos escape freely into space, so the supernova's neutrinos escape as well. A pulse of such neutrinos was detected in February 1987 when a supernova—SN 1987A—blew up in the Large Magellanic Cloud. You may have "seen" one of those neutrinos because more than 1 trillion of them passed through you, and if one happened to collide with an electron in your eyeball, it would have made a tiny flash of light.

Supernova Remnants

Gas ejected by the supernova blast plows through the surrounding interstellar space, sweeping up and compressing other gas that it encounters. The huge, glowing cloud of stellar debris—a **supernova remnant**—steadily expands. At the end of one year, it is 300 billion kilometers (about 0.03 light years) across. At the end of one century it is several light years in diameter, but its expansion slows as it runs into more surrounding gas. Figure 13.20 shows photographs of several supernova remnants of different ages. Notice how ragged the remnants in figures 13.20A and 13.20B look compared with the smoothly ejected bubbles of planetary nebula—evidence that massive stars die more violently than low-mass stars.

One such violent outburst was seen about 1000 years ago by astronomers in China and elsewhere in the Far East. On a date that according to our calendar was July 4, 1054, their records describe a "guest star" that appeared in the evening sky and that was visible even in broad daylight for several weeks. Today, with even a small telescope, you can still see the glowing gases ejected from that dying star. Figure 13.20B is a picture of this perhaps most famous of all supernova remnants, now known as the **Crab Nebula.** In the next chapter we will learn that the star's core survived the blast and lies in the expanding debris.

A supernova explosion marks the death of a massive star, and we chart its complete life in the H-R diagram shown in figure 13.21. We can see that its path through time is far simpler than that of a low-mass star. The massive star begins its life on the upper main sequence as a blue star. As its core hydrogen is consumed, it leaves the main sequence,

FIGURE 13.19
Photographs showing a supernova in the galaxy NGC 4725 in late 1940. The arrow indicates the supernova. Most of the other bright dots are foreground stars in our own galaxy. The top picture shows the galaxy after the supernova has faded. (Courtesy Cal Tech.)

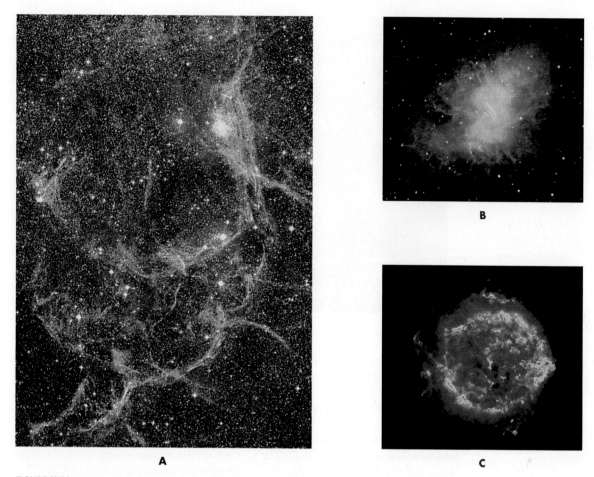

FIGURE 13.20
A few supernova remnants: (A) Color photograph of the Vela supernova remnant. (Courtesy Anglo-Australian Observatory/photographs by David Malin.) (B) Color picture of the Crab Nebula. (Courtesy R. Wainscoat.) (C) Radio image of the Cassiopeia supernova remnant. (Courtesy NRAO/AUI.)

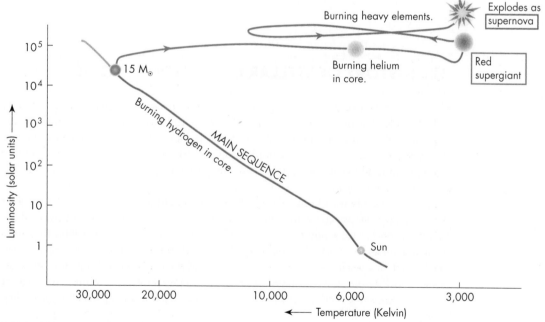

FIGURE 13.21
H-R diagram of a massive star's evolution.

FIGURE 13.22
Photograph of a portion of the Large Magellanic Cloud (a small nearby galaxy) before Supernova 1987A exploded and a second photograph taken during the event. (Courtesy Anglo-Australian Observatory/photograph by David Malin.)

swelling and cooling to become a yellow supergiant while it begins to burn helium in its core. When the helium there is consumed, the star switches to other fuels, all the while contracting and heating its core and expanding and cooling its outer layers to become a red supergiant. Overall, it keeps its luminosity approximately constant as it ages, but some massive stars, depending on how many heavy elements are incorporated in them at their birth, may return partway to the main sequence as they switch to new fuels. The resultant heating may turn them into blue supergiants before they explode as supernovas. Such was the behavior of Supernova 1987A, whose explosion is documented in figure 13.22.

What remains when the supernova remnant dissipates? The core of the star survives either as a ball of neutrons (a neutron star) or perhaps as an even more compressed body—a black hole—as we will discover in the next chapter.

13.8 HISTORY OF STELLAR EVOLUTION THEORIES

The theories we have discussed above about how stars work and evolve are relatively new. For example, Aristotle wrote more than 2000 years ago that stars are heated by their passage through the heavens. He never considered the possibility that they might evolve. In the eighteenth century, the German philosopher Immanuel Kant described stars slightly more scientifically, envisioning the Sun as a fiery sphere, formed when gases gravitated to the center of the solar nebula.

The English physicist Lord Kelvin and the German physicist Hermann von Helmholtz were the first scientists to use the equations of basic physics to arrive at a mathematical description of what we would today call "model stars." In the 1850s and 1860s, they were searching for a mechanism that could explain the Sun's power supply, as discussed in chapter 11. Kelvin and Helmholtz independently solved the mathematical equations for a gas sphere in which pressure is balanced by gravity and showed that a star's core is compressed by the weight of the overlying gas and is much denser than the outer parts of the star. They further showed that, as gravity compresses a star, that compression converts gravitational energy into heat, raising the star's temperature. However, the amount of energy available from gravitational energy is far too small to power it for billions of years.

Astronomers had to await discoveries about the nature of atoms in the early years of the twentieth century before they could solve the mystery of stellar energy. The British astrophysicist Sir Arthur Stanley Eddington, who discovered the mass-luminosity relation, was one of the first scientists to recognize the immense power available in the form of nuclear energy, but at the time, too little was known about how atoms interacted to make models of stars powered by hydrogen burning. Moreover, Eddington discovered that to make a model star, more was required than simply balancing pressure and gravitational forces. He demonstrated that a correct model must also account for the outward leakage of energy from the star's core to its surface, a process that made solving the equations much more difficult.

The equations that govern a star's structure express a complicated set of relations between the star's pressure, density, temperature, gravitational force, energy generation, composition, and luminosity. The equations must balance at each point in the star and yield answers that agree with observations. Solving such complex equations is extremely difficult but became easier in the 1940s, when astronomers began using mechanical calculators. These machines—incredibly primitive compared with a modern computer, were operated by turning a hand crank. Nevertheless, they were faster than pencil-and-paper solutions, and the models so produced revealed more clearly the physical processes in stars. Several months might be needed to make the calculations for a half-dozen models that showed how a main-sequence star could change into a red giant. Then, in the 1950s, when astronomers began to use electronic computers to calculate model stars, only a few minutes were needed to solve the equations of a stellar model. By combining the results of dozens of such models, each of which represented a star with successively more of its precious fuel consumed, astronomers could see how stars evolved. Such work continues to this day, with astronomers still trying to understand the late stages of a star's life, when it is losing mass and burning elements heavier than helium in its core.

13.9 TESTING STELLAR EVOLUTION THEORY

We have seen how astronomers think stars evolve, but what evidence supports these theories? For example, although we now understand that main-sequence stars turn into red giants, as recently as the 1940s some astronomers thought the reverse. The best demonstration that the modern theory is correct comes from comparing the H-R diagrams of real star clusters with theoretically determined H-R diagrams.

We learned in section 13.2 that stars probably form in groups—star clusters. The stars within a cluster are bound together by their mutual gravitational attraction although each star moves along its own orbit. All the stars within a given cluster form at approximately the same time and are therefore nearly the same age. If we look at the stars in a cluster shortly after its birth, when they have all completed their contraction stage from protostars, they will all lie on the main sequence. If we look at some later time, some of the stars will have consumed their core hydrogen and evolved off the main sequence. To check our theory of stellar evolution, we calculate evolutionary tracks, such as those in figures 13.17 and 13.21, for every star on the main sequence to show where each star will be, at say, 10 million years, 100 million years, and so on. The resulting curves show us what the H-R diagram of the entire cluster should look like 10 million years, 100 million years, and so on, after its birth. When we compare such curves with H-R diagrams of actual star clusters, the match is excellent. If our theory of stellar evolution were wrong, the shapes would be unlikely to agree so well. In addition, the models help astronomers understand why so many kinds of stars exist. For example, the models not only offer a natural explanation of main-sequence, red-giant, and white-dwarf stars but also help us see how such totally different objects, such as pulsating variables, planetary nebulas, and supernovas, fit into the scheme of stellar evolution. This success in interpreting such stellar diversity is evidence that the theory is essentially correct. Moreover, that success offers astronomers a way to measure the age of a star cluster.

EXTENDING OUR REACH

MEASURING THE AGE OF A STAR CLUSTER

Astronomers can deduce the age of a star cluster from the shape of its H-R diagram. To understand why its H-R diagram reveals the cluster's age, recall that in a star cluster, all the stars form at approximately the same time. Moreover, at their birth, all the stars are burning hydrogen in their core. Because the main sequence is determined by the location of stars burning hydrogen in their core, all the stars in a newly formed cluster lie on or near the main sequence. Not all the cluster's stars burn their fuel at the same rate, however. Massive stars burn their fuel more rapidly (to maintain their high luminosity) than low-mass stars do. With their

hydrogen used up, high-mass stars leave the main sequence and turn into red giants. The low-mass stars, on the other hand, still have hydrogen to burn and so they remain on the main sequence longer.

Because the high-mass stars in an old cluster turn into red giants (and thus lie off the main sequence), a line in the H-R diagram connecting the position of the cluster's stars bends away to the right, as shown in figure 13.23. The point where that line bends away from the main-sequence is called the **turnoff point**. At the turnoff point, a star is not quite old enough to have used up the hydrogen in its core. That is, its age is just a tiny bit less than its main-sequence lifetime. But *all* stars

in the cluster are the same age, namely, the age of the cluster. To get the cluster's age, we therefore measure the age of the star at the turnoff point by calculating its main-sequence lifetime from its mass and luminosity, as described earlier. The answer we get is the age of the star cluster.

Figure 13.24 shows a series of H-R diagrams for clusters ranging in age from very young to very old (a few million years for the youngest to more than 10 billion years for the oldest). Notice that old clusters have few if any stars on the upper part of the main sequence. Young clusters, on the other hand, do have stars on the upper main sequence.

FIGURE 13.23
The turnoff point in the H-R diagram of a star cluster indicates its age.

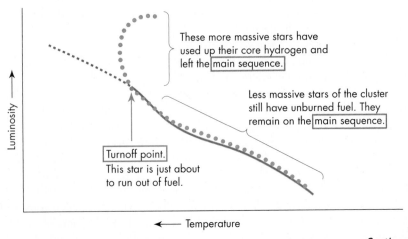

Less massive stars of the cluster still have unburned fuel. They remain on the main sequence.

These more massive stars have used up their core hydrogen and left the main sequence.

Turnoff point.
This star is just about to run out of fuel.

Luminosity ⟶

⟵ Temperature

Continued

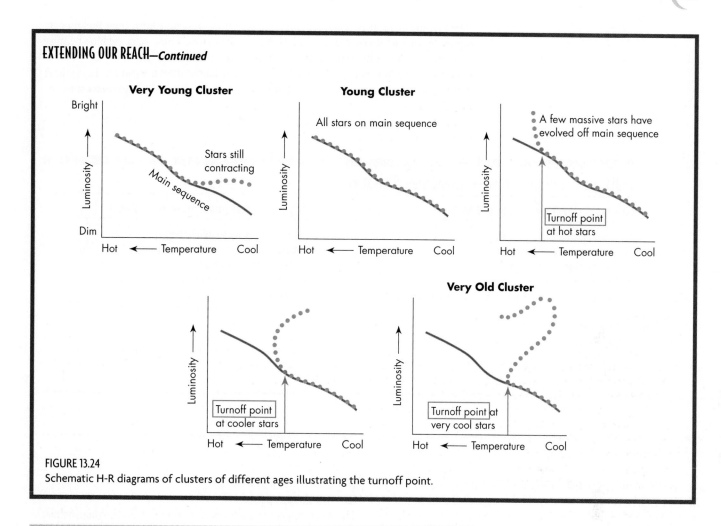

FIGURE 13.24
Schematic H-R diagrams of clusters of different ages illustrating the turnoff point.

SUMMARY

A star forms from interstellar gas drawn together by gravity, which compresses and heats the gas to form a protostar. Further heating causes the core of the protostar to fuse hydrogen into helium. The energy released keeps the core hot and thereby maintains an outward pressure that stops the protostar's collapse and changes it into a main-sequence star. When the hydrogen is exhausted in a star's core, the core shrinks and heats, making the star swell into a red giant.

A low-mass star like the Sun continues to burn hydrogen in a shell during its first red-giant stage but eventually compresses the helium in its core enough to ignite the gas in a helium flash. The energy released expands the core, and the outer layers shrink, turning the star into a yellow giant. As the star burns helium, its core again shrinks, and it once more becomes a red giant. It is so luminous now that its radiation drives off its outer layers to form a planetary nebula shell. The core remains as a small, dense star: a white dwarf.

Because of the tremendous heating from gravitational compression of their cores, high-mass stars more easily ignite fuels heavier than hydrogen and undergo less dramatic changes initially than low-mass stars do. They build in their cores layers of heavy elements: carbon, neon, silicon, and eventually iron. The iron core cannot burn and will collapse, triggering a supernova explosion. The heavy elements produced during the star's life and in the supernova that destroys it are blasted into space, enriching the interstellar medium. The core remains as a ball of neutrons—a neutron star—or may collapse entirely and become a black hole.

The lifetime of a star is dependent on its mass. High-mass stars burn their fuel quickly to supply the energy they need to support their great weight. These stars are blue because they are so hot. Thus blue stars burn out quickly and must be young.

Both high- and low-mass stars pass through a stage late in their lives when they pulsate as yellow giants. Low-mass stars become RR Lyrae variables, whereas high-mass stars become Cepheids.

Table 13.2 summarizes these stages.

TABLE 13.2   Evolution of Low-Mass and High-Mass Stars

Low-mass star (mass less than 5 M☉)	High-mass star (mass more than 5 M☉)
Protostar (gravity supplies energy)	*Protostar* (gravity supplies energy)
Main sequence (hydrogen burns in core)	*Main sequence* (hydrogen burns in core)
Core hydrogen consumed	Core hydrogen consumed
Red giant (hydrogen burns in shell)	*Yellow supergiant* (helium burns in core)
Core contracts, heats, and ignites helium (helium flash)	Star pulsates
Yellow giant (star pulsates)	*Red supergiant* (heavier elements burn in a series of shells in the star's core)
Red giant (helium burns in shell)	Buildup of iron core and its collapse
Outer layers ejected to form:	*Supernova* explosion when core collapses;
Planetary nebula	remnant may be a neutron star or black hole
Core cools to become:	
White dwarf (no energy generation)	

QUESTIONS FOR REVIEW

1. Sketch an H-R diagram of the evolutionary path of the Sun.
2. Sketch an H-R diagram of the evolutionary path of a 10 M☉ star.
3. Describe how a protostar is believed to form.
4. What heats a protostar?
5. What is a bipolar flow?
6. How can we observe protostars? Why are they surrounded by dust and gas?
7. What is a T Tauri star?
8. What is a Bok globule?
9. What determines when a star becomes a main-sequence star?
10. What determines how long a star stays on the main sequence?
11. What makes a star move off the main sequence?
12. Where do main-sequence stars end up as they evolve?
13. Why do high- and low-mass stars evolve differently as they become red giants?
14. Why is it easier for a high-mass star than for a low-mass star to burn helium?

15. What is meant by a pulsating star?
16. Why do stars pulsate?
17. What happens to a solar-mass star when it starts to burn helium in its core? What does it turn into?
18. What is a planetary nebula?
19. What is one explanation for how a low-mass star expels its outer layers to make a planetary nebula?
20. What is left when a planetary nebula dissipates?
21. What makes a high-mass star's core collapse?
22. How do the neutrons form in a massive star's remnant core? What other kind of particle is made along with the neutrons?
23. What is a supernova explosion?
24. What kind of subatomic particles have been observed when a supernova explodes?

THOUGHT QUESTIONS

1. Suppose that stars could mix unburned fuel from their outer layers into their cores. Would that alter the way they evolve? Can you suggest some possible differences?

2. Hold a small rubber ball on top of a basketball, and drop them together toward the floor. What happens to the small ball? Does that help you understand what happens to the outer layers of a supernova as they collapse on the core?

3. Inflate a balloon and carefully measure its size. Put it in the freezer for a few hours. Does it look the same when it is cold? How does this relate to how stars form in cold regions of space?
4. Take a plastic bottle and put a little soapy water in it. Run your finger across the mouth to make a soap film. Now, without breaking the film, run hot water over the bottle. What happens to the soap film? How does this relate to what happens to a star when it is heated?

5. In some very old, dense star clusters, a few blue stars, known as "blue stragglers," are seen. Why are blue stars unexpected in such clusters? If stars that stick together mix their material thoroughly when they collide, how might you explain such blue stars?

PROBLEMS

1. Calculate the main-sequence lifetime of the Sun by first determining the rate at which it burns hydrogen (refer to chapter 11) and then dividing that rate into its core mass. If your answer is in seconds, convert to years given that there are about 3.2×10^7 seconds in a year.
2. Calculate the escape velocity from a red giant's atmosphere (use the formula for escape velocity from chapter 2). Assume that the star's mass is 1 $M_\odot$ and its radius is 100 $R_\odot$ How does this compare with the speed at which a planetary nebula shell is ejected?
3. How long will it take a planetary nebula shell moving at 20 kilometers per second to expand to a radius of one-fourth of a light year?
4. Figure 13.25 (right) shows the H-R diagram of a star cluster. How old is it?

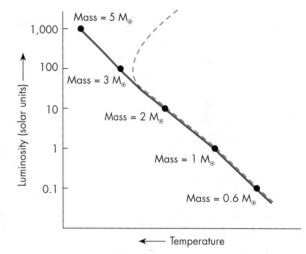

FIGURE 13.25
H-R diagram of a star cluster.

TEST YOURSELF

1. Which of the following sequences correctly describes the evolution of the Sun from young to old?
   (a) White dwarf, red giant, main-sequence, protostar.
   (b) Red giant, main-sequence, white dwarf, protostar.
   (c) Protostar, red giant, main-sequence, white dwarf.
   (d) Protostar, main-sequence, white dwarf, red giant.
   (e) Protostar, main-sequence, red giant, white dwarf.
2. A planetary nebula is
   (a) another term for the disk of gas around a young star.
   (b) the cloud from which protostars form.
   (c) a shell of gas ejected from a star late in its life.
   (d) what is left when a white dwarf star explodes as a supernova.
   (e) the remnants of the explosion created by the collapse of the iron core in a massive star.
3. Stars like the Sun probably do not form iron cores during their evolution because
   (a) all the iron is ejected when they become planetary nebulas.
   (b) their cores never get hot enough for them to make iron by nucleosynthesis.

   (c) the iron they make by nucleosynthesis is all fused into uranium.
   (d) their strong magnetic fields keep their iron in their atmospheres.
   (e) None of the above.
4. As a star like the Sun evolves into a red giant, its core
   (a) expands and cools.
   (b) contracts and heats.
   (c) expands and heats.
   (d) turns into iron.
   (e) turns into uranium.
5. A star whose mass is 5 times larger than the Sun's and whose luminosity is 100 times larger than the Sun's has a main-sequence lifetime about
   (a) 5 times longer than the Sun's.
   (b) 500 times longer than the Sun's.
   (c) 5 times shorter than the Sun's.
   (d) 100 times shorter than the Sun's.
   (e) 20 times shorter than the Sun's.

FURTHER EXPLORATIONS

Bally, John. "Bipolar Gas Jets in Star-Forming Regions." *Sky and Telescope* 66 (August 1987): 94.

Bethe, Hans A., and Gerald Brown. "How a Supernova Explodes." *Scientific American* 252 (May 1985): 60.

Boss, Alan P. "Collapse and Formation of Stars." *Scientific American* 252 (January 1985): 40.

Crosswell, Ken. "Stars Too Small to Burn." *Astronomy* 12 (April 1989): 16.

Crosswell, Ken. "Life and Times of a Star." *New Scientist* 144 (26 November 1994): A1.

de Grasse Tyson, Neil. "Forged in the Stars." *Natural History* 105 (August 1996): 72.

Eicher, David J. "Ashes to Ashes and Dust to Dust." *Astronomy* 22 (May 1994): 40.

Frank, Adam. "Blowing Cosmic Bubbles." *Astronomy* 25 (February 1997): 36.

Frank, Adam. "Starmaker: the New Story of Stellar Birth." *Astronomy* 24 (July 1996): 52.

Hayes, John C., and Adam Burrows. "A New Dimension to Supernovae." *Sky and Telescope* 90 (August 1995): 30.

Henbest, N. "Life of a Star." *New Scientist* 118 (June 2, 1988): 4.

Herbst, William, and George C. Assousa. "Supernovas and Star Formation." *Scientific American* 241 (August 1979): 138.

Jastrow, Robert. *Red Giants and White Dwarfs*. New York: Norton, 1990.

Kaler, James B. "Planetary Nebulae and Stellar Evolution." *Mercury* 10 (July/August 1991): 114.

Kawaler, Steven D., and Donald E. Winget. "White Dwarfs: Fossil Stars." *Sky and Telescope* 74 (August 1987): 132.

Kirshner, Robert P. "The Earth's Elements." *Scientific American* 271 (October 1994): 58. (How the elements formed.)

Lada, Charles J. "Energetic Outflows from Young Stars." *Scientific American* 247 (July 1982): 82.

Meadows, A. J. *Stellar Evolution*. 2nd ed. Oxford, New York: Pergamon Press, 1978.

Morrison, Nancy D., and Stephen Gregory. "What Massive Stars Explode?" *Mercury* 15 (May/June 1986): 77.

Reipurth, Bo, and Steve Heathcote. "Herbig-Haro Objects and the Birth of Stars." *Sky and Telescope* 90 (October 1995): 38.

Rodriguez, L. "Searching for the Energy Source of the Herbig-Haro Objects." *Mercury* 10 (March/April 1981): 34.

Soaker, Noam. "Planetary Nebulae." *Scientific American* 266 (May 1992): 78.

Stahler, Steve W. "The Early Life of Stars." *Scientific American* 265 (July 1991): 48.

Stephens, Sally. "Needles in the Cosmic Haystack." *Astronomy* 23 (September 1995): 50. (About brown dwarfs.)

Tayler, R. "The Birth of Elements." *New Scientist* 124 (December 16, 1989): 25.

Woosley, Stan, and Tom Weaver. "The Great Supernova of 1987." *Scientific American* 261 (August 1989): 32.

KEY TERMS

bipolar flows, 392
Crab Nebula, 404
degenerate, 397
helium flash, 397
instability strip, 399
interstellar cloud, 387
main-sequence lifetime, 395
nucleosynthesis, 402
period-luminosity relation, 400

planetary nebulas, 401
protostar, 390
shell source, 396
supernova, 404
supernova remnant, 404
T Tauri stars, 393
triple alpha process, 397
turnoff point, 408

CHAPTER FOURTEEN

STELLAR REMNANTS: White Dwarfs, Neutron Stars, and Black Holes

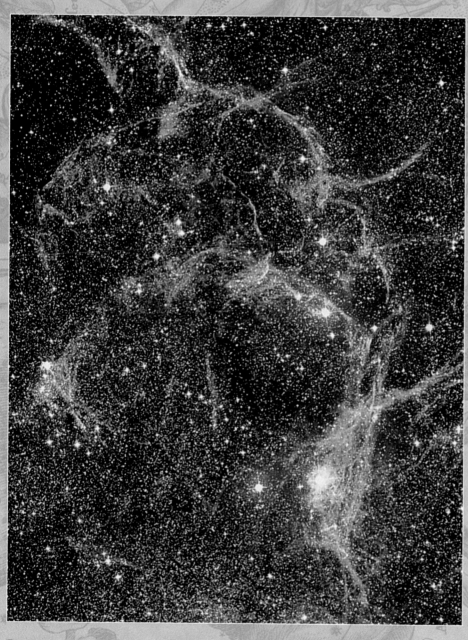

Debris from a supernova explosion (Courtesy Anglo-Australian Observatory/
Photograph by David Malin)

During its life, every star supports itself against gravity by burning nuclear fuel in its core. When its fuel is spent, the star collapses. This fate awaits the Sun 5 billion years from now. Gravity will crush it into a white dwarf—a star about 100 times smaller than the present Sun and roughly the size of the Earth. More massive stars surrender sooner and are more dramatically squeezed to even smaller dimensions, becoming either neutron stars or black holes.

Compact stars, as these three kinds of stellar remnants are known, are the end points of stellar evolution. Because nearly all stars must eventually reach this stage, the galaxy is littered with their shriveled bodies. But unlike stars at earlier stages of evolution, no nuclear fuel makes them shine. They shine—if at all—with heat inherited from their previous state. Their matter, too, is unusual. In crushing compact stars to their tiny dimensions, gravity squeezes them into exotic materials. A white dwarf's matter weighs more than 16 tons per cubic inch. Matter in a neutron star is so compressed that electrons have merged with protons, making the star resemble a giant atomic nucleus. So unmercifully has gravity squeezed the most massive of compact stars that they have collapsed completely, their immense gravity warping the space around them so that no light escapes, making them black holes in space.

Compact, however, does not mean inconspicuous: some of these crushed stars radiate intensely despite having no fuel of their own. Rather, a compact star may "parasitize" a companion star, drawing matter from it that may burn explosively and allow the crushed star to "rise from the dead" as a nova—a "new" star. Moreover, because of a compact star's intense gravity, any material that falls on it releases such immense amounts of gravitational energy that this dead star may be thousands of times brighter than a "live" star like the Sun. But this brightness is bought at a dreadful price. The star's mass may increase to the point that gravity causes it to collapse even further or, for some stars, to explode as a type I supernova, blowing the star to atoms.

14.1 WHITE DWARFS

General Properties, Origin, and Fate

White dwarfs are hot, compact stars whose mass is comparable to the Sun's but whose diameter is about the same as the Earth's (about 0.01 that of the Sun). Their tiny size gives them so small a surface area that they are very dim despite being as hot as about 25,000 K. Even the brightest white dwarfs are about 100 times fainter than the Sun. Their pale light is not supplied by burning fuel, for they have exhausted all fuel supplies. Rather, it comes from residual heat that leaks from the star's interior.

A white dwarf forms when a low-mass star like the Sun nears the end of its life. When the star swells into a red giant, its intense radiation strips off its outer layers to form a planetary nebula shell. This process exposes the star's core, as illustrated in figure 14.1. That core is composed mainly of carbon and oxygen, the end products of its parent star's nuclear burning. The white dwarf's surface is composed of a thin layer of hydrogen and helium, which may initially be as hot as 150,000 K. The core has too little mass and therefore insufficient gravity to contract and heat itself to the ignition temperature of carbon. Unable to burn further, the dying core rapidly cools and shrinks into a white dwarf.

In theory, some white dwarfs could form from very low-mass stars (less than about 0.4 or 0.5 $M_\odot$) without ejecting a planetary nebula shell. Such stars are relatively cool

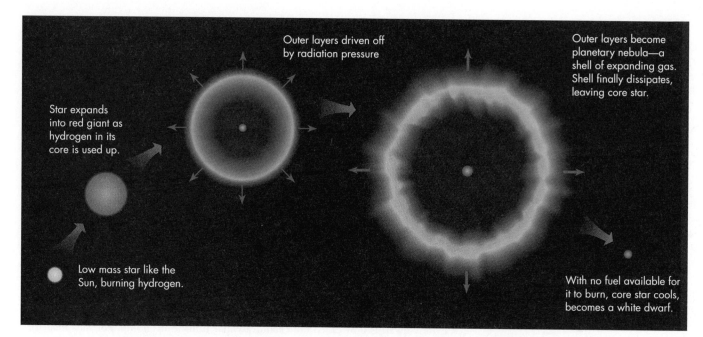

Star expands into red giant as hydrogen in its core is used up.

Outer layers driven off by radiation pressure

Outer layers become planetary nebula—a shell of expanding gas. Shell finally dissipates, leaving core star.

Low mass star like the Sun, burning hydrogen.

With no fuel available for it to burn, core star cools, becomes a white dwarf.

FIGURE 14.1
Origin of a white dwarf. White dwarfs form when a low-mass star like the Sun reaches the end of its life. Having expanded into a red giant, the star loses its outer atmosphere and becomes a planetary nebula. Its core is exposed, and eventually it turns into a white dwarf, a star with no source of nuclear fuel.

throughout, and so their gas, like the relatively cool matter just below the Sun's surface, must carry energy by convection. The convection currents in these small stars extend throughout their spheres, mixing their gas and preventing a core from forming. Moreover, their weak gravity cannot compress the spent fuel to make them red giants. Because low-mass stars evolve so slowly, however, they require longer than the age of the Universe to reach this point in their evolution, and therefore none of these very low-mass stars have turned into white dwarfs yet.

A white dwarf shines by heat inherited from its earlier incarnation. Calculations show that it loses this heat very slowly, taking about 10 million years for the surface to cool to 20,000 K. During this long cooling period, the star grows steadily redder and dimmer, but being dimmer, it loses heat more slowly. Astronomers are uncertain how long it takes a white dwarf to fade to such dimness, but they call such dead stars **black dwarfs**. Cooling white dwarfs are probably very abundant in our galaxy. Astronomers estimate from studies of the number of stars that die each year that perhaps as much as half our galaxy's mass is composed of dead white dwarfs.

Structure of White Dwarfs

Because white dwarfs are very compact and have no fuel supply, their structure differs significantly from that of ordinary stars. Although they are in hydrostatic equilibrium, with pressure balancing gravity, their pressure arises from a peculiar interaction among the electrons that limits the number that can occupy a given volume. This gives white dwarfs an odd property: increasing their mass makes them shrink! Even more crucial, however, white dwarfs must have a mass below a critical limit, or they collapse. To see why, we must consider conditions in a white dwarf's interior.

White dwarfs are very dense, having formed from the shrunken cores of their parent stars. If we divide the mass of a white dwarf by its volume, we find that its density is about 10^6 grams per cubic centimeter—roughly 1 ton per cubic centimeter (about 16 tons per cubic inch). This high density packs the star's atoms so closely that they are separated by less than the normal radius of an electron orbit. If electrons could remain attached to nuclei under these conditions, their orbits would overlap those of neighboring atoms. But electrons can easily escape from their nuclei at such high density and move freely between them, allowing the electrons to interact in a peculiar way.

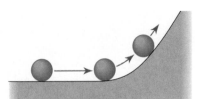

Ball rolling up a hill loses energy and slows down.

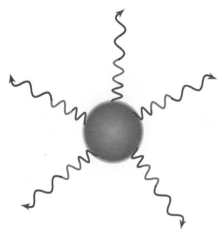

Light moving away from a star loses energy.

FIGURE 14.2

The gravitational redshift. Just as a ball rolling up a hill loses energy, so too does light as it moves away from a star. Light does not slow down, however. Instead, because its energy, E, determines its wavelength, λ, according to the relation $E = \text{Constant}/\lambda$, as the light's energy decreases, its wavelength increases. Longer wavelengths correspond to redder colors, and so we refer to the lengthening of the light's wavelength caused by the star's gravity as the gravitational redshift.

Degeneracy and the Chandrasekhar Limit

A law of subatomic physics called the **exclusion principle** limits the number of electrons that can be put into a given volume. This in turn limits how closely electrons can be squeezed, thereby creating a pressure that, to distinguish it from that of an ordinary gas, is called **degeneracy pressure**. Degeneracy pressure depends only on the gas density, not on its temperature. In an ordinary gas, pressure depends on temperature and density. When such a gas is compressed, both the heat and the higher density contribute to raising its pressure. When a degenerate gas is compressed, it heats up, but the heat does not raise its pressure as it does for an ordinary gas. This makes degenerate gas less "springy" than ordinary gas.

If mass is added to a white dwarf, its radius *shrinks*. The extra mass increases the star's gravity, compressing it. Its pressure rises, but because the star is degenerate, the rise is small and is unable to prevent the star from being squashed into a smaller volume. Squashing the star decreases the distance between particles and makes its gravity larger still, squashing the star even more. In fact, gravity may ultimately overwhelm the star's pressure forces and crush the star completely. Thus too much mass makes a white dwarf collapse, as was demonstrated by theoretical calculations made in 1939 by the Indian astrophysicist Subrahmanyan Chandrasekhar (pronounced /chan-dra-say-car/). Chandrasekhar won the Nobel Prize for physics in 1983 for these calculations and related work. The limiting mass of a white dwarf is now called the **Chandrasekhar limit** in his honor. For stars made of a helium-carbon mix—the presumed composition of many white dwarfs—the limiting mass is about 1.4 M$_\odot$.

Measurement of white dwarf masses made after Chandrasekhar's work was completed show that all white dwarfs are subject to this limit. For those in binary star systems, the masses can be measured using the modified form of Kepler's third law. For isolated white dwarfs, calculations are based on the shift of their spectral lines caused by their immense gravity.

When light escapes from a body, it must work against the force of gravity. The escaping light acts like a ball rolled up a slope: as the ball moves higher, it slows down as it loses energy. Light cannot slow down—it must always travel at the constant speed, c—but it too can lose energy, and it does so. Light's energy, as you have learned, determines its wavelength: lower energy corresponds to longer wavelengths. Longer wavelengths, in turn, correspond to redder colors; thus, as light "climbs" away from a body, the body's gravity stretches the wavelengths and creates a **gravitational redshift**, as shown in figure 14.2. The amount of redshift depends on the star's mass and radius, and from the shift, the white dwarf's mass can be estimated if its radius is known* (from its luminosity and temperature, for example). The greater the redshift, the greater is the mass that causes it.

Spectral lines also permit astronomers to measure the magnetic fields of white dwarfs. We discussed in chapter 11 how the Sun's magnetic field can be measured from the splitting of its spectral lines caused by the Zeeman effect. Similar observations show that some white dwarfs have magnetic fields 10,000 times stronger than sunspots—10^8 times that of the Earth. These immense field strengths occur because as the white dwarf shrinks, it draws its magnetic field lines closer together, amplifying the field.

White Dwarfs in Binary Systems: Novas and Supernovas of Type I

Isolated white dwarfs cool off and disappear from sight, but in binary systems they may heat up and explode. If a white dwarf has a nearby companion, gas expelled from the companion may fall onto the dwarf, as illustrated in figure 14.3. Coming from the companion's outer layers, such gas is generally rich in hydrogen, and it may briefly replenish

---

*To use this method successfully, astronomers need to know the original wavelengths of the spectrum lines. These can be found from the Doppler-shift law if the white dwarf's velocity is known from the motion of a companion star and if the kind of atoms creating the lines can be identified.

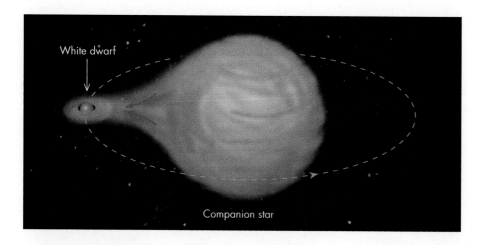

FIGURE 14.3
A white dwarf accreting mass from a binary star companion.

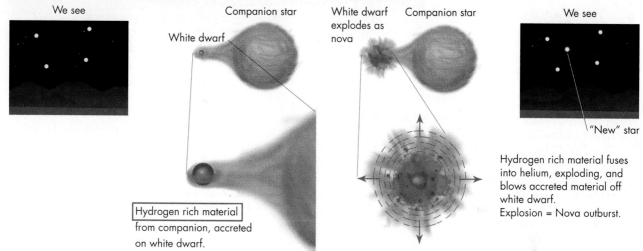

FIGURE 14.4
A white dwarf exploding as a nova. The hydrogen that falls onto the surface of a white dwarf from its companion may suddenly fuse into helium, creating an explosion that makes the star brighten.

the white dwarf's fuel supply. The new fuel forms a layer on the dwarf's surface, where gravity compresses and heats it. The gas layer eventually reaches the ignition temperature for hydrogen, but as we saw in the previous chapter, nuclear burning in a degenerate gas can be explosive. The detonating hydrogen is blasted into space and forms an expanding shell of hot gas, as figure 14.4 shows, that radiates far more energy than the white dwarf itself. Sometimes these stellar explosions are visible to the naked eye. When early astronomers saw such events, they called them **novas,** from the Latin word for "new," because the explosion would make a bright point of light appear in the sky where no star was previously visible.

A nova may erupt repeatedly if it does not accumulate too much mass. If enough gas is added to push it over the Chandrasekhar limit, however, the white dwarf will collapse and may explode. Carbon and oxygen in the collapsing star will be compressed and heated and, although initially too cold to burn, will eventually reach their ignition temperature and begin nuclear fusion. Carbon and oxygen fuse to form silicon (^{12}C + ^{16}O becomes ^{28}Si), which in turn fuses into nickel (2 ^{28}Si becomes ^{56}Ni). The energy released in this burning may be enough to blow the entire star apart, resulting in a **type I supernova***

*Astronomers have recently identified a different form of type I supernova whose origin is uncertain.

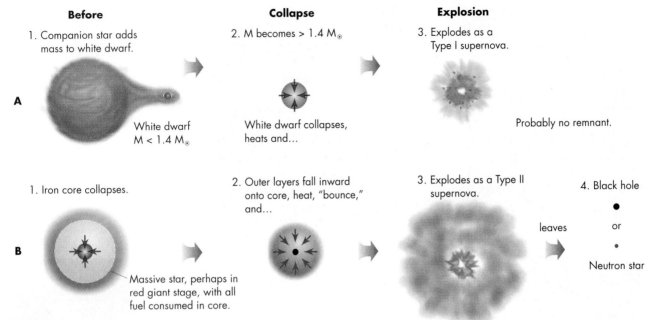

Before

1. Companion star adds mass to white dwarf.

A

White dwarf
$M < 1.4 M_\odot$

1. Iron core collapses.

B

Massive star, perhaps in red giant stage, with all fuel consumed in core.

Collapse

2. M becomes > $1.4 M_\odot$

White dwarf collapses, heats and...

2. Outer layers fall inward onto core, heat, "bounce," and...

Explosion

3. Explodes as a Type I supernova.

Probably no remnant.

3. Explodes as a Type II supernova.

leaves

4. Black hole

●

or

·

Neutron star

FIGURE 14.5
(A) A white dwarf exploding as a type 1 supernova. If too much hydrogen from a companion accumulates on the white dwarf, it may raise its mass to a value above the Chandrasekhar limit (about 1.4 $M_\odot$). A white dwarf more massive than this limit collapses, heats from the resulting compression, and explodes. (B) Formation of a neutron star or black hole by the collapse of the iron core of a massive star.

(fig. 14.5A). The nickel, sprayed into space by the explosion, is highly radioactive and rapidly decays into cobalt (^{56}Co), which finally decays to iron (^{56}Fe), each decay adding additional energy to the outburst.

A type I supernova leaves no remnant star, unlike a type II supernova, which results from the collapse of a massive star's iron core and which leaves a neutron star or a black hole. Type I supernovas leave only a cloud of rapidly expanding gas rich in the leftover carbon and oxygen, plus the silicon, iron, and other elements made during the nuclear burning. If the iron in your blood was not made by the collapse of a massive star's core, it might have been made when a white dwarf met its doom.

14.2 NEUTRON STARS

General Properties and Origin

In 1934, Walter Baade (pronounced /Bah-duh/) and Fritz Zwicky, astrophysicists at Mt. Wilson Observatory and the neighboring California Institute of Technology, respectively, suggested that when a massive star reaches the end of its life, its gravity will crush its core and make the star collapse. The collapse of the core will trigger a titanic explosion, which they called a "supernova"* (see fig. 14.5B). Almost as an afterthought, they speculated that the collapsed core might be so dense that the star's protons and electrons would be driven together and merge into neutrons, forming a **neutron star**.

Astronomers readily accepted Baade and Zwicky's idea that a massive star dies as a supernova—such violent explosions had been observed—but they paid little heed to their suggestion that a neutron star might be born in the blast. Such stars would be incredibly tiny compared even with white dwarfs and were thought to be unobservable. Nevertheless, several physicists calculated the properties of neutron stars, and found that their radii should be about 10 kilometers (about 6 miles) and their masses should be between one and several times that of the Sun. The calculations also showed that, like white dwarfs, neutron stars have a maximum mass, which for them is about 2 to 3 solar masses.

*Referred to as type II supernova.

Pulsars and the Discovery of Neutron Stars

Despite this theoretical model of neutron stars, there was no evidence that they existed, and so Baade and Zwicky's concept of tiny, collapsed stars lay dormant for over 30 years. In 1967, a group of English astronomers led by Anthony Hewish was observing fluctuating radio signals from distant, peculiar galaxies. Jocelyn Bell, a graduate student working with the group, noticed an odd radio signal with a rapid and astonishingly precise pulse rate of one burst every 1.33 seconds. The precision of the pulses led some astronomers to wonder whimsically if they had perhaps stumbled onto signals from another civilization, and informally the signal was known as LGM-1, for little green men # 1. Over the next few months Hewish's group found several more pulsating radio sources, which came to be called **pulsars** for their rapid and precisely spaced bursts of radiation (fig. 14.6).

The discovery of other pulsars convinced astronomers that they were observing a natural phenomenon, and so their explanations focused on peculiar varieties of pulsating stars. Ordinary pulsating stars—for example, Cepheid variables—obey a period-density relation such that each star's pulsation period depends inversely on the square root of its density. That law applies in a generalized form to all pulsating stars. Thus the extreme shortness of pulsar periods indicates that they must be extremely dense objects, and the first pulsars discovered had periods that were consistent with their being white dwarfs. But soon other pulsars were discovered whose periods were much shorter, implying densities far larger than those of white dwarfs, and so astronomers turned back to Baade and Zwicky's ideas about neutron stars.

Neutron stars, according to their theory, would not only be extremely dense (and thus have very short periods), but would also be associated with supernova explosions. Thus, when a very short-period pulsar was found in the Crab nebula supernova remnant, it seemed to confirm Baade and Zwicky's model. But calculations of the pulsation period of neutron stars indicated that they should pulsate much faster than any of the pulsars so far observed. Thus white dwarfs pulsated too slowly and neutron stars pulsated too rapidly to fit the observations. The solution to this puzzle came from the independent work of Franco Pacini and Thomas Gold, astrophysicists at Cornell University.

Even before the discovery of pulsars, Pacini had been studying the link between neutron stars and supernova remnants that had been proposed by Baade and Zwicky, some 40 years earlier. He found that a neutron star remaining in the debris of a supernova explosion could supply large amounts of energy to the gas in its vicinity if it were rapidly spinning and if it had a very strong magnetic field. Rapid rotation and a strong magnetic field were also the keys to Gold's completely unrelated work, which was an attempt to explain not the energizing of supernova debris but the rapidity of pulsar pulses. Gold suggested that a pulsar is a rapidly *spinning*, not pulsing, neutron star. Furthermore, its magnetic field causes it to emit radiation in two narrow beams so that, as the star spins, its beams sweep across space. We see a burst of radiation when the beam points at Earth. Thus a pulsar shines like a cosmic lighthouse whose beams swing around as its lamp

FIGURE 14.6

Pulsar signals recorded from a radio telescope. (Courtesy M. I. Large, University of Sydney, R. N. Manchester, Australia Telescope, CSIRO, and Joseph H. Taylor, Princeton University.)

☀ Approximately how rapidly do these pulsars spin?

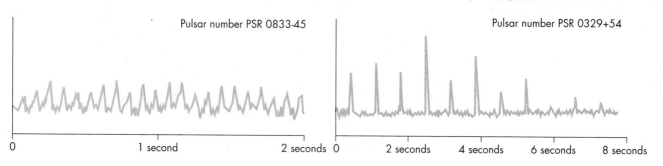

Pulsar number PSR 0833-45

Pulsar number PSR 0329+54

| 0 | 1 second | 2 seconds | 0 | 2 seconds | 4 seconds | 6 seconds | 8 seconds |

rotates, as shown in figure 14.7. But the model worried some astronomers because at least one of the pulsars then known—that in the Crab nebula supernova remnant—rotates 30 times per second. What could make a star spin that fast? For comparison, consider that the Sun spins once a month.

If pulsars are neutron stars, their rapid spin has a natural explanation. When a massive star collapses, forming a neutron star and exploding as a supernova, its core shrinks to a tiny radius—perhaps only 10 kilometers or so—and any rotation it has will be vastly increased. This acceleration results from the "ice-skater effect," which we discussed in describing the formation of the solar system. The ice skater spins faster by drawing his or her arms inward, and thereby turning a slow pirouette into a blur of rotation.

Any rotating body spins faster when it shrinks—a principle called the **conservation of angular momentum**, illustrated in figure 14.8. Angular momentum measures rotational inertia, which is usually denoted by L. For a spherical mass, L is approximately given by the body's mass, M, times its equatorial rotation velocity, V, times its radius, R; or

$$L = MVR$$

The principle of the conservation of angular momentum states that this product must be constant unless some force acts to brake or accelerate the spin. Therefore for a given mass, M, *if we make R smaller, V must increase* to keep the angular momentum constant.

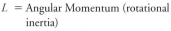

L = Angular Momentum (rotational inertia)
M = Mass of an object
V = Rotation velocity at the object's equator
R = Radius of the object

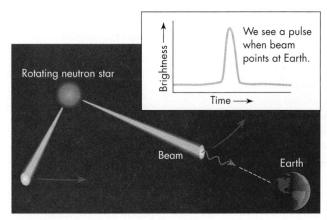

FIGURE 14.7
A pulsar's pulses are like the flashes of a lighthouse as its lamp rotates.

FIGURE 14.8
Conservation of angular momentum spins up a skater and a spinning star. Angular momentum is proportional to a body's mass, M, times its rotation speed, V, times its radius, R. Conservation of angular momentum requires that mathematically MVR = Constant. Thus, if a given mass changes its R, its V must also change. In particular, if R decreases, V must increase, as the ice skater demonstrates by spinning faster as she draws her arms in. Similarly, a rotating star spins faster if its radius shrinks.

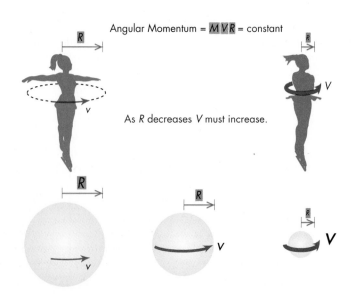

Angular Momentum = MVR = constant

As R decreases V must increase.

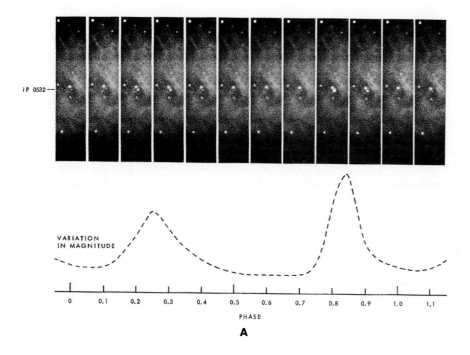

VARIATION IN MAGNITUDE

I P 0532 —

PHASE

A

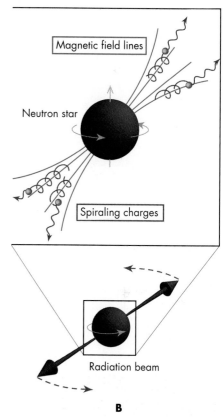

B

FIGURE 14.9

A spinning pulsar generates a powerful electric field, which strips charges from its surface. (A) The rapid rotation of the pulsar in the Crab nebula makes the star appear to turn on and off 30 times per second as its beam of radiation sweeps across the Earth. (B) The charges spiral in the star's magnetic field and emit radiation along their direction of motion. Because the charges stream in a narrow beam confined by the magnetic field, their radiation is also in a narrow beam at each magnetic pole of the star.

Thus shrinking a spinning body (decreasing its radius) makes it spin faster, and a simple calculation using this principle shows that if even a slowly spinning star like the Sun were to shrink to the size of a neutron star, it would rotate several thousand times per second! Such a shrinkage is exactly what happens when the iron core of a massive star collapses. Thus the rapid spin of neutron stars becomes entirely explainable.

Collapse to such a small radius has another effect: it squeezes the star's magnetic field into a denser state, amplifying the field's strength to about 1 trillion times that of the Earth's. Pulsars are thus extremely powerful magnets, and the combination of such a strong magnetic field and the rapid rotation generates the pulsar's radiation beams.

Emission from Neutron Stars

A varying magnetic field creates an electric field—a principle we use here on Earth to generate electricity by spinning magnets in dynamos. Similarly, the spin of a neutron star and its magnetic field generates powerful electric fields. These fields rip positively and negatively charged particles off the star's surface and accelerate them to nearly the speed of light. No metal wires confine the charges; rather, they are channeled by the pulsar's intense magnetic field. We saw in chapter 4 how the Earth's magnetic field directs solar particles toward the poles to create the aurora. The magnetic field of a spinning pulsar, in a like manner, generates two narrow beams of charged particles at the star's magnetic poles.

To explain how the pulsar's particle beams generate radiation, we invoke the physical law that accelerating charges radiate electromagnetic energy. A radio transmitter works according to this principle: electric currents are pulsed through the broadcast antenna, accelerating electrons in it; the accelerated electrons in turn produce the radio waves we detect. In a pulsar, too, the charges radiate as they accelerate, in this case along the magnetic field, spiraling as they go (fig. 14.9B). The pulses of radiation we see are the collective emission from myriads of charges pouring off the neutron star's surface. The spiraling charges radiate electromagnetic energy over a range of wavelengths that depends on their energy. Higher-energy charges generate more energetic—and therefore shorter-wavelength—photons. The radiation is beamed because the charges are traveling along the field lines which, near the star's surface, form a tight cone. Because of this beaming, we see only those neutron stars whose sweeping beams point to Earth; those radiating in other directions are invisible to us. Thus many more neutron stars must exist than the few hundred we have detected so far.

The emission created by the accelerating charges is called **nonthermal radiation** because its properties depend on the charges, acceleration and on the strength of the magnetic field rather than on the temperature of a heated gas. It is also sometimes called **synchrotron radiation**, after a type of atomic particle accelerator in which such radiation was first observed on Earth. For most pulsars, the charges generate mostly low-energy radiation at radio wavelengths. However, in some very young pulsars, the charges are so energetic that they produce radiation across the entire electromagnetic spectrum, including visible light and gamma rays. Figure 14.9A shows the very young pulsar in the Crab nebula, which emits visible light flashes 30 times per second in addition to its radio waves. A high-speed television camera has been used to capture the flashes caused by its rapid spin. Pulsars presumably emit some thermal radiation—ordinary light—from their surfaces, but because they are so tiny, that radiation is very dim and has not yet been detected for sure.

In everyday life spinning objects slow down. Pulsars are no exception. As a pulsar spins, it drags its magnetic field through the particles that boil off its surface into the surrounding space. The magnetic field exerts a force on those particles and speeds them up. But as the particles speed up, they exert a reaction force back on the magnetic field that slows it and the pulsar down. This process is a little like what happens if you lower the beaters of a rapidly spinning electric mixer into a bowl of thick dough: the dough begins to spin, but the beater blades almost stop spinning. Astronomers measure this "spindown" of a pulsar by precisely timing the interval between pulses. Such measurements indicate that the time interval—the period of the spinning pulsar—is lengthening. The slower rotation also limits the energy of the radiation the pulsar emits. Young, rapidly spinning ones emit visible light and radio waves; old, slowly spinning ones generate only radio waves.

Although such observations show that, in general, pulsars are slowing down, some occasionally speed up suddenly. Such jumps in their rotation speed are called **glitches,** and they can tell us about the interior of the pulsar or neutron star. Astronomers study the structure of neutron stars with the same methods used to study ordinary stars; that is, they construct mathematical models of the neutron star. Such models are based on physical laws—for example, the requirement that, in a stable star, gravitational forces must balance pressure forces. From these models, astronomers can deduce the structure of the neutron star's interior—its density and pressure. Such calculations show that neutron stars contain at least three distinct regions: a thin, outer gaseous atmosphere about 1 millimeter thick; a thin, solid (perhaps iron) crust a few hundred meters thick, below the atmosphere; and the bulk of the star, a liquid "sea" of neutrons lying below the crust (fig. 14.10). This liquid has many remarkable properties. For example, it can move with virtually no friction, a condition called **superfluidity**.

A superfluid's lack of friction creates a very odd condition in a neutron star: the core of the star may spin almost independently of its crust. As a result, even though magnetic effects slow the crust's rate of rotation, the interior's spin is not altered. Thus the interior of the star may spin rapidly beneath a more slowly spinning crust. However, as the neutron star's crust spins ever more slowly, the difference between the rotation rates of the crust and interior grows. When the difference in rotation rates reaches a critical value that depends on the properties of the superfluid and the crust, the superfluid alters its behavior and sheds some of its rotational energy, slowing down slightly. That rotational energy has no place to go but into the crust, and as a result the crust speeds up as the core slows down. We observe the crust's acceleration as a glitch, and from such events astronomers can test the correctness of their models of the neutron star's interior.

A neutron star may alter its spin rate in another way, too. When a neutron star is born, its surface is very hot, probably several million K. As the star cools, its crust may shrink and suddenly crack, making the star's radius slightly smaller. We know that if a spinning body shrinks, its rotation speed increases as a result of the principle of the conservation of angular momentum. Thus a sudden shrinkage of a neutron star's crust makes the neutron star suddenly spin faster. Other types of crustal motion are possible on neutron stars. For example, some astronomers have suggested that a pulsar's crust may exhibit a creeping motion somewhat like plate tectonics on Earth. Such crustal motions will also alter a neutron star's rate of rotation.

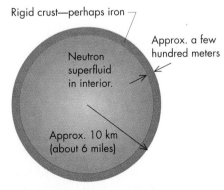

Rigid crust—perhaps iron

Approx. a few hundred meters

Neutron superfluid in interior.

Approx. 10 km (about 6 miles)

FIGURE 14.10
Structure of a neutron star interior.

As the magnetic dipole radiation slows the pulsar, its emission of electromagnetic radiation weakens, and ultimately the star becomes undetectable. Thus a pulsar "dies" by becoming invisible. But just as a white dwarf can be resurrected by a nearby companion star, so too can a pulsar in a binary system be awakened.

Pulsars are not often found in binary star systems, however. The supernova explosion that creates them may expel so much mass that gravity is no longer strong enough to hold the pair of stars together. They rapidly escape from the debris of the explosion that spawned them, helping to explain why so few are seen in supernova remnants. Nevertheless, a few neutron stars do remain bound to companion stars, and they may then become intense sources of x-rays. Moreover, astronomers have calculated that the magnetic dipole radiation generated by a pulsar's spin is so intense at its birth, when it is spinning fastest, that the radiation acts like a rocket engine. The radiation literally pushes the neutron star through space at speeds of hundreds of kilometers per second, making it difficult for its companion to hold onto it. Such speeding pulsars have been detected.

X-Ray Binary Stars

Neutron stars in binary systems sometimes emit intense x-ray radiation. Such stars are called **x-ray binaries**. Astronomers have identified several types of x-ray binaries. In one type, the x-ray emission occurs in intense bursts at irregular time intervals, and so these stars are called x-ray bursters. According to one hypothesis, the x-rays come from gas that falls onto the neutron star's surface, where the star's intense gravity compresses and heats it to millions of degrees. This heating triggers a thermonuclear explosion like those that cause nova outbursts. The exploding gas is so hot that most of its radiation is at x-ray wavelengths, in accordance with Wien's law (the hotter a gas, the shorter the wavelength at which it radiates). Such explosions on the surface of a neutron star emit thousands of times more energy than the Sun, an impressive achievement for a "dead" star.

In another type of x-ray binary, the emission comes in rapid, regular pulses every few seconds or less. The x-ray pulses from these stars are reminiscent of the radio pulses from ordinary pulsars, and so astronomers refer to such objects as "x-ray pulsars." An x-ray pulsar generates its emission from small "hotspots" near its magnetic poles. Gas from a companion star flows in along the field lines, much as gas particles from the Sun stream in toward Earth's magnetic poles, creating the aurora. As the gas falls toward the neutron star, the star's intense gravity accelerates the gas. As it crashes onto the neutron star at the magnetic poles (fig. 14.11), the gas is compressed and heated intensely, causing it to emit x-rays. As the neutron star spins, the hot regions near each pole move into and then out of our field of view, creating the x-ray pulses that we observe.

Because pulsars generally slow down as they age, astronomers were initially puzzled when they discovered some extremely rapidly spinning—and therefore presumably very young—pulsars in several very old star clusters. Some of these seemingly young pulsars rotate approximately 1000 times per second, so that the pulses we receive from them are separated by little more than a thousandth of a second. Astronomers accordingly call them "millisecond pulsars." Millisecond pulsars are distinguished not only by their rapid rotation but also by having companion stars in most cases. In fact, astronomers think they probably all had companions at one time, for the reason outlined below.

If a neutron star has a companion that ejects mass, the neutron star may capture some of the mass by its gravitational attraction. However, the captured mass generally does not fall directly onto the neutron star. Instead, it temporarily goes into orbit around the neutron star, accumulating in a flat, spinning disk that astronomers call an **accretion disk**. The matter in the accretion disk eventually falls onto the neutron star and, when it does, the material carries some of its spinning motion—technically, angular momentum—with it. That spinning motion is imparted to the neutron star when the gas strikes its surface, making the neutron star spin faster. Thus the usual slowdown of an aging pulsar may be reversed if it has a companion star. But what of those millisecond pulsars without companions? What made their rotations speed up? One possibility is that such pulsars originally had companions that have since disappeared. In some cases, the companion may have been lost if the pulsar passed too closely to a neighboring star. Gravitational forces

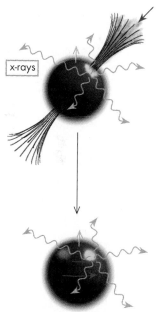

Magnetic field channels gas to small area near poles. Gas heated by impact, making hot spot.

x-rays

As star rotates, hotspot appears and disappears.

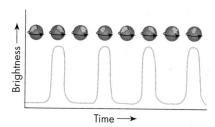

Brightness →

Time →

FIGURE 14.11

Gas falling onto a neutron star follows the magnetic field lines and makes a hotspot on the star's surface, creating x-rays. As the star rotates, we observe x-ray pulses.

from that neighbor may have pulled the pulsar's companion away. A more sinister possibility is that the pulsar destroyed its companion, a process that astronomers believe they may be witnessing in the case of the "black-widow pulsar." This peculiar pulsar has a low-mass companion star that is evaporating under the intense heating created by the pulsar. Given enough time, the companion star may totally evaporate, and future astronomers may see the millisecond pulsar without its original companion.

Very recently, astronomers have discovered that some pulsars may have planetary companions. These odd pulsars exhibit tiny Doppler shifts, similar to those expected from the slight gravitational tug of very low-mass orbiting bodies, ones comparable in mass to planets. The interpretation of these Doppler shifts may be in error, but if such planets prove to be real, they raise many puzzling questions about how planets formed after the supernova explosion that created the pulsar or how they survived the supernova's blast.

14.3 BLACK HOLES

When a star that is initially more massive than about 10 $M_\odot$ reaches the end of its life and collapses, it may create a compact star whose properties differ dramatically from those of white dwarfs or neutron stars. Its greater mass can compress its core so much that pressure is unable to support it, and it totally collapses to form what astronomers call a **black hole** (see fig. 14.5).

Black holes have very strange properties, but to understand them we need to review the concept of **escape velocity,** which we discussed in chapter 2. Escape velocity is the speed a mass needs to avoid being drawn back by another object's gravity. For a body of mass M and radius R, the escape velocity, V, for an object moving away from that body is

V = Velocity necessary to escape an object
M = Object's mass
R = Object's radius

$$V = \sqrt{\frac{2GM}{R}}$$

where G is the gravitational constant and has a value of 6.67×10^{-11} if V is in meters per second, R is in meters, and M is in kilograms. For the Earth, the escape velocity is about 11 kilometers per second; for the Sun, it is about 600 kilometers per second. You can see from the formula that the escape velocity for a body of a given mass will be larger if the body has a small radius. The same mass in a smaller radius creates a larger force of gravity, making that body more difficult to escape. Thus a white dwarf whose mass is the same as the Sun's has an escape velocity much larger than the Sun's because its radius is so small. In fact, because its radius is about 100 times less than the Sun's and because the escape velocity depends on the square root of the radius, its escape velocity is larger by a factor of $\sqrt{100}$ (=10) and equals about 6000 kilometers per second. An object leaving the surface of a white dwarf at a velocity of 6000 kilometers per second would therefore overcome the star's gravity and escape into space. The escape velocity for a neutron star, whose radius is about 10^5 times smaller than the Sun's but whose mass is comparable, is about 180,000 kilometers per second, or approximately half the speed of light. Thus, if a neutron star could be compressed to about one-fourth of its radius, its escape velocity would exceed the speed of light. Such an object would become a black hole.

As long ago as 1783, the English cleric John Michell discussed the possibility that objects with escape velocities exceeding the speed of light might exist. A little more than a decade later, the French mathematician and physicist Pierre Simon Laplace entertained the same idea. Following their logic, we can calculate the radius at which a body of a given mass would become a black hole by equating its escape velocity to the speed of light. That is, $V = c = (2GM/R)^{1/2}$. Notice that only two quantities of the body enter: its mass and its radius. Thus we can solve for the radius at which a body of mass M becomes a black hole and find that

R = Radius at which an object of a
    particular mass becomes a black hole
M = Object's mass
c = Speed of light

$$R = \frac{2GM}{c^2}$$

For a body of one solar mass, this turns out to be about 3 kilometers, or 1.9 miles. In other words, if the Sun could be compressed that small, it would become a black hole.

Not until 1916, with the advent of Einstein's general theory of relativity, however, did astronomers begin to think seriously about black holes. Relativity theory showed that gravity is related to the **curvature of space** and that a black hole is a place where the curvature becomes so extreme that a hole forms. An analogy may help illustrate how gravity and curvature of space are related. Imagine a waterbed on which you have placed a baseball (fig. 14.12). The baseball makes a small depression in the otherwise flat surface of the bed. If a marble is now placed near the baseball, it will roll along the curved surface into the depression. The bending of its environment made by the baseball therefore creates an "attraction" between the baseball and the marble. Now suppose we replace the baseball with a bowling ball. It will make a bigger depression and the marble will roll in further and be moving faster as it hits the bottom. We therefore infer from the analogy that the strength of the attraction between the bodies depends on the amount by which the surface is curved. Gravity also behaves this way, according to the general theory of relativity. According to that theory, mass creates a curvature of space, and gravitational motion occurs as bodies move along the curvature.

We now extend our analogy to black holes by supposing that we remove the bowling ball and put on the waterbed a steel safe. This will immediately rip the plastic and fall through. A marble placed on the torn surface will follow the safe into the hole and disappear. So too a black hole is a "rip" in space where the curvature has become so strong that the structure of space is disrupted.

Needless to say, this simple analogy is only approximate. The mathematically correct treatment of a black hole's structure is very difficult. The German astrophysicist Karl Schwarzschild pioneered such calculations, and the size of a black hole is now called the **Schwarzschild radius**, R_S, in his honor. It turns out, however, that relativity theory gives the same answer for the radius of a black hole that we deduced earlier: that is, the Schwarzschild radius, $R_S = 2GM/c^2 =$ approximately $3\mathcal{M}$ kilometers, where $\mathcal{M}$ is the mass of the body in solar units.

All masses, including the Sun, bend space to some degree, and this curvature can be observed. Far from the Sun, space is "flat," unaffected by the Sun's gravity. As we approach the Sun, however, the curvature increases. If we look at light or radio waves whose paths traverse that curved region, as shown in figure 14.13, the waves' bending can be detected and is found to match precisely the prediction of relativity theory. Thus astronomers *know* that space is curved.

The extreme curvature of a black hole prevents its light from escaping, and so its boundary is sometimes called the **event horizon**. Just as the Earth's horizon blocks our view of what lies beyond it, so too the Schwarzschild radius prevents our seeing into a black hole. All that happens within it is hidden forever from our view. No radiation of any sort, nor any material body—rocket, spacecraft, and so on—can break free of its gravity.

Baseball:
Marble rolls into depression

Bowling ball:
Marble rolls in faster

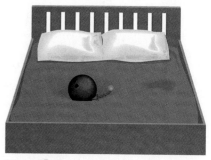

Safe:
Marble disappears into hole

FIGURE 14.12
Objects on a waterbed make depressions analogous to the curvature of space created by a mass. According to the general theory of relativity, that curvature produces the effect of gravity. We can see the similarity by placing a marble at the edge of the depression and watching it roll inward, as if attracted to the body that makes the depression. Bigger bodies make bigger depressions, and so a marble rolls in faster. However, a very big body may tear the waterbed, creating an analog of a black hole.

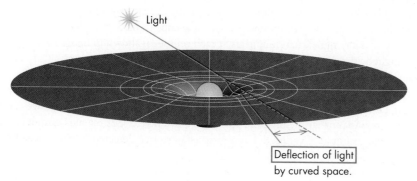

FIGURE 14.13
Curved space around the Sun bends a ray of light passing near the Sun.

Because we cannot observe the interior of a black hole even in principle, there is only a limited number of physical properties we can ascribe to it. For example, it is meaningless to ask what a black hole is made of. When it warps space inside its event horizon, we can no longer observe its interior—it could be made of neutrons, corn flakes, or gerbils. Only the amount of mass is important, not what composes it.

We can measure the mass of a black hole because the mass generates a gravitational field. In fact, the gravitational field generated by a black hole is no different from that generated by any other body of the same mass. For example, if the Sun were suddenly to become a black hole with the same mass it has now, the Earth would continue to orbit it just as it does now: it would not be pulled in. Black holes do not have any special property that gives them abnormally strong gravity except that they are so small that another object can get extremely close to them, which, according to Newton's law of gravity, implies a strong force.

Two other properties of black holes can also make themselves felt beyond the Schwarzschild radius: electric charge and angular momentum. A black hole may become charged if an excess of one kind of charge, say, positive, falls into it.

Angular momentum is much more likely to be a significant property of a black hole. Material nearly always has some angular momentum and, if it becomes a black hole, its angular momentum is retained. As a result, a spinning black hole will not be a sphere. The distance from its center to the event horizon is smaller along the rotation axis than it is perpendicular to it.

The Formation of Black Holes

Not all astronomers are yet completely convinced that black holes exist, but if they do, they are probably made in the supernova explosions that mark the deaths of massive stars. We are virtually certain that such explosions create neutron stars and, as we showed above, it is only a small step in compression from a neutron star to a black hole. It seems a safe extrapolation to suppose that a black hole would form if a massive star had a core whose mass exceeded 5 to 10 $M_\odot$. It remains, however, to be demonstrated by *observation* that black holes exist. Furthermore, how in fact could we observe objects that emit no radiation?

Observing Black Holes

An object that emits no light or other electromagnetic radiation is not easy to observe! But just as you may "see" the wind by its effect on leaves and dust, so too astronomers may be able to see black holes by their effects on their surroundings. Suppose a massive star in a binary system explodes, and its core mass is 10 $M_\odot$, rather than 2 $M_\odot$, so that the core collapses to become a black hole rather than a neutron star. Gas from the companion star may be drawn toward the hole by its gravity, as we know happens for neutron stars that are x-ray bursters. The infalling matter swirls around the black hole and forms an accretion disk whose inner edge lies just outside the hole's Schwarzschild radius, as depicted in figure 14.14. Here, where the disk orbits at nearly the speed of light, turbulence and friction heat the swirling gas to a furious 10 million K, making it emit x-rays and gamma rays.

As the black hole orbits its neighbor, the x-ray–emitting gas may disappear from our view as it is eclipsed by the companion star. An x-ray telescope trained on such a star system will show a steady x-ray signal that disappears at each eclipse, as shown in figure 14.15. Such a signal might be the sign of a black hole, and at least three cosmic x-ray sources bear that signature. But how do we know an x-ray source is not just a neutron star?

Because these x-ray sources are binary stars, we can measure their masses. If x-ray–emitting gas surrounds a body we cannot see but whose mass exceeds 5 $M_\odot$, we can be reasonably confident that the invisible object is a black hole. Neutron stars cannot be that massive.* For example, Cygnus X-1, the first x-ray source detected in the constellation

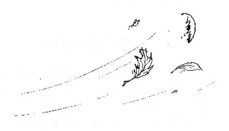

*In choosing 5 $M_\odot$ as the limit, we have allowed some margin above the theoretical maximum mass of a neutron star of about 3 $M_\odot$.

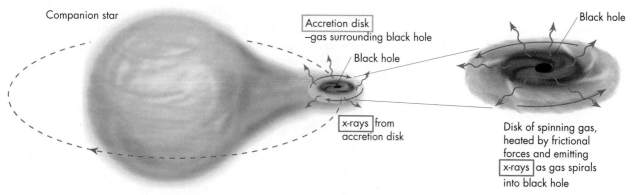

FIGURE 14.14
Black holes may reveal themselves by the x-rays emitted by gas orbiting them in an accretion disk.

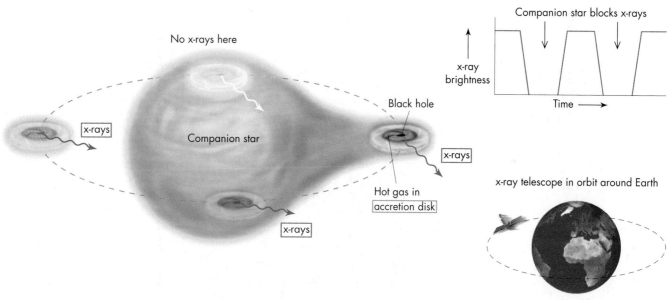

FIGURE 14.15
X-ray emission from gas around a black hole. The signal drops as the hole is eclipsed by its companion star.

Cygnus, the swan, consists of a *B* supergiant star and an invisible companion whose mass—based on an application of the modified form of Kepler's third law, is at least 6 M$_\odot$. Nothing we know of but a black hole can be so massive and yet invisible. An even better candidate is AO620-00, an x-ray—emitting object in the constellation Monoceros, the Unicorn. For A0620-00, the invisible star's mass is probably 16 M$_\odot$.

Recently astronomers have seen in some binary star systems indirect evidence of the event horizon itself. The x-rays from those systems thought to contain a black hole are dimmer than the x-rays from systems thought to contain a neutron star, presumably because matter disappears into the black hole before it has a chance to radiate as much.

Gravitational Waves from Double Compact Stars

In our discussion of black holes, we saw that space near them is curved. If a black hole or other compact star orbits a companion, its motion generates **gravitational waves**. Just as ripples spread away from a stone tossed into a pond, so too gravitational waves spread across space, bending it up and down (fig. 14.16). Scientists are currently building a large gravitational wave detector called LIGO, but astronomical "detectors" offer a shortcut. Early this century, Einstein predicted that rapidly orbiting stars should generate gravitational waves, but his proposal could not then be verified. Even now, astronomers cannot detect gravitational waves directly, but they can observe their effect on the bodies that emit the waves.

Gravitational waves carry energy away from the orbiting stars, making them gradually move together. In 1974, Joseph Taylor and Russell Hulse, then at the University of Massachusetts, discovered a binary pulsar: two neutron stars in orbit around each other. From observations of the pulses, Taylor and Hulse deduced the orbital speed and separation of the neutron stars, and to their delight, the speed and separation changed exactly in the manner predicted by Einstein's general theory of relativity. For this discovery, which has persuaded most astronomers that Einstein was right and that gravitational waves do exist, Taylor and Hulse were awarded the 1993 Nobel Prize in Physics. But collapsed stars may emit even more mysterious radiation than gravitational waves.

Hawking Radiation

In l974, the English mathematical physicist and cosmologist Stephen Hawking predicted that black holes can radiate! Hawking's discovery had been anticipated by Jacob Bekenstein, a physicist who was studying the relation between gravity and thermodynamics, but Hawking was able to show that—astonishingly—a black hole emits blackbody radiation, such as we discussed in chapter 3. In fact, we can use Wien's law, which relates a body's temperature to the wavelength at which it radiates most strongly, to estimate the temperature of a black hole. Hawking calculated that the wavelength of maximum emission is approximately 16 times the black hole's Schwarzschild radius (R_S), a result also found by Bekenstein. Wien's law states that an object's temperature is inversely proportional to the wavelength of the radiation it emits, and so the temperature of a black hole, is equal to a constant divided by R_S. For a solar-mass black hole, this turns out to be about 6×10^{-8} K. This is very cold, but it is not absolute zero. Accordingly, a black hole, like any other body whose temperature is not absolute zero, emits energy in the form of electromagnetic waves, energy now known as **Hawking radiation.** However, for black holes of stellar mass, the amount of Hawking radiation emitted is far too small to be detected at this time. Nevertheless, the existence of such radiation means that black holes are not truly black.

Hawking radiation is created by quantum physical processes that allow energy to escape from black holes despite their intense gravity. Because the only source of energy available to a black hole is its mass, as a black hole "shines" (albeit very dimly), its mass must decrease. In other words, black holes must eventually "evaporate." However, the time it takes for a solar-mass black hole to disappear by "shining itself away" is *very* long: approximately 10^{67} years! This is immensely long—vastly larger than the age of the Universe—but the implications are important: even black holes evolve and "die."

FIGURE 14.16
Gravitational waves generated by a rapidly orbiting collapsed star are ripples in space.

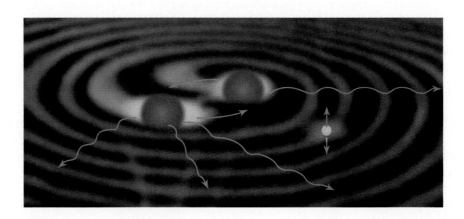

SUMMARY

White dwarfs, neutron stars, and black holes are the remnants of dying stars. A white dwarf forms when a low-mass star expels its outer layers to form a planetary nebula shell and leaves its hot core exposed. The radius of a white dwarf is about the same as that of the Earth. Its matter is degenerate, and its mass must be less than the Chandrasekhar limit, or it collapses. Having no fuel supply, a white dwarf cools and grows dim. If a white dwarf is in a binary system, it may accrete mass from its neighbor and explode either as a nova or as a type I supernova.

A neutron star forms when a massive star's iron core collapses and triggers a supernova explosion. The collapse compresses the core's protons and electrons together to make neutrons, forming a ball of neutrons with a radius of a mere 10 kilometers but which may contain up to about 2 to 3 solar masses. Conservation of angular momentum during its collapse accelerates rotation to about 1000 times per second. This spin and the star's magnetic field generate beams of radiation that sweep across space, making the spinning neutron star a pulsar.

If the massive star's iron core contains more than about 5 $M_\odot$, a black hole, rather than a neutron star, is born. Gravity overwhelms the pressure forces of the core, crushing it and bending space around it so that light cannot escape. The black hole lies invisibly inside its Schwarzschild radius. If it has a companion star, however, we may be able to see x-rays from an accretion disk as gas from the neighbor falls toward the hole.

QUESTIONS FOR REVIEW

1. What are the approximate mass and radius of a white dwarf compared with the Sun?
2. How does a white dwarf form?
3. Can a white dwarf have a mass of 10 solar masses? Why?
4. What is meant by degeneracy pressure?
5. What is a neutron star?
6. What are the mass and radius of a typical neutron star compared with the Sun?
7. How does a neutron star form?
8. How do we observe neutron stars?
9. What is a pulsar? Does it pulsate?
10. Are all neutron stars pulsars? Are all pulsars neutron stars?
11. What creates the beams of radiation seen in pulsars?
12. What is nonthermal radiation?
13. What is a black hole?
14. What is the Schwarzschild radius?
15. What is Hawking radiation?

THOUGHT QUESTIONS

1. Some astronomers have suggested that cooled white dwarfs are made of diamond. Why might it be impractical to mine them?
2. Suppose you jumped into a black hole feet first. What would happen to you as your feet approached its Schwarzschild radius? HINT: Think about tides on the Earth created by the Moon.
3. How would you explain to an 8-year-old child that black holes may exist?
4. Is it surprising that a pulsar is not seen in every supernova remnant? Why?

PROBLEMS

1. Calculate the density of a white dwarf star of 1 solar mass that has a radius of 10^4 kilometers.
2. Calculate the escape velocity from a white dwarf and a neutron star. Assume that each is 1 solar mass. Let the white dwarf's radius be 10^4 kilometers and the neutron star's radius be 10 kilometers.
3. Calculate the Schwarzschild radius of the Sun.
4. Calculate your Schwarzschild radius. How does that compare to the size of an atom? How does it compare to the size of a proton?

TEST YOURSELF

1. Which of the following has a radius (linear size) closest to that of a neutron star?
- (a) The Sun.
- (b) The Earth.
- (c) A basketball.
- (d) A small city.
- (e) A gymnasium.

2. What evidence leads astronomers to believe that they have detected black holes?
- (a) They have seen tiny dark spots drift across the face of some distant stars.
- (b) They have detected pulses of ultraviolet radiation coming from within black holes.
- (c) They have seen x-rays, perhaps from gas around a black hole, suddenly disappear as a companion star eclipses the hole.
- (d) They have seen a star suddenly disappear as it was swallowed by a black hole.
- (e) They have looked into a black hole with x-ray radar telescopes.

3. What causes the radio pulses of a pulsar?
- (a) The star vibrates.
- (b) As the star spins, beams of radio radiation from it sweep through space. If one of these beams points toward the Earth, we observe a pulse.
- (c) The star undergoes nuclear explosions that generate radio emission.
- (d) The star's dark orbiting companion periodically eclipses the radio waves emitted by the main star.
- (e) A black hole near the star absorbs energy from it and re-emits it as radio pulses.

4. If mass is added to a white dwarf,
- (a) its radius increases.
- (b) its radius decreases.
- (c) its density increases.
- (d) it may exceed the Chandrasekhar limit and collapse.
- (e) all of the above except (a).

5. The Schwarzschild radius of a body is
- (a) the distance from its center at which nuclear fusion ceases.
- (b) the distance from its surface at which an orbiting companion will be broken apart.
- (c) the maximum radius a white dwarf can have before it collapses.
- (d) the maximum radius a neutron star can have before it collapses.
- (e) the radius of a body at which its escape velocity equals the speed of light.

FURTHER EXPLORATIONS

Bethe, Hans A., and Gerald Brown. "How a Supernova Explodes." *Scientific American* 252 (May 1985): 60.

Bernstein, Jeremy, "The Reluctant Father of Black Holes." *Scientific American* 274 (June 1996): 80. (About Einstein.)

Charles, Philip A., and R. Mark Wagner. "Black Holes in Binary Stars: Weighing the Evidence." *Sky and Telescope* 91 (May 1996): 38.

Chown, Marcus. "Bright Light, Black Hole." *New Scientist* 150 (8 June 1996): 30.

Croswell, Ken. "What Lies at the Milky Way's Center?" *Astronomy* 23 (May 1995): 32.

Drieger, Walter C. "The Care and Feeding of Black Holes." *Astronomy* 23 (May 1995: 19. (A satire.)

Grahm-Smith, Sir Francis. "Pulsars Today." *Sky and Telescope* 80 (September 1990): 240.

Grahm-Smith, Sir Francis. "The Binary and Millisecond Pulsars." *Contemporary Physics* 33 (May/June 1992): 165.

Greenstein, George. *Frozen Star.* New York: Freundlich Books, 1984.

Herbst, William, and George C. Assousa. "Supernovas and Star Formation." *Scientific American* 241 (August 1979): 138.

Jastrow, Robert. *Red Giants and White Dwarfs.* New York: Norton, 1990.

Jeffries, Andrew D., Peter R. Saulson, Robert E. Spero, and Michael E. Zucker. "Gravitational Wave Observatories." *Scientific American* 256 (June 1987): 50.

Kaler, James B. "Planetary Nebulae and Stellar Evolution." *Mercury* 10 (July/August 1991): 114.

Kaler, James B. "Bubbles from Dying Stars." *Sky and Telescope* 63 (February 1982): 129.

Kawaler, Steven D., and Donald E. Winget. "White Dwarfs: Fossil Stars." *Sky and Telescope* 74 (August 1987): 132.

Kirshner, Robert P. "The Earth's Elements." *Scientific American* 271 (October 1994): 58.

Kurczynski, Peter. "The Mother of All Fireworks." *Mercury* 25 (September/October 1996): 16. (About neutron stars.)

Lewin, Walter H. G. "The Sources of Celestial X-ray Bursts." *Scientific American* 244 (May 1981): 72.

Mackeown, P. Kevin, and Trevor C. Weekes. "Cosmic Rays from Cygnus X-3." *Scientific American* 253 (November 1985): 60.

Murdin, Paul, and Lesley Murdin. *Supernovae.* New York: Cambridge University Press, 1985.

Piran, Tsvi. "Binary Neutron Stars." *Scientific American* 272 (May 1995): 52.

Price, Richard H., and Kip S. Thorne. "The Membrane Paradigm for Black Holes." *Scientific American* 258 (April 1988): 69.

Ruthen, Russell. "Catching the Wave." *Scientific American* 266 (March 1992): 90.

Shaham, Jacob. "The Oldest Pulsars." *Scientific American* 256 (February 1987): 50.

Starrfield, Sumner, and Steven N. Shore. "The Birth and Death of Nova V1974 Cygni." *Scientific American* 272 (January 1995): 76.

Thorne, Kip S., *Black Holes and Time Warps: Einstein's Outrageous Legacy.* New York: W. W. Norton, 1994.

Thorne, Kip S. "A Voyage Among the Holes." *Grand Street* 12 (Winter 1994): 148.

Trimble, Virginia. "Neutron Stars and Black Holes in Binary Systems." *Contemporary Physics* 32 (March/April 1991): 103.

Trimble, Virginia. "White Dwarfs: The Once and Future Suns." *Sky and Telescope* 72 (October 1986): 348.

Weisberg, Joel M., Joseph H. Taylor, and Lee A. Fowler. "Gravitational Waves from Orbiting Pulsars." *Scientific American* 245 (October 1981): 74.

Williams, Robert E. "The Shells of Novas." *Scientific American* 244 (April 1981): 120.

Web Site

"Black Holes and Beyond," from "Science for the Millenium," created at the University of Illinois.
   www.ncsa.uiuc.edu/cyberia/NumRel/BlackHoles.html
This site has many different pages on black holes and relativity. Highly recommended, but not easy.

KEY TERMS

accretion disk, 423

black dwarfs, 415

black hole, 424

Chandrasekhar limit, 416

compact stars, 414

conservation of angular momentum, 420

curvature of space, 425

degeneracy pressure, 416

escape velocity, 424

event horizon, 425

exclusion principle, 416

glitches, 422

gravitational redshift, 416

gravitational waves, 427

Hawking radiation, 428

neutron star, 418

nonthermal radiation, 422

novas, 417

pulsars, 419

Schwarzschild radius, 425

superfluidity, 422

synchrotron radiation, 422

type I supernova, 417

white dwarfs, 414

x-ray binaries, 423

The Milky Way and other Galaxies

As we have journeyed outward from the Earth to the Solar System to the other stars, we have repeatedly seen that astronomical objects occur in groupings created by the force of gravity. Stars too are clustered by gravity into yet larger astronomical systems called galaxies (fig. OV6.1).

A galaxy is an immense swarm of stars held together by their mutual gravity. Our Sun belongs to the Milky Way galaxy, a flattened, disk-shaped system about 80,000 light years in diameter (fig. OV6.2). The Milky Way contains about 100 billion stars, and our Sun lies about two-thirds of the way out from the center of the disk.

FIGURE OV6.1
Photograph of a typical spiral galaxy. (Courtesy Anglo-Australian Observatory/Photographs by David Malin.)

Bulge

Sun

Disk

Halo

Spiral arm

25 kiloparsecs
(approx. 80,000 light years)

FIGURE OV6.2
Sketch of our own Galaxy,
the Milky Way.

Scattered among the stars are vast clouds of dust and gas, tens to hundreds of light-years across (fig. OV6.3). The dust particles are microscopic, but their numbers are so immense that they create a haze that blocks our view of most of our galaxy in visible light. Thus, to understand our galaxy's shape and structure, astronomers must observe at wavelengths that are less affected by the dust, such as infrared and radio waves.

Open Star Cluster   Dark nebula

FIGURE OV6.3
Photograph of an interstellar dust and gas cloud. (Courtesy Anglo-Australian Observatory/Photographs by David Malin.)

Because our galaxy is a thin disk, few gas and dust clouds block our view through its top or bottom. This allows us a relatively clear view in those directions, where we see millions of other galaxies (fig. OV6.4).

FIGURE OV6.4
Picture of distant galaxies made with the Hubble Space Telescope. (This picture, packed with galaxies, shows an area of sky that you could cover with a grain of sand at arm's length!) (Courtesy STSCI.)

Many of these are like the Milky Way in size and shape and are called spiral galaxies because they show spiral "arms" winding out from the central region (fig. OV6.5A). Others lack arms and show a smooth distribution of stars. These systems range from nearly spherical to egg-shaped and are called Elliptical galaxies (fig. OV6.5B). Yet others have no symmetric shape and are called irregular galaxies (fig. OV6.5C). These other galaxies also range widely in size: some are thousands of times less massive than the Milky Way, while others are thousands of times more massive. Why galaxies have such diverse shapes and sizes is a major unsolved puzzle.

A

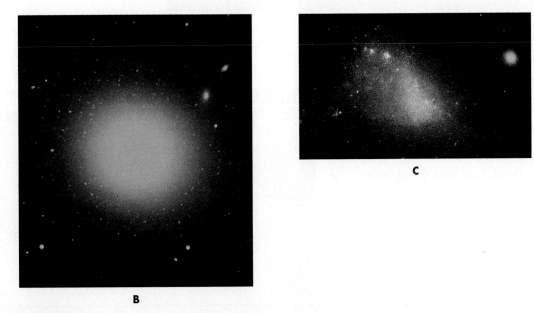

B

C

FIGURE OV6.5
Photographs of (A) a spiral, (B) an elliptical, and (C) an irregular galaxy. (Courtesy Anglo-Australian Observatory/Photographs by David Malin.)

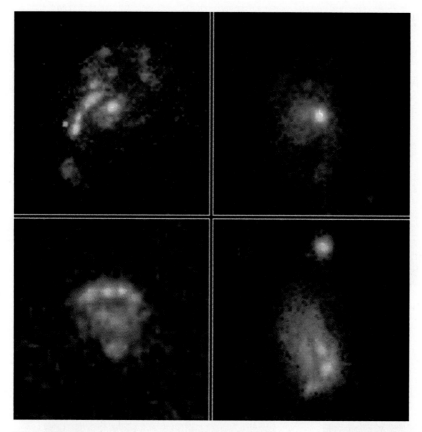

FIGURE OV6.6A
Picture of young blue galaxies taken with the Hubble Space Telescope.
(Courtesy STSCI)

From studies of the Milky Way and other galaxies, astronomers hypothesize that galaxies form from collapsing gas clouds much as the Solar System did, but on a vastly larger scale. As the cloud shrinks, gravity draws the gas within it into smaller clumps. These clumps become stars that shine as hydrogen in their core fuses into helium and heavier elements. The result is a young galaxy filled with stars and leftover gas. Astronomers think they have seen such young galaxies with the Hubble Space Telescope (fig. OV6.6A), but these systems are often small compared to the Milky Way. To form larger galaxies like our own, in the opinion of many astronomers, these young, small galaxies are drawn together by their mutual gravity and gradually merge into bigger galaxies, as we may be witnessing in figure OV6.6B Computer simulations also suggest that merging is important in building galaxies.

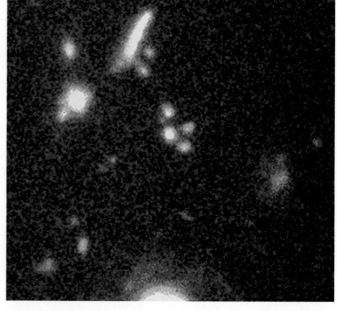

FIGURE OV6.6B
Young galaxies possibly merging to form a larger system. (Courtesy STSCI.)

What determines the final shape of a galaxy is still a mystery. According to one theory, the shape is set by the amount and nature of the motion of the gas from which it formed. Clouds with little internal motion ended up nearly spherical, Others that spun rapidly became flattened (much as our Solar System did) and turned into disklike systems like our own Milky Way.

Most galaxies belong to groups called galaxy clusters (fig. OV6.7). These clusters—a few million light years across—are held together by gravity, just as the galaxies themselves are. Within clusters, galaxies sweep along orbits that carry them back and forth through the cluster.

FIGURE OV6.7
Cluster of galaxies.
(Courtesy NOAO.)

As they move along their paths, galaxies occasionally collide with one another. Such collisions may alter a galaxy's shape (fig. OV6.8). If a small galaxy runs into a bigger one, the bigger galaxy's gravity may tear the smaller one apart. The stars and gas torn free may then fall toward the center of the larger galaxy.

FIGURE OV6.8
Photograph of galaxies colliding.
(Courtesy NOAO and STSCI.)

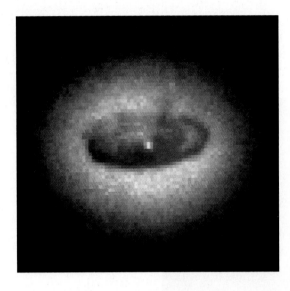

FIGURE OV6.9
Picture of gas surrounding a possible black hole in an active galaxy. (Courtesy STSCI.)

Astronomers have detected hot gas swirling around what they think are enormous black holes in the cores of many galaxies (fig. OV6.9). Such black holes may have a mass nearly a billion times that of our Sun. Material falling into such a huge black hole is heated to millions of degrees and triggers violent activity in the galaxy's core.

Galaxies are so remote (all but a handful are millions of light years or more away from us) that their motion and distribution across the sky traces the structure of the Universe itself. For example, in general, galaxies are moving away from one another (fig. OV6.10). That motion arises from the expansion of space itself, and it "stretches" the wavelength of the light we see from the galaxies, creating a redshift in their light. From the redshift, astronomers deduce the distance to galaxies as well as the age of the Universe.

From its present expansion, astronomers deduce that the Universe was once much smaller, denser, and hotter that it is today. From the rate of expansion, they deduce that the Universe was "born" about 13 billion years ago and that at that time it was packed into a volume about the present size of the Solar System, with a temperature of about 10 trillion K. In a titanic burst of matter and energy called the Big Bang, the Universe began to expand. As it expanded, the Universe cooled, and its subatomic particles assembled into atoms of hydrogen and helium. Further expansion cooled the hot hydrogen and allowed gravity to begin to draw the gas into clouds from which galaxies formed.

FIGURE OV6.10
Sketch illustrating expansion of the Universe.

CHAPTER FIFTEEN
THE MILKY WAY GALAXY

The Summer Milky Way from New Mexico (Chris Cook)

On a clear, moonless night, far from city lights, you can see a pale band of light spangled with stars stretching across the sky (fig. 15.1). The ancient Hindus thought this shimmering river of light in the heavens was the source of the sacred river Ganges. To the ancient Greeks this dim celestial glow looked like milk spilt across the night sky, and so they called it the **Milky Way.** In the seventeenth century, Galileo showed that the Milky Way is millions of stars too dim to be seen as individual points of light. Now, in the twentieth century we know that these stars, along with our Sun, form a huge, slowly revolving disk—our galaxy. The word *galaxy* itself comes from the ancient Greeks and their word for "milk"—*galactos*. Thus "Milky Way" is both the name of the band of light across the night sky and also the name of our galaxy.

A view of the Milky Way on a clear, dark night is one of nature's finest spectacles. Superposed on the dim background glow are most of the bright stars and star clusters that we can see, which all belong to our galaxy. Here and there, however, dark blotches interrupt the glowing backdrop of stars, as you can see in figure 15.2. Ancient Incan astronomers observing the Milky Way from their temple observatories in the Andes gave these dark areas names, just as peoples of the classical world named the star groups. Today we know the dark regions are clouds of dust and gas that give birth to new stars, as we discussed in chapter 13.

FIGURE 15.1
Wide-angle photo of a portion of the Milky Way taken from Mt. Graham in Arizona. The dark ring around the edge of the picture is the horizon. (Courtesy Steward Observatory and NOAO.)

☼ What do you think the whitish blobs along the horizon at right are?

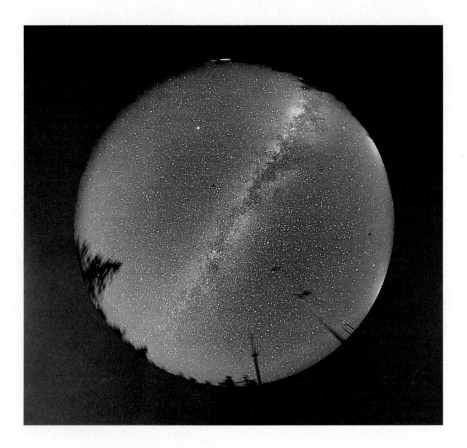

FIGURE 15.2
Dark cloud of dust silhouetted against background stars of Milky Way. (Courtesy Anglo-Australian Observatory; photograph by David Malin.)

If Milky Way were spherical, numerous stars would lie in all directions.

Earth

We would see stars scattered uniformly.

Because the Milky Way is a disk, we see lots of stars when we look in directions lying in the disk, and we see very few stars when we look in directions that lie out of the disk.

We see stars in a concentrated band.

FIGURE 15.3
If the Milky Way had a spherical shape, we would see about the same number of stars in every direction. However, because it is a disk, we see stars concentrated into a band around the sky.

15.1 DISCOVERING THE MILKY WAY

Shape of the Milky Way

Our understanding of the Milky Way as a star system dates to the eighteenth century when Thomas Wright, an English astronomer, and Immanuel Kant, the great German philosopher, independently suggested that the Milky Way is a flattened swarm of stars. They argued that if the Solar System were near the center of a spherical cloud of stars we would see roughly the same number of stars in all directions. Instead, we see vastly more stars in the direction of the Milky Way than in directions perpendicular to it, exactly what we would expect if the Milky Way is a disk rather than a ball of stars, as shown in figure 15.3.

These arguments were refined toward the end of the eighteenth century by the English astronomer Sir William Herschel, who, by counting the number of stars visible in different directions, deduced that the disk was about five times wider than it was thick.

FIGURE 15.4
A copy of Herschel's sketch of the Milky Way made by him in 1784. He based this cross-sectional view on the number of stars he could see in different directions. From this picture he correctly deduced that the Milky Way was wider than it was thick. However, Herschel *incorrectly* concluded that the Sun (the yellow dot) is near the Milky Way's center. He was led astray by his lack of knowledge that dust clouds blot out distant stars and prevent us from seeing our galaxy's true extent.

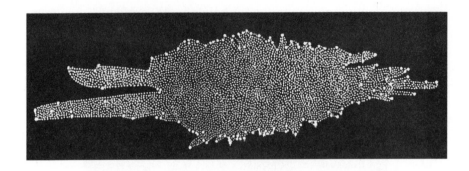

Moreover, because he saw roughly the same number of stars in all directions around the disk, he concluded that the Sun must lie near its middle, as shown in figure 15.4. However, he had no way to measure the distances to the stars (parallax was not discovered until the next century) and so could not determine the Milky Way's size.

Size of the Milky Way

The size of the Milky Way was first measured in the early 1900s by the Dutch astronomer Jacobus C. Kapteyn and the American astronomer Harlow Shapley. Kapteyn measured the distance to nearby stars by parallax and then estimated the distance to more remote stars by their motion. As stars move in their orbits around the galaxy, their positions in the sky change very slowly. Although such motion is not noticeable to the naked eye, it is detectable when pictures taken many years apart are compared. Kapteyn realized that nearby stars change their position faster than more distant stars, just as a ball tossed past your head moves across your field of view more rapidly than one thrown at the far end of a playing field. Using such motions for remote stars and comparing them with the motion of nearby stars whose distance he knew from parallax, he estimated the distance to the remote stars and thus the size of the Milky Way. He concluded that the Milky Way is a disklike system about 20,000* parsecs in diameter, with the Sun located near the center.

Almost immediately after Kapteyn's publication of his model of the Milky Way, Shapley published a model in strong disagreement with it. Shapley argued that the Milky Way was larger than Kapteyn believed and that the Sun was not near the center but about two-thirds of the way out in the disk. Shapley drew his conclusions from a study of globular star clusters—dense groupings of up to a million stars, such as the one shown in figure 15.5. Because these clusters contain so many stars, they are very luminous and can be seen at large distances—across the galaxy and beyond. Moreover, many of them lie above the disk, and so we have a clear view of them.

FIGURE 15.5
A typical globular star cluster.
(Courtesy NOAO.)

---

*In this and the remaining chapters we will be dealing with huge objects at immense distances. Perhaps a quick review of some units and powers-of-10 notation is in order.

Astronomers measure large distances in either light years (ly) or parsecs (pc). One parsec is 3.26 light years and is about 3×10^{13} (30 trillion) kilometers. Recall that 10^{13} stands for 1 followed by 13 zeros.

Galaxy dimensions are so huge that even parsecs are inappropriate for measuring their size, and so astronomers often use kiloparsecs (kpc) for that purpose. One kiloparsec = 1000 parsecs. Later, we will find that an even larger unit, the megaparsec (Mpc) is a more appropriate unit for measuring the distance to galaxies.

Shapley noticed that the globular clusters are not scattered uniformly across the whole sky but are concentrated in the same place that the stars are concentrated, that is, toward the constellation Sagittarius. He hoped therefore that a detailed map of the distribution of the globular clusters would show him the shape of the Milky Way.

To make his map of the clusters, Shapley needed to know their distance from the Sun, distances he could measure by observing variable stars in them. Drawing on the work by Henrietta Leavitt that established the period-luminosity relation for variable stars, Shapley measured the periods of these variables to obtain their luminosities. With their luminosity, their apparent brightness, and the inverse-square law, he could calculate the distance to the variable stars. Knowing the distance to the stars in a cluster then gave him the distance to the cluster and allowed him to plot where each cluster lies in the Milky Way, as shown in figure 15.6. These plots showed that the clusters are distributed through a flattened spheroidal volume about 100 kiloparsecs (roughly 300,000 light years) in diameter. They surround a disk of stars in which the Sun lies about two-thirds of the way from the center, as illustrated in figure 15.6.

The great difference between Shapley's value for the diameter of the Milky Way and the Sun's location and Kapteyn's finding created a major controversy among astronomers. Like so many other controversies, both sides were partially right and partially wrong. For example, Kapteyn deduced a size for the Milky Way that was far too small because he assumed that space was transparent: he did not recognize the dimming effect of interstellar dust. On the other hand, Shapley was correct about the Sun not being at the center, but he greatly overestimated the diameter of the Milky Way, finding a value about three times larger than the one currently accepted. Part of the reason for his error was that Shapley, like Kapteyn, did not recognize the dimming effect of interstellar dust. Thus, Shapley mistook the dimness of the globular clusters to mean that they were very far away. Moreover, he did not realize that he was dealing with two types of pulsating variables: RR Lyrae stars and Cepheids. Leavitt had determined the period-luminosity law for Cepheids, but Shapley was measuring the galaxy's size with RR Lyrae stars. Because RR Lyrae stars are much dimmer than Cepheids, Shapley thought they were much farther away than they really are. Thus he overestimated the size of the Milky Way. Kapteyn came closer to the presently accepted size of our galaxy, but he was wrong about where the Sun lay in it.

These issues were aired in 1920 when the National Academy of Sciences sponsored a debate on the size of the Milky Way and the distance to the spiral nebulas (what we now know to be other galaxies). Shapley defended his position that the Milky Way was huge and that the spiral galaxies were small, nearby objects. He was opposed by Heber D. Curtis of the Lick Observatory in California. The debate did not settle the issues: many additional observations, some depending on techniques and equipment unavailable at the time, were needed to bring a resolution of the problems.

FIGURE 15.6
Schematic version of Shapley's plot of globular clusters from which he inferred the size of the Milky Way and the Sun's location in it. Notice that the clusters fill a roughly elliptical region and that the Sun is *not* at the center of their distribution.

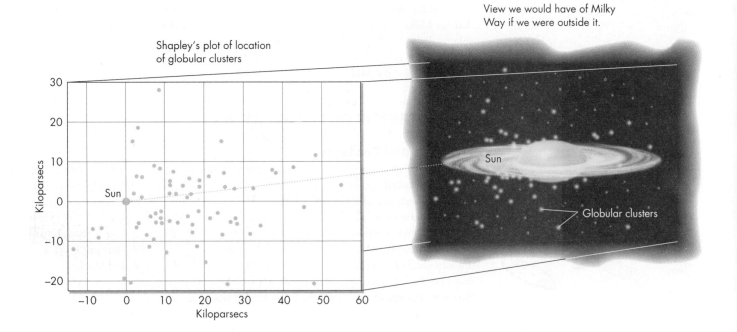

View we would have of Milky Way if we were outside it.

Shapley's plot of location of globular clusters

Sun

Sun

Globular clusters

FIGURE 15.7
Photograph of the spiral galaxy NGC 5236 (also known as M83), illustrating its spiral arms. (Anglo-Australian Observatory; photograph by David Malin.)

Over the last fifty years, astronomers have revised and improved Shapley's model of the Milky Way. Much of that improvement comes from studies of other galaxies. For example, many external galaxies are flat disks with conspicuous spiral arms, as illustrated in figure 15.7. Such arms are extremely hard to detect in our own galaxy because our location in the Milky Way is such that we cannot get an overview of its disk. Nevertheless, given that the Milky Way is a flat disk and that other disk galaxies have spiral arms, it is reasonable to deduce that ours does too. Thus by combining observations of our own galaxy with the general features known to occur in other galaxies, astronomers can assemble a detailed model for the Milky Way.

15.2 OVERVIEW OF THE MILKY WAY

Structure and Contents

The Milky Way consists of three main parts, as illustrated in figure 15.8: a **disk** about 25 kiloparsecs (80,000 light years) in diameter, a surrounding **halo**, and a **bulge** where the central parts of the disk thicken. Within the disk, numerous bright young stars cluster into **spiral arms** that wind outward from near the center. Our Solar System lies between two arms about 8.5 kiloparsecs (roughly 28,000 light years) from the center, with the orbits of the planets tilted by about 60° with respect to the galaxy's disk. This tilt is the reason why the band of the Milky Way on the sky is tipped with respect to the ecliptic. You may find the following analogy helpful in visualizing the scale of our galaxy: if the Milky Way were the size of the Earth, the Solar System would be the size of a large cookie.

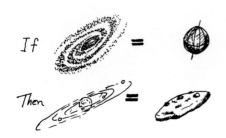

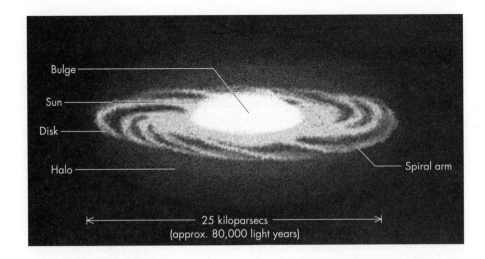

FIGURE 15.8
Artist's sketch of the Milky Way showing the disk, halo, and bulge. The halo surrounds the disk much as a bun surrounds a hamburger.

Mingled with the stars of the Milky Way are huge clouds of gas and dust that amount to nearly 15% of the system's mass. We can see some of these gas clouds by the visible light they emit, whereas others reveal themselves by blocking the light of background stars, as can be seen in figure 15.2. Shapley and others before him were unaware that these clouds contain dust that dims the light from distant stars, making these stars look farther away than they really are. Thus he thought the galaxy was bigger than it actually is. Dust also hides one of the most interesting parts of our galaxy: the **nucleus,** a dense swarm of stars and gas in which perhaps a massive black hole lies. So much dust lies between us and the nucleus that we can "see" it only with radio and infrared telescopes. The visible light it emits is completely absorbed.

Astronomers have not yet fully mapped the outer parts of the Milky Way. Radio telescopes reveal that gas lies there, extending out to nearly 40 kiloparsecs (about 130,000 light years) but in a warped disk that is bent up on one edge and down on the opposite edge. Moreover, for reasons we will discuss later, astronomers believe that our galaxy contains additional, as yet unobserved, dark matter that may extend farther yet.

The flattened shape of its disk implies that the Milky Way rotates (recall how rotation made the solar nebula flat, as we discussed in chapter 7). Astronomers have measured that rotation and find that our Sun and neighboring stars move around the Milky Way with a speed of about 220 kilometers per second, which means that near us the disk makes one complete rotation approximately every 240 million years. As an illustration of this immense time span, our galaxy has made only about one quarter of a revolution since the death of the dinosaurs about 65 million years ago.

The above observations, when combined with physical laws, help astronomers explain the Milky Way's structure. For example, according to Newton's laws of motion, some force must act to deflect the motion of the Sun and other stars into orbits around the center of the Milky Way. That force is supplied by the collective gravitational attraction of the stars and gas within our galaxy that draws them toward its center. In fact, the motion of the Sun around the Milky Way is much like the motion of the Earth around the Sun. That similarity extends to the shape and orientation of the stars' orbits. Just as all the planets move around the Sun in the same direction and lie in nearly the same plane, so too the stars in the disk all move in the same direction along approximately circular orbits lying in nearly the same plane. Moreover, just as planets near the Sun complete their orbits faster than planets farther out, so too stars near the center of the galaxy complete their orbits faster than stars at the edge of the galaxy. Thus, the inner parts of the galaxy spin faster than the outer parts, a phenomenon called "differential rotation," which, as we will see in section 15.5, plays an important role in making the galaxy's spiral arms.

Recently, astronomers have found evidence that the central bulge of our galaxy is not circular. Instead, it has a slightly elongated shape, like a flattened football. If true, the Milky Way should be described as a "barred" spiral (see chapter 16).

Within this swirling cloud of stars, the Sun is nearly lost like a speck of sand on a beach. But unlike grains of sand that touch others, stars are widely separated. For example, near the Sun and throughout most of the disk, stars are typically several light years apart: the Sun's nearest neighbor is 4.2 light years away, a separation like that of pinheads 15 miles apart. Near the core of the Milky Way, stars are packed far more densely, with a separation roughly 1500 times less than that near the Sun, a separation like pinheads at opposite ends of a football field. The density of stars, as measured by the number of stars per cubic light year, is probably more than 10 million stars per cubic light year near the core, but only about 0.003 stars per cubic light year near the Sun. Near the edge of the disk, stars are spread even more thinly, much as our atmosphere thins out as it merges into space. Thus the Milky Way has no sharply defined outer edge.

Mass of the Milky Way and the Number of Stars

Astronomers can estimate the Milky Way's mass from its rotation. On the basis of the gravitational attraction needed to hold the Sun in orbit, the Milky Way must contain about 10^{11} M$_\odot$ of matter—stars and gas. It probably contains much more, however, because its gravitational effects on nearby galaxies imply a mass as large as 5×10^{11} M$_\odot$. Astronomers still have no satisfactory explanation for the large difference between these values of the Milky Way's mass, an indication of how much remains to be learned about our home galaxy.

With so much uncertainty about its mass, it is difficult to estimate the number of stars in the Milky Way. However, if we assume that the average star in our galaxy has a mass comparable to the Sun's (an assumption based on censusing stars, as we will see below), then from the best estimates of the galaxy's total mass, we deduce that it contains about 10^{11} (100 billion) stars, each moving along its own orbit, held in its path by the collective gravity of the other stars.

Age of the Milky Way

We can also use stars to estimate the Milky Way's age. Using the techniques to measure stellar ages described in chapter 13, astronomers calculate that our galaxy's most ancient stars are about 15 billion years old, a number we will take as the approximate age of our galaxy*. During this vast span of time the Milky Way has consumed most of its original gas, converting it into stars. Each year, ever less gas remains, and eventually, about 10 billion years from now, the Milky Way will lack material for making new stars and will begin to dim. Thus even our immense and ancient galaxy has a finite lifetime.

15.3 STARS OF THE MILKY WAY

Stellar Censuses and the Mass Function

The Milky Way contains many types of stars: giants and dwarfs, hot and cold, young and old, stable and exploding. Counts that list all known stars of a given type in a given region of space—stellar censuses—show that all the types described in chapters 12 to 14 occur in our galaxy. By analyzing the types of stars and their relative numbers, astronomers can deduce some of the ancient history of our galaxy, much as archeologists can learn about life in an ancient city from the kinds of buildings it contained. Such studies of stellar populations reveal that despite the wide range of star types, the typical star in the Milky Way is rather like our Sun: small, dim, and cool.

---

*Recent calculations of stellar models and observations of the luminosity of giant stars suggest that the oldest stars may only be 11–13 billion years old. This might mean that our galaxy is also about that old. Despite this uncertainty, we'll continue to use 15 billion years as a round number for the Milky Way's age.

One of the most important statistics that emerges from a stellar census is the number of stars of each mass, technically known as the **mass function.** Because mass is the fundamental property that governs how a star evolves and ages, the evolution of the Milky Way depends strongly on how many stars of each type it contains and how rapidly these stars are born and die. For example, if the Milky Way were composed only of 10 $M_\odot$ stars, it would be very luminous for about 10 million years, and then all its stars would explode as supernovas, leaving a galaxy composed of neutron stars or black holes. On the other hand, if the Milky Way were composed only of 0.1 $M_\odot$ stars, it would never be very bright, but it would shine for several hundred billion years. How long the Milky Way shines also depends on how rapidly stars are born in it. If many stars form each year, our galaxy will quickly deplete the gas from which they form. By dividing the number of stars in the Milky Way by its age, astronomers estimate that currently a few (3 to 5, or so) stars are born each year, though at the time of our galaxy's birth, the birth rate was probably several times higher. This seemingly small number of new star births is nevertheless sufficient to keep our galaxy shining for many billions of years into the future. In fact, our galaxy may shine even longer, because this number ignores the possibility of new stars forming from material ejected by old stars when they die, a form of astronomical recycling.

Studies of the mass function show that stars similar to or smaller than the Sun vastly outnumber more massive stars. The most numerous stars in the Milky Way turn out to be dim, cool, red dwarfs—stars that lie on the main sequence but whose mass is 0.1 to about 0.5 $M_\odot$. Astronomers do not understand yet what determines the distribution of stellar masses, though it probably depends on the conditions in the cloud where the stars form. The average mass for Milky Way stars is about 1 $M_\odot$, though some are significantly more massive (up to 30 or so $M_\odot$), whereas others are much less massive (perhaps as small as 0.08 $M_\odot$). Stars much more massive than about 30 $M_\odot$ are very rare because they are nearly unstable and may break up under the force of their own radiation, as we discussed in chapter 13. Objects less massive than about 0.08 $M_\odot$ are also believed to be very rare because their gravity is too weak to compress them and to create enough heat to ignite nuclear burning.

Low-mass stars are so dim that they are very hard to detect, and so astronomers may simply be overlooking them. Because low-mass stars are hard to observe, they are not the stars we see when we look out at the night sky, despite their abundance. Those visible stars tend to be giants which, although they are rare, we see at great distances because they are so bright. Astronomers refer to this difference between what is seen and what is truly present as a selection effect.

Selection effects can occur whenever one samples a population. For example, when the census bureau counts our population, it must make corrections for people who are homeless or who otherwise might be overlooked. Similarly, astronomers must guard against drawing false conclusions from data that are biased by selection effects.

Two Stellar Populations: Population I and Population II

Hidden in the great diversity of star types described above is an underlying simplicity astronomers first noted in the 1940s. At that time, the 100-inch reflector at Mt. Wilson Observatory, located outside Los Angeles, California, was the largest telescope in the world and the best for observing galaxies. The glow from city lights, however, made it hard to see faint galaxies. But blackouts during World War II darkened the night sky, and Walter Baade, an astronomer at Mt. Wilson, took advantage of the darkness to make a series of photographs of neighboring galaxies. He noticed that stars in these nearby galaxies were segregated by color. Red stars were concentrated in the bulges and halos of the galaxies, whereas blue stars were concentrated in their disks and especially in the spiral arms. To distinguish these groups, Baade called the blue stars in the disk **population I** and the red stars of the bulge and halo **population II** (Pop I and II for short). As he pursued his studies of the stellar populations of other galaxies, he saw that Milky Way stars showed the same division. Furthermore, when he examined the properties of population I and II stars in our galaxy, he found they possessed many striking differences in addition to their color and location.

Population I and population II stars differ in nearly every respect: color, age, location in galaxy, motion, and composition. Population I stars are young (10^6 to a few times 10^9 years old) compared to the galaxy and many of them are blue. They lie in the disk and follow approximately circular orbits, as shown in figure 15.9A. Their composition is like the Sun's, mostly hydrogen and helium, but about 3% of their mass is heavy elements such as carbon and iron.

Population II stars are generally red and old (nearly 10^{10} years old). They lie in the bulge and halo of the galaxy and move along highly elliptical orbits that may be tilted strongly with respect to the disk, as shown in figure 15.9B. Population II stars are also mostly hydrogen and helium, but only a few hundredths percent of their mass is composed of heavy elements, roughly a hundred times less than stars like the Sun. Table 15.1 summarizes some of these properties.

Division of all stars into two broad categories is an oversimplification. For example, the Sun fits in neither category precisely, but it is usually considered a Pop I star because of its high metal content and approximately circular orbit in the disk. To avoid forcing a star into a category in which it does not fit, astronomers sometimes subdivide Pop I and Pop II into *extreme* and *intermediate* Pop I and Pop II. Also, some astronomers use the term "old-disk population" to describe stars like the Sun that are located in the disk but that are not extremely young. In short, stellar populations seem to exhibit a smooth continuum from extremely old to extremely young stars: nevertheless they are still useful for describing both the structure and history of the Milky Way. For example, many astronomers believe that the two different populations show that star formation has not occurred continuously in the Milky Way. Population II stars probably formed in a major burst at the time of the galaxy's birth during its initial collapse (see section 15.8), whereas Population I stars formed much later and continue forming even today. This hypothesis explains the differences between the two populations, as we will discuss in greater detail in section 15.8.

One of the first uses astronomers made of stellar population differences was to map the Milky Way's spiral arms. As Baade noted, the spiral arms of other galaxies are demarcated by their Pop I stars, especially the hot blue ones of spectral classes O and B. It is therefore logical to assume that the same is true for the Milky Way. By measuring the location of O and B stars near the Sun, astronomers can make a crude picture of the spiral arms near us, even though they cannot view the arms from a distance. This method is severely limited, however, because dust in space prevents us from seeing even the brilliant O and B stars if

FIGURE 15.9
Stellar orbits in the Milky Way.
(A) Pop I stars orbit in disk.
(B) Pop II stars orbit in halo. Notice elongation of their orbits.

Pop I stars

Pop II stars

A

B

TABLE 15.1	Properties of Population I and II Stars	
	Pop I	**Pop II**
Age	young to old (10^6 to a few times 10^9 years)	old (about 10^{10} years)
Color	blue (generally)	red
Location	disk and concentrated in arms	halo and bulge
Orbit	approximately circular in disk	plunging through disk
Heavy-element content	high (similar to Sun)	low (10^{-2} to 10^{-3} Sun's)

they are farther from the Sun than about 10,000 light-years. Nevertheless, such maps were the first evidence that we live in a spiral galaxy.

Star Clusters

Some of the Milky Way's stars are bound together gravitationally into groups called **star clusters.** Within a cluster, each star moves along its own orbit about the center of mass of the cluster, held to it by the group's gravity. You can see several such clusters with your naked eye. The most obvious one is the Pleiades, or the Seven Sisters, named for the daughters of the giant Atlas who, in Greek mythology, carried the world on his shoulders. The Pleiades (pronounced /plee-a-dees/) are visible from late August through March as a tiny group of stars north of the V in the constellation of Taurus. With a pair of binoculars you can see that the Pleiades contain many more than the half dozen stars visible to the naked eye. Likewise, by scanning the Milky Way with binoculars you will see many additional clusters dimmer than the Pleiades.

The Milky Way has two main types of star clusters: open clusters and globular clusters. **Open clusters**—the Pleiades, for example—contain up to a few hundred members in a volume with a radius of typically 7 to 20 light years. They are called "open" because their stars are scattered loosely. Sometimes they are also called "galactic clusters" because most of them lie in the disk of the galaxy.

Astronomers think that open clusters form when giant, cold, interstellar gas clouds move into the galaxy's spiral arms. The clouds are compressed by the stronger gravitational field within the arm and collapse, breaking up into hundreds of stars whose mutual gravity binds them into the cluster. Once formed, an open cluster continues to orbit the galaxy in the disk, but over hundreds of millions of years, stars gradually escape from it and so the cluster eventually dissolves. Assuming that open clusters are as abundant elsewhere in the galaxy as within the regions visible from Earth, astronomers estimate that our galaxy contains about 20,000 open clusters, though dust hides most of them from our view. Our own Sun may have originally been a member of such a star group, its companion stars having long since scattered across space.

Within our galaxy's spiral arms extremely young stars sometimes occur in loose groups called **associations** that are a few hundred light-years across. Associations typically spread out from a single large open cluster near their center and may contain other smaller star groupings. Moreover, the stars in associations are usually still mingled with the massive clouds of dust and gas from which they formed.

Globular clusters contain far more stars than open clusters, from a few hundred thousand to several million per cluster. We described them briefly in our discussion of Shapley's study of the shape of the Milky Way. They have larger radii than open clusters have—40 to 160 light years—but the larger number of stars in them creates a stronger gravity than that found in open clusters. This stronger gravity pulls their stars into a denser ball. Astronomers estimate that about 150 to 200 globular clusters orbit the Milky Way, of which about 130 have actually been observed. Figure 15.10 shows pictures of each type of cluster and illustrates the difference between them. Table 15.2 summarizes some properties of open clusters and globular clusters and associations.

TABLE 15.2   Properties of Clusters and Associations

Type	Number of stars	Radius*	Location
Open cluster	few hundred	7 to 20 ly (2 to 6 pc)	along spiral arms
Globular cluster	10^5 to 10^6	40 to 160 ly (12 to 50 pc)	halo and nucleus of galaxy
Associations[†]	5 to 70 O or B stars	65 to 325 ly (20 to 100 pc)	spread along a spiral arm

*Because star clusters do not have sharp edges but instead gradually thin out, quoted dimensions differ substantially.
[†]Astronomers identify several other types of associations as well. For example, T associations are regions with above-average numbers of T Tauri stars.

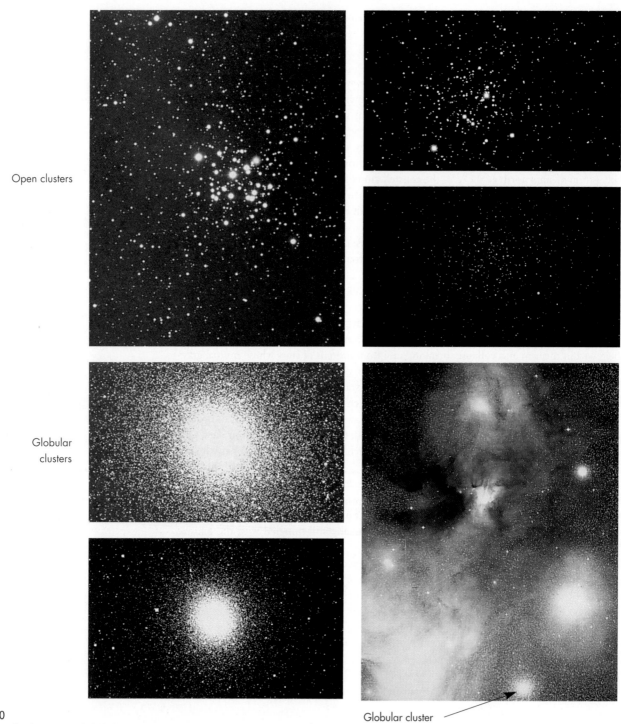

Open clusters

Globular
clusters

Globular cluster

FIGURE 15.10
Star clusters: both open and globular
types. (NOAO; Anglo-Australian
Observatory; photograph by David Malin.)

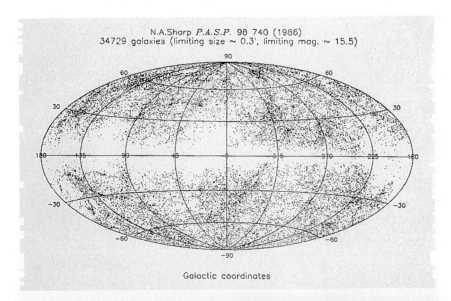

N.A.Sharp *P.A.S.P.* 98 740 (1986)
34729 galaxies (limiting size ~ 0.3', limiting mag. ~ 15.5)

Galactic coordinates

FIGURE 15.14
Dust limits our view in the disk of the galaxy and creates the zone of avoidance. We see no galaxies in the plane of our own galaxy because dust in the disk blocks our view. (Plot of galaxies courtesy NOAO.)

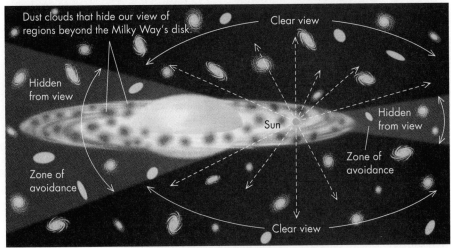

FIGURE 15.15
Infrared photograph of the galaxy made with the satellite COBE. (NASA.)

no galaxies appear is the zone of avoidance and coincides with the plane of the Milky Way. We know today that galaxies do occur in this region because they can be seen with radio and infrared telescopes sensitive to radiation that can penetrate the dust.

Infrared and radio waves pass more readily through dust than visible light does because their wavelengths are much larger than the dust particles. This allows infrared pictures to reveal the structure of the Milky Way far better than visible light pictures can, as illustrated in figure 15.15. The disk and central bulge of our galaxy are easy to see in this wide-angle picture taken from above our atmosphere by the cosmic background explorer satellite (COBE).

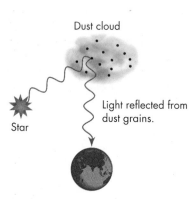

Dust cloud

Light reflected from dust grains.

Star

FIGURE 15.16
Reflection nebula. Light from a star near a dust cloud is reflected from the dust making it visible. (Courtesy Anglo-Australian Observatory; photograph by David Malin.)

 Why is red light not scattered?

The Pleiades

Dust is not all bad;* it may scatter enough light to become visible, thereby creating beautiful **reflection nebulas** (fig. 15.16). Scattered starlight makes the interstellar dust shine just as scattered sunlight makes dust in the air visible. Reflection nebulas are not only lovely, but additional proof that dust exists in space.

Interstellar Gas

Interstellar gas plays several crucial roles in our galaxy. It is the material from which stars form, and it is the repository of matter blown off dying stars. Without interstellar matter to make new stars to replace those that die, the Milky Way would be dark and filled only with stellar corpses. Interstellar gas is also a help to astronomers seeking to map the Milky Way. Dust blocks our view of the Milky Way and makes it difficult for us to observe its structure, but interstellar gas emits radiation that penetrates the dust and allows us to "see" regions of the galaxy that would otherwise be invisible to us. Finally, interstellar gas creates some of the most spectacular structures in our galaxy: huge glowing clouds heated by stars embedded in them, as shown in figure 15.17. An example of such a gas cloud is the Great Nebula in Orion, which is easy to see with a small telescope or even a pair of binoculars. If you look at the middle star in Orion's sword you will see a pale, fuzzy glow created by luminous interstellar gas surrounding a dense star cluster.

Visible Emission from Interstellar Gas. Interstellar gas clouds that emit visible light are called **emission nebulas,** and figure 15.17 illustrates several of the many found within the Milky Way. Emission nebulas give off their own light in visible and other wavelengths, unlike the dust in reflection nebulas, which simply reflects and scatters light from neighboring stars. However, a cloud must be hot to emit visible light, and so such clouds need a source of heat (generally a nearby star), as shown in figure 15.17D. Hot, blue stars are especially effective at heating because they emit large amounts of ultraviolet radiation whose short wavelength is much more energetic than that of visible light. Ultraviolet light

---

*We should also keep in mind that the molecules that comprise the Earth and our bodies came from such dust grains.

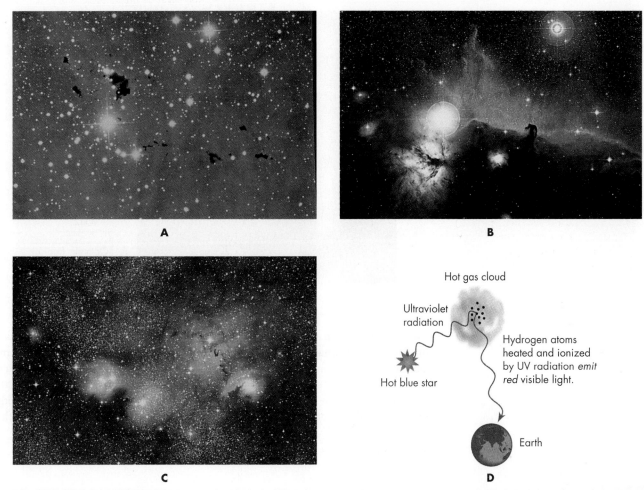

FIGURE 15.17
(A), (B), (C) Emission nebulas. (Courtesy Anglo-Australian Observatory; photographs by David Malin.) Notice the bright (and hot) stars near the glowing gas. (D) These stars emit ultraviolet radiation, which heats and ionizes the nearby hydrogen gas. Electrons dropping from orbit 3 to orbit 2 generate the red light.

whose wavelength is less than 91.2 nm is so energetic that when it is absorbed by hydrogen it tears the electron free of the nucleus, ionizing the gas. In an ionized gas (one in which atoms are missing one or more electrons) the free electrons collide with atoms of oxygen, nitrogen, and other elements and excite them, making them emit light of their own. Eventually the electrons recombine with the ionized hydrogen atoms and emit more radiation, especially the red light of hydrogen alpha, which gives these nebulas their lovely reddish color. Because hydrogen in these hot gas clouds is ionized, they are sometimes called **HII regions** (pronounced /H two/), the H denoting the hydrogen and the II indicating that it is ionized.

HII regions have a temperature of approximately 10,000 K and are heated by the electrons freed by the ionizing ultraviolet radiation. The liberated electrons collide with gas atoms around them and speed them up, increasing their energy and thus their temperature. The freed electrons are themselves a powerful source of radio emission, emission that helps astronomers see HII regions that might otherwise be hidden from us by dust. Because these HII regions are generally located in spiral arms (fig. 15.7 shows them glowing like rubies interspersed with diamonds on a necklace), radio maps of their location can reveal the spiral structure of our galaxy. In fact, maps made by combining radio observations and optical observations of HII regions are the best evidence we have that the Milky Way is a spiral galaxy* (fig. 15.18).

---

*Some astronomers think the Milky Way may be a barred spiral galaxy. That is, its arms wind out from the ends of an elongated nucleus rather than a round nucleus. Barred spiral galaxies are discussed in more detail in the next chapter.

FIGURE 15.18
Map of the Milky Way made by combining radio observations (since radio can penetrate dust and thereby allow one to see all the way across the Milky Way's disk) and optical observations of the position of HII regions. Such regions outline the spiral arms of other galaxies. In our galaxy too they form a spiral pattern. (Courtesy Y. M. and Y. P. Georgelin.)

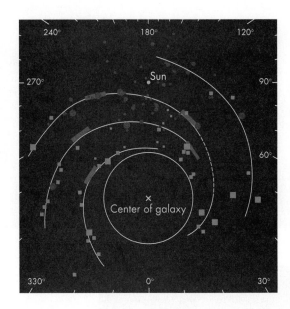

Radio Waves from Cold Interstellar Gas. Although we can easily see the glow from gas clouds that happen to be near hot stars, most interstellar gas is too cold to emit visible light, and we can detect it only with radio telescopes. Clouds more than a few tens of light years from hot stars may be only a few Kelvin (more than 400° below zero on the Fahrenheit scale). Such cold material emits no visible light because it has too little energy to generate visible photons, but it can emit radio waves of very low energy (recall that radio waves carry much less energy than visible light does).

EXTENDING OUR REACH

MAPPING THE MILKY WAY WITH RADIO WAVES

One of the most important sources of radio emission is cold hydrogen atoms,* which radiate at a wavelength of 21 centimeters. This radiation, called **21-centimeter radiation,** arises because subatomic particles such as protons and electrons possess a property, known as "spin," that makes them behave like tiny magnets. Magnets have north and south poles with the property that like poles repel and unlike poles attract. If an electron has the same alignment as the proton it is orbiting—both with north poles "up", for example—the electron will flip over, as shown in figure 15.19. As it flips, the changed magnetic interaction between the electron and the nucleus makes the electron shift to a slightly smaller orbit with

*Cold hydrogen atoms with their electrons attached are called "HI" (pronounced /H one/) to distinguish them from ionized hydrogen atoms, HII.

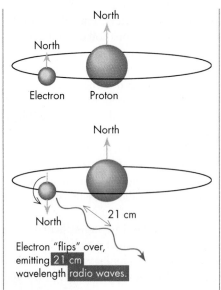

FIGURE 15.19
Radio radiation at 21 centimeters is emitted by cold hydrogen as the electrons "flip."

a slightly lower energy. However, energy must be conserved, and so when the electron shifts to the new orbit, it emits the excess energy as electromagnetic radiation. We discussed in chapter 3 that this process can produce visible light. In cold hydrogen, however, the flip of the electron liberates too little energy to make visible radiation and can create only radio emission at a wavelength of 21 centimeters. The Dutch astronomer Hendrik C. Van de Hulst predicted in 1944 that hydrogen should emit such radiation, and he calculated that its wavelength should be 21 centimeters. But World War II prevented anyone from looking for this radiation, and it was not until 1951 that astronomers detected it from gas in the Milky Way.

The 21-centimeter radiation has proved extremely valuable for studying the Milky Way and other galaxies. Hydrogen is abun-

Continued

TABLE 15.3	Some Interstellar Molecules
H_2	molecular hydrogen
OH	hydroxyl radical
CN	cyanogen radical
CO	carbon monoxide
H_2O	water
HCN	hydrogen cyanide
NH_3	ammonia
HCOH	formaldehyde
HCCH	acetylene
HCOOH	formic acid
CH_3OH	methyl alcohol (methanol)
CH_3CH_2OH	ethyl alcohol (ethanol)

dant in space, and so the signal of the 21-centimeter radiation is strong. From the strength of the signal, astronomers can deduce the amount of hydrogen in an area. If the gas is moving, the wavelength will be Doppler-shifted, from which the gas's speed along our line of sight can be found. Radio observations thereby allow astronomers to map not only where the gas is concentrated in the galaxy, but also how it is moving. Although details in such 21-centimeter maps are very uncertain because of assumptions that must be made about how the gas moves, they nevertheless are extremely useful because the radiation is not absorbed by interstellar dust, thereby allowing astronomers to "see" the entire Milky Way in its radiation. Such maps show that the gas is confined to a thin disk and its distribution is suggestive of a spiral pattern, confirming the picture deduced by other means.

The 21-centimeter radiation is not a perfect tool, however. It is emitted only by hydrogen if its electron is attached. If the hydrogen is heated so that its electron is torn free—that is, the hydrogen is ionized—it produces no 21-centimeter waves. Likewise, very cold hydrogen cannot emit 21-centimeter radiation. At extremely low temperatures (15 K or so), hydrogen atoms bond to one another to make hydrogen molecules and, in its molecular form, hydrogen cannot emit 21-centimeter radiation. However, in the cold, dense regions where hydrogen molecules form, other molecules that do emit radio waves form as well.

More than 100 interstellar molecules have been identified so far, including common compounds such as carbon monoxide (CO), formaldehyde (HCHO), and ethyl alcohol (CH_3CH_2OH), and also more exotic substances. Table 15.3 lists a few of the many interstellar molecules now known. Collisions among neighboring molecules make them spin and give them energy, as shown in figure 15.20, but they quickly radiate that energy away and reduce their spin by emitting radio waves. Each molecule has a distinctive set of wavelengths at which it can emit its rotational energy, just as each atom has a distinctive set of wavelengths (its spectral lines) at which it can radiate energy from electron jumps. For example, carbon monoxide emits at a wavelength of 2.6 and 1.3 millimeters, and ethyl alcohol emits at 3.5 millimeters. Both these molecules emit at many other wavelengths as well.

The strength of the molecular radio emission from a gas cloud depends on the number of molecules it contains, giving astronomers information on its density. If the gas is moving, the wavelength of the emission will be slightly altered by the Doppler shift, displacing and broadening the spectrum lines. From such shifts and broadening, astronomers can deduce the speed of the cloud and its rotation. With that information, they can then map out where in the Milky Way the molecular gas is located and thereby obtain additional information about the galaxy's structure.

FIGURE 15.20
Emission of radio waves by spinning molecules in cold interstellar clouds. Collisions set the molecules spinning. When their rotation speed changes, they emit radio waves. The wavelength emitted depends on the kind of molecule and by how much its rotation speed changes.

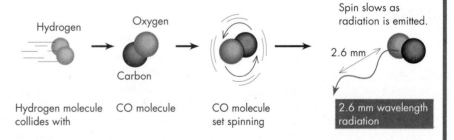

15.5 MOTION OF STARS AND GAS IN THE MILKY WAY

The motion of stars and gas give important clues to the structure of the Milky Way. We saw earlier that in accordance with Newton's first law (the law of inertia), if no force acted on the stars, each would move in a straight line. But a force does act on each star: the gravitational force created collectively by all the other stars and gas. That force acts to draw each star inward toward the center of the galaxy. The result is that each star follows its own orbit around the center of the galaxy, just as each planet moves in orbit about the Sun.

Although all stars within the Milky Way move around its center, the paths followed by stars in the disk and halo are very different, as we discussed in section 15.3. We described there that all the stars in the disk orbit in the same direction and in nearly the same plane, giving the disk its flat shape, just as the planets' motion in approximately the same plane gives the Solar System its flattened shape. On the other hand, stars in the halo and bulge follow orbits that are steeply inclined to the disk's plane (see fig. 15.9).

Motion within the disk also creates the Milky Way's spiral arms. We know that gravity holds a galaxy together, and so we might infer that gravity is also what makes a spiral arm. Gravity does in fact help hold a spiral arm together, but a spiral arm is not a fixed mass of material. Rather, the stars and gas in an arm at one time slowly drift out and are replaced by new stars and gas, according to the density-wave theory of spiral structure. In the **density-wave model,** waves of stars and gas sweep around the galactic disk. The wave is not primarily an up-and-down motion like an ocean wave; rather, it is a place where the density of stars and gas is large compared to the surroundings. This denser region is the wave crest, and we see it as a spiral arm. Spiral waves are not unusual in rotating matter. Batter stirred in a bowl with an electric mixer forms spiral waves, but to better understand how arms form in the Milky Way, consider the following analogy.

As cars move along a freeway, most move at nearly the same speed and keep approximately the same separation. If one car moves slightly slower than the others, however, traffic will begin to bunch up behind it. Cars can pass the slower vehicle but must change lanes to do so. The result is a clump of cars behind the more slowly moving one. But that clump is not composed permanently of the same cars. Cars join the clump from behind, pass the slower moving car, and then leave the clump at its front.

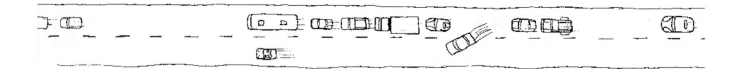

A similar phenomenon happens to stars orbiting in the Milky Way. As stars circle the galaxy, a few closely spaced stars at a given location will make the local gravity slightly stronger than elsewhere in the disk. That excess gravity draws stars into the region so that the clump grows a little. But stars do not remain in the clump. They continue along their orbits and eventually leave the clump. Likewise, gas clouds drift through the clump, initially accelerated inward by the clump's gravity and then slightly slowed as they move out the other side. The result is a wave in the galaxy's disk where stars and gas accumulate before moving through. The accumulation of gas compresses clouds and makes them collide with one another, which in turn makes some of the clouds collapse and form stars. This is the first stage of the star formation process described in chapter 13. The brilliant young *O* and *B* stars so born illuminate the arms and make them stand out against the more ordinary stars of the disk. In fact, most of a galaxy's stars form in its spiral arms, and as we will discover in the next chapter, galaxies without arms generally contain extremely few young stars.

Although the density-wave theory has difficulty explaining the longevity of spiral arms, many observations support the theory. For example, by combining radio and visual pictures of a galaxy, astronomers can see a progression through the spiral arm as gas enters on one side, is compressed, and forms stars, taking a few million years to sweep through the arm. On one side of the arm, you can see gas and young stars; on the other side, older, more evolved stars, as indicated in figure 15.21. Doppler-shift studies of the gas motion also clearly show the gas moving into the arm on one side and leaving on the other.

Although most spiral galaxies probably make their arms this way, another theory has been proposed to explain the very ragged-appearing arms in some galaxies, such as the one illustrated in figure 15.22. This second theory is called **self-propagating star formation,** or SSF, for short.

In the SSF theory, star formation starts at some random point in the disk of a galaxy when a gas cloud collapses and turns into stars. As the stars heat the gas around them and explode as supernovas, they generate a disturbance that makes the surrounding gas clouds collapse and turn into stars. The original stars burn out, but the new stars—formed as the old ones evolve and explode—trigger more gas clouds to collapse and form additional stars. In this fashion the region of star formation spreads across the galaxy's disk much as a forest fire burns outward in a circle through a forest. But unlike a forest where the trees stand still, stars orbit in a galaxy, and those near the center orbit in less time than those farther out. Thus the zone in which star formation occurs is drawn out into a spiral by the

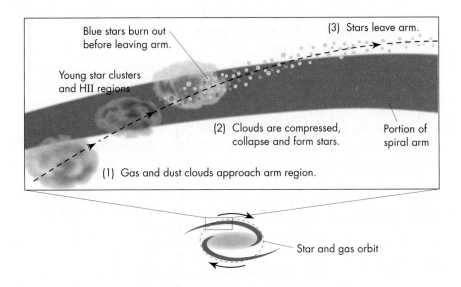

FIGURE 15.21
Sketch of a density wave in a galaxy, showing the progression from dust and gas to stars across the arm.

FIGURE 15.22
Ragged arms in the spiral galaxy M33, perhaps the result of self-propagating star formation. (Courtesy NOAO.)

FIGURE 15.23
Winding of arms in the SSF model. (This sequence was generated by P. Seiden and H. Gerola, *Ap. J. 233*, 56, 1979.)

difference in rotation rate between the inner and outer parts of the disk. The result is spiral arms, as illustrated in figure 15.23.

Astronomers are not certain which theory—density wave or SSF—is correct. Possibly in some galaxies one of the processes is at work, whereas in other galaxies the other is. Alternatively, both mechanisms may operate, or perhaps both hypotheses are wrong! All astronomers can say at this time is that spiral arms might form by either process.

15.6 MEASURING THE MILKY WAY

We have described earlier in this chapter how astronomers measure the basic properties of the Milky Way, such as its radius and mass. In this section we look in greater detail at those methods.

Diameter of the Milky Way

Astronomers have several ways to measure the Milky Way's diameter, but all such methods depend on first knowing how far from the center the Sun is. We described in section 15.1 how Shapley used globular clusters to map out the structure of the galaxy and thereby find its size and where the Sun lies in it. This method remains one of the best techniques for measuring our galaxy's scale. Astronomers measure the distance and direction from the Sun to a large number of globular clusters or some type of star that can be seen for great distances and then plot their positions, as shown in figure 15.24. They next locate the point around which the clusters seem to be most symmetrically located and assume this point is the center of the galaxy. Finally, they measure the distance from that point to the Sun, as shown in the figure. Modern values for this distance turn out to be about 8.5 kiloparsecs (approximately 28,000 light years) on the average.

A more recent method that may eventually give a more accurate distance uses radio observations of cool giant stars. Some such stars emit powerful beams of energy at radio wavelengths analogous to the light emitted by lasers. These stars are called "maser sources." The word *maser* is an acronym for *microwave-amplification by stimulated emission of radiation*. The radiation comes from molecules in a cool shell around the star. Maser sources are especially common in the inner bulge of the Milky Way because of the many red giants there, and with the masers, astronomers can measure the distance from the Sun to the Milky Way's center.

The maser sources, like other stars near the galactic center, move along randomly oriented orbits. Some move along our line of sight and therefore show a Doppler shift, which gives their velocity. To obtain their distance, astronomers assume that, on the average, maser sources moving along the line of sight have the same speed as sources moving across the line of sight. Sources moving across the line of sight show a tiny change in position from one year to the next, and from that change and their speed, astronomers can deduce their distance and therefore the distance to the center of the Milky Way (the point about which they orbit). This method for finding the Sun's distance from the center of the galaxy gives a somewhat smaller value (about 7 kiloparsecs) than is found by other methods, and so astronomers are uncertain which method to believe. Nevertheless, once the Sun's distance from the center is known, the overall size of the Milky Way follows when the distance to objects lying near the galaxy's edge is measured and the two distances are

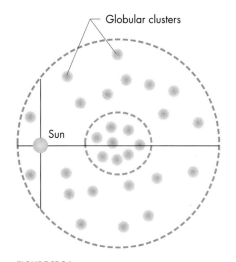

Globular clusters

Sun

FIGURE 15.24
Finding the distance to the center of the Milky Way using globular clusters. Make a scale plot of where the clusters lie with respect to the Sun. Outline the region filled by the clusters and note where they are most concentrated. The center of their concentration is presumably the center of the galaxy. Measure off the distance from the center to the Sun, and from the known scale, determine that distance.

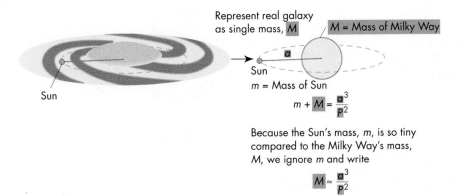

Represent real galaxy as single mass, M

M = Mass of Milky Way

a

Sun

m = Mass of Sun

$$m + M = \frac{a^3}{P^2}$$

Because the Sun's mass, m, is so tiny compared to the Milky Way's mass, M, we ignore m and write

$$M \approx \frac{a^3}{P^2}$$

FIGURE 15.25
Finding the mass of the Milky Way. Represent the entire Milky Way by a single mass, M, around which the Sun (mass m) orbits at a distance a and with an orbital period P. Apply the modified form of Kepler's third law to calculate $M + m$. Because the Sun's mass is so tiny, it can be ignored and we can set $M = a^3/P^2$.

added. From such measures, astronomers believe the Milky Way has a diameter of about 40 kiloparsecs, or even larger, but much of the vast outer part beyond the Sun's orbit is extremely sparsely populated.

Mass of the Milky Way

Astronomers can deduce the Milky Way's mass from its gravitational attraction on matter in and near it. The procedure they use is very similar to the one described in chapter 12 to determine the mass of stars. You may recall that if one star is in orbit around another, we can calculate their combined mass using the modified form of Kepler's third law. Similarly, astronomers can find the mass of the Milky Way by treating it as a scaled-up version of a binary star in which one star, the Sun, for example, orbits under the gravitational influence of the entire galaxy (fig. 15.25). The details are given in the box, "Extending Our Reach."

EXTENDING OUR REACH

MEASURING THE MASS OF THE MILKY WAY

In our calculation we will make use of a result from physics that, if the Milky Way were spherical, only that part of it inside the Sun's orbit would affect the Sun's motion. Moreover, the gravity of that inner part acts as if all the mass interior to the Sun's orbit were packed into a point at the orbit center. The Milky Way is not spherical, but detailed calculations using its true shape differ only slightly from what we will find with our simple model. A more serious difficulty, however, is that our calculation omits all the mass in the Milky Way that lies outside the Sun's orbit. That is, we can assert the Milky Way contains *at least* the amount of mass given by our calculation.

We proceed as follows. We represent the mass of the Milky Way inside the Sun's orbit as a single mass, M, and assume the Sun (with mass m) moves in a circular orbit of radius a about M, as shown in figure 15.25. We now use the modified form of Kepler's third law described in chapter 12 for measuring the masses of binary stars. The modified

form of Kepler's third law states that

$$m + M = a^3/P^2$$

m = Mass of smaller object
M = Mass of larger object
a = Distance between objects
P = Years to complete one orbit

where m and M are the masses of the orbiting bodies in solar masses, P is their orbital period in years, and a is their separation in astronomical units. The orbital period of the sun around the Milky Way is about 240 million years (we will see below how this number is determined). Its orbital radius is about 8.5 kiloparsecs, or about 1.8×10^9 AU. If we insert these values into the equation, we find that the mass of the Sun (m) plus that of the Milky Way (M) is

$$\begin{aligned} m + M &= (1.8 \times 10^9)^3/(2.4 \times 10^8)^2 \\ &= 1.8^3 \times 10^{27}/(2.4^2 \times 10^{16}) \\ &= 10^{11} \, M_\odot \end{aligned}$$

Because m, the mass of the Sun, is so tiny compared with the mass of the galaxy, we can ignore the Sun's mass and conclude that the Milky Way's mass is about $10^{11} \, M_\odot$.

FIGURE 15.26
Schematic rotation curve of the Milky Way. (Based on diagram by Dan P. Clemens, *Ap. J. 295*, 422.)

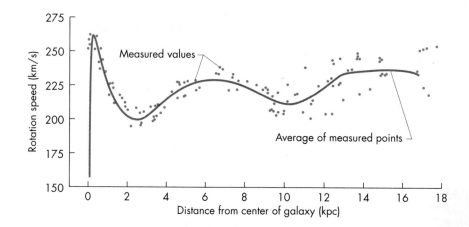

More refined methods build on this same technique but account for the mass of the entire galaxy and use the rotation speed of stars at a variety of distances (the so-called **rotation curve**). The rotation curve, shown in figure 15.26, shows how fast the galaxy spins at each distance from its core. That speed must be exactly sufficient to offset the gravity of the Milky Way that pulls orbiting stars in toward the center. The gravity in turn depends on the Milky Way's mass, and so from the rotation curve and Newton's laws of motion and gravity, astronomers can calculate the mass of our galaxy. The first task therefore is to measure the rotation curve.

To determine the rotation curve, we must remember that we are observing from the Earth, which is traveling with the Sun in the Sun's orbit around the center of the galaxy. Astronomers thus have the same problem that the police in a moving patrol cruiser have in measuring the speed of a traffic violator. That is, to find the speed of the star around the Milky Way, they must measure how fast it moves with respect to the Sun and then add the Sun's motion about the galaxy. Astronomers have several methods for making this measurement.

From many such measurements, astronomers conclude that the Sun is orbiting the galaxy with a velocity of about 220 kilometers per second (about 130 miles per second!). From the Sun's orbital speed, we can calculate its orbital period, P, by dividing the speed, v, into the circumference of its orbit.

That is

$P = $ Sun's orbital period
$R = $ Radius of Sun's orbit
$v = $ Sun's orbital speed

$$P = 2\pi R/v$$

where R is the radius of the Sun's orbit, which we showed earlier is about 8.5 kiloparsecs. Inserting the values of R and v determined above (in appropriate units), we find that P is about 2.4×10^8 years, the number we used earlier to find the Milky Way's mass.

Unfortunately, the measurements we have described above are difficult to do accurately, so the speed of the Sun is uncertain. Because that influences the mass that we calculate for the Milky Way, our galaxy's mass is also uncertain. Moreover, what composes that mass is poorly known. Too few stars and gas clouds are detectable to account for the large value of the mass that is calculated. That is, the galaxy's gravity seems too big for the mass astronomers can see. The Milky Way is not unique, however, in seeming to possess unseen matter. As we shall see in the next chapter, most galaxies seem to contain "dark matter," material that emits no (as yet) detectable radiation but that exerts a gravitational attraction on the stars and gas within the galaxy. The nature of this dark matter is perhaps the greatest mystery in astronomy today.

EXTENDING OUR REACH

MEASURING THE SUN'S SPEED AROUND THE MILKY WAY

The Sun's orbital speed around the Milky Way can be found in several ways. In one method, astronomers take spectra to obtain the Doppler shift of galaxies in the Local Group, the small cluster of galaxies to which the Milky Way belongs. The Doppler shift of the galaxies is created by their own motion and that of the Sun; remember that if either the source or the observer moves, there is a Doppler shift. It turns out that the galaxies move slowly compared with the Sun's motion around the Milky Way, so almost all the Doppler shift is attributable to the Sun's motion. From that Doppler shift astronomers can find the Sun's orbital velocity.

Another method to measure the Sun's orbital velocity uses objects such as globular clusters. Because these clusters fill an approximately spherical volume around the galaxy, they, as a group, presumably revolve only slowly compared with the flat disk. Thus the Sun's speed with respect to the globular clusters is at least approximately its true rotation speed around the Milky Way.

15.7 THE GALACTIC CENTER

The Sun lies at the center of the Solar System and holds the planets in orbit about it. What lies at the center of the Milky Way galaxy? Astronomers think that a dense swarm of stars and gas swirls within the innermost core of our galaxy, and they are actively debating whether a million-solar-mass black hole lies at the very center. The debate arises because numerous interstellar gas and dust clouds lie between us and the core, totally blocking its visible light, and thereby seriously hampering our ability to study it. For that reason, astronomers must observe at radio, infrared, x-ray, and even gamma-ray wavelengths if they wish to study our galaxy's heart.

As long ago as the early 1930s Karl Jansky, an engineer at Bell Telephone Labs in New Jersey, found radio waves coming from the direction of Sagittarius, the location Shapley concluded was the galactic center. Radio waves, as we have seen, can easily penetrate dust clouds. With such waves astronomers can map regions of the galaxy that are otherwise invisible. By the 1950s better radio detectors and bigger antennas showed that the center of the Milky Way has a very complex structure, containing numerous clouds of ionized gas and a small, exceptionally intense source, which came to be known as "Sagittarius A" (Sgr A, for short). This source is not an ordinary cloud of hot hydrogen; much of its radiation is nonthermal emission generated by high-energy electrons spiraling in strong magnetic fields. Moreover, it lies directly at the galaxy's core and may contain a massive black hole, as we will discuss later.

Radio maps reveal many other features toward the Milky Way's center: 5 kiloparsecs from the core, giant clouds of molecular hydrogen and huge regions of ionized hydrogen (HII regions) form a necklace around the center of our galaxy; 3 kiloparsecs from the center, an arc of cold hydrogen sweeps outward through the disk at more than 100 kilometers per second; 1.5 kiloparsecs from the core a vast tilted ring of molecule-rich gas spins. All these features, shown in figure 15.27A, are mere preludes to the amazing structures found at our galaxy's heart. To see these structures, however, astronomers rely on radio telescopes (which show the gas), infrared telescopes (which show the stars), x-ray telescopes (which detect very hot gases), and gamma-ray telescopes (which detect emission lines from electrons and positrons colliding and annihilating each other).

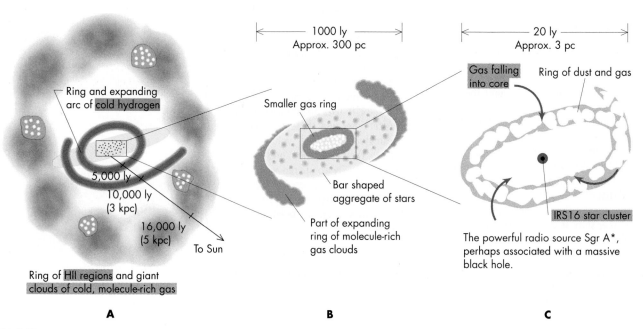

A

Ring and expanding arc of cold hydrogen

5,000 ly

10,000 ly
(3 kpc)

16,000 ly
(5 kpc)

To Sun

Ring of HII regions and giant clouds of cold, molecule-rich gas

B

1000 ly
Approx. 300 pc

Smaller gas ring

Bar shaped aggregate of stars

Part of expanding ring of molecule-rich gas clouds

C

20 ly
Approx. 3 pc

Gas falling into core

Ring of dust and gas

IRS16 star cluster

The powerful radio source Sgr A*, perhaps associated with a massive black hole.

FIGURE 15.27
Artist's representation of the central region of the Milky Way viewed on three different scales. These sketches are based on observations made with radio and infrared telescopes. Details of the structure in the inner part are very uncertain.

Ordinary red giant stars radiate an enormous quantity of infrared energy (in fact, more than their visible-light output). Because infrared energy, like radio waves, can penetrate the dust clouds that veil the galaxy's center, with infrared telescopes astronomers can see the billions of stars hidden there. Maps made with information from such telescopes reveal a giant swarm of stars arranged in an elongated structure roughly 1000 light years across centered on the core of the Milky Way, as illustrated in figure 15.27B. At the center of this swarm, millions of stars are packed into a cubic light year, whereas near the Sun, a similar volume on the average contains less than 1/100 of a star. Almost precisely at the center of the swarm and therefore at the core of the Milky Way lies a dense clump of extremely luminous stars (technically known as IRS16, for Infrared Source #16). Surrounding this dense cluster swirls a lumpy ring of dense dust and gas, as depicted in figure 15.27C. This ring, composed mostly of unusually hot (2000 K) hydrogen molecules, has a diameter of about 20 light-years and a mass of about 30,000 $M_\odot$. It radiates about 2×10^7 $L_\odot$, an astonishing amount of energy for such a small source. Clouds in the ring orbit at an average speed of about 100 kilometers per second, but some move several times faster, implying the gas was violently disturbed within the past 100,000 years. Astronomers therefore think that some energetic event, perhaps a supernova blast, has driven away nearly all the gas from inside the ring. The presence of a few streamers of hot gas moving at high velocity (about 1000 kilometers per second) is additional evidence for a violent event in our galaxy's core. Because the inner edge of the ring is ionized, the implication is that a powerful energy source lies near the ring's center. Moreover, either that source or other as-yet-unseen material must create enough gravitational force to hold the ring together, and calculations show this requires a mass of about 10^6 $M_\odot$. Astronomers find it hard to understand how the dense cluster of stars could contain so large a mass, but the ring encircles another more mysterious source.

Deep within our galaxy's core lies a powerful but incredibly small radio source known as Sgr A* (with the asterisk denoting that it is only part of the larger radio source Sgr A, mentioned earlier). This mysterious object is less than 10 AU in diameter (about the size of Jupiter's orbit), and some of its features are sketched in figure 15.28. Gas swirls around it at hundreds of kilometers per second, from which its mass can be deduced to be at least a few times 10^6 $M_\odot$. If it is a dense star cluster, it should be enormously luminous in visible or infrared radiation, yet it is not. Of all the objects known to astronomers, only a

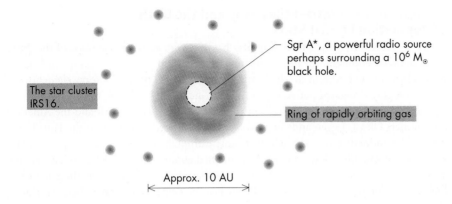

Sgr A*, a powerful radio source perhaps surrounding a 10^6 M$_\odot$ black hole.

The star cluster IRS16.

Ring of rapidly orbiting gas

Approx. 10 AU

FIGURE 15.28
Artist's representation of the very core of our galaxy. The star cluster is IRS16, rich in massive young stars. Within it lies the mysterious radio source Sgr A*, which many astronomers suspect contains a million-solar-mass black hole.

black hole can pack so much mass into so small a volume and not emit much light. Moreover, if this powerful radio source is a several-million-solar-mass black hole, gas falling into it would release the amount of energy needed to heat the gas orbiting it in the ring.

How might such a huge black hole form? We saw in chapter 14 that when a massive star reaches the end of its life, it may explode as a supernova, leaving a black hole as a remnant. Such black holes contain only 10 or so solar masses, but in the central regions of our galaxy, ordinary black holes may grow vastly more massive by drawing in nearby gas with their gravitational attraction. Such gas may be released when stars collide in the crowded environment of the galactic center. Even tidal forces between two passing stars may strip gas from them that may subsequently fall into a black hole. Once a black hole starts growing it will not stop until it runs out of available matter, and a small hole may grow to 10^6 M$_\odot$ in less than 1 billion years.

However, not all astronomers think such a huge black hole exists in the center of the Milky Way. One reason for doubt is the absence of a strong x-ray source in the center of the Milky Way. Recall from chapter 14 that black holes often have gas around them that emits x-rays intensely. Similarly, although there is no dispute about the existence of the strong radio and infrared sources or the small, rapidly rotating ring of dust and gas, what heats and stirs the gas is more controversial. For example, some astronomers think that all the activity we have described above could equally well be produced by a dense, luminous cluster of massive hot stars. A clump of about 100 *O*-type stars would radiate enough energy to heat the gas and dust ring, and such stars might plausibly form as gas falls into the very center of the Milky Way. Sgr A*, however, is difficult to explain in this model unless it is a small-mass black hole, and critics of the star-cluster theory argue if a small black hole is needed along with the cluster to explain what is seen then it is more "economical" to have just a single large hole rather than a small hole and a star cluster. Astronomers are therefore left perplexed: Does our galaxy harbor such a massive black hole at its heart or not? The answer is not known, though we will discover in the next chapter that many other galaxies probably have black holes at their cores.

15.8 HISTORY OF THE MILKY WAY

Formation of our Galaxy

One of the major unsolved problems in astronomy is how galaxies form. Astronomers believe the process is a large-scale version of star formation; that is, a gas cloud collapses under the influence of gravity and breaks up into stars. In galaxy formation, however, we begin with a cloud perhaps a million light years across and containing 100 billion solar masses of gas—one large enough to create billions of stars. Many observations support this hypothesis for the birth of the Milky Way, but perhaps the most important evidence comes from the differing properties of the two stellar populations.

Collapse of the Proto–Milky Way and the Birth of Population I and II Stars

Before we discuss how the Milky Way may have formed, let us review some of the differences between population I and population II stars. Pop I stars are young, and rich in heavy elements, and orbit in the disk. Pop II stars are old, and poor in heavy elements, and orbit on highly elongated paths that are steeply inclined, in general, to the disk. On the basis of these differences, the British astronomer Donald Lynden-Bell and the American astronomers Olin J. Eggen and Allan R. Sandage proposed in 1961 that the Pop II stars, being older, formed early in the galaxy's history during the first stages of collapse of a massive cloud of gas. This cloud must have contained about 10^{11} $M_\odot$ of gas to have enough mass to make the Milky Way, and the gas must have been almost pure hydrogen and helium because no stars yet existed to make the heavier elements. As gravity began shrinking this vast cloud, stars began to condense in the infalling gas. Once formed, they retained the motion of the collapsing gas, motion we see even today in their greatly elongated orbits, as illustrated in figure 15.29A.

This early collapse was very slow, perhaps lasting as much as a few hundred million years. Collapse was in fact so slow that massive stars born in the falling gas consumed all their fuel and exploded as supernovas long before they reached the center of the galaxy. The gas they ejected was enriched by heavy elements made during their evolution and soon mixed with surrounding gas. Stars continued to form as the gas fell inward, but they now contained traces of heavy elements. We see those stars today as population II.

This initial collapse stage probably lasted a few hundred million years, with massive stars forming and dying, adding still more heavy elements to the gas. Star formation was probably not efficient enough to consume all the collapsing gas, and some gas survived to form a rotating disk, as sketched in figure 15.29B. This disk gas, rich in heavy elements from the death of Pop II stars, then began to fragment and form the first population I stars. These stars followed the motion of their parent gas. Pop I stars, therefore, contain more heavy elements than Pop II stars and move along approximately circular orbits lying in the disk, as illustrated in figure 15.29C.

Computer simulations also support a collapse model for the birth of the Milky Way. Figure 15.30 shows that if an unstructured region of gas of the appropriate size and rotation speed collapses it will form a thin disk with spiral arms, surrounded by a halo. Such simulations also hint at the existence of the unseen dark matter mentioned earlier, because computer-model galaxies with dark matter more closely resemble real galaxies than computer models in which the dark matter is not included.

Although the Eggen, Lyden-Bell, and Sandage collapse model explains many of the features of our galaxy, it fails to explain two important properties of its stars. First, according to the model, all population II stars should be about the same age, having formed during the relatively brief period of the galaxy's initial collapse. Observations of Pop II stars, however, shows that they formed over a much longer time span than the model predicts. Second, the model predicts that some stars should have virtually no heavy elements, but no such stars

FIGURE 15.29

(A) Birth of the Milky Way as a gas cloud collapses. A first generation of stars (population II) form. (B) Collapse continues with more population II stars forming and some exploding as supernovas. Heavy elements created by the stars exploding as supernovas are mixed into the gas. Gas not used in making stars settles into a rotating disk. (C) Population I stars form in the disk.

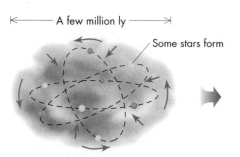

A Slowly rotating protogalactic gas cloud, composed of almost pure hydrogen and helium, begins to collapse under the influence of its gravity.

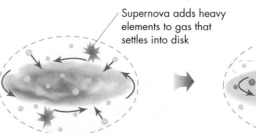

B Some stars form in collapsing gas, some gas collects in disk.

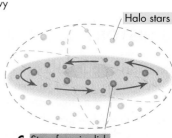

C Stars form in disk

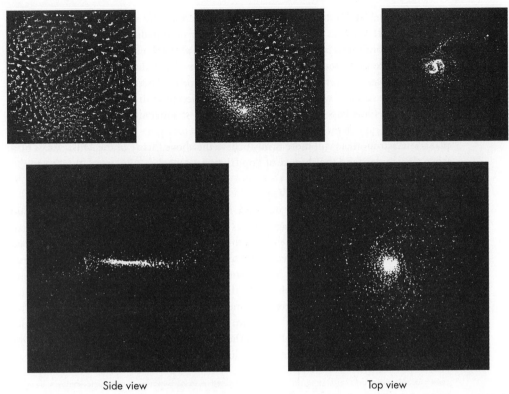

Side view   Top view

FIGURE 15.30
Computer simulation of the birth of a galaxy similar to the Milky Way. The top three frames (A), (B), and (C), show the initial state of the gas and the development of a clump that becomes the galaxy, illustrated in the bottom two frames (D) and (E). Notice the thin disk and spiral arms. (Courtesy Neal Katz, U. of Massachusetts, and James E. Gunn, Princeton.)

have ever been observed. The collapse model also omits a critical feature that many astronomers now believe plays a major role in shaping galaxies: collisions with other galaxies.

We mentioned in Overview VI—and will discuss further in the next chapter—that galaxies collide with one another and sometimes merge. With the Hubble Space Telescope, astronomers have discovered that such collisions and mergers were far more common in the past. How can the Hubble Space Telescope see into the past? Recall that light takes 5 billion years to reach us from a galaxy five billion light years away. Thus we see such distant galaxies as they were in the remote past, when they were young. With this "window" back in time, astronomers see that many young galaxies are smaller than galaxies we see today like the Milky Way. In addition, many more of these young galaxies are colliding with neighbors than galaxies that we see today. Thus, astronomers conclude that rather than forming from a single large collapsing cloud, many galaxies are assembled from smaller ones by collisions and mergers. In fact, the Milky Way itself shows signs of such mergers.

A few years ago, astronomers discovered a dim, dwarf galaxy on the far side of the Milky Way that is nearly completely hidden from our view by intervening dust and gas clouds in our galaxy's disk. This dwarf galaxy is elongated probably because the Milky Way's gravity has stretched it and begun to pull it apart. Moreover, some astronomers believe that two satellite galaxies of the Milky Way—the Large and Small Magellanic Clouds (fig.16.2) are spiraling into our galaxy and will eventually merge with it. Thus, galaxy birth may be a dog-eat-dog event.

Population III

Despite uncertainties, the above picture elegantly explains many of the features of the Milky Way except the puzzling observation mentioned earlier: the collapse theory implies that at least some stars should have formed from pure hydrogen and helium before there was any enrichment of the gas with heavy elements. Why therefore do we see no stars made of only hydrogen and helium? Up to now, all stars found in the Milky Way contain at least a small percentage of heavy elements (typically at least $\frac{1}{1000}$ the amount seen in the Sun).

Astronomers have devised several explanations for why they have so far seen no **population III stars,** as these hypothetical ancient pure hydrogen and helium stars are called. First, conditions in the early galaxy may have allowed only short-lived massive stars to form. For example, the gas may have been too hot or turbulent to make small stars that would have lived long enough for us to see them today. Alternatively, Pop III stars may have had their surface layers contaminated by gas ejected at the death of their more massive brethren. Thus Pop III stars may exist but are masquerading as Pop II stars because they contain heavy elements that were not part of their original makeup. Whatever the explanation, astronomers will more firmly believe the above theory of the Milky Way's birth when they understand the absence of Pop III stars.

The Future of the Milky Way

We have seen that the Milky Way may have formed from the collapse of a giant gas cloud. What of its future? As our galaxy ages, its gas steadily turns into stars, as described in chapter 13. On reaching the end of their lives, these stars return some of that material to space as planetary nebula shells, supernova remnants, and stellar winds. Some of the stars' mass, however, becomes locked up in stellar remnants like white dwarfs, neutron stars, and black holes. Given enough time, therefore, the Milky Way will use up all its gas and be unable to make new stars. At that remote time, billions of years from now, our galaxy will contain only dead stars and surviving dim, low-mass red dwarfs whose lifetime is hundreds of billions of years. Gradually even these dwarfs will burn out, and the Milky Way will fade, slowly spinning in space, a dark disk of stellar cinders.

SUMMARY

Our galaxy, the Milky Way, consists of a disk about 80,000 light years across embedded in a halo about the same size. Together they contain about 10^{11} stars. Near the Sun, stars are a few light years apart. Because the disk rotates, it is made flat, and it contains star groups arranged in a pattern winding out from the center to form spiral arms. Near its center, the galaxy is thicker (several kiloparsecs), forming the galactic bulge, which we can see toward the constellation Sagittarius. The Sun lies about two thirds of the way out in the disk. From Earth we see the Milky Way's disk as a band of dim light stretching across the night sky.

The Milky Way's stars form two main groups—population I and population II—which differ in many ways. Population I stars are found mainly in the disk. They are relatively young, tend to be blue, move in circular orbits, and are rich in heavy elements. Population II stars are found mainly in the halo and bulge. They are old, tend to be red, move in elongated and tilted orbits, and are poor in heavy elements.

About 15% of the galaxy's mass is gas and dust clouds, which orbit with the stars and dim their light. The dust makes it difficult for astronomers to see beyond 10,000 light-years in the disk. The gas, mostly hydrogen and helium, plays a crucial role in the evolution of the Milky Way because it is the source of new stars. Radio emission from the gas, both cold hydrogen emitting 21-centimeter radiation and interstellar molecules, also helps us map the galaxy and see through the dust that obscures our view of its structure.

At the very center of the Milky Way lies the galactic nucleus, a dense collection of stars and gas. Some astronomers think a massive black hole, about 10^6 $M_\odot$, may lie at the center because gas orbiting there moves so rapidly.

Astronomers can measure the diameter of the Milky Way and the Sun's location in it by plotting the position of globular star clusters. They measure its mass by applying the modified form of Kepler's third law to the motion of the Sun and other stars, but such measures give values of the mass that disagree with the value arrived at when stars are simply counted and their masses are added. Astronomers therefore infer that the Milky Way contains dark matter, especially in its outer parts.

The existence of two different stellar populations, I and II, leads astronomers to believe that the Milky Way formed by the collapse of a vast gas cloud about 15 billion years ago. The first stars to form became the population II stars. Some gas did not turn into stars during collapse but instead collected into a rotating disk. From supernova explosions of population II stars, heavy elements became part of the disk of gas. Stars eventually formed from this material to become the population I stars such as our Sun. Eventually the Milky Way will convert all its available gas to stars, and as the surviving stars die, our galaxy will grow dark.

QUESTIONS FOR REVIEW

1. What does the Milky Way look like in the night sky?
2. How do we know our galaxy has a flat disk?
3. Draw a sketch of the Milky Way galaxy and label its major components. Where is the Sun in the Milky Way?
4. Roughly how big in diameter is our galaxy, and how much mass does it contain?
5. How did Shapley deduce the Milky Way's size and the Sun's position in it?
6. What are some differences between Pop I and Pop II stars?
7. How do we know interstellar matter exists?
8. How do we know that some interstellar matter is dust?
9. What is the difference between an emission and a reflection nebula?
10. What are the small reddish blobs along the spiral arms of the galaxy in figure 15.7?
11. How does interstellar dust affect our observations of stars and the Milky Way?
12. What is the zone of avoidance?
13. What evidence makes astronomers believe the Milky Way has spiral arms?
14. What are some ways that we know cold clouds of gas exist in space?
15. How can we determine the Milky Way's diameter?
16. How can we determine the Milky Way's mass?
17. What is meant by dark matter? Why do astronomers conclude the Milky Way may contain such as-yet-unobserved material?
18. What kinds of astronomical objects may lie at the center of the Milky Way?
19. What is the evidence for a black hole at the center of the Milky Way?
20. Describe one model for the origin of the Milky Way. How does this model explain the difference between Pop I and Pop II stars?
21. What is meant by Pop III stars?

THOUGHT QUESTIONS

1. What would the Milky Way look like in the night sky to an observer on a planet at the very edge of the galaxy? What would it look like to an observer at the galactic nucleus?
2. How could we measure the mass of the Milky Way using a distant orbiting globular cluster?
3. Suppose the Sun were a population II star in the halo of the Milky Way. In what ways would that make studying the Milky Way easier?
4. Cut four circles of cardboard whose diameters are 2, 4, 6, and 8 inches. Fasten them together through their centers with a brass paper fastener. Draw a small circle near the edge of the 4-inch disk so that it cuts the other disks. Now turn the two innermost circles. Does this help you see how self-propagating star formation might make spiral arms?

PROBLEMS

1. Given that the Sun moves in a circular orbit of radius 8.5 kiloparsecs around the center of the Milky Way and that its orbital speed is 220 kilometers per second, work out how long it takes the Sun to complete one orbit of the galaxy.
   NOTE: 1 kiloparsec is about 3×10^{16} kilometers, and 1 year is about 3×10^7 seconds.
2. Gas is observed to be orbiting in a circle at 200 km/s about a dark central mass. If the gas's orbit has a radius of 1 parsec, how much mass must be present to hold it in orbit?
3. Suppose the mass in problem 2 were packed into a sphere whose radius is 20 AU. Would it be a black hole?

TEST YOURSELF

1. One way astronomers deduce that the Milky Way has a disk is that they
    (a) see stars arranged in a circular region around the north celestial pole.
    (b) see far more stars along the band of the Milky Way than in other directions.
    (c) see a large dark circle silhouetted against the Milky Way in the Southern Hemisphere.
    (d) see the same number of stars in all directions in the sky.
    (e) None of the above.

2. Astronomers think the Milky Way has spiral arms because
    (a) they can see them unwinding along the celestial equator.
    (b) radio maps show that gas clouds are distributed in the disk with a spiral pattern.
    (c) young star clusters, HII regions, and associations outline spiral arms.
    (d) globular clusters outline spiral arms.
    (e) both (b) and (c) are correct.

3. A young blue star moving along a circular orbit in the disk is a (a) Pop I or (b) Pop II star.

4. The modified form of Kepler's third law allows astronomers to determine the Milky Way's
    (a) mass.
    (b) age.
    (c) composition.
    (d) shape.
    (e) number of spiral arms.

5. Astronomers know that interstellar matter exists because
    (a) they can see it in dark clouds and clouds that absorb light.
    (b) the matter creates narrow absorption lines in the spectra of some stars.
    (c) they can detect radio waves coming from atoms and molecules in the cold gas.
    (d) spacecraft have sampled clouds near Orion.
    (e) All the above except (d).

FURTHER EXPLORATIONS

Benningfield, Damond. "Galaxies Colliding in the Night." *Astronomy* 24 (November 1996): 36.

Berendzen, Richard, Richard Hart, and Daniel Seeley. *Man Discovers the Galaxies.* New York: Science History Publications, 1976.

Binney, James. "The Evolution of Our Galaxy." *Sky and Telescope* 89 (March 1995): 20.

Blitz, Leo. "Giant Molecular Cloud Complexes in the Galaxy." *Scientific American* 246 (April 1982): 84.

Bok, Bart. "The Milky Way Galaxy." *Scientific American* 244 (March 1981): 92.

Croswell, Ken. "The Dark Side of the Galaxy." *Astronomy* 24 (October 1996): 40.

_____. "The Milky Way." *New Scientist* 150 (25 May 1996): S1.

_____. "What Lies at the Milky Way's Center?" *Astronomy* 23 (May 1995): 32.

Elitzer, Moshe. "Masers in the Sky." *Scientific American* 272 (February 1995): 68.

Geballe, Thomas R. "The Central Parsec of the Galaxy." *Scientific American* 241 (July 1979): 60.

Greenberg, J. Mayo. "The Structure and Evolution of Interstellar Grains." *Scientific American* 250 (June 1984): 124.

Gribbin, John. "The Galaxy's Dark Secrets." *New Scientist* 142 (9 April 1994): 26.

Hartquist, T. W., and D. A. Williams. *The Chemically Controlled Cosmos: Astronomical Molecules from the Big Bang to Exploding Stars.* Cambridge, New York: Cambridge University Press, 1995.

Henbest, Nigel, and Heather Couper. *The Guide to the Galaxy.* Cambridge, New York: Cambridge University Press, 1994.

Hoskin, Michael. "William Herschel and the Making of Modern Astronomy." *Scientific American* 254 (February 1986): 106.

Jayawardhana, Ray. "Destination: Galactic Center." *Sky and Telescope* 89 (June 1995): 26.

Knapp, Gillian. "The Stuff Between the Stars." *Sky and Telescope* 89 (May 1995): 20.

Krauss, Lawrence M. "Dark Matter in the Universe." *Scientific American* 253 (December 1986): 58.

Mathewson, Don. "The Clouds of Magellan." *Scientific American* 252 (April 1985): 106.

Morris, Mark. "What's Happening at the Center of Our Galaxy?" *Physics World* 7 (October 1994): 37.

Oort, Jan. "Exploring the Nuclei of Galaxies (Including our Own)." *Mercury* 21 (March/April 1992): 57.

Scoville, Nick, and Judith S. Young. "Molecular Clouds, Star Formation, and Galactic Structure." *Scientific American* 250 (April 1984): 42.

Townes, Charles H., and Reinhard Genzel. "What Is Happening at the Center of Our Galaxy?" *Scientific American* 262 (April 1990): 46.

Trimble, Virginia, and Samantha Parker. "Meet the Milky Way." *Sky and Telescope* 89 (January 1995): 26.

Van den Bergh, Sidney, and James Hesser. "How the Milky Way Formed." *Scientific American* 268 (January 1993): 72.

KEY TERMS

associations, 449

bulge, 444

dark nebula, 452

density-wave model, 457

disk, 444

emission nebulas, 454

globular clusters, 449

halo, 444

HII regions, 455

interstellar matter, 451

mass function, 447

Milky Way, 440

nucleus, 445

open clusters, 449

population I, 447

population II, 447

population III stars, 468

reddening, 452

reflection nebulas, 454

rotation curve, 462

scattering, 452

self-propagating star formation, 459

spiral arms, 444

star clusters, 449

21-centimeter radiation, 457

zone of avoidance, 452

 PROJECTS

Find a spot where the sky is dark and you can see most of the sky. Sketch the path of the Milky Way across the sky, as best you can see it. Do you notice any particularly bright stars along the Milky Way or else- where? Can you see any dark regions along the Milky Way where the number of stars looks small? Mark them on your sketch. Then compare your sketch to one of the star maps inside the covers of this book.

CHAPTER SIXTEEN

GALAXIES

A spiral Galaxy (Courtesy Anglo-Australian
Observatory/Photographs by David Malin.)

A cluster of galaxies (Courtesy Anglo-Australian Observatory/Photographs
by David Malin.)

Beyond the edge of the Milky Way and filling the depths of space are millions of other star systems similar to our own. These remote, immense star clouds are called "external galaxies," and their properties are the subject of this chapter. Many external galaxies are flattened systems with spiral arms, similar to the Milky Way in size and shape. Other external galaxies differ dramatically from our own. For example, many galaxies are not disk-shaped at all. Rather, they are egg-shaped, with their stars distributed in a vast, smooth cloud surrounding a dense central core of stars. Others are neither disks nor smooth elliptical shapes but are completely irregular in appearance.

Galaxies differ in more than shape; they also differ in their contents. Some contain mostly old stars, but others contain predominantly young stars. Some are rich in interstellar matter, but others have hardly any at all. Some galaxies, despite having as much total mass as the Milky Way, have formed far fewer stars, leaving them dim and scarcely visible to us. Others have a small but tremendously powerful energy source at their core that emits as much energy as the entire Milky Way but from a region about 0.001% its size, less than a few light years in diameter. Many of these so-called active galactic nuclei eject gas in narrow jets at nearly the speed of light.

Galaxies are not spread uniformly across the sky. Just as stars often lie in clusters, so too galaxies often lie in galaxy clusters and galaxy clusters themselves lie in clusters of galaxy clusters. These immense groups are not stationary. They are separating from one another, caught up in the expansion of the Universe, like snowflakes carried on the wind.

Astronomers can see all this diversity in galaxies, but as yet they understand only poorly how it arises. Unlike stars, whose structure and evolution are well understood, many aspects of galaxies remain a mystery. Astronomers do not even agree on why some galaxies are disks, whereas others are egg-shaped. Although this lack of knowledge is frustrating to astronomers, it does offer us a chance to see how scientists work, in particular, how astronomers begin with basic observations and go on to construct hypotheses to explain what they observe. For example, many galaxies appear to be undergoing collisions with their neighbors. Could past interactions of this kind account for the different galaxy shapes?

16.1 DISCOVERING GALAXIES

A **galaxy** is an immense cloud of hundreds of millions to hundreds of billions of stars. Each star moves along its own orbit, held within the galaxy by the combined gravitational force of all the other stars. Each galaxy (apart from those few undergoing collisions) is therefore an independent and isolated star system.

Early Observations of Galaxies

All galaxies (except the Milky Way) are extremely distant from Earth, with the nearest being more than 150,000 light-years away. Their great distance makes galaxies look dim to us, and only a few can be seen with the naked eye. For example, on clear, dark nights in autumn and winter in the Northern Hemisphere you can see M31, the Great Galaxy in

Andromeda (fig. 16.1) with the naked eye. It looks like a pale elliptical smudge on the sky. Early observers such as the tenth-century Persian astronomer Al-Sufi noted this object, and it appeared on early star charts as the "Little Cloud." If you live in the Southern Hemisphere, you can easily see with the naked eye the Large and Small Magellanic Clouds, two satellite galaxies of the Milky Way (fig. 16.2). Today astronomers know of millions of galaxies besides the Magellanic Clouds and M31, and they estimate that the visible Universe contains tens of billions more.

The study of other galaxies began in the eighteenth century when the French astronomer Charles Messier (pronounced /mess-yáy/) accidentally discovered many galaxies during his searches for new comets. He noticed a large number of faint, diffuse patches of light, and to avoid confusing them with comets he assigned them numbers and made a catalog of their positions. Although many of Messier's objects have since been identified as star clusters or glowing gas clouds in the Milky Way, several dozen are galaxies. These and the other objects in Messier's catalog are still known by their Messier, or M number, such as M31, mentioned above.

In the nineteenth century, the English astronomer Sir William Herschel, discoverer of the planet Uranus, discovered and cataloged numerous galaxies as he mapped the sky. His list, now known as the *New General Catalog* (NGC, for short) gives many galaxies their names, such as NGC 1275, a peculiar, active galaxy. Herschel's catalog and later catalogs compiled by other astronomers contain the same objects Messier found, and so some galaxies bear two or even more names. For example, M82 is the same galaxy as NGC 3034.

Types of Galaxies

Even early observers noticed that not all galaxies look the same. Lord Rosse discovered that some had a spiral pattern, and others showed no structure at all. This distinction became much clearer in later studies, especially those of Edwin Hubble, an American astronomer working first as a graduate student at the University of Chicago and later at Mt. Wilson in California in the 1920s. Hubble demonstrated that galaxies could conveniently be divided on the basis of their shape into three main types: spirals, ellipticals, and irregulars. The first type, **spiral galaxies,** have two or more arms winding out from the center.

FIGURE 16.1
Photograph of M31, the nearest spiral galaxy to us. (Courtesy George Greany.)

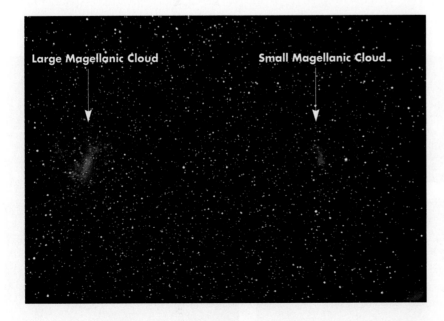

FIGURE 16.2
Photograph of the Large and Small Magellanic Clouds. (Courtesy NOAO.)

We saw in the last chapter that the Milky Way is probably of this type, and figure 16.3 shows photographs of several others.

The second type shows no signs of spiral structure. These galaxies have a smooth and featureless appearance and a generally elliptical shape, as can be seen in figure 16.4. Accordingly, astronomers call them **elliptical galaxies.**

Galaxies of the third major type show neither arms nor a smooth uniform appearance. In fact, they generally have stars and gas clouds scattered in random patches. For this reason they are called **irregular galaxies** (fig. 16.5).

Spiral, elliptical, and irregular galaxies are sometimes denoted by S, E, and Irr respectively. However, in addition to these three main types, astronomers recognize two additional galaxy types closely related to spiral systems. The first of these is **barred spirals,**

FIGURE 16.3
Photographs of typical spiral galaxies:
(A) NGC 5194; (B) NGC 5236; (C) NGC 2997;
(D) NGC 628. (A, C, and D, Courtesy Cal
Tech; B, Courtesy David Malin, Anglo-
Australian Observatory.)

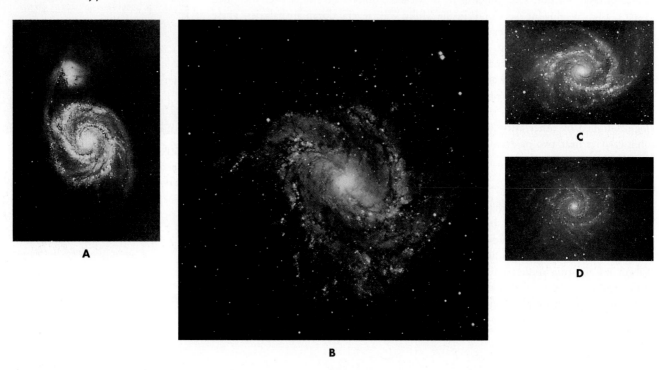

FIGURE 16.4
Photographs of a few elliptical galaxies:
NGC 205, M49, and a color picture of M87.
(Courtesy Cal Tech, NOAO, and Anglo-
Australian Observatory/ David Malin,
respectively.)

which have arms that emerge from the ends of an elongated central region, or bar, rather than from the core of the galaxy (fig. 16.6). It is this bar that gives them their name, and they are denoted SB galaxies to distinguish them from normal spirals. Although Hubble treated SB galaxies as a separate class, astronomers today are less certain that they are a distinct type of galaxy. For example, computer models of galaxies show that most disk-shaped systems will form a bar if they undergo a gravitational disturbance, perhaps as the result of a close encounter with a neighboring galaxy. Even our own Milky Way probably has a weak bar. The bar forms because a gravitational disturbance alters the orbits of the stars, causing them to bunch up in an elliptical pattern that we see as the bar. Some models also suggest that over hundreds of millions of years the orbits spread out again, and the bar dissolves. Thus bars may appear and disappear, and so an ordinary spiral galaxy may change into a barred spiral and then back again.

FIGURE 16.5
Photographs of two irregular galaxies (NGC 6822 and the Small Magellanic Cloud). (Courtesy Anglo-Australian Observatory, photographs by David Malin.)

FIGURE 16.6
Barred spirals: NGC 1530 and NGC 1365. (Courtesy NOAO and Anglo-Australian Observatory/ David Malin, respectively.)

FIGURE 16.7
An S0 galaxy, NGC 1201. (Courtesy J. Young and L. Allen, University of Massachusetts, and J. Kenney, Yale.)

Hubble also identified another kind of galaxy similar to spirals. These are disk systems with no evidence of arms (fig. 16.7), and they are called S0 galaxies. Hubble thought they were intermediate in type between S and E types systems. Today, however, astronomers regard S0 galaxies as ordinary spiral galaxies that have lost their gas. This might happen if a spiral galaxy were to move at high speed through hot, low-density intergalactic gas around it. The impact of such material might strip gas from the ordinary spiral much the way a leaf-blower cleans debris from a sidewalk.

Hubble took his classification system one step further, seeking a pattern that might lead to an understanding of why galaxies exhibit such diversity in their appearance. He noticed that both spiral and elliptical galaxies can be subdivided into additional classes (fig. 16.8) that when properly ordered show a smooth transition from spherical E type of galaxies to flatter E galaxies, to spirals, and to irregulars. Hubble went on to hypothesize that, as a galaxy aged, its type might change from E to S to Irr. This hypothesis seems very plausible if you look at the "tuning fork" diagram that Hubble proposed (fig. 16.9), but

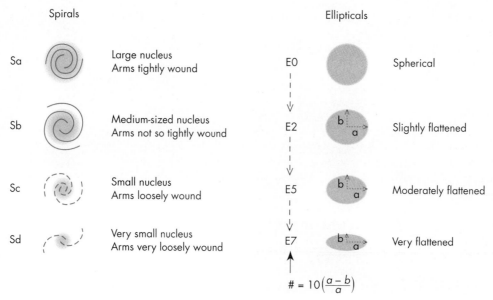

FIGURE 16.8
Sketch of the galaxy subclasses of spiral and elliptical galaxies.

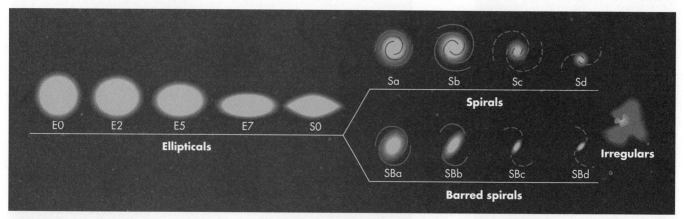

FIGURE 16.9
The Hubble tuning fork diagram.

astronomers now think it is incorrect. Nevertheless, the diagram is still a useful way to organize the galaxy types and subtypes that Hubble described.*

Differences in the Stellar and Gas Content of Galaxies

Since Hubble's time, astronomers have discovered that spiral, elliptical, and irregular galaxies differ, not only in their shape, but also in the type of stars they contain. For example, spiral galaxies contain a mix of young and old stars (Pop I and II), but elliptical galaxies contain mostly old (Pop II) stars. This difference in the kind of stars composing S and E galaxies is understandable in terms of their different gas content. To make young (Pop I) stars, a galaxy must contain dense clouds of gas and dust. Spiral systems have about 15% of their mass in the form of such interstellar clouds, and so they can easily make the young stars of their spiral arms. On the other hand, ellipticals contain much less interstellar matter. In fact, astronomers used to think that the E type of galaxies contained virtually no interstellar gas or dust. However, more recent observations made with x-ray telescopes show that E galaxies often contain very low-density but very hot (10^7 K) gas. Such hot gas does not readily form stars, and so it is no surprise that elliptical galaxies rarely contain young, blue stars like the ones we see in the disks of spiral systems.

Although young, blue stars are rare in elliptical systems, such stars are common in many irregular galaxies. These Irr systems also often contain large amounts of interstellar matter, amounting sometimes to about 50% of their mass. Accordingly astronomers believe that these galaxies are very young and have not yet used up much of their gas in making stars.

Apart from their different star and interstellar matter content, S, E, and Irr galaxies have few other differences to generalize about. For example, if we look at the mass and radius of galaxies, we do not find that all ellipticals are huge and all spirals are small. Instead we discover that elliptical galaxies range enormously in size. Some are not much larger than globular clusters and contain only 10^7 solar masses of material; others are monster galaxies, 10 to 100 times the mass of the Milky Way. This spread is easy to see in figure 16.10, which shows a group of galaxies in which most of the members are elliptical systems.

FIGURE 16.10
Photograph of the central region of the Virgo galaxy cluster, illustrating the range in sizes of elliptical galaxies. (Courtesy NOAO.)

☀ What types of galaxies can you identify in this photograph?

---

*Although Hubble's classification system covers well the most conspicuous galaxy types, it omits two forms that astronomers today think are important: low-surface-brightness galaxies (dim, often large-diameter systems with far less star formation than ordinary galaxies) and blue dwarf systems (perhaps the building blocks of ordinary galaxies).

What fraction of galaxies make up each type? Even that question has no clear answer. If we make a census of the galaxies near the Milky Way, we find that most galaxies are dim dwarf E and Irr systems, sparsely populated with stars. These dim galaxies are totally invisible at greater distances, where we can see only the more luminous galaxies. If we make a census of those luminous objects, we find that the majority of galaxies are spirals, but the percentage depends on whether we look at galaxies in clusters or in more sparsely populated regions. In the former, about 60% of the members are spirals, whereas in the latter that proportion is about 80%.

The diversity of galaxy form and content requires an explanation, and a major part of the astronomical research going on today is devoted to answering the question of why galaxies come in such distinct and different forms. With the Hubble Space Telescope, astronomers have at last been able to observe young galaxies well enough to begin developing a theory of how galaxies form and evolve. As we described in chapter 15 and overview 6, astronomers think that galaxies form by the collapse of gas clouds, much like the process that formed our Sun and other stars, but on a vastly larger scale. As the cloud collapses, stars form within it, evolve, and—if massive enough—explode as supernovas. The supernova debris adds heavy elements to the gas, which are then incorporated into later generations of stars, ultimately to become planets and living beings.

Although astronomers have not witnessed, even with Hubble, the collapse of primordial gas clouds into galaxies, that instrument's ability to see fine details in distant (and thereby ancient galaxies) has allowed astronomers to observe galaxies just past the collapse stage. Surprisingly, these presumably very young galaxies are much smaller than systems like the Milky Way. Moreover, they are far more numerous than ordinary galaxies today. Thus, these young galaxies must somehow disappear. How can a galaxy disappear? Perhaps they grow dim as they use up their gas and can no longer make new stars. Alternatively, they may collide and merge into bigger systems. Deducing from what is currently known, astronomers tend to think that merging is the more likely explanation: it would account both for the formation of large systems like the Milky Way and for the smaller number of galaxies we see today. But a major question still remains: what causes one collapsing gas cloud to become a spiral galaxy, while another becomes an elliptical galaxy?

The Cause of Galaxy Types

Astronomers are still uncertain about what causes the difference between S, E, and Irr galaxies. At one time they considered that the different types might arise from differences in their rotational speeds at the time of their birth. Rotation may play a role in determining a galaxy's type because a rapidly spinning object generally becomes flattened. We saw that this effect creates the equatorial bulge of the Earth and the flat shape of the Solar System. Given this relation between flattening and rotation, it is natural to ask if it occurs in galaxies. In particular, are elliptical galaxies roundish because they rotate slowly, but spirals flat because they rotate fast? When astronomers measure the rotation speed of E-type galaxies, they find that they do rotate but with a slow tumbling motion. On the other hand, the disks of spirals rotate relatively rapidly compared to ellipticals of the same size. Thus the speed of rotation seems to play some role in determining whether a galaxy becomes a spiral or an elliptical. However, a comparison of the rotation speed of elliptical galaxies of different flattening shows little or no relation between rotation velocity and flatness. Thus the rotation speed probably does play a role, but other factors probably do as well. For example, velocity differences of another sort may offer an explanation of the different types of galaxies.

Some computer simulations of galaxy formation show that random motion (that is, motion in random directions) in a collapsing gas cloud can dramatically alter the final shape of the galaxy it forms. According to this model, galaxies that formed from gas clouds with large random motions (fig. 16.11) became ellipticals, whereas those with small random motions became spirals.

Motion, whether rotational or random, may be only one factor in establishing a galaxy's shape. The kind of stars (Pop I or Pop II) they contain may furnish an equally important clue. Recall that E systems contain almost entirely Pop II (old) stars, whereas spirals have many Pop I (young) stars. Perhaps E systems made most of their stars in one brief episode shortly after their birth, using up almost all their gas, but spiral galaxies made their stars over a longer period of time. This difference might lead us to conclude that the rate of star formation plays a role in determining the kind of galaxy. But does the rate of star formation determine the kind of galaxy or does the kind of galaxy determine the rate of star formation? Again, astronomers do not yet know.

In the above discussion of the factors that might affect the type of galaxy formed when a gas cloud collapses, we have said nothing about how spiral arms form. In chapter 15 we discussed some hypotheses about how the Milky Way may have formed its spiral arms. For example, we saw that spiral arms might form as a density wave travels around a galaxy's disk, compressing gas clouds and triggering star formation in a spiral pattern. Astronomers assume that this mechanism probably operates in other galaxies and forms their arms, too. We also discussed, in chapter 15, a different model in which spiral arms form as the galaxy's rotation stretches regions of star formation into elongated arcs. This mechanism also may occur in some other galaxies. Astronomers are still trying to understand which of these models is the more likely explanation, or if either is correct.

Galaxy Collisions and Mergers

The above hypotheses about the origin of the types of galaxies assume that a galaxy's type does not change during the galaxy's lifetime. Over the last few decades, however, astronomers have found evidence that this is not true. Instead, a galaxy can alter its type as a result of collisions and mergers. As galaxies move through space, they occasionally collide with one another. In fact, conditions in a typical cluster of galaxies are such that *most* galaxies will experience several collisions during the course of their existence. When such collisions occur, the stars are essentially unaffected. Because of the vast distances between stars within a galaxy, most of the individual stars pass harmlessly past each other. Perhaps as few as one or two stars will collide directly. Thus galaxy collisions are much like tossing two handfuls of sand together: most of the grains will simply pass by one another without hitting. However, gas and dust clouds in the galaxies will collide, and their impact may

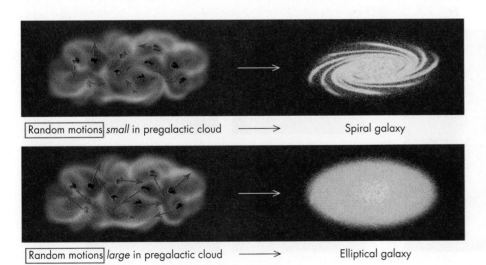

Random motions *small* in pregalactic cloud ⟶ Spiral galaxy

Random motions *large* in pregalactic cloud ⟶ Elliptical galaxy

FIGURE 16.11
Sketch illustrating how random motions may affect the type of galaxy formed when a gas cloud collapses. Large random motions lead to E type of galaxies, whereas small random motions lead to S type of galaxies.

compress the clouds enough to trigger a burst of star formation within them. Such "star-burst galaxies," as these collisionally stimulated galaxies are called, are among the most luminous galaxies known.

During collisions, stars may be shifted into very different orbits. In fact, the gravitational force of one galaxy on the other may fling stars outward in huge luminous arcs. Figure 16.12 shows a computer simulation of such a galaxy collision and a photograph of two real galaxies with just such plumes of stars.

Galaxy collisions can create other bizarre forms, as figure 16.13 illustrates. For example, if a small galaxy collides with a larger spiral system, the small one may "punch a hole"

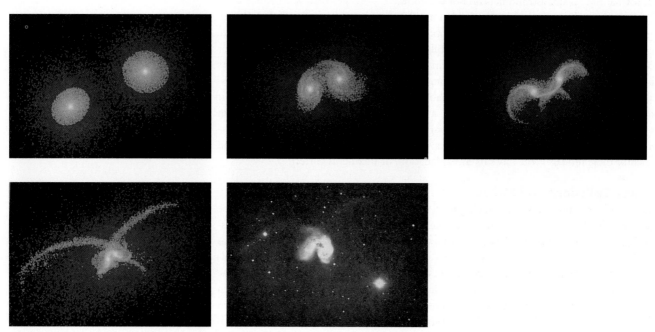

FIGURE 16.12
Computer simulation of two galaxies colliding and photograph of system undergoing just such a collision.
(Courtesy Joshua Barnes, University of Hawaii.)

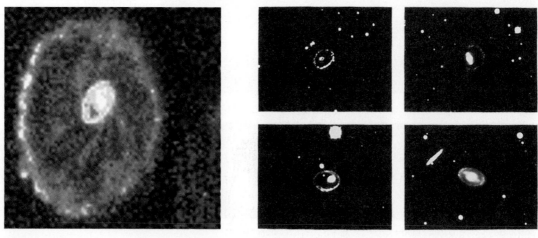

FIGURE 16.13
Several ring galaxies. (Courtesy NOAO.)

through the larger one. Astronomers call the result a ring galaxy. The smaller galaxy does not destroy the stars in the other galaxy's disk. Rather, it makes a hole by shifting the orbits of stars so they no longer pass through the galaxy's central regions.

Because a typical galaxy collision lasts about 100 million years (very brief compared to a galaxy's lifetime) most galaxy collisions lead to nothing more than a temporary disturbance for the colliding systems. However, a collision may sometimes have a more profound effect on the galaxies involved: it may lead to their merger. Figure 16.14 shows a photograph of two galaxies merging. Notice that the merger remnant (as the combined pair is called) looks very much like an elliptical galaxy, though the traces of the individual galaxies still visible indicate that they were originally spirals. Such systems are evidence that some elliptical galaxies may have formed as the result of the collision and merger of two or more spiral galaxies long ago. In fact, some astronomers have proposed that almost all elliptical galaxies formed from such mergers. However, only about 10% of elliptical galaxies show definite evidence now of past mergers. That evidence appears as faint shell-like rings or dense clumps of stars that computer simulations indicate are the debris from the "swallowed" galaxies.

Mergers may in fact affect just about every galaxy sometime during its life. If a large galaxy collides with a small galaxy, the small one is captured and absorbed by the larger one in a process called **galactic cannibalism.** Repeated cannibalism by a galaxy may turn it into a giant, which perhaps explains why some galaxy groups have an abnormally large elliptical galaxy at their center. Cannibalism may also account for much of the activity observed in some galaxies (section 16.4). Even our own galaxy may be devouring smaller neighbors. For example, many astronomers think the Milky Way is in the process of swallowing the Magellanic Clouds. Our neighboring galaxy, M31, also appears to have cannibalized a small galaxy. With the Hubble Space Telescope astronomers have detected a small bright clump of stars near the center of M31 that may be the remains of a galaxy it has absorbed.

The idea described above that collisions and mergers play an important role in shaping galaxies has been strikingly confirmed by pictures of more distant systems made with the space telescope. Such pictures show first of all that distant clusters contain a higher proportion of spirals than nearby clusters do, implying that as clusters age, some spiral galaxies change their form. Second, distant clusters contain more galaxies within a given volume, implying that collisions were more likely in the past than now. Finally, the galaxies seen in the most remote clusters show more signs of disturbance by neighboring galaxies: odd shapes, bent arms, and twisted disks, what some astronomers call "harassment."

FIGURE 16.14
A merger in which two spirals are colliding and forming an elliptical galaxy. (Courtesy F. Schweizer, Carnegie Institution of Washington.)

In all the above discussion of galaxy types and their possible evolution, we have ignored the methods by which the size and mass of a galaxy are found. An even more important point that we have overlooked is how to find the distance to a galaxy. Without knowing how far away a galaxy is, we can learn little of its properties. Thus, before we look further at galaxies, we need to discuss how their properties are measured.

16.2 MEASURING PROPERTIES OF GALAXIES

Much of our ignorance about galaxies can be attributed to their great distance from us, which makes studying them difficult. Exactly how far away from us galaxies lie was for centuries a major controversy among astronomers. For example, in 1755, the German philosopher Immanuel Kant, whose views on the origin of the Solar System were so ahead of their time, suggested that galaxies were remote star systems—Island Universes—similar to the Milky Way. On the other hand, as we saw in the last chapter, the American astronomer Harlow Shapley argued as recently as the 1920s that galaxies were merely small satellites of the Milky Way. The reason for such disagreement comes from the difficulty in *measuring* how far away from us galaxies are.

Galaxy Distances

Galaxies are separated from each other by enormous distances. Even the nearest galaxy to us, the Large Magellanic Cloud, is more than 150,000 light-years from the Sun. Astronomers cannot use parallax to measure such vast distances because the angle by which the galaxy's position changes as we move around the Sun is too tiny to be detected. Astronomers must therefore use other ways to find distances, such as the method of standard candles.

We discussed the method of standard candles in detail in chapter 12. We saw there that if we know the luminosity of a light source, we can use the inverse-square law to deduce its distance from how bright it looks to us. This was the technique we used in chapter 15 to find the size of the Milky Way, and we can apply it here to find the distance to galaxies.

Astronomers use many different astronomical objects as standard candles, but one of the most reliable is the class of luminous, pulsating variable stars known as the "Cepheids" (described in chapter 13). Cepheids have a luminosity about a million times the luminosity of the Sun and, because of their brilliance, they are easy to see in nearby galaxies. Furthermore, because their brightness changes as they pulsate, they are easy to distinguish. Far more important, however, is that the time it takes these variable stars to go from bright to dim to bright again (their period) is related to their luminosity. (Recall our discussion of the period-luminosity, relation in chapter 13). This then allows astronomers to determine a Cepheid's luminosity by simply measuring how long it takes the star to change its luminosity over one cycle. Once astronomers know the variable star's luminosity, they can find its distance from a simple measurement of how bright it appears as seen from Earth and an application of the inverse-square law.

Cepheids are luminous enough that astronomers can use them to measure the distance to nearby galaxies, but, despite their luminosity, Cepheids are invisible in very distant galaxies.* Therefore astronomers use other standard candles such as supergiant stars, planetary nebulas, or supernovas when determining the distance to such remote systems. However, the method's accuracy depends on accurate knowledge of the luminosity of the standard candle employed—for example, supernovas—and such luminosities are still only approximately known. Astronomers therefore have sought other ways to measure the distance to galaxies, as we will discuss below.

In April 1994, English astronomers announced the discovery of a faint, dwarf galaxy lying toward the constellation Sagittarius but on the far side of the Milky Way from us. They estimated that the Sagittarius Dwarf Galaxy (as this tiny star system is called) lies about 50,000 ly from the center of our own galaxy, placing it virtually on the rim of the Milky Way. If these observations are correct, the Sagittarius Dwarf would be our nearest galactic neighbor. It is so close, in fact, that the Milky Way's gravity creates a strong tidal force on the dwarf, stretching it into an elongated shape. That force may ultimately pull the Sagittarius Dwarf apart and draw its stars into the Milky Way.

How could astronomers overlook a galaxy so nearby? The answer is that it lies so close to the plane of the Milky Way that the dwarf's dim light is overwhelmed by the light of foreground stars in our own galaxy. Given the tiny size of this system, many astronomers continue to think of the large Magellanic Cloud as our nearest galactic neighbor. It surely remains the nearest ordinary galaxy to us.

---

*Astronomers can see Cepheids in more remote galaxies with the Hubble Space Telescope than they can using most ground-based instruments, thereby allowing them now to measure the distance to far away systems more accurately.

EXTENDING OUR REACH

MEASURING THE DISTANCE OF A GALAXY USING CEPHEID VARIABLES

Suppose we detect a Cepheid in a galaxy and find that the Cepheid appears 100 million (10^8) times dimmer than a Cepheid of the same pulsation period in the Milky Way. Because the stars have the same period, we assume that they have the same luminosity. Let us denote the brightness and distance of the Cepheid in the Milky Way as B_* and d_*, respectively. Similarly, we will denote the brightness and distance of the Cepheid in the distant galaxy by B_g and d_g. According to the inverse-square law, if two objects have the same luminosity, their apparent brightnesses B_* and B_g, are related to their respective distances, d_* and d_g, by

$$\frac{B_*}{B_g} = \left(\frac{d_g}{d_*}\right)^2$$

$B_* =$ Brightness of Cepheid in Milky Way

$B_g =$ Brightness of Cepheid in distant galaxy

$d_* =$ Distance to Cepheid in Milky Way

$d_g =$ Distance to Cepheid in distant galaxy

To find their distances, we take the square root of both sides, getting

$$\sqrt{\frac{B_*}{B_g}} = \frac{d_g}{d_*}$$

In other words, the square root of the ratio of their brightness equals the ratio of their distances. Because the distant Cepheid in our problem looks 10^8 times dimmer from Earth than the Cepheid in the Milky Way,

$B_*/B_g = 10^8$. Thus, according to the inverse-square law, the ratio of the distance of the Cepheid in the galaxy to the distance of the Cepheid in the Milky Way must be

$$\frac{d_g}{d_*} = \sqrt{\frac{B_*}{B_g}} = \sqrt{10^8}$$
$$= 10^4$$

That is, the Cepheid in the remote galaxy is 10^4 times farther away than the one in our galaxy.

Suppose we know the Cepheid in our own galaxy is 1000 light years away. Then the Cepheid in the other galaxy must be 10^4 times 1000 light-years away. The distant Cepheid, and hence the galaxy in which it is located, is therefore 10^7 light-years distant (fig. 16.15).

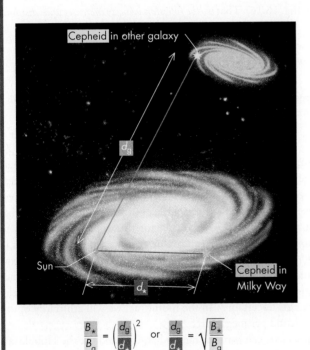

$$\frac{B_*}{B_g} = \left(\frac{d_g}{d_*}\right)^2 \quad \text{or} \quad \frac{d_g}{d_*} = \sqrt{\frac{B_*}{B_g}}$$

FIGURE 16.15
Finding the distance to a galaxy using Cepheid variables as standard candles. B_* and B_g are the apparent brightness of the Cepheid in the Milky Way and in the Cepheid distant galaxy, respectively. Their distances are d_* and d_g, respectively. From a measurement of the brightness ratio of the Cepheids, astronomers can calculate the ratio of their distances and, knowing the distance of the Cepheid in the Milky Way, they can therefore find the distance to the other galaxy.

The Redshift and the Hubble Law

In the 1920s astronomers discovered that galaxies are moving away from one another. This discovery came from the spectra of distant galaxies. The spectrum of a galaxy is the spectrum of all its component stars added together. If a galaxy is moving either toward or away from us, its spectral lines will be Doppler-shifted (recall our discussion of the Doppler shift in chapters 3 and 12). Motion away from us lengthens the wavelength of the lines and produces a so-called redshift. The Doppler-shift formula can then be used to find the galaxy's velocity from the shift of the spectral lines.

The redshift of galaxies is not a true Doppler shift. Rather, it is caused by the expansion of the Universe carrying galaxies away from us, as we will see in the next chapter.

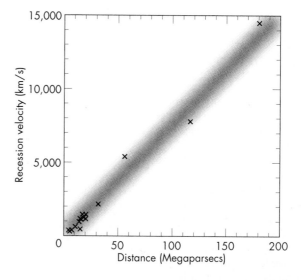

When astronomers first made such observations around 1911, they were astonished to discover that nearly all galaxies were moving away from the Milky Way. That is, nearly all galaxies show a redshift in their spectral lines. The only exceptions are a few nearby galaxies, whose motion is influenced by the gravitational attraction of the Milky Way and its neighbors. More astonishing, the velocity of the galaxy, its **recession velocity,** is larger for galaxies that are farther away (fig. 16.16). That is, *the recession velocity increases with distance.*

Vesto Slipher discovered that galaxies are moving away from each other in 1911, but the details of how the velocity changed with distance were worked out by Hubble.

In 1920, Edwin Hubble discovered that a simple formula related the recession velocity, V, and the distance, D. He found that

$$V = HD$$

where H is a constant. Because of his discovery, this relation is called the **Hubble law,** and H is called the **Hubble constant.**

The value of the Hubble constant depends on the units used to measure V and D. Astronomers generally measure V in kilometers per second and D in **megaparsecs,** where 1 megaparsec (abbreviated Mpc) is a million parsecs (= 3.26 million light-years). In these units, H is somewhere between 50 and 100 kilometers per second per megaparsec. Here we will use H = 75 kilometers per second per megaparsec, but you should be aware that astronomers disagree about the value of H. The reason for this controversy is that, to determine H, one must know *both* the distance and velocity of at least a few galaxies in order to calibrate the law. H is then simply the average value of each calibration galaxy's velocity divided by its distance. However, as we have seen above, measuring the distance to even a nearby galaxy accurately is difficult. Observations with the Hubble Space Telescope may eventually resolve this uncertainty. In the meantime, we will arbitrarily opt to use H = 75 kilometers per second per megaparsec.

Once a value for H is known, we can turn the method around and use the Hubble law to find the distance to a galaxy. The method is as follows: take a spectrum of the galaxy whose distance you want. From the spectrum, we measure the Doppler shift of the spectral lines. From the Doppler-shift formula, we calculate the recession velocity. Finally, we use that velocity in the Hubble law to find the distance, as we show below.

We begin by writing out the Hubble law

$$V = HD$$

To find D, we divide both sides by H, obtaining

$$D = V/H$$

V = Recession velocity of galaxy
D = Distance of galaxy
H = Hubble constant

Finally, we insert the measured value of V and our choice of H and solve for the distance. For example, suppose we find from a galaxy's spectrum that its recession velocity is

60,000 kilometers per second. We find its distance by inserting this value for V in the expression above, and then divide by H.

$$D = (60{,}000 \text{ km/s})/H$$

For H = 75 kilometers per second per megaparsec, the value of D is then

$$D = \frac{60{,}000 \text{ km/s}}{75 \text{ km/s/Mpc}} = 800 \text{ Mpc}$$

Thus the galaxy is 800 megaparsecs away from us.

In applying this method, care must be used about the units. When we divide V in kilometers per second by H in kilometers per second per megaparsec, the answer we get will be in megaparsecs. Thus the 800 found in the above example is not in parsecs or light-years but in megaparsecs. Moreover, the Hubble law applies only to galaxies that are reasonably far away. The velocity of nearby galaxies is strongly affected by the gravitational attraction of their neighbors, and so the method fails when applied to such systems.

With the distance to a galaxy known, we can now measure its radius and mass as we explain below.

Measuring the Diameter of a Galaxy

Astronomers measure the diameter of a galaxy using the same method we discussed in chapter 1 for measuring the diameter of the Moon or the Sun. There we showed that an object's true diameter, d, is given by

$$d = \frac{2\pi A D}{360}$$

A = Angular size of object
d = Diameter of object
D = Distance to object

EXTENDING OUR REACH

OTHER WAYS TO MEASURE A GALAXY'S DISTANCE

Partly as a result of the controversy about what value of the Hubble constant to use in the Hubble law, astronomers have sought other ways to measure the distances to galaxies. One new method is similar to the simple observation that you have made if you look closely at a picture printed in a newspaper or comic book. Holding such a picture near your eyes, you can see that it consists of many tiny separate dots. As you move the picture farther from your eyes, the dots disappear and merge together to form the picture. When astronomers look at a picture of a galaxy, it too consists of many dots, each dot being one of its stars. In galaxies near us, we can easily see the separate spots of its more luminous stars. However, in more distant galaxies, the stars blend together into a uniform blur of light. Careful measurement of the graininess of a galaxy's picture can therefore give an indication of its distance.

Another method widely used for finding galaxy distances is based on the standard-candle method but uses the galaxies themselves as the standard candle. This method relies on the observation that the rotation velocity of a spiral galaxy is related to its luminosity. In particular, the more luminous a galaxy, the more rapidly it rotates. We might anticipate this result because if a galaxy rotates very rapidly it must be very massive to hold itself together gravitationally. Moreover, other things being equal, a massive galaxy contains more stars and thus emits more light than a small galaxy. Thus, the faster a galaxy rotates, the more luminous it is.

Astronomers use this relation to find a galaxy's distance by first obtaining a spectrum of the galaxy at optical or radio wavelengths. If the galaxy is rotating, one side of it is moving toward us and the other side is moving from us, producing a Doppler shift to the red on one side and to the blue on the other. From the Doppler shift law, astronomers can find the galaxy's luminosity and its apparent brightness in the inverse-square law to calculate the galaxy's distance. This procedure, known as the "Tully-Fisher Method" after the two astronomers who devised it, is currently one of the best ways to measure a galaxy's distance.

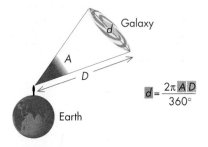

FIGURE 16.17
Measuring the diameter of a galaxy from its angular diameter.

$$d = \frac{2\pi A D}{360°}$$

where A is its angular size on the sky, and D is its distance (fig. 16.17). The units of d are the same as D; for example, in measuring galaxy sizes, we might express both in light years or in parsecs.

To illustrate the method, we will measure the diameter of M31, a close neighbor of the Milky Way. M31 is a spiral galaxy and so, when we speak of its diameter, we are referring to the diameter of its disk. Even that is somewhat ambiguous because galaxies do not have sharp edges, the stars just thin out near the edge, and it is hard to say exactly where the galaxy ends and intergalactic space begins.

The angular diameter, A, of M31 is about 5° and its distance, D, is about 670,000 parsecs. Using the expression for d above and inserting these quantities, we find that the diameter, d, of M31 is

$$d = \frac{2\pi A D}{360}, \text{ or}$$

$$d = \frac{2\pi \times 5 \times 670,000}{360}$$

$$= \text{about } 58,500 \text{ pc}$$

Measuring the Mass of a Galaxy

Once we know the diameter of a galaxy, we can find its mass by using the modified form of Kepler's third law, as we discussed in the last chapter, where we measured the mass of the Milky Way by this technique. Recall that the modified form of Kepler's third law relates a body's mass to the distance and period of an object orbiting around it. Thus, if we observe a star at the outer edge of the galaxy and we measure its orbital period (how long it takes the star to circle the galaxy) and its distance from the galaxy's center, we can calculate the galaxy's mass.

From the above we see that to measure a galaxy's mass we must first measure the orbital speed of stars or gas within it. We can do this by taking a spectrum of the galaxy because the motion of its stars causes a Doppler shift in the spectral lines we observe. From that shift we can determine the star's rotation speed around the galaxy. From the rotation speed and star's distance from the center of the galaxy, we find its orbital period by dividing the circumference of its orbit by its rotation speed. Finally, we put the star's orbital period and distance from the center of the galaxy into the modified form of Kepler's third law and solve for the galaxy's mass. In actually using this method, astronomers do not get a spectrum of a single star, but rather the spectra of groups of stars or gas clouds moving together around the galaxy. Moreover, the spectra do not need to be at visible wavelengths. Many of the most accurate observations of galaxy masses come from radio telescope observations of the Doppler shift.

The above method is an excellent and accurate way to measure a galaxy's mass, yet a puzzling discrepancy arises when astronomers compare that mass with the mass of the visible material in the galaxy. Almost without exception, the mass calculated from the modified form of Kepler's law is *always* larger than the mass of stars and interstellar matter detectable by optical, infrared, and radio telescopes. Thus galaxies seem to contain matter that cannot be seen, what astronomers therefore call "dark matter."

16.3 DARK MATTER

Dark matter is the material that astronomers believe accounts for the discrepancy between the mass of a galaxy as found from the modified form of Kepler's third law and the mass observable in the form of stars and gas. The amount of dark matter required to resolve this discrepancy is very large, amounting in some galaxies to 10 times their visible mass. That is, when we look at a galaxy we may in some cases be seeing only one-tenth of its matter.

The strongest evidence that dark matter exists comes from galaxy-rotation curves. We discussed such curves in chapter 15 when we described the rotation of the Milky Way, but let us review them here briefly. A galaxy's rotation curve is a plot of the orbital velocity of the stars and gas moving around it at each distance from its center (curve A in fig. 16.18). The rotation curve is useful for studying dark matter because the orbital velocity depends on the mass, visible or not, that lies between the star and the galaxy's center. We can understand how that dependence arises by using the formula for orbital velocity that we developed in chapter 2 or by simply thinking about what keeps a star in orbit.

According to Newton's first law, if an object does not move in a straight line (for example, a star moving in a circular orbit around a galaxy), a force must be acting on it. We know that for a star orbiting in a galaxy, that force is supplied by gravity. We also know that the strength of the gravitational force depends on the mass of the attracting body (in this case, the galaxy). All other things being equal, a greater mass therefore implies a greater force. It turns out, however, that to a good approximation, *only the material between the star's orbit and the galaxy's center contributes* to the gravitational force. Stars near the center of a galaxy will therefore have only a small force acting on them because there is little mass between them and the center. Because the force acting on them is small, such stars cannot be orbiting very rapidly, or they would have flown out of the galaxy. On the other hand, stars orbiting far from the center should also feel a small gravitational force acting on them from the galaxy. They feel the effect of the galaxy's full mass, but the gravitational force on them is nevertheless weak because they are so far from the main mass (recall that the force of gravity grows rapidly weaker at greater distances from a mass). These outermost stars must also be orbiting slowly if they are not to fly out of the galaxy. Thus both stars near the center and stars far from the center should have low orbital velocities. Accordingly, the galaxy's rotation curve should have a peak, as indicated in curve B of figure 16.18.

Detailed mathematical models bear out this analysis and show that a galaxy's rotation curve should start at small velocities near the core. The curve should then rise to higher velocities at middle distances and finally drop to lower velocities again far from the center. However, when astronomers *measure* the rotation curves of galaxies, they essentially *never* find this behavior. The rotation curves do rise near the center, but they then flatten out and virtually never drop off to low velocities again. What can explain such a major discrepancy between theory and observation? If our laws of physics are correct, only one thing can explain such behavior: galaxies must contain large amounts of unseen mass in their outer parts. That unseen mass exerts a gravitational force on distant stars, holding them in orbit despite their large orbital velocity. Moreover, the amount of unseen matter needed to account for the observed rotation curves is huge (many times the galaxy's luminous mass). In fact, such studies indicate that most galaxies, our own Milky Way included, may be embedded in a huge halo of dark matter vastly more massive and larger than the "visible" galaxy.

What can this dark matter be? Astronomers are baffled. They can rule out ordinary dim stars because such objects would emit at least some infrared radiation that would be detectable. They can rule out cold gas because that would be detectable by radio telescopes. They can rule out hot gas because that would be detectable by optical, radio, or x-ray telescopes. Objects they cannot rule out are tiny planetesimal-sized bodies, extremely low-mass cool stars, dead white dwarfs or neutron stars, black holes, or subatomic particles such as neutrinos. Other possibilities include subatomic particles predicted by theory but not observed as yet in the laboratory, such as photinos, or the strangely named, WIMPS, standing for *weakly interacting, massive particles*. The nature of the dark matter remains one of the central mysteries in astronomy today. Moreover, as we will see later in this chapter and in the next one as well, dark matter is not confined to galaxies. It also permeates galaxy clusters and perhaps the Universe itself.

We should not be surprised that something as vast as our Universe still has mysteries. In fact, dark matter is but one of many aspects of galaxies that astronomers do not fully understand. Another mystery about galaxies is the violent activity in the centers of many of them.

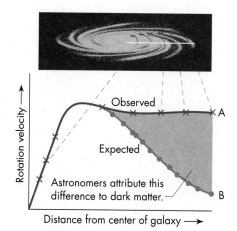

FIGURE 16.18
A schematic galaxy-rotation curve. The line with dots represents the curve expected if the galaxy's mass comes only from its luminous stars. The line with crosses represents the observed curve, implying that the galaxy contains dark matter.

16.4 ACTIVE GALAXIES

Active galaxies are galaxies whose centers (nuclei) emit abnormally large amounts of energy from a tiny region in their core. Not only is the emitted radiation intense, but it also usually fluctuates in intensity. Moreover, in many instances active galaxies exhibit intense radio emission and other activity outside their main body. At least 10% of all known galaxies are active, and they can be divided into three overlapping classes: radio galaxies, Seyfert galaxies, and quasars. The centers of these active systems are sometimes called "AGN's" for *a*ctive *g*alactic *n*uclei.

Radio Galaxies

Radio galaxies, as their name indicates, emit large amounts of energy in the radio part of the spectrum. They are generally elliptical galaxies, and their radio emission differs in two important ways from that of ordinary galaxies. First, it comes not so much from the main body of the galaxy, but from the core and from regions outside the galaxy on either side of it, as you can see in figure 16.19A. This figure was made by aiming a radio telescope at the galaxy and recording those places where the radio signal was strong. The strength is indicated by color, with blue being weak emission and red being relatively strong emission.

The second important feature of the radio emission from such galaxies is its strength, amounting in some cases to millions of times more radio energy than normal galaxies emit. You can get some sense of these differences by comparing figure 16.19A with Figure 16.20, which shows a map of the radio emission from the galaxy M51 (a system similar to the Milky Way). The radio energy is particularly strong along the spiral arms but also comes from the entire body of the galaxy. Even so, the radio emission from M51 is puny compared to that of a radio galaxy.

The radio-emitting regions visible outside the radio galaxy in figure 16.19 are called "radio lobes." These lobes, often located symmetrically on opposite sides of the galaxy, may span hundreds to millions of light-years. Although most of the radio emission comes from the lobes, some comes from within the visible body of the galaxy, primarily from a tiny region at its core. The core source is typically less than a light month across, which is a tiny fraction of the several-thousand-light-year diameter of the central region of the galaxy.

What causes this intense radio emission? Astronomers can tell from its spectrum that the emission is synchrotron radiation, created by electrons traveling at nearly the

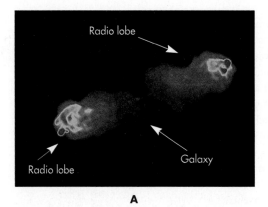
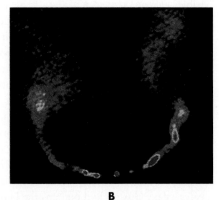

A   B

FIGURE 16.19
(A) False color map of the radio emission from the radio galaxy, Cygnus A. Notice the two radio lobes on either side of the radio galaxy. (B) A similar map of a different radio galaxy (NGC 1265) showing how its motion through gas around it has swept the radio lobes back. (Courtesy NRAO/AUI.)

speed of light and spiraling around magnetic field lines. We discussed this mechanism in chapter 14, where we described radio emission from pulsars. Pulsars generate high-speed electrons at their surface, but radio galaxies generate them in their core. The electrons are then shot out of the core in narrow **jets** (such as the one shown in fig. 16.21) that generally point at the radio lobes. Streaming away from the end of the jet, the electrons eventually collide with the gas around the galaxy. There they spread out to form the lobes, all the while emitting synchrotron radiation. Although the lobes are often on opposite sides of the radio galaxy, they are sometimes swept into arcs or plumes, presumably the result of the radio galaxy's motion through the gas around it (fig. 16.19B). We will discuss below one hypothesis of how the jet is formed, but before we do, we need to describe the other types of active galaxies: Seyfert galaxies and quasars.

Seyfert Galaxies

A **Seyfert galaxy** is a spiral galaxy whose nucleus is abnormally luminous (figure 16.22). These peculiar galaxies are named for the American astronomer Carl Seyfert, who first drew attention to their optical peculiarities. The core luminosity of a typical Seyfert galaxy

FIGURE 16.20
False color map of the radio emission at 21 centimeters from an ordinary spiral galaxy, M51. Notice how clearly the radio intensity shows the spiral arms. (Courtesy NRAO/AUI.)

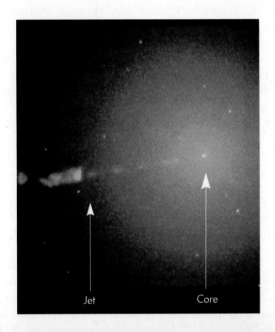

Jet   Core

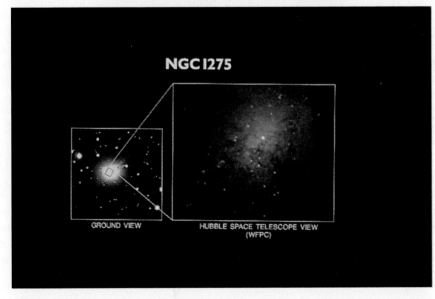

NGC 1275

GROUND VIEW   HUBBLE SPACE TELESCOPE VIEW (WFPC)

FIGURE 16.21
Photograph of the jet in the core of the radio galaxy M87. (Courtesy NASA.)

FIGURE 16.22
Photograph of the Seyfert galaxy NGC 1275 showing the galaxy and a magnified view of its core as observed by the Hubble Space Telescope. (Courtesy HST.)

is immense, amounting to the entire radiation output of the Milky Way, but coming from a region far less than a light-year across. Moreover, the radiation is at many wavelengths: optical, infrared, ultraviolet, and x-ray. However, despite its immense luminosity, the radiation from a Seyfert galaxy's core fluctuates rapidly in intensity, sometimes changing appreciably in a few minutes.

In addition to the intense source of radiation in their nuclei, Seyfert galaxies also contain gas clouds moving at speeds of about 10,000 kilometers per second. In some instances, the core is ejecting gas in small jets. This rapidly moving gas and the small, bright nuclear region make Seyfert galaxies similar to radio galaxies and, in fact, some Seyfert galaxies are radio galaxies. Nevertheless, most Seyfert systems are spiral galaxies, whereas most strong radio galaxies are ellipticals. However, the similarity in their activity is suggestive of a similar cause, a process that probably also acts in quasars, as we shall see below.

Quasars

Quasars are extremely luminous, extremely distant, active galaxies. Some are powerful radio sources, similar to the radio galaxies. Others are similar both to radio and Seyfert galaxies in that they are ejecting hot gas from their centers, as deduced from the extremely strong emission lines in their spectra. (Recall our discussion in chapter 3 that a hot, low-density gas produces a spectrum consisting of emission lines.)

Quasars get their name by contraction of the term *quasi-stellar radio source,* where *quasi-stellar* (meaning "almost star-like") refers to their appearance in the first photographs taken of them (figure 16.23), in which they look like stars (that is, like points of light). Early photographs also revealed that some quasars have small wispy clouds near them or tiny jets of hot gas coming from their cores. However, more recent images reveal that quasars often lie in faint, fuzzy-looking objects that appear to be ordinary galaxies.

The jets observed in some quasars (and at least one Seyfert galaxy) exhibit a very curious property: they contain clumps that appear to move faster than the speed of light. Astronomers refer to such seeming faster-than-light motion as superluminal motion. According to relativity theory, no material object can travel faster than light, but the theory does allow objects to *appear* to move faster than light if their true speed is only slightly less than the speed of light. Thus apparent superluminal motions are additional evidence for extremely high-speed motions in active galaxies.

Pictures of quasars lack detail because these objects are so far away. We deduce their huge distance from the huge redshift of their spectra (recall that a galaxy's distance can be found from its redshift using the Hubble law). In fact, quasars have the largest redshifts known for any kind of astronomical object. On the basis of its redshift, the most distant one yet observed is about 10 billion light-years from Earth. To be visible at such an immense distance, an object must be immensely luminous, and a typical quasar turns out to be about 1000 times more luminous than the Milky Way. Yet, despite this huge power output, quasars, like Seyfert galaxies, fluctuate in brightness. Slow changes occur over months (fig. 16.24), but short-term flickerings occur in times as brief as hours. Rapid changes of this sort give clues to the size of the emitting object, as discussed in the box on the next page.

However, astronomers need not rely solely on light variations to determine the size of an active galaxy's core. Observations with radio interferometer telescopes show that the active cores are extremely small (only a fraction of a light-year, in some cases).

Cause of Activity in Galaxies

Because all three kinds of active galaxies (radio galaxies, Seyfert galaxies, and quasars) share many features in common (for example, tiny cores ejecting gas at tremendous speed), astronomers have sought a model for their activity that could explain all three. In fact, astronomers now realize that in addition to the three classes of active galaxies that we have described above, many other kinds of active galaxies exist. For example, BL Lacertae

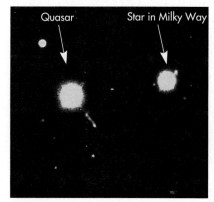

FIGURE 16.23
Photograph of the quasar 3C 273 ("3C" stands for *Third Cambridge Catalog of Radio Sources.* (Courtesy NOAO.)

☀ Do you see the "jet"?

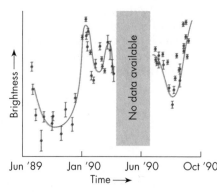

FIGURE 16.24
Plot of the light variation of a quasar. (Courtesy Tom Balonek, Colgate University.)

objects (BL Lac, for short) are luminous objects that are probably distant radio galaxies in which the jet of ejected matter is pointing almost straight at the Solar System. BL Lac objects, like quasars, look starlike and, in fact, BL Lac was originally believed to be a peculiar variable star within the Milky Way.

Any successful model for active galaxies must explain how such a small central region can emit so much energy over such a broad range of wavelengths. For example, at least one quasar is a powerful source of both radio waves and gamma rays. Astronomers therefore hypothesize that the core must contain something very unusual. No ordinary single star could be so luminous. No ordinary group of stars could be packed into so small a region. One kind of object that is very small but that can also emit intense radiation is the accretion disk around a black hole. We saw in chapter 14 that such accretion disks can emit intense x-ray radiation. Could the center of an active galaxy contain an immense black hole?

EXTENDING OUR REACH

MEASURING THE DIAMETER OF ASTRONOMICAL OBJECTS BY USING THEIR LIGHT VARIABILITY

Brightness changes give an important clue to the size of a quasar's or active galaxy's core. Suppose we have a ball of glowing material that we can turn on and off. Suppose further that when we turn it on, it begins to glow everywhere at the same time, and when we turn it off, it stops glowing everywhere at the same time. If we now look at this ball as it begins to glow, light from its nearest part, A, will reach us before light from its farthest part, B (fig. 16.25). An exactly similar effect occurs with sound produced from a lightning stroke. The lightning stroke lasts only fractions of a second, but the sound it generates lasts many seconds. The reason the sound is drawn out for such a long time is that sound travels relatively slowly (only about 1000 feet/second). Thus, if the stroke is 5000 feet long (a typical size for a bolt of lightning), sound from the near part reaches us 5 seconds before sound from the far part does. So, too, even though the ball begins to glow everywhere on its surface at exactly the same moment, the light we see does not turn on instantly. Instead, it gradually increases in brightness over a time interval equal to the time required for light to travel across the ball. That is, if the ball has a diameter of one light-second (the distance light travels in 1 second = 3×10^5 kilometers), the light reaching an observer does not reach full brilliance in less than 1 second. Likewise, if it is a light-month across (about 10^{12} kilometers) and it is turned off instantly, it does not dim away to darkness in less than 1 month.

From this we can conclude that if a quasar varies in brightness in a time as short as a few days (and assuming simultaneous illumination and dimming) then the emitting region can only be a few light-days across. A core a few light-days across is tiny compared to the size of a typical galaxy (100,000 light-years across). In fact, such a core is about 10 million times smaller than a galaxy, or about 10 AU in diameter, about the size of Saturn's orbit around the Sun.

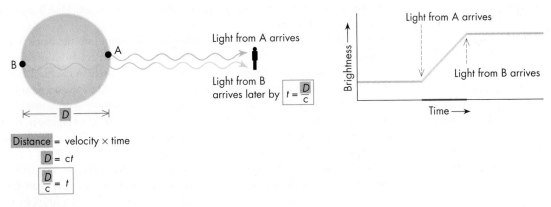

Distance = velocity × time

$D = ct$

$\dfrac{D}{c} = t$

Light from A arrives

Light from B arrives later by $t = \dfrac{D}{c}$

FIGURE 16.25

Sketch illustrating that when a light source turns on, we see the light from its near side before we see the light from its far side, and so we do not *see* its light turn on instantly. For the objects that we encounter in everyday life, the time delay is so short we do not notice it. However, for astronomical-sized bodies the effect can create delays ranging from days to months, from which we can deduce that the emitting body is light-days or light-months across.

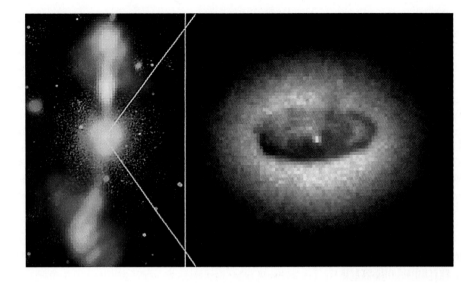

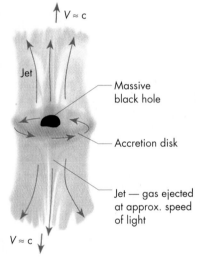

FIGURE 16.27
Sketch of jets formed by an accretion disk.

With Hubble, astronomers have found evidence that many more galaxies than previously thought contain huge black holes in their cores, implying that massive black holes may exist in most galaxies.

Could the radiation we observe from the core be hot gas in an accretion disk around such a black hole? This is precisely the model for active galaxies most widely favored by astronomers today. According to this model, active galaxies have in their cores a black hole about the size of the Earth's orbit around the Sun that contains a hundred million or more solar masses. Around the black hole swirls an accretion disk of gas with a radius of tens to hundreds of AU. Gas within the disk spirals toward its center, drawn inward by the black hole's gravity, ultimately to be swallowed. Orbiting under the black hole's fatal attraction, the gas is heated to incandescence by frictional forces in the disk and by the gravitational energy released as it falls inward.

Figure 16.26, a picture made with the Hubble Space Telescope, supports this model. The picture shows just such a disk in the central regions of the radio galaxy NGC 4261. Gas falling into the disk releases gravitational energy, which heats the material to temperatures perhaps as hot as several million K. Although most of the mass ultimately falls into the black hole, some material from the accretion disk boils off into space. If a strong magnetic field exists in the disk, it may channel such escaping gas into narrow jets from the central regions of the top and bottom surfaces of the disk (fig. 16.27), as shown by computer simulations.

Left to itself, the accretion disk would gradually drain into the black hole and activity would cease. But the disk may be replenished with matter from at least two sources. Sometimes stars within the galaxy may approach too closely to the black hole and be torn apart by its tidal forces. Their gaseous debris then falls into the accretion disk as a new source of material. Alternatively, from time to time collisions with neighboring galaxies may add new material as the active galaxy cannibalizes smaller neighbors. Each such event gives the accretion disk a new burst of energy, causing the active galaxy's core to increase in brightness.

How might such a massive black hole form? We saw in chapter 15 that our own galaxy may harbor a black hole at its center, albeit a much smaller one than the one we are describing here. According to one theory, our own galaxy's black heart formed in several steps. First, a massive star in the central region of the Milky Way reached the end of its life, exploded as a supernova, and formed a black hole perhaps as small as 5 $M_\odot$. Next, this small black hole grew as interstellar matter happened to fall into it. As gas fell in, the hole's mass and radius increased, making it easier for the hole to attract and swallow yet more material. Eventually the hole became large enough to swallow entire stars, at which point it grew rapidly to a million or more solar masses. During its growth, the hole was surrounded by an accretion disk, and our Milky Way was probably an active galaxy. As the Milky Way aged, its black hole consumed the gas around it, and the activity subsided to the low level we see today. Other galaxies may have had more mass at their cores to begin with, or they may have gained mass by collisions with small, neighboring galaxies. This extra mass thereby allowed their central black holes to grow far larger than the black hole in the Milky Way. In either case, we expect that the central black hole is active mainly in young galaxies during the time the hole is growing in mass.

Why are none of the galaxies close to us as active as quasars? The youth of quasars may be the key. Because quasars are so distant, we see such objects the way they were in the distant past, not the way they are "now." For example, when we look at a quasar that is 2 billion light years away, we are seeing light that has taken 2 billion years to reach us. Thus we are seeing the quasar as it was 2 billion years ago. The fact that most quasars are billions of light years away implies that the phenomenon causing them occurred billions of years ago when the Universe was younger. Objects that once were quasars may surround us, but their activity has long since died away.

Some astronomers find the black-hole theory of active galaxies incredible and have proposed that quasars are not truly at the immense distance implied by their redshift and are in reality much closer. Astronomers favoring the black-hole model counter with the question, "What then causes their immense redshift and why do they look like Seyfert galaxies, which we *know* are far away from Earth?" Attempts to explain these extraordinary objects by ordinary processes have generally failed. For example, at one time astronomers thought that the activity was produced by a single star of 10^8 solar masses that had managed to form in the center of the active galaxy. Another idea, also rejected, was that some galaxies might experience a chain reaction of supernovas. In this picture, one star blowing up would trigger the explosion of a neighbor, and so forth. Astronomers have also considered more exotic explanations. For example, a few scientists proposed a new class of objects called "white holes." These would be the opposite of black holes in the sense that they would spew material out into space rather than draw it in. Although certain mathematical theories allow the existence of white holes, no observational evidence for them exists, and therefore using them to explain active galaxies is hardly more plausible than using black holes. Whether another theory involving less exotic objects will supersede the black-hole model remains to be seen. Moreover, despite the reliance of the model on an object as peculiar as a black hole, the model is consistent with observations of active galactic nuclei and quasars. But consistency is not proof, and the history of astronomy is littered with the wrecks of consistent models (Ptolemy's epicycles, for example).

16.5 QUASARS AS PROBES OF INTERGALACTIC SPACE

Although interesting in their own right, quasars are also useful to astronomers as probes of intergalactic space. Because quasars are so distant, their light has a high probability of passing through other galaxies or intergalactic gas clouds on its way to Earth. Even if the intervening object is too dim to see, its matter may imprint its signature on the quasar's light, allowing us to detect these otherwise invisible objects.

For example, many quasars show absorption lines in their spectra that have very different Doppler shifts from those of the emission lines in the quasars themselves. These absorption lines are created by cool gas clouds lying between us and the quasar, much as similar lines in stars are created by cool interstellar clouds, as we discussed in chapter 15. The cool gas responsible for the absorption lines may be within galaxies, or it may lie in clouds that have not yet formed into galaxies. Matter between us and a quasar may also affect its light in an even more remarkable way, creating a gravitational lens.

Gravitational Lenses

About half a dozen quasars seem to have nearby companion quasars with essentially identical spectra but slightly different brightnesses and shapes. The existence of two so nearly identical objects so close together is unlikely, and astronomers now think that the "companions" are actually images of a single quasar created by a gravitational lens.

FIGURE 16.28
(A) Sketch of a gravitational lens. (B) The arcs in this picture of the galaxy cluster Abell 2218 are the image of a distant galaxy distorted by the gravitational lens effect created by the galaxy cluster. (Courtesy STSCI.)

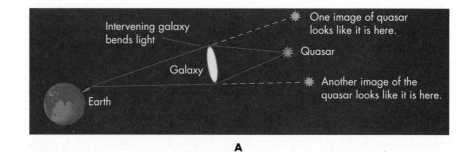

A

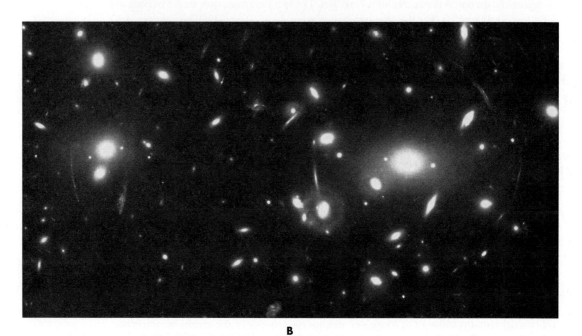

B

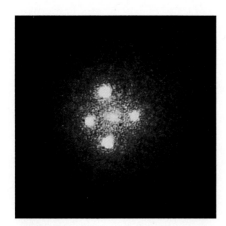

FIGURE 16.29
An Einstein cross, a complex image of a quasar created by a gravitational lens. (Courtesy NASA.)

An ordinary lens forms an image because light bends as it passes through the lens's curved glass. A **gravitational lens** forms an image because light bends as it passes through the curved space around a massive object such as a galaxy. The galaxy's gravitational force bends the space around it so that light rays that would otherwise travel off in other directions and never reach the Earth are bent so that they do, as depicted in figure 16.28A. The existence of gravitational lenses thus supports Einstein's theory that massive bodies can bend space, as we described in chapter 14 in the discussion of black holes. Although we cannot yet directly see the extreme bending of space at a black hole, we can "see" the more modest bending that creates gravitational lenses. Moreover, we will learn in the next chapter that such bending may occur on such an enormous scale that it affects the shape of the Universe.

Gravitational lenses may create forms other than multiple images of quasars. For example, figure 16.28B shows arcs created when light from a distant galaxy was bent into elongated shapes as it passed through a galaxy cluster. Likewise, figure 16.29 shows a picture made with the Hubble Space Telescope of an "Einstein cross." Scientists studying gravitational lenses predicted many years before the example in figure 16.29 was discovered that such a pattern should exist.

EXTENDING OUR REACH

DARK MATTER AND GRAVITATIONAL LENSES

Although the most dramatic gravitational lenses are created by distant galaxies, astronomers have recently detected such lenses within or near the Milky Way. If a star or even a planet passes in front of a background star as seen from Earth, the gravitational force of the foreground object bends the light of the background star and focuses it, causing the light we see to briefly brighten. With special telescopes that monitor regions of the sky with large numbers of stars, astronomers have discovered about a dozen such brightenings that they believe are gravitational lenses. Even though the foreground object is itself too dim to see, astronomers can estimate it mass from the brief lens effect it creates. Moreover, from a statistical analysis of the number of such brightenings, astronomers can deduce the number of objects creating the lens effect. From the evidence so far available, the objects creating the lens effect are low-mass stars. These objects—referred to as MACHO's (for *m*assive *c*ompact *h*alo *o*bjects)—are invisible by ordinary means but may contribute a significant part of the dark matter we described in section 16.3. Their detection by gravitational lensing may thereby resolve part of the dark matter puzzle.

FIGURE 16.30
Photograph of the Coma cluster of galaxies. (Courtesy Cal Tech.)

16.6 GALAXY CLUSTERS

We have mentioned several times earlier in this chapter that galaxies are often found in groupings called **galaxy clusters**. A galaxy cluster is a group of galaxies held together by their mutual gravity. Each galaxy moves along its own orbit, just as within a galaxy each star moves on its own orbit. Galaxy clusters are huge. A typical cluster is more than several million light-years across and may contain anywhere from a handful to a few thousand member galaxies (fig. 16.30).

Astronomers are still uncertain about how galaxy clusters form. Does a single large gas cloud break up into galaxies forming a cluster at one stroke, the way gas clouds in the Milky Way break up into stars to form a star cluster? Or do galaxies form as separate, independent systems that are then drawn together by their mutual gravitational attraction to form the cluster much later? Once again, the ability to "see back in time" by looking out in space offers astronomers clues. The galaxies in nearby clusters are generally fairly smoothly spread within the clusters. On the other hand, galaxies in remote clusters sometimes have a clumpy distribution in the cluster, as if they have recently joined it and have not had time to spread themselves smoothly.

The Local Group

Our own galaxy, the Milky Way, belongs to a very small cluster called the **Local Group**. The Local Group contains about thirty members, as sketched in figure 16.31. Its three largest members are spiral galaxies: the Milky Way, M31, and M33. The smaller members include the satellite galaxies orbiting M31 and the Milky Way, the most famous of which are the Large and Small Magellanic Clouds, which we have referred to many times before.

The Magellanic Clouds (Figs. 16.2 and 16.32) are about one-tenth the size of the Milky Way. They appear to be dwarf irregular galaxies, but infrared photographs of the large cloud show that it has a faint, barlike structure. Both galaxies seem to be approaching the Milky Way in a very elliptical orbit and are being pulled into a distorted shape by the Milky Way's gravity.

FIGURE 16.31
Artist's view of the Local Group.

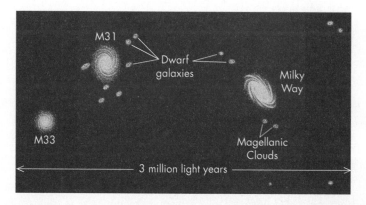

Local Group

A — 3 million light years →

M31, Dwarf galaxies, Milky Way, Magellanic Clouds, M33

A

B

FIGURE 16.32
Photographs of (A) the Small Magellanic Cloud and (B) the Large Magellanic Cloud. (Courtesy Anglo-Australian Observatory; photographs by David Malin.)

What are the irregular reddish regions visible in the pictures?

The Local Group is particularly interesting to astronomers not only because it is our home galaxy cluster, but also because it allows us to study the "demographics" of galaxies. That is, we can see the true population of galaxies in this region of space "close" to home. For example, most of the galaxies in the Local Group are faint, small "dwarf" galaxies. Figure 16.33 shows a typical dwarf galaxy. It is a ragged, disorganized group of stars with little or no interstellar gas. Such galaxies are too dim to be visible except in clusters close by. Thus without the Local Group to let us see the true abundance of these small systems, we might grossly underestimate their true numbers. Moreover, the Local Group's small size helps us appreciate the immensity of the giant galaxy clusters that contain hundreds and in some cases thousands of members.

Rich and Poor Galaxy Clusters

The largest groups of galaxies are called **rich clusters** because they contain hundreds to thousands of member galaxies (fig. 16.34). The great mass of such a galaxy cluster draws its members into an approximately spherical cloud with the most massive galaxies near the center.

FIGURE 16.33
The dwarf galaxy Leo I, a member of the Local Group of galaxies. (Courtesy Anglo-Australian Observatory; photograph by David Malin.)

FIGURE 16.34
The Hercules Cluster, a rich cluster of galaxies. (Courtesy NOAO.)

☀ What types of galaxies can you identify in this photograph? Notice the interacting galaxy pair near the center. Suppose one of those galaxies is 100,000 light years in diameter. What is the approximate diameter of the part of the cluster visible in this picture?

In contrast, some galaxy clusters contain relatively few members (sometimes only a dozen or so). These so-called **poor clusters** have so little mass that the gravitational attraction they create is much weaker than that of the rich clusters. Thus their member galaxies are not held in so tight a grouping, and they generally have a ragged, irregular appearance with no central concentration. The Local Group (fig. 16.31) is a typical poor cluster.

Not only do rich and poor clusters differ in their number of galaxies, but they also tend to contain different types of galaxies. Rich clusters contain mainly elliptical and S0 galaxies. Moreover, the few spiral galaxies that they contain tend to be found in the outer parts of the cluster. On the other hand, poor clusters tend to have a high proportion of spiral and irregular galaxies. Spirals are rare in the inner regions of rich clusters because of galaxy collisions. In the core of a rich cluster the galaxies are close together, and so collisions between them are frequent. As we saw earlier, collisions may change spiral galaxies into ellipticals. Thus, although the rich clusters may have once contained many spirals, today they contain few, and those are the lucky ones that escaped running into a neighbor. The many collisions in the crowded inner part of these clusters also create giant elliptical galaxies by "cannibalism," as we described in section 16.1.

Rich and poor clusters differ in yet another way. Observations with x-ray telescopes show that rich clusters often contain large amounts (10^{12} to 10^{14} $M_\odot$) of extremely hot gas that emits x-rays. In fact, they sometimes contain several times more hot gas than they do stars. Astronomers are not certain where that gas originates. Some of it probably comes from within the cluster. For example, when a star explodes as a supernova, some of the gas it ejects may leave the star's galaxy and end up in the intergalactic space within the cluster. On the other hand, much of the gas in rich clusters contains very few heavy elements, an indication that it is primordial material left over from the time when galaxies formed.

In addition to the x-ray-emitting gas, galaxy clusters also contain large amounts of the dark matter we discussed earlier. In fact, some of the first evidence for dark matter came from observations of the motion of galaxies within clusters. Just as stars within galaxies seem to orbit too rapidly for the gravitational force that can be attributed to their luminous mass, so too galaxies in clusters orbit too rapidly for their luminous mass. The amount of dark matter in some galaxy clusters is so immense that it amounts to perhaps 100 times the amount of luminous material.

Superclusters

The great mass of galaxy clusters creates gravitational interactions between them and holds them together into larger structures called "superclusters." A **supercluster** contains a half dozen to several dozen galaxy clusters spread throughout a region of space tens to hundreds of millions of light-years across. That is, a supercluster is a cluster of galaxy clusters. For example, the Local Group is part of the Local Supercluster, which contains the Virgo galaxy cluster and more than a dozen others spanning nearly 100 million light-years (about 30 megaparsecs). It is sketched in figure 16.35.

FIGURE 16.35
Artist's view of the Local Supercluster.

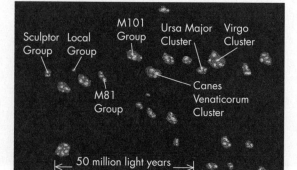

Local Supercluster

50 million light years
(Approx. 15 Mpc)

Superclusters have irregular shapes and are themselves often part of yet larger groups. Figure 16.36 shows a map of the galaxies that form such a cluster of superclusters. This immense flattened system of galaxies recently found lying about 330 million light-years (100 megaparsecs) away from us stretches across a region about 500 × 200 × 16 million light years (170 × 60 × 5 megaparsecs). Because of its relative thinness, astronomers call this system of galaxies the "Great Wall."

The immensity of the Great Wall implies that it contains an enormous mass, which in turn creates gravitational forces that alter the motion of galaxies near it. In fact, such gravitational forces may be pulling on the Local Group, though the forces appear to come not from the Great Wall but from a nearer and as-yet-unseen system called "the Great Attractor."

The Great Attractor is probably an assemblage of superclusters about 130 million light-years (40 megaparsecs) away from us and about 260 million light-years (80 megaparsecs) in diameter. It may contain about 3×10^{16} M., the equivalent of 10,000 or more galaxies like the Milky Way. Although we cannot see it, we can "feel" it because its gravitational force may be drawing galaxies around it (including the Local Group) toward its center. This motion causes deviations in the Hubble law for galaxies near the Great Attractor, which is how it was discovered. Those deviations give astronomers rough clues to its mass and size.

Superclusters are the frontier of the Universe at this time, and little is known of their properties or formation. However, one feature is clear: they often appear to form chains or shells surrounding regions nearly empty of galaxies. These latter spaces are called *voids* and, although they are not totally empty, the few galaxies they contain are small and dim. Whether dark matter exists within them is not yet known. At the vast distances involved, even the largest telescopes now available allow us to see only the brightest galaxies. Thus clusters of superclusters and voids mark the end of the structure of the Universe that we can currently see. The next step up in scale takes us to the entire Universe itself.

Recently, astronomers have found evidence for this clustering of superclusters in other directions as well. Our Universe seems to be "lumpy," even on extremely large scale.

FIGURE 16.36
Plot of the location of many distant galaxies showing the huge, narrow collection of galaxies called the "Great Wall." (Courtesy M. J. Geller and J.P. Huchra; Smithsonian Astrophysical Observatory.)

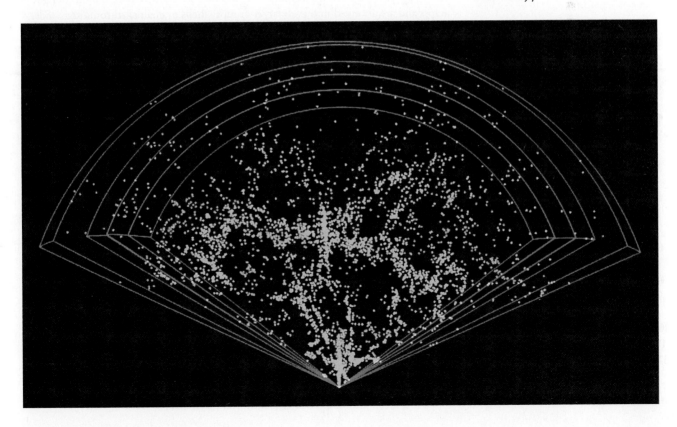

SUMMARY

Astronomers classify galaxies into three main groups: spirals, ellipticals, and irregulars. Spirals are disklike, with arms of young stars and gas winding out from the nucleus. They contain a mixture of population I and II stars, and about 15% of their mass is in the form of interstellar gas and dust. Elliptical galaxies are spherical or egg-shaped and contain hardly any population I stars because they have essentially no dense interstellar matter from which new stars could form. Irregular galaxies lack a symmetry to their shape but are often extremely rich in interstellar matter and thus contain many young, population I stars.

Galaxies form from the collapse of primordial gas clouds. Astronomers do not yet understand why some clouds become spiral and others become elliptical galaxies. Perhaps the initial rotation speed and the rate at which the collapsing gas cloud forms stars play a role. If so, ellipticals may develop from slowly rotating clouds that create stars rapidly, whereas spirals may develop from more rapidly spinning clouds that flatten to a disk. Alternatively, random motions in the primordial clouds may determine a galaxy's type.

Many galaxies have extremely small but very energetic and luminous nuclei. These include radio and Seyfert galaxies as well as quasars. According to one currently popular theory, the activity results from matter falling in toward a giant black hole at the galactic center. For the theory to work, the hole must have a mass of about 100 million solar masses packed into a region not much larger than the Solar System. An accretion disk forms around the hole as matter falls in, and hot gas boils off the disk, forming jets and clouds of radio-emitting gas. The amount of energy generated may dwarf the light output from the galaxy's stars and, for a quasar, outshine the Milky Way by a factor of a thousand.

Galaxies are usually grouped into clusters a few million light-years across. Small (poor) clusters show little symmetry and tend to contain spiral galaxies. The Milky Way belongs to a poor cluster called the Local Group, which contains about thirty member galaxies. Rich clusters contain thousands of galaxies, which are primarily ellipticals. Clusters themselves appear to be clustered into superclusters that are many tens of millions of light years across and are arranged into even larger groups (shells or chains) surrounding voids.

QUESTIONS FOR REVIEW

1. What are the three main types of galaxies?
2. How do the basic galaxy types differ in shape, stellar content, and interstellar matter?
3. Why is it not surprising that elliptical galaxies contain only old stars?
4. Which of the basic galaxy types has the smallest proportion of Pop I stars?
5. What happens to the stars when two galaxies collide?
6. Why are galaxy collisions of interest?
7. How do astronomers measure the distance to nearby galaxies?
8. What is the Hubble law?
9. Why are astronomers uncertain about the value of the Hubble constant?
10. How do astronomers measure the mass of a galaxy?
11. What is meant by dark matter?
12. What are the three main types of active galaxies?
13. Make a sketch of a radio galaxy illustrating its structure and where its radio emission comes from. Why do astronomers call these objects "radio" galaxies?
14. What process generates the radio emission in radio galaxies?
15. What is a Seyfert galaxy?
16. What is a quasar?
17. Why do some astronomers think that quasars are very distant?
18. How is it known that active galaxies have small core regions?
19. What mechanism has been suggested to power active galaxies?
20. How might a large black hole form in a galaxy's core?
21. What is a gravitational lens?
22. What are some differences between rich and poor clusters?
23. What is the Local Group?
24. What is the Local Supercluster?
25. What is a void?

THOUGHT QUESTIONS

1. Hubble proposed that his tuning-fork diagram might represent evolutionary changes with E galaxies gradually changing to S galaxies. What evidence based on the amount of interstellar matter and kind of stars in E and S galaxies indicates that this is very unlikely?

2. If you wanted to look at yourself as a baby, how far out in space would you need to be? (Assume you have a sufficiently powerful telescope and that you can travel instantly to any distance.) How does this relate to astronomers studying young galaxies?

PROBLEMS

1. A Cepheid variable in a nearby galaxy looks 10^6 times fainter than an identical Cepheid in the Milky Way. The Cepheid in the Milky Way is 1000 parsecs away. How far away is the nearby galaxy?

2. A galaxy has a recession velocity of 30,000 kilometers per second. What is its distance in megaparsecs?

3. A galaxy has a rotation speed of 200 kilometers per second at a distance from its center of 10 kiloparsecs. What is its mass within this distance?

4. A galaxy has an angular diameter of 0.1 degree. Its distance is 10 megaparsecs. What is its true diameter?

TEST YOURSELF

1. A large galaxy contains mostly old (Pop II) stars spread smoothly throughout its volume, but it has little dust or gas. What type galaxy is it most likely to be?
   (a) Irr
   (b) S
   (c) SB
   (d) E
   (e) All of the above are possible.

2. A galaxy's spectrum has a redshift of 30,000 kilometers per second. If the Hubble constant is 75 kilometers per second per megaparsec, how far away from Earth is the galaxy?
   (a) 10^6 megaparsecs
   (b) 1000 megaparsecs
   (c) 20 megaparsecs
   (d) 50 megaparsecs
   (e) 400 megaparsecs

3. The Local Group
   (a) contains about thirty member galaxies.
   (b) is a poor cluster.
   (c) is the galaxy cluster to which the Milky Way belongs.
   (d) is a rich cluster.
   (e) is all of the above except (d).

4. A spiral galaxy has a small bright central region, and its spectrum shows that it contains hot, rapidly moving gas. It is most likely a _____ galaxy.
   (a) barred spiral
   (b) Seyfert
   (c) radio
   (d) BL Lac
   (e) quasar

5. Astronomers believe that dark matter exists because
   (a) they can detect it with radio telescopes.
   (b) the outer parts of galaxies rotate faster than expected on the basis of the material visible in them.
   (c) the galaxies in clusters move faster than expected on the basis of the material visible in them.
   (d) it is the only way to explain the black holes in active galaxies.
   (e) both (b) and (c) are correct.

FURTHER EXPLORATIONS

Barnes, Joshua, Lars Hernquist, and Francois Schweizer. "Colliding Galaxies." *Scientific American* 265 (August 1991): 40.

Bartusiak, Marcia. "Giving Birth to Galaxies." *Discover* 18 (February 1997): 58.

_____. "The New Dark Age of Astronomy." *Astronomy* 24 (October 1996): 36.

Bartusiak, Marcia. *Through a Universe Darkly: A Cosmic Tale of Ancient Ethers, Dark Matter, and the Fate of the Universe.* New York: HarperCollins Publishers, 1993.

_____. "What Makes Galaxies Change." *Astronomy* 25 (January 1997): 36.

Beardsley, Tim. "Galactic Gushers." *Scientific American* 276 (February 1997): 26.

Blanford, Roger D. , Mitchell C. Begelman, and Martin J. Rees. "Cosmic Jets." *Scientific American* 246 (May 1982): 124.

Bothon, Gregory D. "The Ghostliest Galaxies." *Scientific American* 276 (February, 1997): 56.

Burns, Jack O., and R. Marcus Price. "Centaurus A: The Nearest Active Galaxy." *Scientific American* 249 (November 1983): 56.

Crosswell, Ken. "How Far to Virgo? [Hubble Constant]" *Astronomy* 23 (March 1995): 48.

Djorgovski, S. George, "Fires at Cosmic Dawn [Role of Quasars in Galactic Evolution]." *Astronomy* 23 (September 1995): 36.

Dressler, Alan. "The Large-Scale Streaming of Galaxies." *Scientific American* 257 (September 1987): 46.

_____. "Observing Galaxies through Time." *Sky and Telescope* 82 (August 1991): 126.

Dressler, Alan Michael. *Voyage to the Great Attractor: Exploring Intergalactic Space.* New York: Knopf/Random House, 1994.

Eicher, David J. "Candles to Light the Night [On the Size of the Universe]." *Astronomy* 22 (September 1994): 32.

Ferris, Timothy. *Galaxies.* San Francisco: Sierra Club Books.

Finkbeiner, Ann. "Active Galactic Nuclei." *Sky and Telescope* 84 (August 1992): 138.

Ford, Holland, and Zlatan I. Tsvetanov. "Massive Black Holes in the Hearts of Galaxies." *Sky and Telescope* 91 (June 1996): 28.

Freedman, Wendy L. "The Expansion Rate and Size of the Universe." *Scientific American* 267 (November 1992): 54.

Gallagher, Jay, and Jean Keppel. "Seven Mysteries of Galaxies." *Astronomy* 22 (March 1994): 38.

Gorenstein, Paul, and Wallace Tucker. "Rich Clusters of Galaxies." *Scientific American* 239 (November 1978): 110.

Harris, William E. "Globular Clusters in Distant Galaxies." *Sky and Telescope* 81 (February 1991): 148.

Hodge, Paul. "Our New! Improved! Cluster of Galaxies [The Local Group]." *Astronomy* 22 (February 1994): 26.

Impey, Chris. "Ghost Galaxies of the Universe [Low-Surface-Brightness Galaxies]." *Astronomy* 24 (June 1996): 40.

Lake, George. "Cosmology of the Local Group." *Sky and Telescope* 84 (December 1992): 613.

_____. "Understanding the Hubble Sequence." *Sky and Telescope* 83 (May 1992): 515.

Macchetto, F. Ouccin, and Mark Dickinson. "Galaxies in the Young Universe." *Scientific American* 276 (May 1997): 92.

Miller, Richard H. "Experimenting with Galaxies." *American Scientist* 80 (March/April 1992): 152.

Naeye, Robert. "Ghosts in the Cosmic Machine: The Answer to the Dark Matter Mystery May Lie in the Murky Realm of Ghostly Elementary Particles." *Astronomy* 24 (October 1996): 48.

Price, Jill S., and Karen Ann Caldwell. "Galaxies That Go Bump in the Night." *Mercury* (July/August 1995): 23.

Silk, Joseph, Alexander S. Szalay, and Yakov B. Zel'dovich. "Large Scale Structure of the Universe." *Scientific American* 249 (October 1983): 72.

Strom, Richard G., George K. Miley, and Jan Oort. "Giant Radio Galaxies." *Scientific American* 233 (August 1975): 26.

Trefil, James. *The Dark Side of the Universe.* New York: Charles Scribner's Sons, 1988.

Van den Bergh, Sidney. "The Age and Size of the Universe." *Science* 258 (16 October 1992): 421.

Veilleux, Sylvain, Gerald Cecil, and Jonathan Bland-Hawthorn "Colossal Galactic Explosions." *Scientific American* 274 (February 1996): 98.

West, Michael. "Galaxy Clusters: Urbanization of the Cosmos." *Sky and Telescope* 93 (January 1997): 30. See also related articles in same issue.

Web Sites

www.seds.org/messier
Contains many galaxy images and descriptions of them.

KEY TERMS

active galaxies, 490
barred spirals, 476
dark matter, 488
elliptical galaxies, 476
galactic cannibalism, 483
galaxy, 474
galaxy clusters, 497
gravitational lens, 496

Hubble constant, 486
Hubble law, 486
irregular galaxies, 476
jets, 491
Local Group, 498
megaparsecs, 486
poor clusters, 500

quasars, 492
radio galaxies, 490
recession velocity, 486
rich clusters, 499
Seyfert galaxy, 491
spiral galaxies, 475
supercluster, 500

 PROJECTS

If the season allows, use a pair of binoculars or small telescope to find M31, the great galaxy in Andromeda. What shape does it appear to have? Use the information in the text to estimate how long it took the light you are seeing to reach you.

CHAPTER SEVENTEEN

COSMOLOGY

Distant galaxies as seen by the Hubble Space Telescope. (Courtesy of STSCI.)

Cosmology is the study of the structure and evolution of the Universe as a whole. Cosmologists seek answers to questions such as, How big is the Universe? What shape is it? How old is it? How did it form? What will happen to it in the future?

Given our insignificant size in the cosmos, such questions may seem futile or even arrogant, but most cultures have tried to answer them, and many of the early attempts have become part of humanity's religious heritage. For example, the *Bible,* in the book of Genesis (1 : 1–3), states that

> In the beginning God created the heaven and the earth; and the earth was without form and void; and darkness was upon the face of the deep and the spirit of God moved upon the face of the waters. And God said, "Let there be light," and there was light. And God saw the light, that it was good.

These simple but powerful words, rich in symbols, give one picture of the birth of the Universe. But despite their power, they offer few specifics of what our Universe was like at its birth. In this final chapter, we will learn how contemporary scientists view the creation and evolution of our Universe, beginning with the observational discoveries that the Universe is expanding and that it is filled with a very low-energy background radiation. This expansion and radiation imply that the Universe was born about 15 billion years ago in a hot, dense, violent burst of matter and energy called the **Big Bang.** Within the last few decades, cosmologists have sought to extend our knowledge of the Universe to before the Big Bang and have discovered that the Universe may have been born in even more turbulent events known as the inflationary stage, a time when the entire Universe that we see today may have fitted in a volume smaller than a proton.

17.1 OBSERVATIONS OF THE UNIVERSE

In the early years of this century, most astronomers believed that the Universe consisted of the Milky Way galaxy and a number of small companions. Immanuel Kant, the German natural philosopher, suggested in the eighteenth century that the Milky Way was merely one of a multitude of galaxies, but his views were not widely accepted until the 1920s, when astronomers showed that other galaxies were millions of light-years from the Milky Way and comparable to it in size. About the same time, astronomers such as Hubble showed that galaxies were moving away from one another—in short, the Universe is expanding.

Distribution of Galaxies

We begin our study of the structure of the Universe by looking at the night sky, shown in figure 17.1A. This wide-angle view shows the distribution of over one million galaxies as seen looking up out of the disk of our own galaxy, the Milky Way. Around the edge of the picture, you can see only a few spots of light because dust in the Milky Way (the zone of avoidance) hides the galaxies in that part of the sky. But the central region of the picture shows two important clues about the structure of the Universe. First, in no matter what direction you look (ignoring the dust layer of the Milky Way) you see approximately the same number of galaxies (fig.17.1B). Second, the galaxies are not spread smoothly over the sky, but are clumped into loose groups.

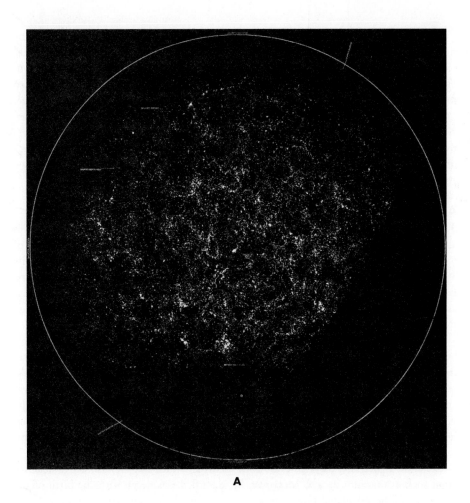

A

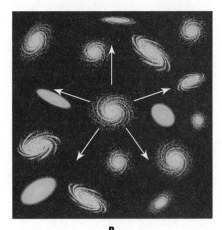

B

FIGURE 17.1
(A) A computer-generated picture showing the location of a million galaxies. (Courtesy Seldner, M., Siebers, B., Groth, E. J., and Peebles, P. J. E., A. J. 82, 4. "New Reduction of the Lick Catalog of Galaxies.") (B) A sketch illustrating the uniformity of the Universe. Roughly the same number of galaxies occur in all directions.

Because we see approximately the same number of galaxies in all directions,* we conclude that galaxies are spread approximately smoothly throughout the Universe like raisins in a fruitcake. That smooth distribution (apart from the "lumps" of galaxy clusters and superclusters, as discussed in chapter 16) is also reflected in the motion of the galaxies.

Motion of Galaxies

In chapter 16 we learned that nearly all galaxies have their spectral lines redshifted, and their shift to a longer wavelength implies that most of the galaxies are moving away from us. Moreover, the redshift increases with distance, D, in such a way that the recession velocity, V, of a galaxy obeys the Hubble law:

$$V = HD$$

where H is the Hubble constant.

The motion of galaxies away from us is caused by the expansion of the Universe. This expansion is the result of space itself expanding. It is *not* like a bomb burst flinging fragments in all directions from a single central location. Instead it is like the motion of buttons glued on a balloon when the balloon is inflated (fig. 17.2). Or, returning to the fruitcake analogy, it is like a giant, raisin-filled fruitcake expanding as it bakes.

---

*This is in sharp contrast to how stars are distributed in our Milky Way. We see its stars concentrated in a narrow band from which we deduce that our galaxy is a disklike system.

FIGURE 17.2
Representations of the expansion of
the Universe.

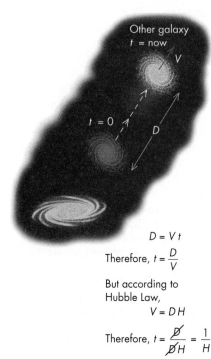

$$D = V t$$

Therefore, $t = \dfrac{D}{V}$

But according to
Hubble Law,

$$V = D H$$

Therefore, $t = \dfrac{\cancel{D}}{\cancel{D} H} = \dfrac{1}{H}$

FIGURE 17.3
Estimating the age of the Universe
from the recession of one galaxy from
another galaxy.

The motion of both the buttons on the balloon and the raisins in the cake is caused by the motion of the medium in which they are set. So too for galaxies: space itself is expanding, and galaxies are carried apart by that expansion, not by any motion of their own. However, although space expands, the objects themselves—galaxies, stars, planets, and so forth—do not. Gravity and other forces maintain such material bodies at a constant size. On the other hand, light waves *are* altered by the expansion of space, and that expansion creates the redshift of the galaxies that we observe. As a light wave from a galaxy moves through space, it is stretched as the space expands, just as a kinked curl of carrot in a piece of Jell-O is stretched if the ends of the Jell-O are pulled on.

Age of the Universe

The above model of the Universe as an expanding space filled uniformly with galaxies allows cosmologists to estimate its age and to calculate conditions near the time of its birth. Alexander Friedmann, a Russian scientist, made such calculations in the 1920s, but his work attracted little attention. Not until 1927, when Abbé Georges Lemaître, a Belgian cosmologist and priest, independently made similar calculations, did astronomers appreciate that conditions at the birth of the Universe could be deduced from what is observed today.

Lemaître argued that since galaxies are moving apart now they must in the past have been closer together. He went on to propose that at one time in the distant past all matter in the Universe was packed into a region approximately the size of the present Solar System. With so much mass in such a relatively small volume, the density of the Universe was enormous. In fact, at that remote time in the past, the density was so immense that the protons and neutrons that now compose the far-flung stars and galaxies were touching, just as today the protons and neutrons within a single atom touch. For that reason, Lemaître called this dense stage of the early Universe the Primeval Atom.

Lemaître argued that the Primeval Atom "decayed" much like an ordinary radioactive atom such as uranium decays. Moreover, he identified this decay with the birth of the Universe and he calculated how long ago that occurred. Astronomers no longer accept Lemaître's theory of the birth of the Universe, but most think that it was created from a highly compressed state in the distant past, and they measure its age from that time.

We can estimate the age of the Universe as follows. We know that the galaxies are currently moving apart. If we know their present separation and their speed and, if we further assume that their speed has been constant, we can determine (see box below) how long ago their motion began (fig. 17.3), a time we identify as the age of the Universe.

Lemaître made a similar but more detailed calculation that included the effect of the gravitational attraction of the Universe on itself. The effect of that attraction is to slow the motion of the galaxies (that is, their speed in the past was faster than what we see today). Because their speed was greater then than it is now, the galaxies must have begun their movement apart more recently than we deduce from their present speed. In other words, the Universe is younger than our simple model based on a constant speed implies.

Lemaître found that the effect of the Universe's gravitational attraction on itself was to make its age $(2/3)/H$. Astronomers today recognize that the age they deduce for the Universe depends on the amount of matter it contains. However, many astronomers think that Lemaître overestimated the importance of gravity in his calculation and that, in fact, our simple model of *no* gravitational effects is nearer the truth. Thus, we will take the age of the Universe, t, to be $t = 1/H$. With the above uncertainties in mind, we can now evaluate what that age is.

To evaluate t in years, we must know H, the Hubble constant. However, we cannot simply insert the value of H in the expression for t because H is expressed in units of kilometers/second/megaparsec. If we were to use H in these units we would get an answer for the age of the Universe in equally odd units. We must therefore express H in units of 1/seconds by converting megaparsecs to kilometers so that the distance units in H cancel out. One megaparsec is 3×10^{19} km, so if the Hubble constant is 75 km/second/Mpc

EXTENDING OUR REACH

ESTIMATING THE AGE OF THE UNIVERSE

To calculate the age of the Universe, t, we proceed as follows: let two representative galaxies be a distance, D, apart and separating with a velocity, V. For simplicity we will assume that V has remained constant. Then in the time t, the age of the Universe, they will have separated by a distance given by

$$D = Vt$$

D = Distance between two galaxies

V = Velocity at which galaxies are separating

t = Age of the universe

We can find how long ago they began moving apart by dividing both sides by V.

This gives

$$t = \frac{D}{V}$$

Thus, in this simple model, the age of the Universe is D/V. The Hubble law, however, relates a galaxy's velocity to its distance by $V = DH$. Therefore, substituting this expression for V into the expression for the age of the Universe, we find

$$t = \frac{D}{V}$$
$$= \frac{D}{DH}$$
$$= \frac{1}{H}$$

where H is the Hubble constant.

(the value deduced from observations made with the Hubble Space Telescope*) this gives

$$H = 75 \text{ km/second} / (3 \times 10^{19} \text{ km}) = 2.5 \times 10^{-18} \text{ seconds}^{-1}.$$

Therefore, $1/H = 4 \times 10^{17}$ seconds. Because there are about 3×10^7 seconds per year, $1/H = 1.3 \times 10^{10}$ years. Thus we find that the age of the Universe is about 13 billion years.

Many astronomers are uncomfortable with this value for the age because it conflicts with the ages of some astronomical objects. For example, they feel very confident that the Earth is about 4.5×10^9 years old, making our planet safely younger than the Universe. However, the ages estimated for the oldest stars and globular clusters are about two to three times older, or about 10^{10} years. If it turns out that H is as large as 100 km/second/Mpc, or if further work shows that gravity has slowed the expansion of the Universe significantly so that the age should be $(2/3)/H$ rather than just $1/H$, then the age of the Universe is less than the age of its stars. Such a clearly impossible result would imply a serious error in either our theory of stars or our theory of the Universe.

 ## RE-MODELING

AGE DISCREPANCY

Two separate, recent discoveries may resolve this discrepancy. First, astronomers have used the *Hipparchos* satellite (see chapter 12) to measure the distance to a few Cepheid variable stars directly by parallax. They find that the Cepheids are about 10% farther away than previously thought. This implies that Cepheids are brighter than previously believed and, because they are used to measure the distances to galaxies, galaxies too must be more remote. This in turn implies a smaller value for the Hubble constant and thus a greater age for the Universe.

Second, if Cepheids are in fact brighter than previously thought, astronomers have been wrong about the brightness of other luminous stars. If these objects are also brighter (more luminous), they must burn their fuel faster than thought before, implying that the stars are younger than previously believed. At the same time, astronomers studying the evolution of giant stars have discovered several possible flaws in the earlier models that would lead to higher luminosities for the stars, in agreement with the observations. These new discoveries—by implying that stars are younger and the Universe is older than thought—help resolve the age discrepancy. Whether future work will bear this out remains to be seen.

*Recent Hubble Space Telescope measurements suggest H may be closer to 65 km/second/Mpc.

FIGURE 17.4
We cannot even in theory see beyond the cosmic horizon. Light takes more time than the age of the Universe to reach us from there.

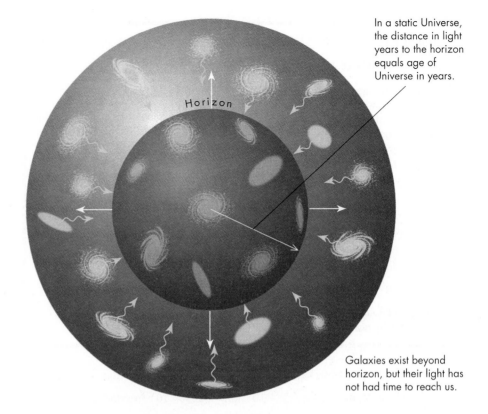

In a static Universe, the distance in light years to the horizon equals age of Universe in years.

Horizon

Galaxies exist beyond horizon, but their light has not had time to reach us.

The Cosmic Horizon

The age of the Universe limits the distance we can see. Because light requires time to move from one point to another, we can see no farther than the distance light travels in a time span equal to the age of the Universe. If the Universe did not expand, calculating the distance that light travels in that time span would be simple. For example, if the Universe were not expanding and were 13 billion years old, we could see nothing farther than 13 billion light years away (fig. 17.4). The Universe may well extend beyond that distance, but the maximum distance that light can travel in the Universe's age therefore defines a **cosmic horizon**. Astronomers call the space within the horizon the visible Universe.

However, because the Universe is expanding, it turns out that there is no simple expression for the distance to the cosmic horizon. Nevertheless, the existence of a cosmic horizon has another use.

The Size of the Universe

Because the cosmic horizon defines the size of the visible universe, astronomers sometimes use the distance to the horizon as a rough measure of the **radius of the Universe.** As we saw above, the distance to the cosmic horizon is related to the distance light travels in the age of the Universe. If for simplicity we ignore the Universe's expansion, and we let D be the distance to the horizon, then $D = ct$ where c is the speed of light and t is the age of the Universe. We can insert values for c and t and multiply them to find D. However, it is much easier to make use of the principle that, by definition, light travels one light year in one year. Therefore in the 13-billion-year age of the Universe, it travels 13 billion light years. Thus the size of the visible Universe is approximately 13 billion light years.

The cosmic horizon describes how far we can see in principle. It takes no account of whether our instruments can in fact detect light from sources that far away. How far can astronomers in fact see? The most distant object known at this time is a quasar receding from us at 94% the speed of light. From its redshift and the Hubble law, its distance turns out to be about 10 billion light years. The horizon therefore still lies far beyond the reach of today's telescopes, though the Hubble Space Telescope may ultimately allow astronomers to see nearly to the cosmic horizon.

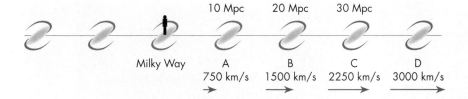

FIGURE 17.5
A line of galaxies illustrating that in a Universe that obeys a Hubble law of expansion (V = HD), an observer on any galaxy sees the same type of law.

Are We at the Center of the Universe?

The recession of distant galaxies from the Milky Way often leads to the mistaken belief that the Milky Way is at the center of the Universe. Because space itself is expanding, no matter where you are in the Universe galaxies will be moving away from you. A simple example illustrating this is shown in figure 17.5, showing galaxies distributed along a line. Each galaxy is separated from its nearest neighbor by 10 megaparsecs. Suppose the Milky Way is the third galaxy in the line and that galaxies A, B, and C obey the Hubble law, with each receding at a velocity given by $V = 75D$. That is, A recedes from us at 750, B at 1500, and C at 2250 kilometers per second. Next, suppose we could communicate with an alien on galaxy C and ask what it sees. It would see us receding at 2250 kilometers per second, A receding at $2250 - 750 = 1500$ kilometers per second, and B receding at $2250 - 1500 = 750$ kilometers per second. Thus the alien would see galaxies receding from it exactly the same way we see galaxies receding from us. Furthermore, the galaxies seen by the alien will obey the alien's version of the Hubble law.

A similar argument can be made for an observer on any of the galaxies. Thus, for a Hubble law in which the velocity increases proportional to the distance, *every observer, no matter where in the Universe he/she/it is, will see the Universe expanding the same way*. The Universe has no preferred "center" of expansion if it expands uniformly. Astronomers describe this lack of a preferred location as the cosmological principle. It is a statement of cosmic modesty: there are no special positions. We are not at the center of the Universe, nor is anyone else. In fact, the notion of a "center" to the Universe is meaningless, as we will discuss further in section 17.3.

Olbers' Paradox

The Universe's finite age can also be demonstrated by the simple experiment of stepping outside at night and noting that the sky is dark. This relation between the darkness of the night sky and the structure of the Universe is known as **Olbers' paradox,** after the German astronomer Heinrich Olbers, who in 1823 presented an argument that if the Universe extends forever and has existed forever, the night sky should be bright. The paradox arises from following reasonable premises (or so it seemed at that time) to a clearly nonsensical conclusion. The argument is as follows:

Suppose you stand in a small grove of trees and look out between them to the surrounding landscape. If the grove is small, your line of sight will be blocked by trees in a few directions but will pass between the trees in most directions (fig. 17.6A–D). If the grove is larger, more distant trees in it will block your view in what were previously clear gaps. Then, no matter what direction you look, your line of sight is intercepted by a tree.

Now suppose that, rather than looking between trees, you look out into space through the galaxies and stars that compose the Universe. If space extends sufficiently far and is populated with galaxies and stars such as we see around us in the Local Group, then no matter in what direction you look, your line of sight will ultimately intercept a star just as in a sufficiently large forest your line of sight ultimately hits a tree. Therefore the sky should be covered with stars, each glowing brilliantly, with no dark spaces between them. Even though such distant stars are very faint to us, there are so many of them that their collective brightness should be very large. In other words, the night sky should not be dark. This argument must be fallacious because the most obvious observation one can make about the sky is that it is dark at night. That is the paradox: the night sky should be bright, but it is not. Where therefore is the error in our reasoning?

A In a small grove of trees, only a few block your view. Lots of space to view between trees.

B In a larger forest, more distant trees block your view. No open space visible between trees.

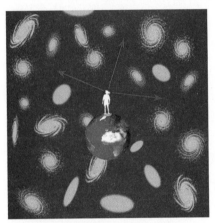

C In a small group of stars and galaxies, lots of dark sky shows between them.

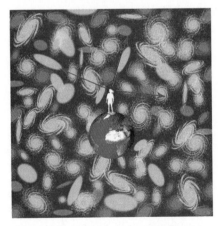

D In a sufficiently large group of stars and galaxies, no dark sky shows between them. That is, the sky is bright.

FIGURE 17.6

(A) An observer in a small grove of trees can see out through gaps between them to the surrounding countryside. (B) In a deeper woods, distant trees block the gaps so that no matter in what direction you look your line of sight ends on a tree. (C) So too in a universe that extends forever and has existed forever your line of sight will always end on a star. Thus, the disks of stars completely cover the sky and leave no dark spaces between them. Therefore, the night sky should be bright.

The paradox is avoided if there are no stars beyond some distance. That is, just as we can see out of the woods if there are only a few trees, so too we will see a dark night sky if there is only a limited number of stars. The paradox is also avoided if we account for the time it takes light from the stars to reach us. Light from a star 1 billion light-years away takes a billion years to get here. Thus, to see such light requires that the Universe has existed long enough for the light to arrive. That is, if stars have not been shining for billions of years, the night sky will be dark. A final way to avoid the paradox is that light from remote stars is redshifted by the expansion of the Universe. The redshift reduces the light's energy (recall that longer-wavelength radiation carries less energy) and thereby reduces the brightness of the most distant stars and galaxies. Thus an expanding Universe can make the night sky dark.

Although the paradox bears Olbers' name, more than two centuries earlier, Kepler pointed out that in an infinite Universe the night sky should be bright and he therefore concluded the Universe was finite in size. Today, astronomers think that the finite *age* of the Universe implies a finite cosmic horizon, making the night sky dark. Stars have simply not existed long enough for their radiation to flood space and make it bright. Moreover, they will never be able to make the night sky bright because they will burn up all their fuel first. The darkness of the night sky, however, results from the limits of our senses. If we could see radio waves the night sky would glow uniformly, as we shall see below.

The Cosmic Background Radiation

Lemaître's idea that the young Universe was extremely dense led George Gamow, a Russian astrophysicist who fled to the United States before World War II, to realize that it must also have been extremely hot. His argument was based on the observation that compression heats a gas, and so the enormous compression of the early Universe must have heated it to at least 10 billion Kelvin. Gamow went on to note that conditions in the young Universe would have been similar to those in the core of stars, and so he theorized that heavy elements might have been created in the young Universe just as they are in stars. Gamow proposed that at its birth the Universe was composed only of neutrons, which rapidly decayed into protons and electrons. Some of the protons then fused to form helium and heavier elements. Gamow's ideas were developed further by two of his collaborators, the American physicists Ralph Alpher and Robert Hermann. In 1948 they predicted that as the young Universe expanded and cooled from its hot dense state the radiation it emitted would survive to the present, though in altered form. To understand how that alteration occurred, we need to look at the properties of the radiation in greater detail.

According to Wien's law, a dense gas radiates most intensely at a wavelength determined by its temperature, as discussed in chapter 3. In particular, hot matter radiates most strongly at short wavelengths. Because the early Universe was so hot—perhaps 10 trillion Kelvin for a very brief time—the wavelength of maximum emission turns out to be very short, less than a millionth of a nanometer. Such short-wavelength radiation, however, is typical only in the young, hot, dense Universe. As space expands, it stretches the wavelength of radiation traveling through it, increasing its wavelength. Recall the analogy used earlier of a carrot curl in a block of stretched Jell-O. Calculations show that as the Universe expanded to its present size, the wavelength of the radiation emitted at its birth was stretched from fractions of a nanometer to millimeters, a radio wavelength. Thus Alpher and Hermann showed that the Universe today should be filled with faint, long-wavelength radiation created when the Universe was very young. Their prediction was confirmed in 1965 when Arno Penzias and Robert Wilson of the Bell Telephone Laboratories accidently discovered cosmic microwave radiation having just that property.

Penzias and Wilson were trying to identify sources of background noise on telephone satellite links and noticed a weak radio signal with the unusual property that its strength was constant no matter in which direction of the sky they pointed their detector. They soon demonstrated that the signal came not from isolated objects such as stars or galaxies but rather from all space. Initially mystified by the signal, they soon discovered that scientists at Princeton University had repeated Alpher, Hermann, and Gamow's calculations and were searching for the predicted radiation. Once aware of that work, Penzias and Wilson realized that the background interference they had found was radiation created in the young Universe. Their discovery of the **cosmic background radiation,** as it is now called, won for them the Nobel Prize in physics in 1978.

Figure 17.7 shows measurements of the intensity of the cosmic background radiation (CBR, for short) and demonstrates that it has a perfect blackbody spectrum, as theory predicts. Moreover, the radiation is most intense at about 1 millimeter (10^6 nanometers), in the microwave part of the radio spectrum. From this wavelength and using Wien's law ($\lambda = 3 \times 10^6 / T$, where λ is the wavelength where the emission is strongest and T is the temperature) the temperature that describes the radiation turns out to be about 3 K, within a few degrees of the value predicted nearly 20 years earlier. A more precise measurement of λ shows that the temperature that describes the CBR is 2.7 K, only a little warmer than absolute zero, the lowest temperature anything can have.

To understand how expansion of the Universe has cooled the CBR, we again use the relation between a body's temperature and the wavelength at which it radiates most intensely: Wien's law. We know from Wien's law that cool objects emit most strongly at long wavelengths and hot objects emit most strongly at short wavelengths. We described earlier that as the Universe expands, it stretches the wavelengths of the radiation and makes them longer (fig. 17.8). Thus, the radiation becomes more like that emitted by a cool body

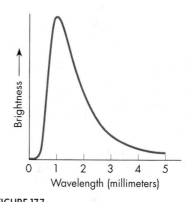

FIGURE 17.7
Spectrum of the cosmic background radiation. The shape of the spectrum matches perfectly a blackbody with a temperature of 2.7 K. Notice the wavelength at which it peaks. (Based on curve supplied by NASA/Goddard Spaceflight Center: COBE Science Working Group.)

FIGURE 17.8
As space expands, it stretches radiation moving through it, making the wavelength longer. Because longer wavelengths are associated with cooler objects, the stretching of the radiation has the effect of cooling it.

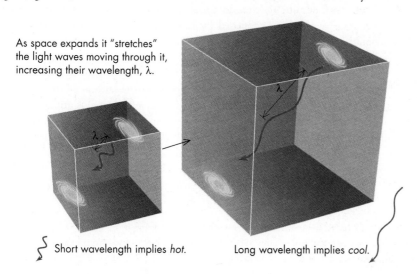

As space expands it "stretches" the light waves moving through it, increasing their wavelength, λ.

Short wavelength implies *hot*.   Long wavelength implies *cool*.

rather than a hot one. We therefore say that the radiation has cooled from the high temperature at which it was born in the early Universe to the mere 2.7 K we observe today.

It is ironic that the existence of this radiation—such important proof that the Universe was born in a hot dense state—was originally only a minor part of Alpher, Hermann, and Gamow's theory of the young Universe. Their original goal was to explain the formation of the heavy elements. Today, however, we know that most heavy elements are made in stars as they evolve. Nevertheless, the idea that conditions in the early Universe might allow the synthesis of elements is still important because it allows astronomers to determine conditions in the early Universe in yet another way.

Composition of the Oldest Stars

Gamow's hypothesis that some helium formed in the young Universe is supported by more recent theory as well as by observations. According to modern theory, however, the young Universe consisted of protons and electrons (that is, hydrogen), rather than neutrons, as Gamow believed. Nevertheless, his basic idea is correct: in its youth the Universe was hot and dense enough for nuclear reactions to fuse some of the protons into helium. Moreover, current calculations show that the amount of helium formed depends on how hot and dense the Universe was and how long those conditions lasted. These conditions are determined by how rapidly the Universe expands. Too rapid an expansion cools the Universe quickly and allows only a little helium to form. Too slow an expansion leaves the Universe hot for so long that too much helium forms. Current estimates of the expansion rate imply that about 24% of the matter should be transformed into helium, an amount in good agreement with what is observed in old stars in the Milky Way and other galaxies.* Thus the amount of helium in old stars further supports the theory that at its birth the Universe was hot and dense. Similar measurements of the amount of deuterium (^{2}H) and the light element lithium also indicate the high temperature and density of the young Universe.

Conclusions Deduced from the Basic Observations of the Universe

The recession of galaxies and their uniform distribution leads us to conclude that the Universe was born in a cataclysmic event about 15 billion years ago. At that time the Universe must have been much hotter and denser than now. Its temperature can be calculated to have been about 10^{13} K—about 10 trillion Kelvin—and its density must have been about 10^{10} grams per cubic centimeter. Moreover, all the matter we now see in the Universe must have been compressed into a volume about the size of the present Solar System.

Confirmation that this hot dense state existed comes from (1) the cosmic background radiation—the radiation created during that hot stage but cooled by the Universe's subsequent expansion—and (2) from the hydrogen/helium mix found in old stars. The high temperature and rapid expansion from a hot, dense state led the English astronomer Sir Fred Hoyle to jokingly refer to this theory of the birth of the Universe as the "Big Bang," a name that has stuck despite his sarcasm. Hoyle himself has long championed an alternative "steady-state" cosmology, proposed by him and two other British astronomers, Hermann Bondi and Thomas Gold, in the early 1950s. In the steady-state universe, new matter forms continually out of nothing to create new galaxies in the empty, expanding space between older galaxies, as depicted in figure 17.9. This keeps the average density of the Universe constant so the Universe on the average looks the same at all times. This unchanging appearance (which gives the steady-state Universe its name) is unlike that of the Big Bang theory where space becomes progressively emptier as the galaxies move apart.

Although the steady-state theory was popular in the 1950s, few astronomers now accept it because it necessitates creating new material out of nothing. Moreover, it requires a special process to create the cosmic background radiation, whereas the Big Bang theory explains it as a natural consequence of the Universe expanding from a hot, dense state.

---

* Most stars today contain more than 24% helium because they formed from material enriched with helium from earlier generations of stars.

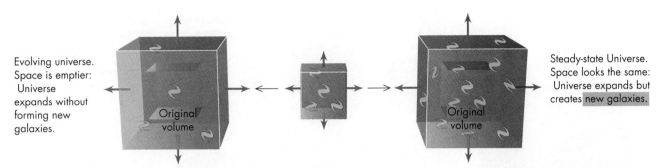

Evolving universe. Space is emptier: Universe expands without forming new galaxies.

Original volume

Original volume

Steady-state Universe. Space looks the same: Universe expands but creates new galaxies.

FIGURE 17.9
In the steady-state theory, new galaxies must form in between old ones. Otherwise, space will grow more and more empty as the Universe expands, thereby changing its appearance, a violation of the premise that it is in a steady state.

Another difficulty with the steady-state theory arises from its prediction that the overall appearance of the Universe should not change in time, a prediction that can be tested. But how can astronomers today determine whether the Universe looks different today from the way it looked in the past? They can look into the past by looking out to great distances in space. For example, when we look at an object 4 billion light-years away from us, its light has taken 4 billion years to reach us, and so we see it the way it was 4 billion years ago. Thus looking out in space is roughly equivalent to looking back in time. Therefore, by comparing the appearance of the Universe at great distances with the Universe near at hand, astronomers can see if it has changed.

The best way yet developed for comparing the Universe long ago with the Universe today is to count the number of quasars in a volume near the Milky Way and in a similar volume far away. Quasars are ideal for this test because they are so luminous that they can be seen at great distances, or equivalently at times in the remote past. When astronomers make such counts, they discover that quasars are very rare near the Milky Way but are much more common at great distances, that is, long ago. Thus quasars were more abundant in the past than they are now, and so we can infer that the Universe has changed, in contradiction to the hypothesis of the steady-state theory. Proponents of the steady-state theory, however, say that inferring the distance of quasars from their redshifts, as is done, is invalid and therefore any evidence about the age of the Universe based on the number of quasars is also invalid. But only a few astronomers adopt this point of view. Most accept that quasars were more common in the past and that the Universe has changed in time. However, popularity does not make a theory right. The Big Bang hypothesis has its own flaws, as we will discover later in this chapter, and many astronomers believe it is accepted too uncritically.

17.2 EVOLUTION OF THE UNIVERSE: OPEN OR CLOSED?

The Universe is currently expanding. But what of its future? Will it expand forever, or will it sometime stop expanding and contract? This question is fundamental to the nature of the Universe, for according to the Big Bang theory all the hydrogen in the Universe was created in the Big Bang. Thus, if the Universe expands forever, its stars will one by one consume their hydrogen and burn out, leaving the Universe a black, empty space—a dismal prospect. However, recollapse of the Universe means that all the objects and atoms (and us) within it will be compressed into what some astronomers have called the "Big Crunch." The Big Crunch might lead to another Primeval Atom, perhaps leading to a new Big Bang and the birth of another universe. Such a universe might oscillate, expanding and collapsing repeatedly.

Astronomers use the terms "open" and "closed" to describe these two different outcomes. An **open universe** expands forever. A **closed universe** stops expanding and collapses again. But there is a third possibility as well: the expansion may be so precisely tuned that the expansion speed becomes zero when the Universe has reached infinite size—a case called a **flat universe.**

A simple analogy illustrates these three types of universes. Suppose you throw a stone into the air. As the stone moves upward, it is slowed by the Earth's gravitational force,

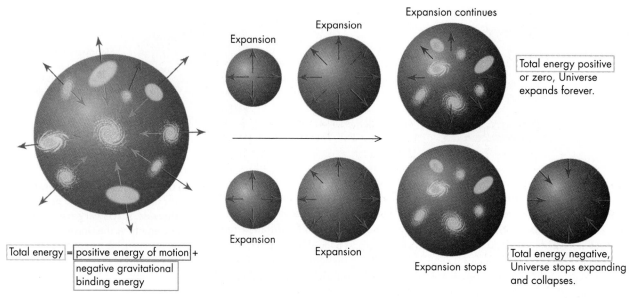

Total energy = positive energy of motion + negative gravitational binding energy

FIGURE 17.10
A sketch of how the energy of the Universe determines its behavior.

which eventually stops it and makes it fall back to the ground. On the other hand, if you could throw the stone faster than the escape velocity, it would rise forever. The expansion of the Universe behaves similarly. The gravitational force between galaxies slows their separation, but if the force is weak, the retardation is small and does not stop the expansion. If the gravitational force is sufficiently strong, however, expansion will stop and the Universe will collapse. Thus two very different fates are possible for the Universe: "open" and "closed." A flat universe is an intermediate case. Its behavior is like rolling a ball up a rounded hill so that it stops exactly on reaching the top and stays perched there.

Open, closed, and flat universes differ not only in whether they expand forever, but also in their energy content. It turns out that an open universe has a positive total energy; a closed universe has a negative total energy; and a flat universe has zero total energy. Thus the energy of the Universe determines its fate (fig. 17.10), and to determine whether the Universe is open or closed, we must compute its energy.

The Universe has two main forms of energy: a positive energy of its expansion (kinetic energy) and a negative energy of its gravitational binding force. Rapid expansion creates a large positive-energy contribution, whereas a large mass creates a large negative gravitational contribution. Adding them together gives the total energy and determines whether the Universe is open or closed. Thus, whether the Universe is open or closed can, in theory, be found by comparing its expansion rate to its mass. However, rather than dealing with the mass of the Universe, it is easier to work with its density (the amount of mass contained within a given volume). The reason is simple: astronomers can measure the density of the Universe but not its mass. To measure its mass, they would need to observe all of space. To measure the density, however, requires measuring the mass of some small part of space and then assuming the remainder is similar.

The Density of the Universe

To find the density of the Universe, astronomers choose a volume of space and count the galaxies in it (fig. 17.11). Next, they measure the mass of each galaxy, add them up to find the total mass, and divide by the volume.

To determine whether the Universe is open or closed, astronomers compare the observed density (as deduced above) with a theoretically calculated **critical density.** If the actual density of the Universe is greater than the critical density, the Universe is closed; if it is less, the Universe is open. The critical density, ρ_c, depends on the Universe's rate of expansion. Mathematically, $\rho_c = 3H^2/(8\pi G)$ where H is the Hubble constant, and G is the gravitational constant. The Hubble constant affects the critical density because a large gravitational force, such as that created by a high density, is needed to stop a rapidly expanding Universe.

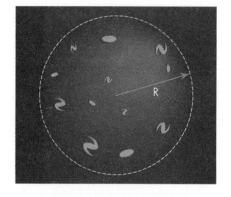

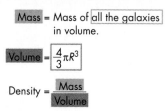

Mass = Mass of all the galaxies in volume.

$$\text{Volume} = \frac{4}{3}\pi R^3$$

$$\text{Density} = \frac{\text{Mass}}{\text{Volume}}$$

FIGURE 17.11
Calculating the density of the Universe by dividing the mass of a region by its volume.

EXTENDING OUR REACH

MEASURING THE DENSITY OF THE UNIVERSE

To measure the local density, astronomers might choose the Local Group, which contains three large galaxies and about two dozen smaller ones. The total mass of the Local Group is about 10^{12} solar masses ($M_\odot$), which we can convert to grams if we multiply by the number of grams in a solar mass (2×10^{33} grams). This then gives a mass for the group of about 2×10^{45} grams.

Next, that mass is divided by its volume, assumed to be a sphere whose radius is the distance from the center of the Local Group to the nearest galaxy cluster, which turns out to be about 3 megaparsecs ($= 9 \times 10^{24}$ centimeters). Using the formula for the volume of a sphere ($V = 4/3 \pi R^3$) gives a volume for the Local Group of about 3×10^{75} cubic centimeters. Dividing that volume into the mass (2×10^{45} grams) gives a density of about 7×10^{-31} grams per cubic centimeter. A similar calculation for a larger volume of the Universe gives a slightly smaller value of about 2×10^{-31} grams per cubic centimeter. This incredibly tiny density, far less than that of interstellar space, is equivalent to about 1 atom in 10 cubic meters.

When the value of H determined by observation (75 kilometers per second per megaparsec) is used, the critical density is about 10^{-29} grams per cubic centimeter, the equivalent of about one hydrogen atom per cubic meter. This is much larger than our observationally measured density, and so we infer that the Universe is open: there is insufficient matter in it to halt its expansion. In fact, for the Universe to be closed, the Universe needs about 25 times more material than astronomers see in visible stars and gas: the observed density of stars and gas is too small to stop expansion. Many astronomers are dismayed by this result because some theories of the Universe's structure and evolution imply that its density should exactly equal the critical density. However, if these theories are correct, how can the density discrepancy be reconciled? One possibility is that the Universe contains large quantities of unobserved dark matter, as we discussed in the last chapter. There we saw that astronomers have postulated such dark matter to explain the rapid rotation of the outer parts of spiral galaxies and motions within galaxy clusters. The amount of dark matter required to explain that discrepancy is approximately the amount needed to close the Universe, an agreement that suggests that dark matter may really exist. Astronomers, however, do not know what the dark matter is. It might be some combination of black holes, dead low-mass stars, subatomic particles such as neutrinos, or theoretically predicted particles such as WIMPs (mentioned in the last chapter) for which no laboratory evidence yet exits. On the other hand, theories implying that the density should equal the critical value may be wrong, and dark matter may be only wishful thinking. Its reality remains unproved.

In summary, we have deduced that our Universe is expanding and was born in the Big Bang. Whether it continues to expand or eventually stops and contracts depends on its rate of expansion and its density. The observed density of the Universe is too small to stop expansion, implying that the Universe is open and will expand forever. If it contains large amounts of dark matter, however, the Universe may be closed and, as we shall discover next, whether the Universe is "open" or "closed" also gives information about its shape.

17.3 THE SHAPE OF THE UNIVERSE

What shape is the Universe? To answer this fundamental question we need to discuss **curved space.** Astronomers know space can be curved, as we saw in chapter 14 in discussing black holes and in chapter 16 where we discussed gravitational lenses. How can we picture such curvature for the Universe? Our home planet Earth offers a simple example. Athletic fields and parking lots appear flat and, if we ignore hills, the surface of the Earth certainly *looks* flat. We know, however, that if we could walk in a "straight" line along this

"flat" surface, we would return to our starting point. We are therefore like an ant that accidentally crawls into a straw on a kitchen counter. The unbent straw represents a flat, uncurved space, and so, if the ant crawls in a straight line, it will reach the straw's end and emerge onto the counter again. However, suppose we pick up the straw with the ant still in it and bend the straw so that its ends meet. Now the ant could crawl forever and would never reach the straw's end.

A similar situation exists for a Universe curved around on itself. Such a space looks as if it extends forever in all directions, but if we could travel in a spaceship along a "straight" line we would eventually return to our starting position. Thus, we would have moved through space forever without coming to an end point. A universe with this property is said to have **positive curvature** and is "closed" because it has no "outside." We can infer that the Earth has a "positively curved surface" because if we travel in a straight line on it we return to our starting point. Positively curved surfaces have other properties as well: "parallel" lines meet when extended, and the sum of the interior angles of a triangle drawn on them is greater than 180°, as illustrated in figure 17.12A.

The Universe might have other types of curvature, however. It might be flat (that is, have no curvature) or have **negative curvature.** Flat space is what most people picture when they think of space. In it, parallel lines do not meet, and a triangle's interior angles add up to exactly 180°, as illustrated in figure 17.12B. Negatively curved space is harder to visualize, but you can think of it as being bent into a saddle-like shape, as shown in figure 17.12C. Such curvature corresponds to bending space one way along one line and the opposite way along another line. In negatively curved space, the sum of the interior angles of a triangle is less than 180°.

Can we measure the curvature of space and determine whether it is positive, negative, or flat? In theory, yes, but in practice not yet. One such method is to count the number of galaxies within a fixed angle at progressively larger distances (fig. 17.13A). If galaxies are uniformly spread through space, then in flat space their total number increases as the cube of the distance. In positively curved space the number increases less rapidly than the cube of the distance; in a negatively curved space the number increases more rapidly than the cube of the distance. You can see why this occurs by looking at figure 17.13B. In a flat universe, represented by the plane, the lines defining the angle are straight and spread apart uniformly. In a positively curved space, the lines defining the angle spread apart initially but then converge at great distances. Thus the number of galaxies they enclose at a given distance decreases when the distance is very large. In a negatively curved space, the saddle shape, the lines defining the angle spread apart more rapidly than they do on the flat surface, and so the number of galaxies between the lines increases rapidly. Figure 17.13C shows the resulting differences in the number of galaxies counted. Unfortunately, astronomers cannot yet make accurate counts on such dim objects for the curvature effects to be clearly evident.

What causes the curvature? The answer is gravity, as worked out by Einstein in his general theory of relativity. Einstein showed that the mass that composes the Universe bends its space just as the mass concentration in a black hole bends space there. In fact, comparison of the Universe with a black hole may be doubly significant if the Universe is closed because a closed universe is equivalent to a black hole: it has enough mass to bend space around on itself and "close off" access to the outside.

FIGURE 17.12
Figures illustrating how the shape of a surface determines the geometry of a triangle on it. (A) The interior angles of a triangle on the surface of the Earth add up to more than 180°. (B) The interior angles of a triangle on a flat surface add up to 180°. (C) The interior angles of a triangle on a saddle-shaped surface add up to less than 180°.

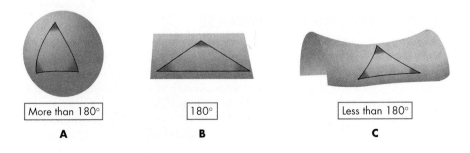

More than 180°   180°   Less than 180°

A   B   C

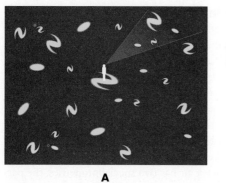
A

Positively curved
universe

Flat universe
B

Negatively curved
universe

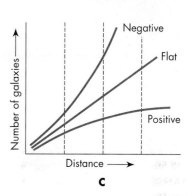
C

FIGURE 17.13
Sketches illustrating how the number of galaxies at a given distance can give clues to the shape of the Universe. (A) We count the number of galaxies lying within a fixed angle at each distance. (B) Notice how the lines defining that angle are straight on the flat surface but approach again after spreading apart on the spherical (positive curvature) surface. On the saddle-shaped (negative curvature) surface, the lines spread ever farther at greater distances. (C) A plot of how the number of galaxies should change in a flat, positively curved, and negatively curved universe.

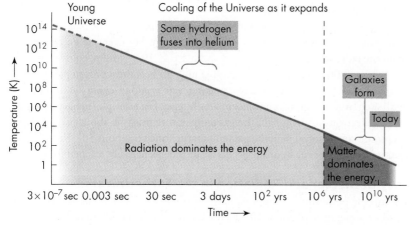

FIGURE 17.14
A plot of how the temperature of the Universe changes with time according to the Big Bang theory.

17.4 THE ORIGIN OF THE UNIVERSE

Where did the Universe come from and what was it like in its early stages? Cosmologists study that remote time by extrapolating the present conditions in the Universe backward in time. That is, they use the laws of physics and present conditions to infer what the Universe was like at the time of its birth. We have already described one important conclusion of such extrapolations: the early Universe was hot and dense (as confirmed by the cosmic background radiation). Figure 17.14 shows how its temperature has changed over the last 15 billion years.

The Universe's enormous temperature in its youth, about 10^{13} K, implies that it may have had a very simple structure. At such a temperature, the Universe was gaseous with its matter and radiation mingled in a manner unlike their sharp distinction today. The mingling occurred because, in a high-temperature gas, radiation can be so energetic that it can be converted into matter. Thus, in the early Universe, matter and radiation behaved almost as a single entity.

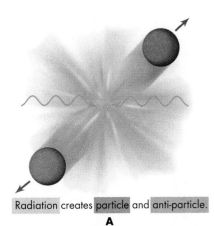

Radiation creates particle and anti-particle.

A

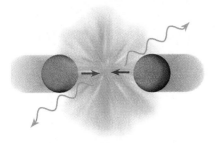

Particle and antiparticle annihilate, creating radiation.

B

FIGURE 17.15
(A) The energy of electromagnetic radiation can be converted into mass with the formation of a particle and its antiparticle. (B) The collision of a particle and its antiparticle leads to their annihilation and conversion of their mass back into energy.

Radiation, Matter, and Antimatter in the Early Universe

Einstein's equation, $E = mc^2$ equates an amount of energy, E, to an amount of mass, m, times the speed of light, c, squared. It tells us that stars can obtain their energy from mass lost when, for example, hydrogen is converted to helium, as discussed in chapters 11 and 13. In the early Universe, the implication of this energy-mass relation was reversed: a quantity of energy E may have turned into particles whose mass equaled E/c^2. Scientists know that matter can be created this way: they can observe high-energy radiation (gamma rays) forming particles in laboratory experiments. Given that energy can change into matter, it seems likely that high-energy radiation in the early Universe might do the same. Such transformations of energy into mass, however, always create a *pair* of particles, as illustrated in figure 17.15A. Moreover, the particle pair has two special properties: the particles must have opposite charge (or no charge), and one of the pair must be made of ordinary matter whereas the other must be made of what is called **antimatter**.

We briefly discussed antimatter in chapter 11 where we saw that the electron has an antimatter companion particle, the positron, with identical mass, but opposite electric charge. Protons also have a corresponding antiparticle called the anti-proton, and all other subatomic particles are similarly paired.

Antimatter has the important property that a particle and its antiparticle are annihilated on contact, leaving only high-energy radiation, as depicted in figure 17.15B. Thus radiation (electromagnetic energy) can become matter, and matter can become radiation. In the early Universe this shifting between mass and energy happened continually as particles formed and were annihilated.

Energetic radiation cannot turn into just any pair of particles: the pair's combined mass multiplied by c^2 must be less than the radiation's energy. For example, to become an electron pair, the radiation needs an energy of about 1.6×10^{-13} joules, equivalent to a wavelength of about 0.0012 nanometers (a gamma ray). To become a proton pair the radiation needs roughly 1800 times more energy, making it a very high-energy gamma ray. Radiation with such large energies occurs only in very hot gas. For example, to have enough energy to make an electron-positron pair, the gas temperature must be about 6×10^9 Kelvin. To make a proton-anti-proton pair, the gas temperature must be about 10^{13} Kelvin. Such high temperatures do not occur in ordinary stars, but they were easily reached in the Big Bang.

History of Matter and Radiation in the Early Universe

Astronomers hypothesize that approximately 1 microsecond after the Big Bang the Universe had a temperature of about 10^{13} K, hot enough for its radiation to be converted into not only protons and anti-protons, but even quarks and antiquarks, the particles that make up protons and neutrons. At that time, the Universe we now see was packed into a volume smaller than the Earth's orbit around the Sun. This volume, an inferno jammed with quarks and antiquarks created from the high-energy radiation, began to expand at nearly the speed of light. As space expanded, this dazzling ball of matter and radiation cooled. The cooler radiation was no longer energetic enough to create new quark-antiquark pairs. Thus, as quarks annihilated antiquarks, they were no longer recreated and most of the quarks therefore disappeared from the Universe. This massacre was not total, however. Theories of subatomic particles predict that a tiny excess of quarks, literally, *one in a billion,* survived with no antiquark to annihilate. These survivors then combined with one another to make the protons and neutrons of the Universe around us. Without that tiny imbalance in the annihilation of quarks, our Universe would have been essentially empty. Every particle would have an antiparticle with which it could combine and be annihilated into radiation, leaving the Universe filled only with the cosmic background radiation and containing no matter at all.

Space, by then filled with ordinary matter and radiation, continued to expand, and for about 5 more seconds was still hot enough to create electron-positron pairs. At the end of that brief interval (about the time needed to take a deep breath), the Universe cooled to a few billion degrees and its creation of matter ceased.

The next landmark in the history of the early Universe occurred about 3 minutes after the Big Bang when the expansion of space had cooled the Universe to a few hundred million

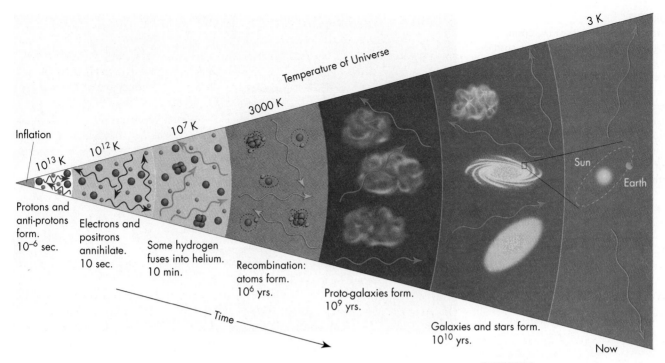

Inflation

10^{13} K

10^{12} K

10^{7} K

3000 K

Temperature of Universe

3 K

Protons and anti-protons form. 10^{-6} sec.

Electrons and positrons annihilate. 10 sec.

Some hydrogen fuses into helium. 10 min.

Recombination: atoms form. 10^{6} yrs.

Proto-galaxies form. 10^{9} yrs.

Galaxies and stars form. 10^{10} yrs.

Sun

Earth

Now

Time

FIGURE 17.16
A sketch depicting the history of matter and radiation in the early Universe.

degrees. Such conditions are similar to those in the core of a very hot star, and just as the star can fuse protons into helium, so did the Universe. Roughly one quarter of the Universe's protons were changed into helium at that time. That helium, mixed with the more abundant hydrogen, permeates all space, explaining why nearly all stars have the same proportion of hydrogen to helium in their outer layers—71% H and 27% He—as discussed earlier.

For the next half million years little happened in the young Universe apart from further expansion and cooling. Eventually the temperature dropped to about 10,000 K, and electrons and protons had low enough energy to bind together to form hydrogen atoms. This period (known as the recombination era because of the combination of electrons with nuclei) marks the time when the Universe's radiation and matter ceased to interact strongly. Before recombination, electrons and radiation collided violently enough to exchange energy and maintain the same temperature. However, when electrons became attached to nuclei and formed hydrogen and helium atoms, their interactions with radiation became much weaker because the atoms have no overall electric charge. Thus, before recombination, the Universe behaved as a single blend of matter and radiation, but after recombination, its matter and radiation acted as separate ingredients.

The changing behavior of the early Universe's matter and radiation is summarized in figure 17.16. The diagram shows how the temperature of the Universe has changed as the Universe has expanded and cooled and changed from quarks to atoms of relatively cool hydrogen and helium. One vital stage remains to link that ancient time to the present: the formation of galaxies.

The Formation of Galaxies

From the ages of the stars within galaxies, astronomers deduce that the Milky Way and other galaxies are at least 10 billion years old. Galaxies must therefore have formed soon after the recombination of protons and electrons in the early Universe. In considering the results of theoretical calculations, astronomers have come to believe that gravity pulled gas clouds into protogalaxies, much the way protostars formed. But the details remain obscure. For example, the amount of matter observed in the Universe exerts too feeble a gravitational attraction for this process to have been able to form galaxies in the time available. Many astronomers therefore think that additional matter must be present to create a stronger gravity to enhance galaxy formation. This apparent need for nonluminous, or

FIGURE 17.17
A map of the brightness of the cosmic background radiation made by the cosmic background explorer (COBE) satellite. Notice the slight patchiness of the brightness. Each pink patch may represent a "lump" of matter from which groups of galaxies ultimately grew. The patches were approximately one half billion light years across when they emitted the radiation. (Courtesy NASA Goddard Spaceflight Center: COBE Science Working Group.)

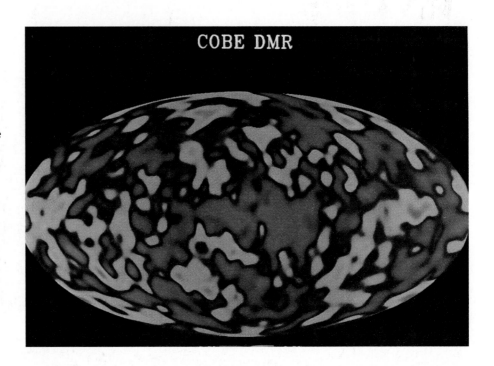

dark, matter to aid galaxy formation strengthens astronomers' suspicion that unseen dark matter might also play a role in closing the Universe.

If dark matter exists, its gravity should draw it into clumps, which then might pull in ordinary matter to form galaxies. According to this theory, galaxies form more readily in areas where dark matter has already clumped. This theory, called "biased galaxy formation," predicts that the regions rich in dark matter should also be rich in galaxies, whereas regions poor in dark matter should have few or no galaxies. The former regions are identified as the galaxy chains and sheets, and the latter empty regions are the voids, discussed in chapter 16. Thus the theory accounts for the distribution of galaxies and voids. However, as yet, no definitely young galaxies or uncondensed gas have been detected in such voids despite intensive searching. A more serious problem is that most theories of galaxy formation predict that the hot gas in clumps of matter that become galaxies should interact with the cosmic background radiation and make its brightness vary across the sky in a patchy fashion. Astronomers detect such patchiness (fig. 17.17), but it is very faint, leaving them with at least two major problems in current theories of galaxy formation. First, they can find very few young galaxies and have so far seen none in their formation stage, and second, the birth of galaxies seems to have affected the cosmic background radiation much less than expected.

Galaxy formation is, however, only one of many unsolved mysteries of the young Universe. For example, where did the original matter and energy of the Big Bang come from?

17.5 THE INFLATIONARY UNIVERSE

The classic Big Bang theory allows cosmologists to understand the Universe back to about 1 microsecond after its birth. But what happened earlier? Current theories suggest that during that first microsecond the Universe was even hotter, denser, and more energetic than we have yet described. Cosmologists hypothesize that before the Big Bang, the visible Universe was packed into a volume smaller than a proton and that its temperature was extremely hot. It was so hot, in fact, that forces such as gravity acted very differently from the way they do now. For example, some theories imply that, under such energetic conditions, gravity, rather than being a force of attraction, would be a *repulsive* force creating a violent expansion, which cosmologists call **inflation.**

Cosmologists hypothesize that the inflation stage of the early Universe lasted a mere 10^{-43} second, during which time space expanded explosively in size, its radius swelling by a factor of 10^{25}. If the period at the end of this sentence expanded by a similar factor, it would end up the size of the Local Group of galaxies. During that expansion, the Universe cooled, and the forces within it evolved to their present form. The inflationary stage ended with the Universe an incandescent fireball that became what we previously called the Big Bang. Thus the Big Bang originated from the inflationary stage, just as the Universe we see around us originated from the Big Bang.

Inflationary models of the first moments of the Universe mark the frontier of our understanding of the cosmos and offer tentative explanations of some of its most mysterious features. For example, some of these theories suggest that the Universe may have been created literally from nothing. Some imply the existence of other universes forever separated from our own. Still others suggest that space has more dimensions than the four we are familiar with: three of space and one of time. In fact, according to some versions of these theories, space has as many as 10 or 11 dimensions. Inflationary models, however, also deal with more prosaic questions such as, "Why is the energy of expansion so nearly in balance with the collective gravitational force between galaxies?" "Why (at least as far as we can tell) is the Universe all ordinary matter with essentially no antimatter?" "Why is the cosmic background radiation so smooth but the distribution of matter so lumpy?" Perhaps the most intriguing aspect of these theories is that they hypothesize links between the Universe at large, the properties of subatomic particles, and the fundamental forces of nature. These theories are extremely complex and not well understood. Nevertheless, because they deal with some amazing properties of the Universe, we will sketch below some of these startling ideas.

EXTENDING OUR IMAGINATION

NEW IDEAS IN COSMOLOGY

Grand Unified Theories

Inflationary models are based on theories that attempt to unify the basic forces of nature: gravity, the electromagnetic/weak force, and the strong force. We discussed gravity and the electromagnetic force in chapters 2 and 3. The weak component of the electromagnetic force is responsible for radioactive decay and, until recently, was treated as a separate force. Today, physicists generally refer to it as the *electroweak force* to emphasize the concept of unification. The strong force's most direct effect is to bond quarks into protons and neutrons. However, it also creates the nuclear "glue," the force between protons and neutrons that links them together to form the nucleus of atoms, as mentioned briefly in the Introduction and chapter 11. In the Universe today, each force dominates at a different scale. Gravity dominates at cosmic scales. Electromagnetic forces dominate at the scale of atoms and molecules. The nuclear force dominates in the nucleus of atoms. In the early Universe, however, the strong and

electromagnetic/weak forces may have been blended into a single effect, a grand unified force. Such hypotheses are therefore called **grand unified theories,** or GUTs, for short. Some theories indicate that gravity may also have been part of the unification, which, if true, leads to the satisfying simplification that a single force controlled all aspects of the early Universe.

According to theory, unification of the forces results from the unimaginably high temperature of the early Universe. If the Universe really was compressed into the tiny volume implied by inflationary theories, its temperature must have been not trillions of degrees but trillions of times hotter than trillions of degrees—about 10^{27} K. At such enormous temperatures, the contents of the Universe were incredibly energetic, and high energy alters how matter responds to the forces. For example, a B-B flicked at a pane of glass with your finger (low energy) bounces off the glass as if repelled. If it is shot from a pellet gun (high energy), however, it pierces the glass. Thus the repelling effect "vanishes" at high energy. So, too, in the high-energy environ-

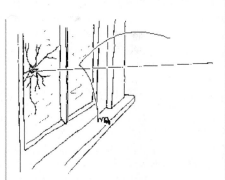

ment of the early Universe, forces may have behaved very differently from the way they do today in our low-energy world, making their unification a possibility. The forces stayed unified only while the Universe was extremely hot (10^{27} K). Even a tiny amount of expansion cooled the early Universe enough to destroy unification by a process called "symmetry breaking."

Symmetry breaking is a little like the change that takes place in a thick solution of sugar dissolved in water. As time passes, solid crystals of sugar form in the liquid medium. Similarly, in the early Universe, an

Continued

EXTENDING OUR IMAGINATION—*Continued*

initially unified force separates into a mixture of forces in a medium known as the **false vacuum** The false vacuum is called "false" in part because its energy is not zero. It also has the bizarre property that it has a pressure that is *negative*. Negative pressure means that if the false vacuum expands, it does not cool as ordinary matter does. Instead, it releases energy. Even more strange, however, the negative pressure leads to negative gravity: a gravity that *repels*. According to Newton's law of gravity, all mass exerts a gravitational force. According to Einstein's theory of relativity, mass and energy are equivalent, that is, $E = mc^2$. Thus we might expect that energy in any form can produce a gravitational force. Just such a possibility is developed further in Einstein's general theory of relativity. In particular, according to the general theory of relativity, the heat energy responsible for generating pressure can exert a gravitational attraction. In ordinary stars, the energies are too small for this effect to be important and pressure behaves in its usual way. However, in the false vacuum the energies are vastly larger and, because pressure in the false vacuum is negative, so too is the gravity it creates. That negative gravity is a repulsive force that can make the false vacuum expand and, as it expands, the energy it releases makes its negative gravity even stronger, so it expands even faster.

Regardless of the force that causes inflation in the early Universe, energy is required to drive the expansion. That energy comes from the separation of the previously unified forces into their more familiar separate forms. Gravity separates first from the electroweak and strong forces; later the electroweak and strong forces separate. We saw earlier that energy can produce matter, for example, an electron-positron pair. Similarly, the energy released as the fundamental forces separate also creates matter, and the negative gravity of the false vacuum drives that matter and space apart. As you might imagine, the details of this process are extremely complex. Nevertheless, cosmologists calculate that the energy released as the forces separate in 10 kilograms of false vacuum could create the mass of the entire Universe. In releasing this energy, moreover, the Universe swells from a microscopic size far smaller than a proton to the size of a grape (about 1 centimeter) and

thereby generates the Big Bang, as discussed earlier. Figure 17.18 summarizes how inflation may have occurred.

An analogy may help explain where the energy liberated by symmetry breaking comes from. When water freezes, a kind of symmetry change occurs as the water molecules align themselves into a solid crystal. Energy is released in this process as the molecules give up their energy of motion and lock into the rigid grid of the ice. That energy is sometimes used to protect delicate fruit crops from frost damage. Fruit growers spray trees with water as the air temperature nears freezing. The congealing water then liberates energy to the fruit or blossoms it coats, preventing the crop itself from freezing. The protection is only good for light freezes, but it illustrates the reality of energy liberation caused by symmetry changes.

Inflation also helps explain other important features of the Universe. For example, according to the theories on which the inflationary model is built, the symmetry breaking should create matter with a slight imbalance between the number of particles of ordinary matter and antimatter. We saw earlier that such an imbalance is critical because without it, matter and antimatter might totally annihilate each other leaving a universe devoid of matter and containing only radiation.

Another aspect of the Universe nicely explained by inflation is the smoothness of the cosmic microwave background. In the original Big Bang theory, the visible Universe was initially contained in a volume about 10 centimeters (4 inches) in diameter. Although small, the Universe expands so quickly from this size that its radiation and matter cannot mix smoothly. Thus theories based on the standard Big Bang predict that the brightness of the cosmic background radiation should be uneven. In the inflationary universe, however, the visible Universe was originally one trillionth the size of a proton. The incredible smallness of such a volume allowed radiation and the forces acting at that time to thoroughly mix and smooth the matter and energy. That smoothness was preserved during the early stages of expansion and is retained to this day in the extreme uniformity of the 2.7 K cosmic background radiation. This radiation is not perfectly uniform, however. The satellite COBE (Cosmic Background Explorer) has detected tiny differences in the intensity of the background radiation in different directions, as can be seen in figure 17.17. Those differences may arise from clumpiness of the gas in the very young Universe, but cosmologists are still uncertain about their origin.

The huge expansion associated with inflation may also explain why the gravitational energy that binds the Universe is

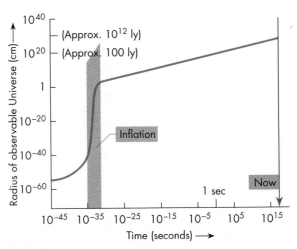

FIGURE 17.18
A sketch showing how the radius of the Universe suddenly inflates.

Continued

EXTENDING OUR IMAGINATION—*Continued*

approximately equal and opposite to its energy of expansion. We said earlier that if the Universe contains substantial quantities of dark matter, it may be closed. That is, the density may just barely exceed the critical density. Inflationary theories imply that these two densities are almost exactly equal. If the density of the Universe was even a tiny bit less than the critical density, it would make gravity too weak to stop expansion. Then, during the period of inflation, the Universe would have expanded so rapidly that galaxies by now would be enormously far apart and the Universe would appear virtually empty. Conversely, if the density was even a tiny bit greater than the critical density, it would make gravity strong enough to stop expansion, and by now the Universe would have recollapsed. Thus the fact that the Universe has neither recollapsed nor expanded into a nearly empty state implies just the precise balance between gravity and expansion energies predicted by the inflationary theory.

Finally, although inflation smooths out the *early* and large-scale irregularities, it actually helps smaller irregularities to develop in later stages into galaxies. To form galaxies, irregularities must become relatively more dense than their surroundings. The more the surrounding density decreases, the more the density in local regions may grow relative to that background. Inflation of the Universe makes the average density drop enormously compared with what occurs in standard Big Bang universes. Thus the first steps toward the formation of galaxies and their intervening voids may have been taken when the Universe was smaller than an atom.

Other Universes?

The inflationary universe has yet another curious property: It allows adjacent regions of space to be completely separated from each other with no possible contact between them. Such isolated regions might well be considered separate universes, independent of ours and completely unobservable by us. Furthermore, according to inflationary theories, these other universes may be either expanding or contracting.

Inflation creates such sections of space much as bubbles form in a pot of boiling water. As water begins to boil, regions of the liquid suddenly change from liquid to gas (steam) and begin expanding. We see that region of steam as a bubble. Some bubbles expand, rise to the surface, and burst, whereas others collapse. So too in inflationary theories, entire universes may form and dissolve, entirely unknowable to us (fig. 17.19). It would be amazing indeed if all the wonderful intricacy and beauty of our Universe is merely a single bubble in an even vaster sea of space, a space that we can never see.

FIGURE 17.19
A sketch depicting bubbles that might become separate universes forming in an inflating universe.

 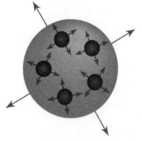 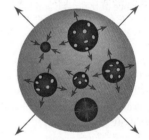

"Bubbles" grow.   Expansion cools Universe.   "Bubbles" become separate universes?

SUMMARY

Because distant galaxies are receding from the Milky Way, most astronomers believe that the Universe in its youth was much smaller, hotter, and denser than it is today. By extrapolating the present recessional motion of galaxies back in time, astronomers deduce that the Universe was created approximately 10 to 15 billion years ago in the Big Bang, a release of energy that initiated the expansion of space and the creation of matter. That burst of energy cooled as space expanded. At the time of the Big Bang, the temperature of the radiation was about 10^{13} K, but the expansion of space has stretched the wavelengths of the radiation, cooling it to about 3 K. That cool radiation has been detected and is called the "cosmic microwave background." It is additional evidence for a hot dense beginning of the Universe.

If the expansion continues forever, the Universe is said to be open. If the expansion stops and the Universe recollapses at some time in the distant future, the Universe is said to be closed. Whether the Universe is open or closed depends on its density and rate of expansion. If the density is less than a critical value, not enough mass is present for gravity to stop the cosmic expansion, and the Universe is therefore open. Observations, in fact, imply that this is the case, and so unless large quantities of dark matter exist, the Universe is open.

Astronomers are currently studying the state of the Universe before the Big Bang. Some theories suggest that the Universe may have originated from a peculiar state known as the "false vacuum," in which gravity is a repulsive force. While in this false vacuum state, the visible Universe was trillions of times smaller than a proton, but the negative gravity occurring then "inflated" the Universe in an almost unimaginably brief time, cooling it, and releasing huge amounts of energy, which became the Big Bang.

QUESTIONS FOR REVIEW

1. Why do astronomers believe the Universe is expanding?
2. What is meant by a cosmic horizon?
3. What is Olbers' paradox?
4. What is the cosmic background radiation? What is its origin?
5. Why do astronomers believe the early Universe was hot and dense?
6. What is meant by the age of the Universe? How old is it? How is its age found?
7. How was helium formed from hydrogen in the early Universe?
8. What explanation do astronomers offer for why the relative abundance of hydrogen and helium is uniform?

9. Is our cosmic horizon the same as that of beings in another galaxy? Why?
10. What is meant by inflation in cosmology?
11. What is the Big Bang? What are some of its properties?
12. What problems of the Big Bang theory are resolved by inflationary models?
13. What is meant by open and closed universes?
14. How do the shapes of open, closed, and flat universes differ? How does their energy content differ?
15. What is "unified" in grand unified theories?
16. What is meant by dark matter?

THOUGHT QUESTIONS

1. M31, the great galaxy in Andromeda and a member of the Local Group of galaxies, is approaching us. Does this mean that the Hubble law is wrong? Why?
2. Suppose the Hubble constant were discovered to be five times larger than is now believed. How would that affect our estimate of the age of the Universe?

3. Does it bother you that the Universe may expand forever?
4. Why do most stars, including our own Sun, contain more helium than the 24% formed during the Big Bang?

TEST YOURSELF

1. From what evidence do astronomers deduce that the Universe is expanding?
   (a) They can see the disks of galaxies getting smaller over time.
   (b) They see a redshift in the spectral lines of distant galaxies.
   (c) They detect cosmic background x-ray radiation.
   (d) They can see distant galaxies dissolve, pulled apart by the expansion of space.
   (e) All of the above.
2. Why do astronomers think the Universe was much hotter at its birth than it is now?
   (a) They detect cosmic microwave background radiation.
   (b) The amount of helium observed in all stars implies that the early Universe was hot enough to fuse some of its hydrogen into helium.

   (c) They detect intense x-ray emission from gas in galaxy clusters.
   (d) All of the above.
   (e) Only (a) and (b).
3. What is meant by inflation in the early Universe?
   (a) The force of gravity suddenly grew stronger in the distant past.
   (b) Protons expanded to the size of stars, which was how our Sun formed.
   (c) The Universe increased dramatically in size in an extremely brief period of time.
   (d) The number of galaxies that we see at large distances is much greater than the number we can see near us.
   (e) The diameter of distant galaxies is much greater than the diameter of galaxies near us.

4. What is meant by the cosmic background radiation?
   (a) It is radiation from distant quasars.
   (b) It is radiation from hot gas in intergalactic space.
   (c) It is radiation from the first stars formed when the Universe was young.
   (d) It is radiation created during the early days of the Universe.
   (e) It is the explanation of Olbers' paradox.

5. What is meant by "open" and "closed" universes?
   (a) An open universe expands forever.
   (b) An open universe has no beginning.
   (c) A closed universe has no beginning.
   (d) A closed universe expands to a maximum size and then contracts.
   (e) Both (a) and (d).

FURTHER EXPLORATIONS

Brush, Stephen G. "How Cosmology Became a Science." *Scientific American* 267 (August 1992): 62.

Davies, Paul. "Everyone's Guide to Cosmology." *Sky and Telescope* 81 (March 1991): 250.

———. "The First One Second of the Universe." *Mercury* 21 (May/June 1992): 82.

———, ed. *The New Physics.* Cambridge, New York: Cambridge University Press, 1989.

Gribbin, John. *In Search of the Big Bang: Quantum Physics and Cosmology.* Toronto, New York: Bantam Books, 1986.

Gulkis, Samuel, Philip M. Lubin, Stephan S. Meyer, and Robert F. Silverberg. "The Cosmic Background Explorer." *Scientific American* 262 (January 1990): 132.

Guth, Alan H. "The Inflationary Universe." *Scientific American* 250 (May 1984): 116.

Halliwell, Jonathan J. "Quantum Cosmology and the Creation of the Universe." *Scientific American* 265 (December 1991): 76.

Hawking, Steven W., and Roger Penrose. "The Nature of Space and Time." *Scientific American* 275 (July 1996): 60.

Horgan, John. "Universal Truths." *Scientific American* 263 (October 1990): 108.

Kinney, Anne L. "Fourteen Billion Years Young." *Mercury* 25 (March/April 1996): 29.

Krauss, Lawrence M. "Dark Matter in the Universe." *Scientific American* 255 (December 1986): 58.

MacRobert, Alan. "Beyond the Big Bang." *Sky and Telescope* 65 (March 1983): 211.

Osterbrock, Donald E., Joel A. Gwinn, and Ronald S. Brashear. "Edwin Hubble and the Expanding Universe." *Scientific American* 269 (July 1993): 84.

Pagels, Heinz R. *Perfect Symmetry.* New York: Simon and Schuster, 1985.

Peebles, P., E. James, David N. Schramm, Edwin L. Turner, and Richard G. Kron. "The Evolution of the Universe." *Scientific American* 271 (October 1994): 52.

Powell, Corey S. "The Golden Age of Cosmology." *Scientific American* 267 (July 1992): 17.

———. "Mirroring the Cosmos." *Scientific American* 265 (November 1991): 112.

Ronan, Colin A., ed. *The Universe Explained: The Earth-Dweller's Guide to the Mysteries of Space.* New York: H. Holt, 1994.

Schramm, David N. "Dark Matter and Cosmic Structure." *Sky and Telescope* (October 1994): 28.

Silk, Joseph. *The Big Bang: Revised and Updated.* San Francisco: W. H. Freeman and Co., 1989.

Silk, Joseph, Alexander S. Szalay, and Yakov B. Zel'dovich. "Large Scale Structure of the Universe." *Scientific American* 249 (October 1983): 72

Trefil, James S. *The Moment of Creation.* New York: Scribner, 1983.

KEY TERMS

Life in the Universe

The evolution of the Universe from Big Bang to galaxies, stars, and planets has along the way created us. This evolutionary path from raw quarks to thinking beings who seek to understand how they fit into this grand scheme—what eighteenth-century scientists called "the great chain of being"—is one of many marvels of the Universe and is the theme of this section.

In earlier chapters we described astronomical evolution and how the Earth and Sun fit into the cosmos. Now we turn to life on Earth and how it may have formed and developed. We will see that the existence of life on our own planet is suggestive of the possibility of its existence elsewhere, and so we will also speculate about whether we are alone, leading to the question, "Do other intelligent civilizations exist?" Astronomers are strongly divided about such possibilities. Some argue that such life is highly probable, whereas others argue that it is highly improbable. Thus, one goal of this section is to describe the evidence that goes into such judgments.

We will begin our discussion by examining the long history of life on Earth, looking at the factors thought to play a role in its development here. We will also discuss the likelihood of similar conditions elsewhere, discovering that planets suitable for life are probably numerous. In addition, we will describe how some astronomers have been motivated by such arguments to search for radio signals from other civilizations. Finally, we will speculate about the role of life in the history of our planet—the so-called Gaia hypothesis—and whether our existence may even say something about the nature of the Universe itself.

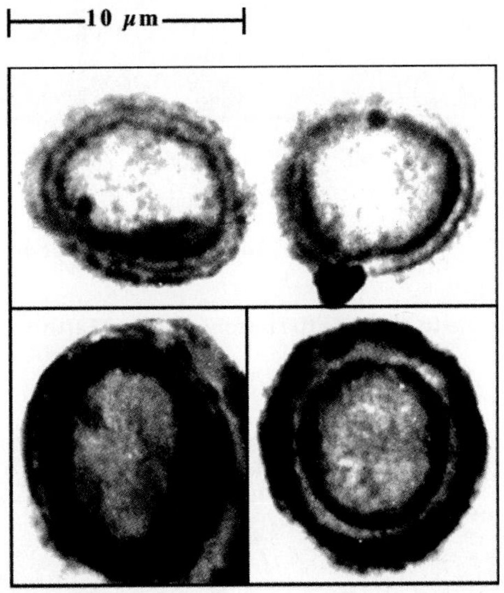

Life on Earth
History of Life on Earth

Life has existed on Earth for nearly the entire history of our planet. For example, fossil algae and bacteria such as illustrated in figure E3.1 occur in rocks 3.8 billion years old. In fact, one of the most ancient life forms, algal colonies called "stromatolites," still exist in shallow, warm, salty bays such as those along the coast of western Australia.

Given that Earth formed about 4.5 billion years ago and that its surface was probably molten rock for almost 1 billion years, life began remarkably quickly. How quickly, we can perhaps appreciate from figure E3.2,

FIGURE E3.1
Photographs of fossil bacteria. (Andrew H. Knoll. Raven and Johnson, Biology, ed.3. Mosby, 1992)

which shows a time line for the history of life on Earth. This line shows the epoch at which various life forms first appear in the fossil record. Their dates have been measured by radioactivity in their associated rock, as described in chapter 4, or other means. Along with the ages, the figure also illustrates some of these ancient life forms.

Scientists conclude from such evidence that only extremely simple life forms such as algae and single-celled animals existed for roughly three fourths of Earth's history. Then, approximately 600 million years ago, more complex forms such as invertebrates developed, and about 500 million years ago, only a little more than one-tenth of the Earth's lifetime, shells and crustaceans appeared. Mammals and dinosaurs both appeared roughly 250 million years ago, but the latter were wiped out about 65 million years ago in the great Cretaceous extinction, caused perhaps by a totally chance event—an asteroidal collision, as described in chapter 10. Hominids, our immediate ancestors, appeared roughly 5.5 million years ago, but our own species, *Homo sapiens,* evolved only about 500,000 years ago. Earth has therefore existed as a planet roughly 10,000 times longer than we have as a species. Thus, if all Earth's history were compressed into a year, humans would appear only in the last hour, and civilization in the last few minutes.

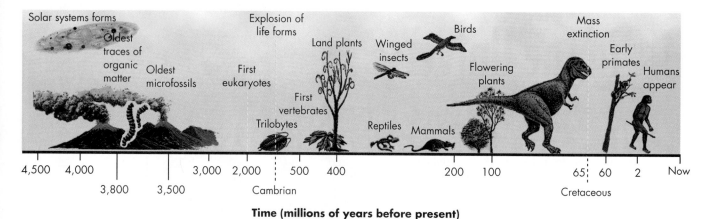

FIGURE E3.2
A time line to illustrate the history of life on Earth.

FIGURE E3.3
Photographs of some of Earth's life forms. (Visuals Unlimited/Tom Stack & Assoc./Animals Animals, Earth Scenes.)

Unity of Living Beings

Life on Earth shows a dazzling variety of forms from butterflies to birds, worms to whales, and mushrooms to maples, as shown in figure E3.3. Despite that variety, all living beings on Earth have an amazing underlying unity of structure, reproduction, and metabolism. For example, all living organisms use the same kinds of atoms for their structure and function (hydrogen, oxygen, carbon, nitrogen, and iron, for example). These atoms are not only widely used, but they are also abundant throughout the Universe. Thus life on Earth gen-

erally is very economical in its use of the same chemical substances in widely differing organisms. Moreover, life tends to draw on the materials with which our planet is most richly endowed: we are made primarily of the same substances that make up our ocean and atmosphere.

The chemical elements that compose living things are linked into long-chain molecules for their structure and function. The primary structural units of many of these chain molecules are about 20 amino acids. **Amino acids** are organic molecules containing hydrogen, carbon, nitrogen, and oxygen atoms arranged in a certain

FIGURE E3.4
Diagrams illustrating the structure of several amino acids. (Raven and Johnson, Biology, ed. 2. Mosby. 1989)

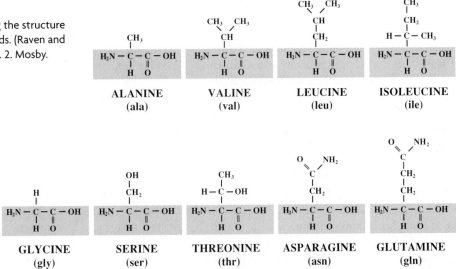

ALANINE (ala) VALINE (val) LEUCINE (leu) ISOLEUCINE (ile)

GLYCINE (gly) SERINE (ser) THREONINE (thr) ASPARAGINE (asn) GLUTAMINE (gln)

pattern (fig. E3.4). A few amino acids also contain one or more sulfur atoms.

Moving up to the next level of complexity, we find that all living things use amino acids as the structural units of more complicated molecules called **proteins.** Protein molecules give living things their structure. For example, the scales of reptiles and fish are composed of the protein keratin, and the stiff cartilage of our own bodies is composed of the protein collagen. Proteins not only give cells their structure, but also supply their energy needs. That energy supply reveals yet another underlying unity among living things: all cells use the same kind of molecule, adenosine triphosphate (abbreviated as ATP) to supply energy for action and growth.

This unity of structure and energy supply is echoed in the reproduction of life forms. All single-celled organisms divide to create new cells; all multicelled organisms produce egg cells, which divide to create new cells. Moreover, in all cases "parents" pass on genetic information to their offspring by the same molecule, **DNA.**

DNA stands for deoxyribose nucleic acid. It is a massive molecule consisting of two extremely long chains of smaller molecules wound around each other and linked by molecules called "base pairs." Thus DNA has the appearance of a twisted ladder with the role of the connecting rungs played by the base pairs (fig. E3.5). When a cell reproduces, the DNA separates into two individual chains. Each chain then adds a new base pair at each point along its length to match the one previously present on the other chain. Thus each strand makes a copy of the other so that after division and replication a new DNA molecule exists identical to the original one.*

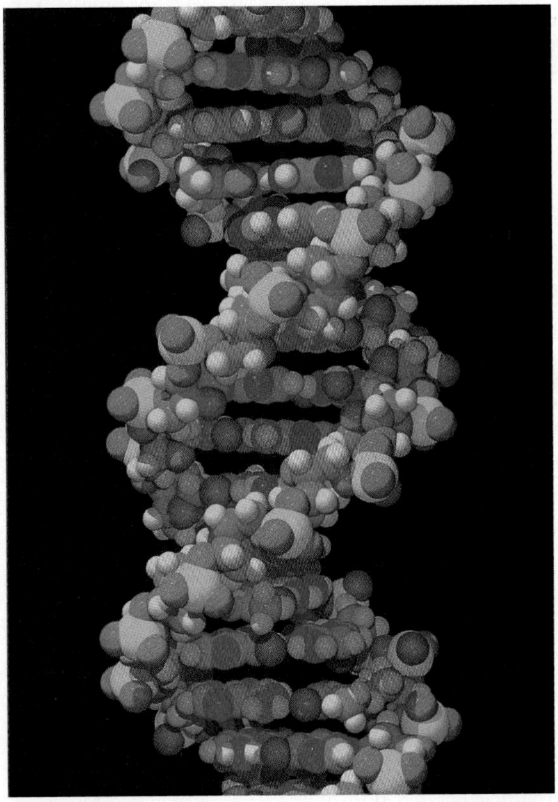

---

*Copying errors—mutations—sometimes occur. Most are harmful and lead to the organism's death. Nonlethal ones make evolution possible.

FIGURE E3.5
A diagram illustrating the structure of the DNA molecule. (Ken Eward/BioGrafx)

That all terrestrial organisms use DNA for their reproduction gives life a truly remarkable unity.

Unity can also be seen in the way higher organisms function. For example, plants derive their food from carbohydrates that they manufacture during photosynthesis. The chlorophyll they use in this process is structurally very similar to the hemoglobin that transports oxygen in the blood of animals. These two molecules differ primarily in that chlorophyll is built around a magnesium atom, whereas hemoglobin is built around an iron atom. Likewise, cellulose, the structural material of plants, has a molecular structure very similar to chitin, the structural material in insect bodies and crustacean shells.

Finally, living systems exhibit a unity in their functions, such as the chemical processes that move biologically important molecules into and out of cells.

Deductions from the Unity of Life and the Time Line

What conclusions can we draw from the chemical similarities of living beings and their ancient history as deduced from the time line? First, the chemical similarity almost inescapably suggests that all life on Earth had a common origin. Second, the antiquity of life suggests that under suitable conditions life develops rapidly from complex molecules. Thus we might infer that wherever such conditions occur, so too will life.

We should treat these conclusions with great caution, however, for they are based on only a single case: life on Earth. For example, if the first time you hit a golf ball you make a lucky hole-in-one, it does not mean that you will do so again anytime soon. Thus the fact that life exists on Earth may say next to nothing about its chances elsewhere. Supposing, however, we do choose to speculate. What can we say about the origin of life and its likelihood elsewhere?

Origin of Life on Earth

Most scientists today think that terrestrial life originated from chemical reactions among complex molecules present on the young Earth. This belief dates back to at least the time of Charles Darwin, but it was strongly bolstered in 1953 by an experiment performed by Stanley Miller—then a graduate student at the University of Chicago—and his professor, Harold Urey. Miller and Urey filled a sterile glass flask with water, hydrogen, methane, and ammonia—gases thought to be present in the Earth's early atmosphere. They then passed an electric spark through this mixture. The spark generated both visible and ultraviolet radiation, which triggered reactions in the gas/water mixture. At the end of a week, they analyzed the mixture, finding in it a variety of organic molecules as well as five of the amino acids used to make proteins. They therefore concluded that at least some of the ingredients necessary for life could have been generated spontaneously on the early Earth.

The success of Miller and Urey's experiment motivated other researchers to perform similar but more complex experiments. For example, the U.S. chemist Sydney Fox took complex organic molecules such as those made by Miller and Urey and by repeatedly heating and cooling them in water was able to create short strands of proteins called "proteinoids." More interestingly, however, Fox discovered that these proteinoids spontaneously formed droplets reminiscent of cells. These droplets have no power to reproduce, nor do they show any of the complex structure commonly found in living cells, but they do demonstrate that organic material can spontaneously give itself structure and wall itself off from its environment, a step probably taken by the first living things.

Even the energy needs of living things can be produced in Miller-Urey type of experiments. For example, the Sri Lankan-American biochemist Cyril Ponnamperuma and the U.S. astronomer Carl Sagan showed that ATP can be readily synthesized from compounds similar to those generated by Miller and Urey, demonstrating one additional step in creating living from nonliving material.

Why then do we not observe similar processes occurring today on Earth? One reason is that our atmosphere now contains oxygen, which quickly attacks organic molecules, breaking them down. Thus the absence of oxygen earlier in the Earth's history—an absence born out by analysis of ancient rocks—was probably crucial for the development of life. Another reason was given by Charles Darwin in his speculations on life's origin: today's living things immediately consume such organic material.

Even without the success of the Miller-Urey experiment in creating organic material, scientists studying the origin of life on Earth are confident that such material would be present on the young Earth. One source of organic material is interstellar matter. We discussed in chapter 15 that many interstellar clouds contain a rich mix of organic molecules, including many that are precursors to amino acids. Because the Solar System and Earth formed from such clouds, complex organic molecules must have been present on or near the Earth from the time of its birth. Even if these molecules were destroyed during the Earth's formation, fresh molecules may have been carried to its surface by planetesimals toward the end of the heavy bombardment of its surface by these planetesimals. In fact, carbonaceous chondritic meteors—probably surviving fragments of the ancient planetesimals—are rich in organic material and some even contain amino acids, as we saw in chapter 10. Finally, some scientists have demonstrated experimentally that the compression and heating of gases resulting from planetesimal bombardment may be as fruitful as the electric discharges used by Miller and Urey to form complex organic materials. Thus astronomers are confident that organic molecules were common on the ancient Earth.

On the other hand, if one assumes that organic matter existed on the early Earth (whether created by lightning, ultraviolet radiation, or planetesimal bombardment), many steps are needed to take it from amino-acid-filled droplets in a primordial ocean to living cells. For example, the structures created must be able to reproduce, a feature not yet seen in any origin-of-life experiments.

Cells replicate today using the double-chained molecule DNA. Some researchers have speculated, however, that the earliest life forms used the slightly simpler molecule, ribose nucleic acid, or RNA for short. In fact, biochemists have shown that if the proper chemicals are present (especially certain enzymes) RNA can replicate in the absence of living cells. Moreover, they have also discovered several other molecules far simpler in structure than RNA that can replicate but, so far, only in a "mechanical" sense. That is, if such molecules are put in a solution containing the necessary raw materials, they will make copies of themselves. In fact, one particular mix of molecules not only replicates, but also can "mutate" into a slightly better replicator. Even so, far more is needed to produce life as we know it. For example, simple life forms must be able to form cells that work cooperatively.

Origin of Complex Organisms

The original cellular life on Earth was probably far simpler than the cells we see today. For example, the earliest cells probably had no nucleus, a condition found now only in bacteria and certain algae. **Prokaryotes,** as these cells without nuclei are known, may accidentally have merged with other cells, a merger that benefited both cells by allowing them to grow and reproduce more rapidly. For example, a cell good at storing energy might have combined with a cell good at reproducing, thereby creating a more complex cell. From such simple beginnings, **eukaryotes** (cells with nuclei) probably evolved and, after them, multicelled organisms. Today, most cells have a complex membrane that holds them together and allows food to enter and waste products to leave. Moreover, most cells other than bacteria and some algae have a nucleus in which genetic information is stored and, in addition, mitochondria, which are tiny bodies within the cell that convert food into energy that the cell utilizes.

The greater complexity of multicelled organisms would in some cases allow them to exploit resources unavailable to simpler beings. For example, an organism that can move from nutrient-poor to nutrient-rich environments may be able to reproduce more prolifically than a stationary creature. Such reproductive advantages combined with changing environments have led to the incredible diversity of life that we see around us today. In fact, such reproductive advantage (natural selection) and random mutations are crucial to evolution and the development of higher life forms, as Charles Darwin explained in 1859 in his monumental work *On the Origin of Species by Means of Natural Selection.*

The processes we have discussed above, though hypothetical, persuade many scientists that life originated on Earth spontaneously. Not all scientists, however, accept this idea. Some favor the idea that life originated elsewhere and was transported to Earth either accidentally or deliberately.

Panspermia

The belief that terrestrial life descended from organisms created elsewhere in the Universe is sometimes called **panspermia.** According to the theory, simple life forms (perhaps bacteria) from some other location drifted from their place of origin across space to Earth. For example, bacteria spawned on some distant planet may have stuck to tiny dust particles that were lofted by winds or volcanic eruptions to the top of the planet's atmosphere. From there, they were subsequently lost to space, where, driven by radiation pressure from stars or stellar winds, they might drift almost anywhere. Although space is highly hostile to advanced life forms, simple organisms may be able to survive long passages through it if embedded in dust grains or otherwise protected. Eventually the dust may have arrived in the Solar System, where it was captured by the Earth's gravity. Sifting down to the surface, the dust infected the virgin Earth, bringing life to our planet.

Panspermia has a certain appeal in that it suggests our roots spring from elsewhere in the Universe. The theory was very popular early in this century, but critics point out that it does not really simplify the problem of the origin if life, it just shifts it elsewhere. Moreover, it further requires getting the life to Earth, a perilous voyage even for a bacterium. For these reasons the panspermia theory has gradually fallen into disfavor.

Life Elsewhere in the Universe

Humans have speculated for thousands of years about life elsewhere in the Universe. For example, about 300 B.C. the ancient Greek philosopher Epicurus wrote in his "Letter to Herodotus" that "there are infinite worlds both like and unlike this world of ours." Epicurus went on to add that "in all worlds there are living creatures and plants." Likewise, the Roman scholar Lucretius wrote about 50 B.C. that "it is in the highest degree unlikely that this Earth and sky is the only one to have been created." Such views were not universal, however. For example, Plato and Aristotle argued that the Earth was the only abode of life, a view that prevailed through most of the Middle Ages. Today, however, influenced by science-fiction movies, television, and books, many people find no difficulty at all in believing that extraterrestrial life exists. In fact, supermarket tabloids make silly claims almost weekly about encounters with aliens. If we look at the facts, however, there is not a shred of evidence that life exists elsewhere. This does not mean there is no extraterrestrial life; it simply means we do not *know* if there is. Thus, with no evidence, what can we say meaningfully about life elsewhere in the Universe?

Are We Alone?

Do other life forms exist in the Universe? Is there somewhere hundreds of light-years from Earth a "student alien" reading a book discussing the possibility of life on planets such as Earth? Scientists do not know, and, in fact, they are strongly divided into two groups. One group of scientists (let us call them "many-worlders") believes millions of planets with life exist in the Milky Way and many of these have advanced civilizations. The other group (let us call them "loners") argues that we are the only intelligent life in the galaxy. We will discuss below how these two radically different points of view arise.

Arguments for Many Worlds

Many-worlders argue as follows: Earth-like planets are common. Life has formed on Earth. Therefore it would be surprising if life did not exist elsewhere. In fact, even if only a small fraction of such planets have life, many advanced civilizations probably exist. Such arguments may actually vastly *underestimate* the number of habitable planets because life forms elsewhere may be able to live in environments intolerable to terrestrial organisms. To see how such numbers are deduced, let us begin by estimating the number of Earthlike planets in the Milky Way.

We start by asking how many Sunlike stars are in the Milky Way. We limit ourselves to such stars because luminous blue ones burn their fuel so rapidly that they will burn out and explode as supernovas before life has much chance to evolve, whereas dim red stars are so cool that planets must be very close to them for conditions to be warm enough for a terrestrial type of life. Given those restrictions on star types, we turn to censuses of stars, which show that stars like the Sun make up about 10% of our galaxy's stars, for a total of about 10^{10} such bodies.

Of these 10^{10} stars, we next ask, How many of them have Earthlike planets? Our own system has two planets out of a total of nine that might harbor life as we know it: Earth and Mars. If we are conservative and say that other systems might not be quite so lucky, we might therefore choose the probability of a life-sustaining planet around a solar type of star to be 1 in 10, or 10%. Taking 10% of the 10^{10} Sunlike stars leaves us with 10^9 habitable planets.

How many of these actually have life on them? That will depend on how easy it is for life to form, a probability that the Miller-Urey experiments indicate may be high but that some astronomers have thought to be very low indeed. In fact, the British astrophysicist Fred Hoyle compared the likelihood of life arising spontaneously to the likelihood of a box containing several hundred tons of aluminum being shaken and by chance assembling itself into a 747 jumbo jet. Although many scientists find such a comparison silly, it illustrates the divergence of opinion on the likelihood of life forming. The difficulty in assessing that probability is that we have only the single case of life here as our basis. If, however, we use Earth as the pattern, the rapidity with which life developed on Earth and the Miller-Urey experiment's success in making molecular precursors to life indicate that the chance of life starting is fairly high, say, 1 in 100 rather than 1 in a million. Given that probability, we end up with $10^9 \times 0.01 = 10^7 = 10$ million worlds with life in our galaxy.

We now go to the next step and ask, On how many of these life-bearing planets do technical civilizations form? The chances of higher life evolving may be very large or very small. Again scientists have no idea. Moreover, even if life succeeds in starting, perhaps it will be annihilated by asteroidal impact or, if it develops a high technology and is careless or thoughtless, life may destroy itself by pollution, nuclear war, or some other disaster. Suppose we are very conservative and say that if life forms, it has a 1 in 10,000 chance of developing a technical civilization. Thus we find that the number of advanced civilizations in our galaxy is $10^7 \times 0.0001 = 1000$. When we further consider that there are billions of other galaxies, it seems very likely that we are not alone.

The above argument was first presented by the U.S. astronomer Frank Drake, who pointed out that one can estimate the number of civilizations by multiplying together the probability of each condition necessary for it, a result now called the "Drake equation." Unfortunately, the accuracy of the probabilities is completely unknown. Perhaps, for example, Earthlike planets are far rarer than we have estimated. Perhaps evolution on other planets produces life forms that are so totally unlike those of Earth that we cannot recognize them as being alive. Furthermore, even if life forms similar to us do evolve elsewhere, the time period during which we could communicate with them might be so narrow that the opportunity passes. For example, if a civilization existed on Mars even as recently as 100 years ago, we could not have communicated with them by radio, and if in the meantime they destroyed themselves somehow, the chance would have been lost forever. Thus the answer given by the Drake equation is filled with uncertainties. Nevertheless, even with very pessimistic estimates, the equation suggests that *many* civilizations other than ours exist. Given their existence, we might then ask, How close is the nearest one? We can estimate the distance as follows.

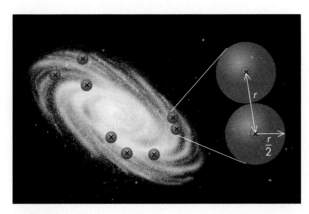

FIGURE E3.6
A sketch illustrating how to estimate the distance between civilizations in the Milky Way.

Suppose the Milky Way contains N civilizations scattered randomly throughout its volume, as illustrated in figure E3.6. We can calculate their average separation, r, as follows. Imagine drawing a sphere of radius $r/2$ around each civilized world. Then N such spheres must fill the volume of the Milky Way. That is,

$$\frac{4\pi}{3}(r/2)^3 N = \frac{4\pi R^3}{3}$$

where, for simplicity, we assume that the Milky Way is a sphere of radius R. We can now cancel $4\pi/3$ to obtain

$$\frac{Nr^3}{8} = R^3$$

Moving the N and 8 to the right-hand side of the equation gives

$$r^3 = 8R^3/N$$

Taking the cube root of both sides gives

$$r = 2R/(N)^{1/3}$$

To evaluate r and thereby find how far away our neighbors are, we set R equal to the Milky Way's radius (taken to be 40,000 light years) and $N = 1000$, the value we deduced for the number of technical civilizations. These choices then give a value for r of 8000 light years. Thus, even if such a large number of civilizations exist in our galaxy, they are all extremely far from us. Such immense separations also make it very unlikely that we will be able to exchange information with other civilizations.

Loners

Loners suggest we are the only advanced life in the Milky Way. Their argument is based on the rapidity with which our species has spread across Earth and the potential for its rapid spread across the Milky Way. Loners claim that once a civilization masters space flight, it will take it at most a few thousand years to colonize planetary systems near their home system and that within a few million years they will colonize the entire galaxy. Moreover, they assume that once such colonization occurs, a civilization will quickly contact other civilizations, either by direct means or by radio communication. Because no other civilization has contacted us, it must mean that no other civilization is out there anywhere near us. Thus we are probably alone in the Milky Way, or nearly alone with at most only a handful of other civilizations. However, this argument, like that of the many-worlders for a multiplicity of civilizations, rests on several unproved assumptions. For example, it assumes that (1) most civilizations are driven to colonize, (2) they seek rather than avoid contact, and (3) they have mastered interstellar flight and communication.

Searching for Life Elsewhere

Is there any evidence for life elsewhere in the Universe? Not really, although in the summer of 1996 two groups of scientists suggested that traces of organic chemicals that they detected in meteorites from Mars may indicate that life once existed there (see chapter 8 for more details). The absence of hard evidence may seem discouraging, but as the British astronomer Sir Martin Rees has said, "Absence of evidence is not the same as evidence of absence." Even if the Universe were teeming with life, we have so far examined only a tiny portion of it: some half dozen spots on the Moon, two on Mars (by robot space-landers), and some asteroidal fragments picked up on Earth. The Moon's lack of atmosphere and water, even water locked up in minerals, makes it so inhospitable that astronomers did not expect to find life there. Mars seemed more promising because its environment, though harsh, is more like the Earth's than that of any other planet, and laboratory experiments showed that some terrestrial organisms (bacteria and lichens) could in fact survive Martian conditions, even though they did not reproduce or grow. Moreover, photographs show that its environment was less harsh in the past. For example, Mars once had flowing water on its surface, an indication that a warmer climate and denser atmosphere existed long ago. Nevertheless, the two *Viking* spacecraft that landed there in 1976 found no trace of life or even of organic carbon compounds, although, as we mentioned earlier, some Martian meteorites do contain organic carbon compounds.

Radio Searches

Given the failure to find life near to us, and the difficulty of directly sampling distant locales, a few astronomers have searched for radio signals from other civilizations (fig. E3.7). Such signals might either be broadcast to us deliberately, or be communications directed elsewhere that we might simply "overhear."

FIGURE E3.7
One of the radio telescopes used to search for signals from other civilizations. (Courtsey of Seth Shostak.)

Astronomers began "listening" for extraterrestrial radio signals in the 1960s with Project Ozma, which monitored radio emission from several nearby star systems, hoping to detect emission from other than natural processes. More recent projects, called **SETI** (*Search for Extraterrestrial Intelligence*), use receivers that automatically scan millions of radio wavelengths to search for signals from other civilizations. In fact, recent searches will tune through the equivalent of two billion radio channels.

Searching for signals is not easy because even if they exist, they are probably very weak and buried in cosmic static. Thus astronomers use considerable care when choosing the wavelengths to monitor. For example, at very long wavelengths interstellar gas in the galaxy emits strongly and will overwhelm all but the most powerful signals. On the other hand, at very short wavelengths, molecules in our atmosphere block the signals. It turns out that the optimum wavelengths are near 21 centimeters, the wavelength emitted by neutral hydrogen in the cold gas clouds of our galaxy. This is also near the wavelength of OH (hydroxyl) molecules, which emit strongly at 18 centimeters. These two wavelengths are very distinctive and any other intelligent civilization would probably have discovered their significance and have equipment that could receive or broadcast there. Thus several searches have been made for signals in this wavelength region, which is sometimes called the "waterhole" because it is bracketed on one side by emission from H and on the other by emission from OH. However, Ozma and all subsequent searches have proved fruitless so far.

One possible drawback to such searches is that other civilizations might not be transmitting signals. Like us, they may be listening. However, if other civilizations are like ours, they inadvertently broadcast a wide range of signals. For example, television stations and defense radar transmit extremely powerful signals that would be detectable many light-years from Earth. Moreover, on at least one occasion, a ceremony in 1974 celebrating the renovation of the huge Arecibo radio telescope in Puerto Rico, astronomers deliberately broadcast a signal toward the globular cluster M13. This cluster is more than 20,000 light years away from us, so the signal is still on its way.

The many difficulties associated with radio searches have led scientists to consider other ways to search for signs of extraterrestrial life. One interesting method involves looking for unexpected gases in the atmosphere of a planet, gases that might be produced by living things.

In the late 1960s James Lovelock, a British chemist, noted that several of the gases making up the Earth's atmosphere would disappear if Earth had no living beings to replenish them. For example, oxygen is very reactive and readily combines with surface rock. Lovelock went on to argue that the presence of such gases in a planet's atmosphere is an indication of life there. Thus, rather than searching for signals or sending robot spacecraft to explore a planet's surface (fig. E3.8), astronomers might find evidence of life by analyzing the light from a planet's atmosphere. Unfortunately, we cannot yet make such observations for any planets other than those in our own Solar System. However, Lovelock's idea has since led to a very different speculation.

The Gaia Hypothesis

In 1974, Lovelock and the U.S. microbiologist Lynn Margulis suggested that life creates a single "larger entity" with a planet, a "symbiosis" of life and planet, which they called the **Gaia hypothesis**. They chose this name from the Greek goddess of the Earth, Gaia (pronounced "Guy-uh" in this context, but "Jee-uh," elsewhere).

According to the Gaia hypothesis, life does not merely respond to its environment but actually alters its

FIGURE E3.8
An artist's view of a Martian Rover, controlled from Earth and designed to explore the surface of Mars.
(Michael Carrol)

planet's atmosphere and temperature to make it more hospitable. As we discussed in chapter 4, plant life has almost certainly done just that on Earth. For example, by photosynthesis, plants have created an oxygen-rich atmosphere on Earth that shields them from dangerous ultraviolet radiation. Similarly, plants may alter a planet's temperature through the greenhouse effect. If the planet is too cold, plant metabolism is slowed, and less CO_2 is converted to oxygen. The greater CO_2 abundance warms the planet and enhances plant growth. Conversely, if the planet gets too warm, plant growth runs amok and produces lots of oxygen, reducing the CO_2 abundance and therefore also reducing the greenhouse warming.

Lovelock and Margulis have also proposed that many other facets of the environment, such as humidity, salinity of the oceans, sea level, and even plate tectonics, may be controlled by living organisms so as to optimize their reproduction. Such an intimate linkage between the living and nonliving is a very appealing idea. It almost makes the planet "alive." But is the Gaia hypothesis correct? Some aspects certainly are, but many scientists are skeptical of its more extreme claims (for example, that life can control plate tectonics). Nevertheless, living beings do exert a remarkable control over their local environment, and many unsuspected links between life and the physical world surely remain to be found. One such possible connection has its roots in cosmology, a connection called the "anthropic principle."

The Anthropic Principle

In 1961, Robert Dicke, a physicist at Princeton University, wrote a paper concerning some remarkable cosmological coincidences. Dicke noted, for example, that the age of the Universe is very close to the lifetime of stars like the Sun and went on to argue that this "coincidence" is in fact not remarkable at all. Rather, it is a *necessity* if cosmologists are to exist to note it. Dicke pointed out that life requires elements such as carbon, silicon, and iron that are made in massive stars. Thus, for life to form, enough time must pass for massive stars to evolve and make the heavy elements and then eject them into space with a supernova explosion. Moreover, additional time is needed for the ejected material to be incorporated back into interstellar clouds and form new stars. Only with this second generation of stars does life become likely. Dicke therefore concluded that intelligent life is impossible in a Universe only a few thousand years old. He likewise argued that life requires stars, and so it can exist only in a Universe young enough that stars have not all burnt out. Thus the very existence of an intelligent observer in the Universe requires that the age of the Universe fall within certain limits.

In 1974, the English physicist Brandon Carter took Dicke's idea further and proposed what he called the **anthropic principle**. According to the anthropic principle, "what we can expect to observe must be restricted by the conditions necessary for our presence as observers." Since Carter's work, many scientists have shown how "fine-tuned" the Universe must be for life to exist. Thus, no matter how unlikely some aspects of the Universe may appear to us, if those aspects are necessary for life, then that is what we will observe. For example, we might argue how truly marvelous it is that conditions on Earth are just right for life. According to the anthropic principle, the rude reply is, "Of course conditions are just right for life. If they were not, there would be no life here to do the marveling."

Summary

Life has existed on Earth for at least 3.8 billion years, a goodly portion of our planet's existence. That long duration and the underlying similarities of all life forms in terms of their structure, energy supply, and methods of reproduction indicate that life had a common origin relatively soon after Earth's formation. Experiments such as Miller and Urey's show that the precursors to life can be easily created under natural conditions. Thus many astronomers believe that life has formed in numerous other places in the Universe. All searches for life, however, have found nothing, but given the vast distances between stars, that negative result is not surprising.

Once life develops on a planet, it may subtly alter its environment to its own ends, a proposition known as the "Gaia hypothesis." Such alterations may be detectable more readily than signals from other civilizations and may therefore be signatures of extraterrestrial life.

Questions for Review

1. What is the earliest evidence for life on Earth?
2. How old are the earliest life forms thought to be?
3. What features of life are suggestive of a common origin?
4. What is meant by panspermia?
5. What is the Drake equation?
6. Why is 21 centimeters believed to be a good wavelength region to search for signals from other civilizations?
7. What is meant by the Gaia hypothesis?
8. What is meant by the anthropic principle?

Thought Questions

1. Would you be upset if it turned out that we were the only intelligent life in the Universe? Why or why not?
2. Suppose astronomers did detect signals from another civilization. What effect do you think it would have on our society? What effect would it have on you?
3. Do you think it is a good idea to send out signals in the hope that aliens might detect them? Suppose, for example, they were more technically advanced than us and could do to us what we have done to less developed societies on Earth.
4. What message would you send to an alien civilization?
5. Use a method similar to the Drake equation to estimate the number of people getting a haircut at some given moment. That is, multiply the probabilities of each such event times the number of people in, for example, the United States.

Test Yourself

1. Evidence that life on Earth is very ancient comes from
   (a) fossil algae in rocks more than 3 billion years old.
   (b) fossils buried thousands of feet below the surface.
   (c) fossil algae in rocks a few million years old.
   (d) the discovery of silicon-based life forms in ancient rocks.
   (e) the discovery of silicon-based life forms in a Miller-Urey type of experiment.
2. A planet on which life forms create an atmosphere that makes the planet more hospitable for life is an example of
   (a) the Drake equation.
   (b) the Gaia hypothesis.
   (c) SETI.

(d) the anthropic principle.

(e) the Miller-Urey experiment.

3. The Miller-Urey experiment demonstrated that

(a) very simple bacteria can be created from the chemicals present in the atmosphere of the ancient Earth.

(b) Earth's early atmosphere was too harsh for life to have formed until recently.

(c) conditions on the early Earth were suitable for the creation of many of the complex organic molecules found in living things.

(d) if simple life forms landed on the early Earth, they could have survived.

(e) early life forms were silicon-based.

4. According to the anthropic principle

(a) we are the only civilization in the Universe.

(b) there are 10^{12} civilizations more advanced than us in the Milky Way.

(c) human beings are the highest form of life.

(d) the Universe could not differ radically from its present properties, or we would not be here to observe it.

(e) the greenhouse effect is beneficial to us.

5. The Drake equation attempts to determine

(a) the conditions under which life originated on Earth.

(b) the optimum wavelength for communicating with extraterrestrials.

(c) the age of life on Earth.

(d) the number of other technically advanced civilizations.

(e) the lifetime of our own civilization.

Further Explorations

Chown, Marcus. "Is There Anybody Out There? *New Scientist* 152 (23 November 1996): 32.

Crowe, M. J. *The Extraterrestrial Life Debate:* 1750–1900. Cambridge, New York: Cambridge University Press, 1986.

Davies, John. "Searching for the Alien Earth." *New Scientist* 146 (13 May 1995): 24.

Diamond, Jared. "Alone in a Crowded Universe." *Natural History* 99 (June 1990): 30.

Dick, S. J. *Plurality of Worlds.* Cambridge, New York: Cambridge University Press, 1982.

Dick, S. J. *The Biological Universe.* Cambridge, New York: Cambridge University Press. 1996.

Drake, Frank, and Dava Sobel. *Is Anyone Out There? The Scientific Search for Extraterrestrial Intelligence.* New York: Delacorte Press, 1992.

Gould, Stephen J. "The Evolution of Life on the Earth." *Scientific American* 271 (October 1994): 84.

Gould, Stephen Jay., "War of the Worldviews." *Natural History* 105 (December 1996): 22.

Horgan, John. "In the Beginning." *Scientific American* 264 (February 1991): 116.

Knoll, A. H. "End of the Proterozoic Eon." *Scientific American* 265 (October 1991): 64.

Lazio, T. Joseph W., and James M. Cordes. "Pulsars. Planets, and Genetics." *Mercury* 24 (March/April 1995): 23. See also related articles in same issue.

Lovelock, James E. *The Ages of Gaia:* a *Biography of Our Living Earth.* New York: Norton, 1988.

Lovelock, James E. *Gaia, a New Look at Life on Earth.* Oxford, New York: Oxford University Press, 1979.

Lovelock, James E., and Lynn Margulis. "The Gaia Hypothesis." *Tellus* 26 (1974): 1.

Naeye, Robert. "SETI at the Crossroads." *Sky and Telescope* 84 (November 1992): 507.

Naeye, Robert. "OK, Where Are They?" *Astronomy* 24 (July 1996): 36.

Orgel, Leslie E. "The Origin of Life on the Earth." *Scientific American* 271 (October 1994): 76.

Rood, Robert T., and James S. Trefil. *Are We Alone?* New York: Charles Scribner's and Sons. 1982.

Sagan, Carl. "The Search for Extraterrestrial Life." *Scientific American* 271 (October 1994): 92.

Sheehan, William. "The Red Planet's Colorful Past." *Astronomy* 25 (March 1977): 44.

Shklovski, I. S., and Carl Sagan. *Intelligent Life in the Universe.* New York: Dell Publishing Co., 1966.

Shostak, Seth. "Listening for Life." *Astronomy* 20 (October 1992): 26.

Weinberg, Steven. "Life in the Universe." *Scientific American* 271 (October 1994): 44.

Key Terms

amino acids, 530

anthropic principle, 538

DNA, 531

eukaryotes, 533

Gaia hypothesis, 536

panspermia, 533

prokaryotes, 533

proteins, 531

SETI, 536

Answers to "Test Yourself"

Introduction	1–b; 2–c; 3–e; 4–c; 5–e
Chapter 1	1–d; 2–b; 3–a; 4–e; 5–a
Essay 1	1–c; 2–b; 3–a; 4–d; 5–e
Chapter 2	1–e; 2–d; 3–d; 4–a; 5–c
Chapter 3	1–e; 2–c; 3–d; 4–e; 5–a
Chapter 4	1–e; 2–a; 3–d; 4–c; 5–d
Chapter 5	1–d; 2–c; 3–d; 4–d; 5–a
Chapter 6	1–d; 2–c; 3–d; 4–d; 5–b
Essay 2	1–b; 2–c; 3–e; 4–e; 5–a
Chapter 7	1–c; 2–c; 3–e; 4–d; 5–a
Chapter 8	1–d; 2–b; 3–c; 4–d; 5–a
Chapter 9	1–c; 2–b; 3–a; 4–d; 5–d
Chapter 10	1–e; 2–d; 3–b; 4–d; 5–d
Chapter 11	1–d; 2–c; 3–a; 4–d; 5–d
Chapter 12	1–d; 2–e; 3–e; 4–a; 5–b
Chapter 13	1–e; 2–c; 3–b; 4–b; 5–e
Chapter 14	1–d; 2–c; 3–b; 4–e; 5–e
Chapter 15	1–b; 2–e; 3–a; 4–a; 5–e
Chapter 16	1–d; 2–e; 3–e; 4–b; 5–e
Chapter 17	1–b; 2–e; 3–c; 4–d; 5–e
Essay 3	1–a; 2–b; 3–c; 4–d; 5–d

Appendix

POWERS-OF-TEN NOTATION

Powers-of-ten notation is a shorthand method for expressing and working with very large or very small numbers. With this method, we express numbers as a few digits times ten to a power, or exponent. The power indicates the number of times that ten is multiplied by itself. For example, $100 = 10 \times 10 = 10^2$. Similarly $1,000,000 = 10 \times 10 \times 10 \times 10 \times 10 \times 10 = 10^6$. Note that we do not always need to write out $10 \times \ldots \ldots$ Instead, we can simply count the zeros. Thus, 10,000 is 1 followed by 4 zeros, so it is 10^4.

We can write 1 and 10 itself in powers-of-ten notation as well: $1 = 10^0$ and $10 = 10^1$.

To write a number like 300, we break it into two parts, $3 \times 100 = 3 \times 10^2$. Similarly, we can write $352 = 3.52 \times 100 = 3.52 \times 10^2$.

We can also write very small numbers (numbers less than 1) using powers of ten. For example, $0.01 = 1/100 = 1/10^2$. We can make this even more concise, however, by writing $1/10^2$ as 10^{-2}. Similarly, $0.0001 = 10^{-4}$. Note that for numbers less than 1, the power is 1 more than the number of zeros.

We can write a number like 0.00052 as $5.2 \times 0.0001 = 5.2 \times 10^{-4}$.

Suppose we want to multiply numbers expressed in powers of ten. The rule is simple: we add the powers. Thus, $10^3 \times 10^2 = 10^{3+2} = 10^5$. Similarly $2 \times 10^8 \times 3 \times 10^7 = 2 \times 3 \times 10^8 \times 10^7 = 6 \times 10^{15}$. In general, $10^a \times 10^b = 10^{a+b}$.

Division works similarly, except that we subtract the exponents. Thus $10^5/10^3 = 10^{5-3} = 10^2$. In general, $10^a/10^b = 10^{a-b}$.

The last operations we need to consider are raising a power-of-ten number to a power and taking a root. In raising a number to a power, we multiply the powers. Thus $(10^3)^4 = 10^{3 \times 4} = 10^{12}$. Care must be used if we have a number like $(4 \times 10^2)^3$. Both the 4 and the 10^2 are raised to a power so the result is $4^3 \times (10^2)^3 = 64 \times 10^{2 \times 3} = 64 \times 10^6$.

Taking a root is equivalent to raising a number to a fractional power. Thus, the square root of a number is the number to the $\frac{1}{2}$ power. The cube root is the number to the $\frac{1}{3}$ power, and so forth. For example, $\sqrt{100} = (100)^{1/2} = (10^2)^{1/2} = 10^1 = 10$.

SOME USEFUL FORMULAS

$$\text{Distance} = \text{velocity} \times \text{time } (D = vt)$$
$$\text{Circumference of circle} = 2\pi r$$
$$\text{Area of a circle} = \pi r^2$$
$$\text{Area of a sphere} = 4\pi r^2$$
$$\text{Volume of a sphere} = \frac{4\pi r^3}{3}$$

SOLVING DISTANCE, VELOCITY, TIME (D, v, t) PROBLEMS

A large number of problems in this book (and in science in general) involve the motion of something. In such problems, we often know two of the three quantities (D, v, t) and want to know the third. For example, we have something moving at a speed v and want to know how far it will travel in a time t. Or we know that something travels with a speed v

and want to find out how long it takes for the object to travel a distance D. We all can solve such problems in our heads if the motion involves automobiles. For example, if it is 150 miles to the city and we travel at 50 miles per hour how long does it take to get there? Or how far can we drive in 2 hours if we are traveling at 45 miles per hour? Because we all do such problems routinely, you might find it easier to think of astronomical problems in terms of cars. Regardless of your approach, the method of solution is simple.

Begin by making a simple sketch of what is happening. Draw some arrows to indicate the motion. Label the quantities known and put a question mark beside the things you want to find.\

Then write out the basic relation $D = vt$. If you want to find D and know v and t, just multiply them for the answer. If you want to find the time and are given v and D, solve for t by dividing both sides by v to get $t = D/v$. If you want the velocity, divide D by t. In some problems, the motion may be in a circle of radius r. In that case, the distance traveled will be related to the circumference of the circle, $2\pi r$. For such cases, you may need to use the expression $D = 2\pi r$.

In most problems you will find that it is extremely helpful to write the units of the quantities in the equation. For example, suppose you are asked how long it takes to travel 1500 km at a velocity of 30 kilometers per hour. First write out $D = vt$. Then rewrite it as

$$D/v = t$$

Next insert the quantities so that

$$t = \frac{1500 \text{ km}}{30 \text{ km/hr}} = 50 \text{ hr}$$

Note that the units of km cancel out and leave us with units of hours, as the problem requires.

APPENDIX TABLE 1 Physical and Astronomical Constants

Physical Constants

velocity of light (c)	$= 2.99792458 \times 10^8$ m/s = approximately 3×10^5 km/s
gravitational constant (G)	$= 6.67259 \times 10^{-11}$ m^3-kg^{-1}-s^{-2} (or newton $-$ m^2 $-$ kg^{-2})
Planck's constant (h)	$= 6.62608 \times 10^{-34}$ joule - s
mass of hydrogen atom (M$_H$)	$= 1.6735 \times 10^{-27}$ kg
mass of electron (M$_e$)	$= 9.1094 \times 10^{-31}$ kg
Stefan-Boltzmann constant (σ)	$= 5.6705 \times 10^{-8}$ watts - m^{-2} - deg^{-4}
constant in Wien's law ($\lambda_m T$)	$= 0.28978$ cm/deg

Astronomical Constants

astronomical unit (AU)	$= 1.495978706 \times 10^{11}$ m $\approx 1.5 \times 10^8$ km
light-year (ly)	$= 9.4605 \times 10^{15}$ m $= 9.4605 \times 10^{12}$ km $= 6.324 \times 10^4$ AU
parsec (pc) $= 3.26$ ly	$= 3.085678 \times 10^{16}$ m $= 3.085678 \times 10^{13}$ km $= 206,265$ AU
sidereal year	$= 365.256366$ days $= 3.155815 \times 10^7$ s
mass of Earth	$= 5.974 \times 10^{24}$ kg
mass of Sun (M$_\odot$)	$= 1.989 \times 10^{30}$ kg
equatorial radius of Earth	$= 6,378.0$ km
radius of Sun (R$_\odot$)	$= 6.96 \times 10^8$ m $= 6.96 \times 10^5$ km
luminosity of Sun (L$_\odot$)	$= 3.83 \times 10^{26}$ watts

APPENDIX TABLE 2   Conversion Between American and Metric Units*

Length

1 km	=	1 kilometer	=	1000 meters	=	0.6214 mile
1 m	=	1 meter	=	1.094 yards	=	39.37 inches
1 cm	=	1 centimeter	=	0.01 meter	=	0.3937 inch
1 μm	=	1 micrometer	=	10^{-6} meter	=	10^{-4} cm = 3.93×10^{-5} inch
1 nm	=	1 nanometer	=	10^{-9} meter	=	10^{-7} cm
1 mile	=	1.6093 km				
1 inch	=	2.5400 cm				

Mass

1 metric ton	=	10^6 grams	=	1000 kg	=	2.2046×10^3 lb
1 kg	=	1000 grams	=	2.2046 lb		
1 g	=	1 gram	=	0.0022046 lb	=	0.0353 oz
1 lb	=	0.4536 kg				
1 oz	=	28.3495 g				

* You will find it easier to use the metric system if you remember the following prefixes that denote very small to very large quantities:

nano	=	10^{-9}	=	1 billionth
micro	=	10^{-6}	=	1 millionth
milli	=	10^{-3}	=	1 thousandth
kilo	=	10^3	=	1 thousand
mega	=	10^6	=	1 million
giga	=	10^9	=	1 billion

APPENDIX TABLE 3   Physical Properties of the Planets

Name	Radius (Eq) (Earth Units)	Radius (Eq) (km)	Mass (Earth Units)	Mass (kg)	Average Density (g/cm³)
Mercury	0.382	2,439	0.055	3.30×10^{23}	5.43
Venus	0.949	6,051	0.815	4.87×10^{24}	5.25
Earth	1.00	6,378	1.00	5.97×10^{24}	5.52
Mars	0.533	3,397	0.107	6.42×10^{23}	3.93
Jupiter	11.19	71,492	317.9	1.90×10^{27}	1.33
Saturn	9.46	60,268	95.18	5.69×10^{26}	0.71
Uranus	3.98	25,559	14.54	8.68×10^{25}	1.24
Neptune	3.81	24,764	17.13	1.02×10^{26}	1.67
Pluto	0.181	1,160	0.00216	1.29×10^{22}	2.05

APPENDIX TABLE 4   Orbital Properties of the Planets

Name	Distance from Sun* (AU)	Distance from Sun* (10⁶ km)	Period Years	Period Days	Orbital Inclination†	Orbital Eccentricity
Mercury	0.387	57.9	0.2409	(87.97)	7.00	0.206
Venus	0.723	108.2	0.6152	(224.7)	3.39	0.007
Earth	1.00	149.6	1.0	(365.26)	0.00	0.017
Mars	1.524	227.9	1.8809	(686.98)	1.85	0.093
Jupiter	5.203	778.3	11.8622	(4,332.59)	1.31	0.048
Saturn	9.539	1,427.0	29.4577	(10,759.22)	2.49	0.056
Uranus	19.19	2,869.6	84.014	(30,685.4)	0.77	0.046
Neptune	30.06	4,496.6	164.793	(60,189)	1.77	0.010
Pluto	39.53	5,900	247.7	(90,465)	17.15	0.248

* Semimajor axis of the orbit.
† With respect to the ecliptic.

APPENDIX TABLE 5   Satellites of the Solar System*

Planet Satellite	Radius† (km)	Distance from Planet (10^3 km)	Orbital Period (days)	Mass (10^{20} kg)	Density (g/cm³)
EARTH					
Moon	1,738	384.4	27.322	734.9	3.34
MARS					
Phobos	13 × 11 × 9	9.38	0.319	1.3×10^{-4}	2.2
Deimos	8 × 6 × 5	23.5	1.263	1.8×10^{-5}	1.7
JUPITER					
Metis	20	127.96	0.295	9×10^{-4}	—
Adrastea	13 × 10 × 8	128.98	0.298	1×10^{-4}	—
Amalthea	135 × 82 × 75	181.3	0.498	8×10^{-2}	—
Thebe	55 × ? × 45	221.9	0.6745	1.4×10^{-3}	—
Io	1,821	421.6	1.769	893.3	3.57
Europa	1,565	670.9	3.551	479.7	2.97
Ganymede	2,634	1,070.0	7.155	1482	1.94
Callisto	2,403	1,883.0	16.689	1076	1.86
Leda	~8	11,094	238.7	4×10^{-4}	—
Himalia	92.5	11,480	250.6	8×10^{-2}	—
Lysithea	~18	11,720	259.2	6×10^{-4}	—
Elara	~38	11,737	259.7	6×10^{-3}	—
Ananke	~15	20,200	631R	4×10^{-4}	—
Carme	~20	22,600	692R	9×10^{-4}	—
Pasiphae	~25	23,500	735R	1.6×10^{-3}	—
Sinope	~18	23,700	758R	6×10^{-4}	—
SATURN					
Pan	10	133.570	0.574	4.2×10^{-6}	—
Atlas	19 × ? × 14	137.64	0.602	1.6×10^{-4}	—
Prometheus	74 × 50 × 34	139.35	0.613	5×10^{-3}	—
Pandora	55 × 44 × 31	141.7	0.629	3.4×10^{-3}	—
Epimetheus	69 × 55 × 55	151.42	0.694	5.6×10^{-3}	—
Janus	97 × 95 × 77	151.47	0.695	2.0×10^{-2}	—
Mimas	210 × 197 × 193	185.52	0.942	0.370	1.17
Enceladus	256 × 247 × 244	238.02	1.370	1.2	1.24
Tethys	536 × 528 × 526	294.66	1.888	6.17	1.26
Calypso	15 × 8 × 8	294.66	1.888	4×10^{-5}	—
Telesto	15 × 13 × 8	294.67	1.888	6×10^{5}	—
Dione	559	377.4	2.737	10.8	1.44
Helene	18 × ? × 15	377.4	2.737	1.6×10^{-4}	—
Rhea	764	527.04	4.518	23.1 1.33	
Titan	2,575	1,221.85	15.945	1,345.5 1.88	
Hyperion	180 × 140 × 112	1,481.1	21.277	0.28 —	
Iapetus	718	3,561.3	79.331	15.9 1.21	
Phoebe	115 × 110 × 105	12,952	550.4R	0.1 —	

* Note: Authorities differ substantially on many of the values above. "R" means orbit is retrograde.
† a × b × c values for the Radius are the approximate lengths of the axes for irregular moons.

APPENDIX TABLE 5   Satellites of the Solar System—Continued*

Planet Satellite	Radius (km)	Distance from Planet (10^3 km)	Orbital Period (days)	Mass (10^{20} kg)	Density (g/cm^3)
URANUS					
Cordelia	13	49.75	0.336	1.7×10^{-4}	—
Ophelia	16	53.77	0.377	002.6×10^{-4}	—
Bianca	22	59.16	0.435	7×10^{-4}	—
Cressida	33	61.77	0.465	2.6×10^{-3}	—
Desdemona	29	62.65	0.476	1.7×10^{-3}	—
Juliet	42	64.63	0.494	4.3×10^{-3}	—
Portia	55	66.1	0.515	1×10^{-2}	—
Rosalind	29	69.93	0.560	1.5×10^{-3}	—
Belinda	33	75.25	0.624	2.5×10^{-3}	—
Puck	77	86.00	0.764	5×10^{-3}	—
Miranda	$240 \times 234 \times 233$	129.8	1.413	0.66	1.26
Ariel	$581 \times 578 \times 578$	191.2	2.520	13.5	1.65
Umbriel	584.7	266.0	4.144	11.7	1.44
Titania	788.9	435.8	8.706	35.2	1.59
Oberon	761.4	582.6	13.463	30.1	1.50
NEPTUNE					
Naiad	29	48.2	0.296	1.4×10^{-3}	—
Thalassa	40	50.0	0.312	4×10^{-3}	—
Despina	74	52.5	0.333	2.1×10^{-3}	—
Galatea	79	62.0	0.396	3.1×10^{-2}	—
Larissa	$104 \times ? \times 89$	73.6	0.554	6×10^{-2}	—
Proteus	$218 \times 208 \times 201$	117.6	1.121	0.6	—
Triton	1,352.6	354.59	5.875R	214	2.0
Nereid	170	5,588.6	360.125	0.31	—
PLUTO					
Charon	635	19.6	6.38718	17.7	1.83

* Note: Authorities differ substantially on many of the values above. "R" means orbit is retrograde.

APPENDIX TABLE 6   Meteor Showers

Shower	Dates	Hourly Rate	Radiant R.A.[†]	Radiant Dec.[†]	Associated Comet
Quadrantids	Jan. 2–4	30	$15^h\,24^m$	50°	?
Lyrids	Apr. 20–22	8	$18^h\,4^m$	33°	1861 I
η Aquarids	May 2–7	10	$22^h\,24^m$	0°	Halley
δ Aquarids	July 26–31	15	$22^h\,36^m$	−10°	?
Perseids	Aug. 10–14	40	$3^h\,4^m$	58°	Swift-Tuttle
Orionids	Oct. 18–23	15	$6^h\,20^m$	15°	Halley
Taurids	Nov. 1–7	8	$3^h\,40^m$	17°	Enke
Leonids	Nov. 14–19	6	$10^h\,12^m$	22°	Temple-Tuttle
Geminids	Dec. 10–13	50	$7^h\,28^m$	32°	Phaeton (an asteroid)

† R.A. = right ascension; Dec. = declination.

APPENDIX TABLE 7 The Constellations*

Constellation Name	Genitive	Description	Abbreviation	Position in Sky† R.A.	DEC.
Andromeda	Andromedae	The Princess of Ethiopia	And	1^h	+40°
Antlia	Antliae	The Air Pump	Ant	10^h	−35°
Apus	Apodis	The Bird of Paradise	Aps	16^h	−75°
Aquarius	Aquarii	The Water Bearer	Aqr	23^h	−15°
Aquila	Aquilae	The Eagle	Aql	20^h	+5°
Ara	Arae	The Altar	Ara	17^h	−55°
Aries	Arietis	The Ram	Ari	3^h	+20°
Auriga	Aurigae	The Charioteer	Aur	6^h	+40°
Boötes	Boötis	The Herdsman	Boo	15^h	+30°
Caelum	Caeli	The Sculptor's Chisel	Cae	5^h	−40°
Camelopardus	Camelopardalis	The Giraffe	Cam	6^h	+70°
Cancer	Cancri	The Crab	Cnc	9^h	+20°
Canes Venatici	Canum Venaticorum	The Hunting Dogs	CVn	13^h	+40°
Canis Major	Canis Majoris	The Big Dog	CMa	7^h	−20°
Canis Minor	Canis Minoris	The Little Dog	CMi	8^h	+5°
Capricornus	Capricorni	The Sea Goat	Cap	21^h	−20°
Carina	Carinae	The Keel of Argo‡	Car	9^h	−60°
Cassiopeia	Cassiopeiae	The Queen of Ethiopia	Cas	1^h	+60°
Centaurus	Centauri	The Centaur	Cen	13^h	−50°
Cepheus	Cephei	The King of Ethiopia	Cep	22^h	+70°
Cetus	Ceti	The Sea Monster (Whale)	Cet	2^h	−10°
Chamaeleon	Chamaeleontis	The Chameleon	Cha	11^h	−80°
Circinus	Circini	The Compasses	Cir	15^h	−60°
Columba	Columbae	The Dove	Col	6^h	−35°
Coma Berenices	Comae Berenices	Berenice's Hair	Com	13^h	+20°
Corona Australis	Coronae Australis	The Southern Crown	CrA	19^h	−40°
Corona Borealis	Coronae Borealis	The Northern Crown	CrB	16^h	+30°
Corvus	Corvi	The Crow	Crv	12^h	−20°
Crater	Crateris	The Cup	Crt	11^h	−15°
Crux	Crucis	The Southern Cross	Cru	12^h	−60°
Cygnus	Cygni	The Swan	Cyg	21^h	+40°
Delphinus	Delphini	The Dolphin	Del	21^h	+10°
Dorado	Doradus	The Swordfish	Dor	5^h	−65°
Draco	Draconis	The Dragon	Dra	17^h	+65°
Equuleus	Equulei	The Little Horse	Equ	21^h	+10°
Eridanus	Eridani	The River	Eri	3^h	−20°
Fornax	Fornacis	The Furnace	For	3^h	−30°
Gemini	Geminorum	The Twins	Gem	7^h	+20°
Grus	Gruis	The Crane	Gru	22^h	−45°
Hercules	Herculis	Hercules	Her	17^h	+30°
Horologium	Horologii	The Clock	Hor	3^h	−60°
Hydra	Hydrae	The Water Snake (female)	Hya	10^h	−20°
Hydrus	Hydri	The Water Snake (male)	Hyi	2^h	−75°
Indus	Indi	The American Indian	Ind	21^h	−55°
Lacerta	Lacertae	The Lizard	Lac	22^h	+45°
Leo	Leonis	The Lion	Leo	11^h	+15°

* Note: Constellations with declinations between −50° and −90° are below the horizon from most of the United States.
† R.A. = right ascension; Dec. = declination.
‡ Carina, Puppis, Pyxis, and Vela were once a single large constellation called Argo, the Ship.

APPENDIX TABLE 7   The Constellations*—Continued

Constellation Name	Genitive	Description	Abbreviation	Position in Sky†	
				R.A.	DEC.
Leo Minor	Leonis Minoris	The Little Lion	LMi	10^h	$+35°$
Lepus	Leporis	The Hare	Lep	6^h	$-20°$
Libra	Librae	The Scales	Lib	15^h	$-15°$
Lupus	Lupi	The Wolf	Lup	15^h	$-45°$
Lynx	Lyncis	The Lynx	Lyn	8^h	$+45°$
Lyra	Lyrae	The Lyre	Lyr	19^h	$+40°$
Mensa	Mensae	Table (Mountain)	Men	5^h	$-80°$
Microscopium	Microscopii	The Microscope	Mic	21^h	$-35°$
Monoceros	Monocerotis	The Unicorn	Mon	7^h	$-5°$
Musca	Muscae	The Fly	Mus	12^h	$-70°$
Norma	Normae	The Square	Nor	16^h	$-50°$
Octans	Octantis	The Octant	Oct	22^h	$-85°$
Ophiuchus	Ophiuchi	The Serpent Bearer	Oph	17^h	$0°$
Orion	Orionis	The Hunter	Ori	5^h	$+5°$
Pavo	Pavonis	The Peacock	Pav	20^h	$-65°$
Pegasus	Pegasi	The Winged Horse	Peg	22^h	$+20°$
Perseus	Persei	The Hero	Per	3^h	$+45°$
Phoenix	Phoenicis	The Phoenix	Phe	1^h	$-50°$
Pictor	Pictoris	The Easel	Pic	6^h	$-55°$
Pisces	Piscium	The Fishes	Psc	1^h	$+15°$
Piscis Austrinus	Piscis Austrini	The Southern Fish	PsA	22^h	$-30°$
Puppis	Puppis	The Stern of Argo‡	Pup	8^h	$-40°$
Pyxis	Pyxidis	The Compass of Argo‡	Pyx	9^h	$-30°$
Reticulum	Reticuli	The Net	Ret	4^h	$-60°$
Sagitta	Sagittae	The Arrow	Sge	20^h	$+10°$
Sagittarius	Sagittarii	The Archer	Sgr	19^h	$-25°$
Scorpius	Scorpii	The Scorpion	Sco	17^h	$-40°$
Sculptor	Sculptoris	The Sculptor	Scl	0^h	$-30°$
Scutum	Scuti	The Shield	Sct	19^h	$-10°$
Serpens	Serpentis	The Serpent	Ser	17^h	$0°$
Sextans	Sextantis	The Sextant	Sex	10^h	$0°$
Taurus	Tauri	The Bull	Tau	4^h	$+15°$
Telescopium	Telescopii	The Telescope	Tel	19^h	$-50°$
Triangulum	Trianguli	The Triangle	Tri	2^h	$+30°$
Triangulum Australe	Trianguli Australis	The Southern Triangle	TrA	16^h	$-65°$
Tucana	Tucanae	The Toucan	Tuc	0^h	$-65°$
Ursa Major	Usrae Majoris	The Great Bear	UMa	11^h	$+50°$
Ursa Minor	Ursae Minoris	The Little Bear	UMi	15^h	$+70°$
Vela	Velorum	The Sails of Argo‡	Vel	9^h	$-50°$
Virgo	Virginis	The Maiden	Vir	13^h	$0°$
Volans	Volantis	The Flying Fish	Vol	8^h	$-70°$
Vulpecula	Vulpeculae	The Fox	Vul	20^h	$+25°$

* Note: Constellations with declinations between $-50°$ and $-90°$ are below the horizon from most of the United States.

† R.A. = right ascension; Dec. = declination.

‡ Carina, Puppis, Pyxis, and Vela were once a single large constellation called Argo, the Ship.

APPENDIX TABLE 8   The Brightest Stars

Star	Name	Apparent Visual Magnitude	Spectral Type	Luminosity Class	Distance (ly)	Absolute Visual Magnitude
α CMa A	Sirius	−1.47	A1	V	8.7	1.4
α Car	Canopus	−0.72	F0	Ib/II	98	−3.1
α Cen	Rigel Kentaurus	−0.01	G2	V	4.3	4.4
α Boo	Arcturus	−0.06	K2	III	36	−0.3
α Lyr	Vega	0.04	A0	V	26.5	0.5
α Aur	Capella	0.05	G8	III	45	−0.6
β Ori	Rigel	0.14	B8	Ia	900	−7.1
α CMi	Procyon	0.37	F5	IV/V	11.4	2.7
α Ori	Betelgeuse	0.41	M2	Iab	520	−5.6
α Eri	Achernar	0.51	B3	V	118	−2.3
β Cen AB	Hadar	0.63	B1	III	490	−5.2
α Aql	Altair	0.77	A7	IV/V	16.5	2.2
α Tau A	Aldebaran	0.86	K5	III	68	−0.7
α Cru	Acrux	0.90	B2	IV	260	−3.5
α Vir	Spica	0.91	B1	V	220	−3.3
α Sco A	Antares	0.92	M1	Ib	520	−5.1
α PsA	Fomalhaut	1.15	A3	V	22.6	2.0
β Gem	Pollux	1.16	K0	III	35	1.0
α Cyg	Deneb	1.26	A2	Ia	1,600	−7.1
β Crux	Beta Crucis	1.28	B0.5	IV	490	−4.6

APPENDIX TABLE 9   The Nearest Stars

Name	Distance (ly)	Spectral Type	Absolute Visual Magnitude	Apparent Visual Magnitude
Sun		G2	4.83	−26.8
Proxima Centauri	4.23	M5.5V	15.5	11.09
α Cen A	4.35	G2V	4.4	0.01
B	4.35	K0V	5.7	1.34
Barnard's Star	5.98	M4V	13.2	9.55
Wolf 359	7.80	M6V	16.6	13.45
Lalande 21185	8.23	M2V	10.5	7.47
Luyten 726-8 A	8.57	M5.5V	15.3	12.41
B	8.57	M6V	16.1	13.2
Sirius A	8.57	A1V	1.5	−1.43
B	8.57	DA2 (white dwarf)	11.3	8.44
Ross 154	9.56	M3.5V	13.1	10.47
Ross 248	10.33	M5.5V	14.8	12.29
ε Eridani	10.67	K2V	6.2	3.73
Ross 128	10.83	M4V	13.5	11.12
Luyten 789-6 ABC	11.08	M5V	14.7	12.33
Groombridge 34 A	11.27	M1.5V	10.4	8.08
B	11.27	M3.5V	13.4	11.07
ε Indi	11.29	K5Ve	7.0	4.68
61 Cygni A	11.30	K5V	7.5	5.22
B	11.30	K7V	8.3	6.03
BD + 59° 1915 A	11.40	M3V	11.2	8.90
B	11.40	M3.5V	12.0	9.68
τ Ceti	11.40	G8Vp	5.8	3.50
Procyon A	11.41	F5IV-V	2.7	0.38
B	11.41	DA (white dwarf)	13.0	10.7
Lacaille 9352	11.47	M1.5V	9.6	7.34
GJ 1111	11.83	M6.5V	17.0	14.79
GJ 1061	12.06	M5.5V	15.2	13.03

Source: Data taken from "Our Nearest Celestial Neighbors," by Joshua Roth and Roger W. Sinnot, *Sky and Telescope* 92 (October 1996), p. 34. Compiled by Roger W. Sinnott.
Notes: BD stands for Bonner Durchmusterung catalog and GJ for Gliese-Jahreiss catalog; *p* indicates a peculiar spectrum; *e* indicates that the star shows emission lines in its spectrum.

APPENDIX TABLE 10   Properties of Main-Sequence Stars*

Spectral Type	Luminosity ($L_\odot$)	Temperature (K)	Mass ($M_\odot$)	Radius ($R_\odot$)
O5	790,000	44,500	60	12.0
B0	52,000	30,000	17.5	7.4
B5	830	15,400	5.9	3.9
A0	54	9,520	2.9	2.4
A5	14	8,200	2.0	1.7
F0	6.5	7,200	1.6	1.5
F5	3.2	6,440	1.3	1.3
G0	1.5	6,030	1.05	1.1
G5	0.79	5,770	0.92	0.92
K0	0.42	5,250	0.79	0.85
K5	0.15	4,350	0.67	0.72
M0	0.08	3,850	0.51	0.6
M5	0.01	3,240	0.21	0.27
M8	0.001	2,640	0.06	0.1

* Note: Authorities differ substantially on many of the above values, especially at the upper and lower mass values. Note also that the values are generally not consistent with the Stefan-Boltzmann Law.

The data in the above tables come from many sources, including the following:

Lang, Kenneth R. *Astrophysical Data: Planets and Stars.* Berlin, New York: Springer-Verlag, 1992.
NOTE: Be sure to get a sheet of corrected errors, however.
Landolt-Bornstein: *Zahlenwerte und Funktionen aus Naturwissenschaften und Technik/Numerical Data and Functional Relationships in Science and Technology.* Neue Serie/New Series. Group VII: Astronomy, Astrophysics, and Space Research, Vols. 1–2, Astronomy and Astrophysics. Berlin, New York: Springer-Verlag, 1965–1982.
Donald Y. Yeomans of JPL for satellite dimensions/masses.

Glossary

a

absorption the process in which light or other electromagnetic radiation gives up its energy to an atom or molecule. For example, ozone in our atmosphere absorbs ultraviolet radiation.

absorption-line spectrum a spectrum showing dark lines at some narrow color regions (wave-lengths). The lines are formed by atoms absorbing light, which lifts their electrons to higher orbits.

acceleration a change in an object's velocity (either its speed or its direction).

accretion addition of matter to a body. Examples are gas falling onto a star and asteroids colliding and sticking together.

accretion disk a nearly flat disk of gas or other material held in orbit around a body by its gravity. The radiation from the hot gas in such disks allows us to infer the existence of black holes.

active galaxy a galaxy whose central region emits abnormally large amounts of electromagnetic radiation from a small volume. Examples are radio galaxies, Seyfert galaxies, and quasars.

AGN an active galactic nucleus. The core of an active galaxy.

alpha particle a helium nucleus: two protons plus (usually) two neutrons.

amino acid a carbon-based molecule used by living organisms to build protein molecules.

angstrom unit a unit of length used in describing wavelengths of radiation and the sizes of atoms and molecules. One angstrom = 10^{-10} meters.

angular momentum a measure of an object's tendency to keep rotating and to maintain its orientation. Mathematically it depends on the object's mass, M, radius, r, and rotational velocity, v, and is proportional to Mvr.

angular size measure of how large an object *looks* to you. It is defined as the angle between lines drawn from the observer to opposite sides of an object. For example, the angular diameter of the Moon is about $\frac{1}{2}^{\circ}$.

annular eclipse an eclipse in which the body in front does not completely cover the other. In an annular eclipse of the Sun, a bright ring of the Sun's disk remains visible around the black disk of the Moon. We therefore see a ring (annulus) of light around the Moon.

anthropic principle the principle that the properties we observe the Universe to possess are limited to those that make our existence possible.

antimatter a type of matter that, if brought into contact with ordinary matter, annihilates it, leaving nothing but energy. The positron is the antimatter analog of the electron. The antiproton is the antimatter analog of the proton. Antimatter is observed in cosmic rays and can be created from energy in the laboratory.

aphelion the point in an orbit where a body is farthest from the Sun.

apollo asteroids those asteroids whose orbits cross the Earth's.

association a loose grouping of young stars and interstellar matter, generally consisting of several star clusters.

asteroid a small, generally rocky, solid body orbiting the Sun and ranging in diameter from a few meters to hundreds of kilometers.

asteroid belt a region between the orbits of Mars and Jupiter in which most of the Solar System's asteroids are located.

astronomical unit (AU) a distance unit based on the average distance of the Earth from the Sun.

atmospheric window a wavelength band in which our atmosphere absorbs little radiation. For example, on Earth the visible window ranges from about 300 to 700 nm, allowing the light we can see with our eyes to pass through the atmosphere.

atom a submicroscopic particle consisting of a nucleus and orbiting electrons. The smallest unit of a chemical element.

aurora the light emitted by atoms and molecules in the upper atmosphere. This light is a result of magnetic disturbances and appears to us as the northern or southern lights.

autumnal equinox the autumn equinox in the northern hemisphere. Fall begins on the autumnal equinox, which is on or near September 23.

azimuth a coordinate for locating objects on the sky. Azimuth is the angle measured from North along the horizon to the point below the object.

b

barred spiral galaxy a galaxy in which the spiral arms wind out from the ends of a central bar rather than from the nucleus.

biased galaxy formation a theory that galaxies form primarily in regions with a high density of cosmic matter.

Big Bang the event that, according to many astronomical theories, created the Universe. It occurred about 15 billion years ago and generated the expanding motion that we observe today.

binary star two stars in orbit around each other, held together by their mutual gravity.

bipolar flow the narrow columns of high-speed gas ejected by a protostar in two opposite directions.

blackbody an object that is an ideal radiator when hot and a perfect absorber when cool. It absorbs all radiation that falls upon it, reflecting no light; hence, it appears black. Stars are approximately blackbodies. The radiation emitted by blackbodies obeys Wien's Law and the Stefan-Boltzmann Law.

black dwarf a collapsed star that has cooled to the point where it emits little or no visible radiation.

black hole an object whose gravitational attraction is so strong that its escape velocity equals the speed of light, preventing light or any radiation or material body from leaving its "surface."

BL Lac object a type of active galaxy named for the peculiar galaxy BL Lac. These objects generally are strong radio sources, and their visible light varies rapidly and erratically.

blueshift a shift in the wavelength of electromagnetic radiation to a shorter wavelength. For visible light, this implies a shift toward the blue end of the spectrum. The shift can be caused by the motion of a source of radiation toward the observer or by the motion of an observer toward the source. For example, the spectrum lines of a star moving toward the Earth exhibit a blueshift. *See also* Doppler shift.

Bode's law a numerical expression for the approximate distances of most of the planets from the Sun.

Bok globule small, dark, interstellar cloud, often approximately spherical. Many globules are the early stages of protostars.

brown dwarf a star that has a mass too low for it to begin nuclear fusion.

bulge the dense, central region of a spiral galaxy.

c

carbonaceous chondrite a type of meteorite containing many tiny spheres (chondrules) of rocky or metallic material stuck together by carbon-rich material.

CCD charged-coupled device: an electronic device that records the intensity of light falling on it. CCD's have replaced film in most astronomical applications.

celestial equator an imaginary line on the celestial sphere lying exactly above the Earth's equator. It divides the celestial sphere into northern and southern hemispheres.

celestial pole an imaginary point on the sky directly above the Earth's North or South pole.

celestial sphere an imaginary sphere surrounding the Earth. Ancient astronomers pictured celestial objects as attached to it.

cepheid a class of yellow-giant pulsating stars. Their pulsation periods range from about a day to about 70 days. Cepheids can be used to determine distances. *See also* standard candle.

Chandrasekhar limit the maximum mass of a white dwarf. Approximately 1.4 solar masses. Named for the astronomer who first calculated that such a limit exists.

chondritic meteorite a meteorite containing small spherical bodies called chondrules.

chondrule a small spherical body embedded in a meteorite.

chromosphere the lower part of the Sun's outer atmosphere that lies directly above the Sun's visible surface (photosphere).

closed universe a model of the Universe in which expansion eventually stops and reverses into contraction, so that the Universe collapses. *See also* critical density.

cluster a group of objects (stars, galaxies, and so forth) held together by their mutual gravitational attraction.

CNO cycle/process a reaction involving carbon, nitrogen, and oxygen (C, N, and O) that fuses hydrogen into helium and releases energy. The process begins with a hydrogen nucleus fusing with a carbon nucleus. Subsequent steps involve nitrogen and oxygen. The carbon, nitrogen, and oxygen act as catalysts and are released at the end of the process to start the cycle again. The CNO cycle is the dominant process for generating energy in main-sequence stars that are hotter and more massive that the Sun.

coma the gaseous atmosphere surrounding the head of a comet.

comet a small body in orbit around the Sun, consisting of a tiny, icy core and a tail of gas and dust. The tail forms only when the comet is near the Sun.

compact stars very dense stars whose radii are much smaller than the Sun's. These stars include white dwarfs, neutron stars, and black holes.

conjunction the appearance of two astronomical objects in approximately the same direction on the sky. For example, if Mars and Jupiter happen to appear near each other on the sky, they are said to be in conjunction. *Superior conjunction* refers to a planet that is approximately in line with the Sun but on the far side of the Sun from the Earth. *Inferior conjunction* refers to a planet that lies approximately between the Sun and the Earth.

conservation of angular momentum a principle of physics stating that the angular momentum of a rotating body remains constant unless forces act to speed it up or slow it down. Mathematically, conservation of angular momentum states that MVR is a constant, where M is the mass of a body moving with a velocity V in a circle of radius R. One extremely important consequence of this principle is that if a rotating body shrinks, its rotational velocity must increase.

conservation of energy a principle of physics stating that energy is never created or destroyed, although it may change its form. For example, energy of motion may change into energy of heat.

constellation a grouping of stars on the night sky. Astronomers divide the sky into 88 constellations.

continuous spectrum a spectrum with neither dark absorption nor bright emission lines. The intensity of the radiation in such a spectrum changes smoothly from one wavelength to the next.

convection the rising and sinking motions in a liquid or gas that carry heat upward through the material. Convection is easily seen in a pan of heated soup on a stove.

convection zone the region immediately below the Sun's visible surface in which its heat is carried by convection.

Coriolis effect a deflection of a moving object caused by its motion across the surface of a rotating body. The Coriolis effect makes storms on Earth spin, generates large-scale wind systems, and creates cloud belts on many of the planets.

corona the outer, hottest part of the Sun's atmosphere.

coronal hole a low-density region in the Sun's corona. The solar wind may originate in these regions.

cosmic background radiation (CBR) the radiation that was created during the Big Bang and that permeates all space. At this time the temperature of this radiation is 2.73 K.

cosmic horizon the maximum distance one can see out into the Universe at a given time. The horizon lies at a distance in light years equal to the age of the Universe in years.

cosmic rays extremely energetic particles (protons, electrons, and so forth) traveling at nearly the speed of light. Some rays are emitted by the Sun, but most come from more distant sources, perhaps exploding supernovas.

cosmological principle the hypothesis that on the average the Universe looks the same to every observer, no matter where he or she is located in it.

cosmology the study of the structure and evolution of the Universe.

Crab nebula a supernova remnant in the constellation Taurus. Astronomers in ancient China and the Far East saw the supernova explode in A. D. 1054. A pulsar lies in the middle of the nebula.

crater a circular pit, generally with a raised rim, and sometimes with a central peak. Crater diameters on the Moon range from centimeters to several hundred kilometers. Most craters on bodies such as the Moon are formed by the impact of solid bodies, such as asteroids.

critical density the density necessary for a closed Universe. If the density of the Universe exceeds the critical density, the Universe will stop expanding and collapse. If the density is less than the critical density, the Universe will expand forever.

crust the solid surface of a planet, moon, or other solid body.

curvature of space the bending of space by a mass, as described according to Einstein's General Theory of Relativity. Black holes bend the space around them, curving it so that the region within the black hole is cut off from the rest of the Universe. The Universe too may be curved in such a way as to make its volume finite.

curved space space that is not flat. *See also* curvature of space.

d

dark matter matter that emits no detectable radiation but whose presence can be deduced by its gravitational attraction on other bodies.

dark nebula a dense cloud of dust and gas in interstellar space that blocks the light from background stars.

daughter atoms the atoms produced by the decay of a radioactive element. For example, uranium decays into lead. Thus lead atoms are daughter atoms.

declination one part of a coordinate system for locating objects in the sky north or south of the celestial equator. Declination is analogous to latitude on the Earth's surface.

degeneracy pressure the pressure created in a dense gas by the interaction of its electrons. Degeneracy pressure does not depend on temperature.

degenerate gas an extremely dense gas in which the electrons and nuclei are tightly packed. The pressure of a degenerate gas does not depend on its temperature.

density the mass of a body or region divided by its volume.

density-wave theory of spiral structure a theory to account for the spiral arms of galaxies. According to the theory, waves traveling through the disk of a galaxy sweep stars and interstellar gas into a spiral pattern.

differential gravitational force the difference between the gravitational forces exerted on an object at two different points. The effect of this force is to stretch the object. Such forces create tides and, if strong enough, may break up an astronomical object. *See also* Roche limit.

differentiation the separation of different previously mixed materials inside a planet or other object. This is the same separation that occurs when a heavy material, for example iron, settles to the planet's core, leaving lighter material on the surface.

diffraction a disturbance of light (or other electromagnetic waves) as it passes through an opening or around an obstacle. Diffraction limits the ability to distinguish fine details in images.

disk the flat, round portion of a galaxy. The Sun lies in the disk of the Milky Way.

dispersion the spreading of light or other electromagnetic radiation into a spectrum. A rainbow is an example of dispersion of light caused by raindrops.

Doppler shift the change in the observed wavelength of radiation caused by the motion of the emitting body or the observer. The shift is an increase in the wavelength if the source and observer move apart and a decrease in the wavelength if the source and observer approach. *See also* redshift and blueshift.

dwarf a small dim star.

dynamo a physical process for generating magnetic fields in astronomical bodies. In many cases the process involves the generation of electric currents from an interaction between rotation and convection.

e

Earth's magnetic field the magnetism generated by the Earth. The magnetic field is what exerts the magnetic force on a compass needle.

eclipse the blockage of light from one astronomical body caused by the passage of another between it and the observer. The shadow of one astronomical body falling on another. For example, the passage of the Moon between the Earth and Sun can block the Sun's light and cause a solar eclipse.

eclipse seasons the times of year, separated by about 6 months, when eclipses are possible. At any given eclipse season, both a solar eclipse and lunar eclipse generally occur.

eclipsing binary a binary star pair in which one star periodically passes in front of the other, totally or partially blocking the background star from view as seen from Earth.

ecliptic the path followed by the Sun around the celestial sphere. The path gets its name because eclipses can only occur when the Moon crosses the ecliptic.

electromagnetic force the force arising between electrically charged particles or between charges and magnetic fields. Forces between magnets are a special case of this force. This force holds electrons to the nucleus of atoms, makes moving charges spiral around magnetic field lines, and deflects a compass needle.

electromagnetic spectrum the assemblage of all wavelengths of electromagnetic radiation. The spectrum includes the following wavelengths, from long to short: radio, infrared, visible light, ultraviolet, x-rays, and gamma rays.

electromagnetic wave a wave consisting of alternating electric and magnetic energy. Ordinary visible light is an electromagnetic wave, and the wavelength determines the light's color.

electron a low-mass, negatively charged subatomic particle. Electrons orbit the atomic nucleus but may at times be torn free. *See also* ionization.

electroweak force one of the fundamental forces of nature. At one time it was considered to be two separate forces: electromagnetism and the weak force. The weak force causes radioactive decay.

element a fundamental substance, such as hydrogen, carbon, or oxygen, that cannot be broken down into a simpler chemical substance. Approximately 100 elements occur in nature.

ellipse a geometric figure related to a circle, but elongated in one direction.

elliptical galaxy a galaxy in which the stars smoothly fill an ellipsoidal volume. Abbreviated E galaxy. The stars in such systems are generally old (Pop II).

emission the production of light, or more generally, electromagnetic radiation by an atom or other object.

emission nebula a hot gas cloud in interstellar space that emits light.

emission-line spectrum a spectrum consisting of bright lines at certain wavelengths separated by dark regions in which there is no light.

energy a quantity that measures the ability of a system to do work or cause motion.

energy level any of the numerous levels that an electron can occupy in an atom, roughly corresponding to an electron orbit.

epicycle a fictitious, small, and circular orbit superimposed on another circular orbit and proposed by early astronomers to explain the retrograde motion of the planets.

equator the imaginary line that divides the Earth (or other body) symmetrically into its northern and southern hemispheres. The equator is perpendicular to a body's rotation axis.

equinox the time of year when the number of hours of daylight and night are approximately equal. The spring and fall (vernal and autumnal) equinoxes mark the beginning of the spring and fall seasons.

escape velocity the speed an object needs to move away from another body in order not to be pulled back by its gravitational attraction. Mathematically, the escape velocity, v, is defined as

$$\sqrt{\frac{2GM}{R}}$$

where M is the body's mass, R is its radius, and G is the gravitational constant.

eukaryotes cells with nuclei. Most cells in current terrestrial organisms have nuclei and are thus eukaryotes.

Evening Star any bright planet, often Venus, seen low in the western sky after sunset.

event horizon the location of the "surface" of a black hole. An outside observer cannot see in past the event horizon.

excited the condition in which the electrons of an atom are not in their lowest energy level (orbit).

exclusion principle the condition that no more than two electrons may occupy the same energy state in an atom. This limitation leads to degeneracy pressure.

f

falsecolor picture/photograph a depiction of an astronomical object in which the colors are not the object's real colors. Instead, they are colors arbitrarily chosen to represent other properties of the body, for example, the intensity of radiation, which we cannot see.

false vacuum a state of the early Universe.

fission the splitting of a single body, such as an atom, into two or more pieces.

flare an outburst of energy on the Sun. *See also* solar flare.

flat universe a universe that extends forever with no curvature. Its total energy is zero.

fluorescence the conversion of ultraviolet light (or other short-wavelength radiation) into visible light.

focus (1) one of two points within an ellipse used to generate the elliptical shape. The Sun lies at the focus of each planet's elliptical orbit. (2) a point in an optical system in which light rays are brought together. The location where an image forms in such systems.

frequency the number of times per second that a wave vibrates.

fundamental forces the three basic forces of nature: gravitation, the electroweak force, and the strong force. According to some modern theories (*see also* GUTs), these three forces are different forms of a single, more fundamental, unified force.

Gaia hypothesis the hypothesis that life does not merely respond to its environment but actually alters its planet's atmosphere and temperature to make the planet more hospitable. For example, by photosynthesis, plants have created an oxygen-rich atmosphere on Earth, which shields the plants from dangerous ultraviolet radiation. Gaia is pronounced / Guy-uh/ in this context.

galactic cannibalism the capture and merging of one galaxy into another.

galaxy a massive system of stars held together by their mutual gravity. Typical galaxies have a mass between about 10^7 and 10^{13} solar masses. Our galaxy is the Milky Way.

galaxy cluster a group of galaxies held together by their mutual gravitational attraction. The Milky Way belongs to the Local Group galaxy cluster.

Galilean satellites the four moons of Jupiter discovered by Galileo: Io, Europa, Ganymede, and Callisto.

geocentric centered on the Earth. Many of the earliest attempts to describe the Solar System were geocentric in that they supposed that the planets moved around the Earth rather than around the Sun.

giant a star of large radius and large luminosity.

glitches abrupt charges in the pulsation period of a pulsar, perhaps the result of adjustments of its crust.

globular cluster a dense grouping of old stars, containing generally about 10^5 to 10^6 members. They are often found in the halos of galaxies.

globule *see* Bok globule.

grand unified theories (GUTs) theories that propose that in the early Universe all the fundamental forces behaved as a single force.

granulation texture seen in the Sun's photosphere. Granulation is created by clumps of hot gas that rise to the Sun's surface.

grating a piece of material that creates a spectrum by reflecting light from or passing it through many very fine and closely spaced parallel lines.

gravitational lens an object that bends space (and thereby the light passing through the space) by its gravitational attraction and focuses the light to create an image of a more distant object. *See also* curvature of space.

gravitational redshift the shift in wavelength of electromagnetic radiation (light) created by a body's gravitational field as the radiation moves away from the body. Only extremely dense objects, such as white dwarfs, produce a significant redshift of their radiation.

gravitational waves a wavelike bending of space generated by the acceleration of massive bodies.

gravity the force of attraction that is between two bodies and is generated by their masses.

greatest elongation the position of an inner planet (Mercury or Venus) when it lies farthest from the Sun on the sky. Mercury and Venus are particularly easy to see when they are at greatest elongation. Objects may be at greatest eastern or western elongation according to whether they lie east of the Sun or west of the Sun.

greenhouse effect the trapping of heat by a planet's atmosphere, making the planet warmer than would otherwise be expected. Generally, the greenhouse effect operates if visible sunlight passes freely through a planet's atmosphere, but the infrared radiation produced by the warm surface cannot escape readily into space.

GUTs *see* grand unified theories.

h

halo the approximately spherical region surrounding spiral galaxies that contains mainly old stars, such as the globular clusters. The halo may also contain large amounts of dark matter.

Hawking Radiation radiation that black holes are hypothesized to emit as a result of quantum effects. This radiation leads to the extremely slow evaporation of black holes.

heliocentric centered on the Sun. Used to describe models of the Solar System in which the planets orbit the Sun.

helium flash the beginning of helium fusion in a low-mass star. The fusion begins explosively and causes a major readjustment of the star's structure.

Herbig-Haro objects clumps of gas seen near some young stars.

highlands the old, heavily cratered regions on the Moon.

horizon the line separating the sky from the ground. *See also* cosmic horizon.

H-R diagram a graph on which stars are located according to their temperature and luminosity. Most stars on such a plot lie along a diagonal line, called the main sequence, which runs from cool, dim stars in the lower right, to hot, luminous stars in the upper left.

Hubble constant the multiplying constant H in the Hubble law, $v = HD$. The reciprocal of the Hubble constant (in appropriate units) is approximately the age of the Universe.

Hubble law a relation between a galaxy's distance D, and its recession velocity, v, which states that more distant galaxies recede faster than nearby ones. Mathematically, $v = HD$, where H is the Hubble constant.

hydrogen burning nuclear fusion of hydrogen into helium. It is not "burning" like ordinary fire, but is instead the transformation of one kind of atom into another accompanied by the release of energy.

hydrostatic equilibrium the condition in which pressure and gravitational forces in a star or planet are in balance. Without such balance, bodies will either collapse or expand.

hypothesis an explanation proposed to account for some set of observations or facts.

HII region a region of ionized hydrogen. HII regions generally show a luminous pink/red glow and often surround luminous, hot, young stars.

i

ideal gas law a law relating the pressure, density, and temperature of a gas. This law states that the pressure is proportional to the density times the temperature. Also called the *perfect gas law*.

inclination the tilt angle of an astronomical object or its orbit.

inertia the tendency of an object at rest to remain at rest and of a body in motion to continue in motion in a straight line at a constant speed. *See also* mass.

inferior planet a planet whose orbit lies between the Earth's orbit and the Sun. Mercury and Venus are inferior planets.

inflation the rapid expansion of the early Universe by an enormous factor.

infrared a wavelength of electromagnetic radiation longer than visible light but shorter than radio waves. We cannot see these wavelengths with our eyes, but we can feel many of them as heat. The infrared wavelength region runs from about 1 nm to 1 mm.

inner core the innermost part of a planet, also called the *solid core*. The Earth's inner core is a mixture of solid iron and nickel.

inner planet a planet orbiting in the inner part of the Solar System. Sometimes taken to mean Mercury, Venus, Earth, and Mars.

instability strip a region in the H-R diagram in which stars pulsate.

interferometer a region consisting of two or more telescopes connected together to work as a single instrument. Used to obtain a high resolving power, the ability to see small-scale features. This device operates at radio, infrared, or visible wavelengths.

interstellar cloud a cloud of gas and dust in between the stars. Such clouds may be many light years in diameter.

interstellar grain microscopic solid dust particles in interstellar space. These grains absorb starlight, making distant stars appear dimmer than they truly are.

interstellar matter matter in the form of gas or dust in the space between stars.

inverse-square law (1) any law in which some property varies inversely as the square of the distance, d. Mathematically, as $1/d^2$. (2) the law stating that the apparent brightness of a body decreases inversely as the square of its distance.

ionization the removal of one or more electrons from an atom, leaving the atom with a positive electric charge. Under some circumstances, an extra electron may be attached to an atom, in which case the atom is described as negatively ionized.

ionized a condition in which the number of an atom's electrons does not equal the number of its protons. Typically, this means the atom is missing one or more electrons.

irregular galaxy a galaxy lacking a symmetric structure.

j

jets narrow streams of gas ejected from any of several types of astronomical objects. Jets are seen near protostars and in many active galaxies.

jet stream a narrow stream of high-speed wind that blows in the atmosphere of a planet. Such winds occur on Earth and many other planets.

joule a unit of energy. One joule per second equals one watt.

Jovian planet one of the giant, gaseous planets: Jupiter, Saturn, Uranus, and Neptune. The name *Jovian* was chosen because Jupiter's structure is representative of the other three.

k

Kepler's three laws laws that describe the motion of planets around the Sun. The first law states that planets move in elliptical orbits with the Sun off center at a focus of the ellipse. The second law states that a line joining the planet and the Sun sweeps out equal areas in equal times. The third law relates a planet's orbital period, P, to the semimajor axis of its elliptical orbit, a. Mathematically, the law states that $P^2 = a^3$, if P is measured in years and a in astronomical units.

Kirkwood gaps regions in the asteroid belt with a lower-than-average number of asteroids. Some of the gaps result from the gravitational force of Jupiter removing asteroids from orbits within the gaps.

Kuiper belt a region from which some comets come. The region extends from the orbit of Neptune, past Pluto, and into the Oort cloud.

l

law of gravity a description of the gravitational force exerted by one body on another. The gravitational force is proportional to the product of their masses and the inverse square of their separation. If the masses are M and m and their separation is d, the force between them, F, is $F = GMm/d^2$, where G is a physical constant.

light electromagnetic energy.

light-year a unit of distance equal to the distance light travels in one year. A light-year is roughly 10^{13} km, or about 6 trillion miles.

liquid core the molten interior of a planet, also called the *outer core*.

lobes regions lying outside the body of a radio galaxy where much of its radio emission comes from.

Local Group the small group of about 30 galaxies to which the Milky Way belongs.

Local Supercluster the cluster of galaxy clusters in which the Milky Way is located. The Local Group is one of its member clusters.

luminosity the amount of energy radiated per second by a body. For example, the wattage of a light bulb defines its luminosity. Stellar luminosity is usually measured in units of the Sun's luminosity (approximately 4×10^{26} watts).

lunar eclipse the passage of the Earth between the Sun and the Moon so that the Earth's shadow falls on the Moon.

m

Magellanic clouds two small companion galaxies of the Milky Way.

magnetic field a representation of the means by which magnetic forces are transmitted from one body to another.

magnetic lines of force fictitious lines used to visualize the orientation and strength of a magnetic field.

magnetosphere the extreme upper regions of the Earth's atmosphere in which the motion of the gas is controlled by the Earth's magnetic field.

magnitude a unit for measuring stellar brightness. The smaller the magnitude, the brighter the star.

main sequence the region in the H-R diagram in which most stars, for instance the Sun, are located. The main sequence runs diagonally across the H-R diagram from cool, dim stars to hot, luminous ones. Stars on the main sequence fuse hydrogen into helium in their cores. *See also* H-R diagram.

main-sequence lifetime the time a star remains a main sequence star, fusing hydrogen into helium in its core.

mantle the solid, outer part of a planet. This part is immediately below the crust.

mare a vast, smooth, dark, and congealed lava flow filling a basin on the Moon and on some planets. Maria often have roughly circular shapes.

maria plural of mare.

maser an intense radio source created when excited gas amplifies some background radiation. "Maser" stands for *m*icrowave *a*mplification by *s*timulated *e*mission of *r*adiation.

mass a measure of the amount of material an object contains. A quantity measuring a body's inertia.

mass function the number of stars of a given mass range in some group.

mass-luminosity relation a relation between the mass and luminosity of stars. Higher-mass stars have higher luminosity.

Maunder minimum the time period from about A.D. 1600 to 1740 during which the Sun was relatively inactive. Few sunspots were observed during this period.

megaparsec a distance unit equal to 1 million parsecs and abbreviated Mpc.

metal astronomically, any chemical element more massive than helium. Thus carbon, oxygen, iron, and so forth are metals.

meteor the bright trail of light created by small solid particles entering the Earth's atmosphere and burning up. A "shooting star."

meteor shower an event in which many meteors occur in a short space of time, all from the same general direction in the sky. The most famous shower is the Perseids in mid-August.

meteorite the solid remains of a meteor that falls to the Earth.

meteoroid the technical name for the small, solid bodies moving within the Solar System. When a meteoroid enters our atmosphere and heats up, the trail of luminous gas it leaves is called a meteor. When the body lands on the ground, it is called a meteorite. ("A meteoroid is in the void. A meteor above you soars. A meteorite is in your sight.")

method of standard candles *See* standard candle.

mid-Atlantic ridge an underwater mountain range on Earth created by plate-tectonic motion and running approximately north-south down the center of the Atlantic Ocean.

Milky Way galaxy the galaxy to which the Sun belongs. Seen from Earth, the galaxy is a pale, milky band in the night sky.

Miller-Urey experiment an experimental attempt to stimulate the conditions under which life might have developed on Earth. Miller and Urey discovered that amino acids and other complex organic compounds could form from the gases that are thought to be present in the Earth's early atmosphere, if the gases are subjected to an electric spark or ultraviolet radiation.

millisecond pulsar a pulsar whose rotation period is about a millisecond.

minute of arc a measure of angle equal to one-sixtieth of a degree.

model a theoretical representation of some object or system.

molecule two or more atoms bonded into a single particle, such as water, H_2O (two hydrogen atoms bonded to one oxygen) or carbon dioxide, CO_2 (one carbon atom bonded to two oxygen atoms).

Morning star any bright planet visible in the eastern sky before dawn.

n

nanometer a unit of length equal to 1 billionth of a meter (10^{-9} meters) and abbreviated nm. Wavelengths of visible light are several hundred nanometers. The diameter of a hydrogen atom is roughly 0.1 nm.

neap tide an abnormally low tide occurring when the Sun's and Moon's gravitational effects on the ocean partially offset each other.

nebula cloud in interstellar space.

negative curvature a form of curved space sometimes described as being open in that it has no boundary. Negative curvature is analogous to a saddleshape.

neutrinos tiny neutral particles with little or no mass and immense penetrating power. These particles are produced in great numbers by the Sun as it fuses hydrogen into helium and also by some supernova explosions.

neutron a subatomic particle of nearly the same mass as the proton but with no electric charge. Neutrons and protons comprise the nucleus of the atom.

neutron star a very dense, compact star composed primarily of neutrons.

Newton's first law of motion a body continues in a state of rest or uniform motion in a straight line unless made to change that state by forces acting on it. *See also* inertia.

Newton's second law of motion $F = ma$. In words, the amount of acceleration, a, that a force, F, produces depends on the mass, m, of the object being accelerated.

Newton's third law of motion when two bodies interact, they exert equal and opposite forces on each other.

nonthermal radiation radiation emitted by charged particles moving at high speed in a magnetic field. The radio emission from pulsars and radio galaxies is nonthermal emission. More generally, nonthermal means "not due to high temperature."

north celestial pole the point on the celestial sphere sky directly above the Earth's north pole. Objects on the sky appear to circle around this point.

North Star any star that happens to lie very close to the north celestial pole. Polaris has been the North Star for about 1 years, and it will continue as such for about another 1 years, at which time a star in Cepheus will be nearer the north celestial pole.

nova a process in which a surface layer of hydrogen builds up on a white dwarf and then fuses explosively into helium. Nova explosions do not destroy the star and may be recurrent.

nuclear force the force that holds protons and neutrons together in the atomic nucleus. Also called the *strong force*.

nuclear fusion the binding of two light nuclei to form a heavier nucleus with some nuclear mass converted to energy. For example, the fusion of hydrogen into helium. This process supplies the energy of most stars and is commonly called "burning" by astronomers.

nucleosynthesis the formation of elements, generally by the fusion of lighter elements into heavier ones. For example, the formation of carbon by the fusion of three helium nuclei.

nucleus the core of an atom around which the electrons orbit. The nucleus has a positive electric charge and comprises most of an atom's mass.

o

obscuration the blocking of light of background stars by interstellar matter.

occultation the covering up of one astronomical body by another. For example, the Moon passes directly between Earth and the planets from time to time, covering them up and causing their occultation.

Olbers' paradox an argument that the sky should be bright at night because of the light from many distant stars and galaxies.

Oort cloud a vast region in which comet nuclei orbit. This cloud lies far beyond the orbit of Pluto.

opacity the blockage of light or other electromagnetic radiation by matter.

open cluster a loose cluster of stars, generally containing a few hundred members.

open universe a universe that expands forever.

opposition the configuration of a planet when it is opposite the Sun in the sky. If a planet is in opposition, it rises when the Sun sets and sets when the Sun rises.

orbit the path in space followed by a celestial body; also a description of an electron's position in an atom.

outer core the molten interior of a planet; also called the *liquid core*.

outer planet a planet whose orbit lies in the outer part of the Solar System. Jupiter, Saturn, Uranus, Neptune, and Pluto are outer planets.

ozone a form of oxygen consisting of three oxygen atoms bonded together. Its chemical symbol is O_3. Because it absorbs ultraviolet radiation, ozone in our atmosphere shields us from the Sun's harmful ultraviolet radiation.

P

panspermia a theory that life originated elsewhere than on Earth and came here across interstellar space either accidently or deliberately.

parallax the shift in an object's position caused by the observer's motion. A method for finding distance based on that shift.

parsec a unit of distance equal to about 3.26 light years (3.09 × 10^{13} km) defined as the distance at which an observer sees the maximum angle between the Sun and the Earth to be one arc second.

perfect gas law a law relating the pressure, density, and temperature of a gas. It states that the pressure is proportional to the density times the temperature. This is also called the *ideal gas law*.

perihelion the point in an orbit closest to the Sun.

period the time required for a repetitive process to repeat. For example, orbital period is the time it takes a planet or star to complete an orbit. Pulsation period is the time it takes a star to expand and then contract back to its original radius.

period-density relation a relation that states that the period of a variable star is inversely proportional to the square root of its average density. That is, high-density stars pulsate more rapidly than low-density stars.

period-luminosity relation a relation stating that the longer the period of a pulsating variable star, the more luminous it is.

phases the changing illumination of the Moon or other body that causes its apparent shape to change. The following is the cycle of lunar phases: new, crescent, first quarter, gibbous, full, gibbous, third quarter, crescent, new.

photo dissociation the breaking apart of a molecule by intense radiation.

photon a particle of visible light or other electromagnetic radiation.

photosphere the visible surface of the Sun. When we look at the Sun in the sky we are seeing its photosphere.

planet a body in orbit around a star.

planetary nebula a shell of gas ejected by a low-mass star late in its evolutionary lifetime. Planetary nebulas appear as a glowing gas ring around a central star.

planetesimal one of the numerous small, solid bodies that, when gathered together, form a planet.

plate tectonics the hypothesis that the crust of the Earth (or some other planet) is divided into large regions (plates) that move very slowly over the planet's surface. Interaction between plates at their boundaries creates mountains and activity such as volcanoes and earthquakes.

polarity the property of a magnet that causes it to have a north and south pole.

poor cluster a galaxy cluster with a small number of members. The galaxy cluster to which the Milky Way belongs, the Local Group, is a poor cluster.

population (Pop) I the younger stars, some of which are blue, that populate a galaxy's disk, especially its spiral arms.

population (Pop) II the older, redder stars that populate a galaxy's halo and bulge.

population (Pop) III a hypothetical stellar population consisting of the first stars that formed in a galaxy, composed of only hydrogen and helium.

positive curvature bending of space leading to a finite volume. A space that is "closed." A universe with positive curvature is analogous to a spherical shape.

precession the slow change in direction of the pole (rotation axis) of a spinning body.

pressure the force exerted by a substance such as a gas on an area divided by that area. That is, pressure × area = force.

prokaryotes cells without nuclei. Presumably the first life forms on Earth were prokaryotes.

prominence a cloud of hot gas in the Sun's outer atmosphere. This cloud is often shaped like an arch.

proton a positively charged subatomic particle. One of the constituents of the nucleus of the atom.

proton-proton chain the nuclear fusion process that converts hydrogen into helium in stars like the Sun and thereby generates their energy. This is the dominant energy-generation mechanism in cool, low-mass stars.

protostar a star still in its formation stage.

pulsar a spinning neutron star that emits beams of radiation that happen to sweep across the Earth each time the star spins. We observe the radiation as regularly spaced pulses.

pulsate to expand and contract regularly. For example, pulsating variable stars swell and shrink in a predictable, regular fashion.

q

quantized the property of a system that allows it to have only discrete values.

quasar a peculiar galaxy characterized by a large redshift, high luminosity, and an extremely small, active core. Quasars are the most luminous and most distant objects known to astronomers.

r

radial velocity the velocity of a body along the line of sight. That is, the part of its motion directly toward or away from the observer.

radiant the point in the sky from from which meteors in showers appear to come. *See also* meteor showers.

radiation pressure the force exerted by radiation on matter.

radiative zone the region inside a star where its energy is carried outward by radiation (that is, photons).

radio galaxy a galaxy, generally an elliptical system, that emits abnormally large amounts of radio energy. *See also* lobes.

radioactive decay the breakdown of an atomic nucleus by the emission of subatomic particles.

radioactive element an element that undergoes radioactive decay and breaks down into a lighter element.

rays long, narrow, light-colored markings on the Moon or other bodies that radiate from young craters. Rays are debris "splashed" out of the crater by the impact that formed it.

recession velocity the velocity of an external galaxy (or other object) away from the Sun.

recombination the reattachment of one (or more) electrons to an atom following its removal.

reddening the alteration in a star's color as seen from Earth as the star's light passes through an intervening interstellar dust cloud. The dust preferentially scatters the blue light from the beam, leaving the remaining light redder.

red giant a cool, luminous star whose radius is much larger than the Sun's.

redshift a shift in the wavelength of electromagnetic radiation to a longer wavelength. For visible light, this implies a shift toward the red end of the spectrum. The shift can be caused by a source of radiation moving away from the observer or by the observer moving away from the source. For example, if a star is moving away from Earth, its spectrum lines exhibit a redshift. *See also* Doppler shift.

reflection nebula an interstellar cloud in which the dust particles reflect starlight, making the cloud visible.

reflector a telescope that uses a mirror to collect and focus light.

refraction the bending of light when it passes through one substance and enters another.

refractor a telescope that uses a lens to collect and focus light.

regolith the surface rubble of broken rock on the Moon or other solid body.

resolving power the ability of a telescope or instrument to discern fine details. Larger-diameter telescopes have greater (that is, better) resolving power.

resonance a condition in which the repetitive motion of one body interacts with the repetitive motion of another so as to reinforce the motion. Sliding back and forth in a bathtub to make a big splash is an example.

retrograde motion the drift of a planet westward against the background stars. Normally planets shift eastward because of their orbital motion. The planet does not actually reverse its motion. The change in its direction is caused by the change in the position from which we view the planet as the Earth overtakes and passes it.

rich cluster a galaxy cluster containing hundreds to thousands of member galaxies.

rifting the breaking apart of a continental plate.

right ascension a coordinate for locating objects on the sky, analogous to longitude on the Earth's surface. Measured in hours and minutes of time.

rilles narrow canyons on the Moon or other body.

ring galaxy a galaxy in which the central region has an abnormally small number of stars, causing the galaxy to look like a ring. Caused by the collision of two galaxies.

Roche limit the distance from an astronomical body at which its gravitational force breaks up another astronomical body.

rotation axis an imaginary line through the center of a body about which the body spins.

rotation curve a plot of the rotation velocity of the stars or gas in a galaxy at different distances from its center.

RR Lyrae stars a type of white, giant, pulsating variable star with a period of about one day or less. They are named for their prototype star, RR Lyrae.

S

satellite a body orbiting a planet.

scattering the random redirection of a light wave or photon as it interacts with atoms or dust particles.

Schwarzschild radius the radius of a black hole. The distance from the center of a black hole to its event horizon.

scientific method the process of observing a phenomenon, proposing a hypothesis on the basis of the observations, and then testing the hypothesis.

scintillation the twinkling of stars.

seeing a measure of the steadiness of the atmosphere during astronomical observations. Under conditions of bad seeing, fine details are difficult to see. Bad seeing results from atmospheric irregularities moving between the telescope and the object being observed.

seismic waves waves generated in the Earth's interior by earthquakes. Similar waves occur in other bodies. Two of the more important varieties are S and P waves. The former can travel only through solid material; the latter can travel through either solid or liquid material.

selection effect an unintentional selection process that omits some set of the objects being studied and leads to invalid conclusions about the objects.

self-propagating star formation a model that explains spiral arms as arising from stars triggering the birth of other stars around them. The resulting pattern is then drawn out into a spiral by the galaxy's rotation.

semimajor axis half the long dimension of an ellipse.

SETI *S*earch for *E*xtra*t*errestrial *I*ntelligence. Some such searches involve automatic "listening" to millions of radio frequencies for signals that might be from other civilizations.

Seyfert galaxy variety of active galaxy with a small, abnormally bright nucleus containing hot gas. Named for the astronomer Carl Seyfert, who first drew attention to these objects.

shell source a region in a star where the nuclear energy generation occurs around the core rather than in it.

shepherding satellite one or more satellites that by their gravitational attractions prevent particles in a planet's rings from spreading out and dispersing. Saturn's F-ring is held together by shepherding satellites.

short-period comet a comet whose orbital period is shorter than 200 years. For example, Halley's comet has a period of 76 years.

sidereal day the length of time from the rising of a star until it next rises. The length of the Earth's sidereal day is 23 hours 56 minutes.

sidereal period the time it takes a body to turn once on its rotation axis or to revolve once around a central body, as measured with respect to the stars.

sidereal time a system of time measurement based on the motion of stars across the sky rather than the Sun.

silicates material composed of silicon and oxygen, and generally containing other substances as well. Most ordinary rocks are silicates. For example, quartz is silicon dioxide.

solar cycle the cyclic change in solar activity, such as sunspots and solar flares.

solar day the time interval from one sunrise to the next sunrise or from one noon to the next noon. That time interval is not always exactly 24 hours but varies throughout the year. For that reason, we use the mean solar day (which, by definition, is 24 hours) to keep time.

solar eclipse the passage of the Moon between the Earth and the Sun so that our view of the Sun is partially or totally blocked. *See also* total eclipse.

solar flare a sudden increase in brightness of a small region on the Sun. This flare is caused by a magnetic disturbance.

solar nebula the rotating disk of gas and dust from which the Sun and planets formed.

solar nebula hypothesis the hypothesis that the Solar System formed from a rotating cloud of gas and dust, the solar nebula.

Solar System the Sun, planets, their moons, and other bodies that orbit the Sun.

solar wind the outflow of low-density, hot gas from the Sun's upper atmosphere. It is partially this wind that creates the tail of a comet by blowing dust and gas away from the comet's immediate surroundings.

solid core the inner iron/nickel core of the Earth or other planet. Despite its high temperature, the core is solid because it is under great pressure. Also called the *inner core.*

solstice (winter and summer) the beginning of winter and summer. Astronomically the solstice occurs when the Sun is at its greatest distance north (June) or south (December) of the celestial equator.

south celestial pole the imaginary point on the celestial sphere directly over the Earth's South Pole.

spectral class an indicator of a star's temperature. A star's spectral class is based on the appearance of its spectrum lines. The fundamental classes are, from hot to cool, *O, B, A, F, G, K,* and *M.*

spectrograph a device for making a spectrum.

spectroscopic binary a type of binary star in which the spectrum lines exhibit a changing Doppler shift as a result of the orbital motion of one star around the other.

spectroscopy the study and analysis of spectra.

spectrum electromagnetic radiation (for example, visible light) spread into its component wavelengths or colors. The rainbow is a spectrum produced naturally by water droplets in our atmosphere.

spicule a hot, thin column of gas in the Sun's chromosphere.

spiral arm a long, narrow region containing young stars and interstellar matter that winds outward in the disk of spiral galaxies.

spiral galaxy a galaxy with a disk in which its bright stars form a spiral pattern.

spring tide the abnormally large tides that occur at new and full moon.

standard candle a type of star or other astronomical body in which the luminosity has a known value, allowing its distance to be determined by measuring its apparent brightness and applying the inverse-square law: for example, Cepheid variable stars, supernovas, and so forth.

star a massive, gaseous body held together by gravity and generally emitting light. Normal stars generate energy by nuclear reactions in their interiors.

star cluster a group of stars numbering from hundreds to millions held together by their mutual gravity.

steady-state theory a model of the Universe in which the overall properties of the Universe do not change with time.

strong force the force that holds protons and neutrons together in the atomic nucleus. Also called *nuclear force*.

subatomic particles particles making up an atom, such as electrons, neutrons, and protons, or other particles of similar small size.

subduction the sinking of one crustal plate where it encounters another.

sunspot a dark, cool region on the Sun's visible surface created by intense magnetic fields.

supercluster a cluster of galaxy clusters. Our Milky Way belongs to the Local Group, a small galaxy cluster that is but one of many galaxy groups making up the Local Supercluster.

superfluidity a condition in which a fluid has no friction (technically, the absence of viscosity).

superior planet a planet orbiting farther from the Sun than the Earth. Mars, Jupiter, Saturn, Uranus, Neptune, and Pluto are superior planets.

supernova an explosion marking the end of some star's evolution. Astronomers identify two main kinds of supernova: Type I and II. Type I occurs in a binary system in which one star is a white dwarf. The explosion is triggered when mass from a companion star falls onto the white dwarf raising its mass above the Chandrasekhar limit and causing the star to collapse. Collapse heats the white dwarf so that its carbon and oxygen fuse explosively, destroying the star and leaving no remnant. Type II probably occurs when a massive star's iron core collapses. A Type II supernova leaves either a neutron star or a black hole, depending on the mass of the collapsing core.

supernova remnant the debris ejected from a star when it explodes as a supernova. Typically this material is hot gas, expanding away from the explosion at thousands of kilometers or more per second.

surface gravity the acceleration caused by gravity at the surface of a planet or other body.

synchronous rotation the condition that a body's rotation period is the same as its orbital period. The Moon rotates synchronously.

synchrotron radiation a form of nonthermal radiation emitted by charged particles spiraling at nearly the speed of light in a magnetic field. Pulsars and radio galaxies emit synchrotron radiation. The radiation gets its name because it was first seen in synchrotrons, a type of atomic accelerator.

t

tail the plume of gas and dust from a comet. The plume is produced by the solar wind and radiation pressure acting on the comet. The tail points away from the Sun and gets longer as the comet approaches perihelion.

terrestrial planet a rocky planet similar to the Earth in size and structure. The terrestrial planets are Mercury, Venus, Earth, and Mars.

tidal braking the slowing of one body's rotation as a result of gravitational forces exerted on it by another body.

tidal bulge a bulge on one body created by the gravitational attraction on it by another. Two tidal bulges form, one on the side near the attracting body and one on the opposite side.

tides the rise and fall of the Earth's oceans created by the gravitational attraction of the Moon. Tides also occur in the solid crust of a body and its atmosphere.

total eclipse an eclipse in which the eclipsing body totally covers the other body. Only at a total solar eclipse can we see the Sun's corona.

transit the passage of a planet directly between the observer and the Sun. At a transit we see the planet as a dark spot against the Sun's bright disk. From Earth, only Mercury and Venus can transit the Sun.

triangulation a method for measuring distances. This method is based on constructing a triangle, one side of which is the distance to be determined. That side is then calculated by measuring another side (the base line) and the two angles at either end of the base line.

triple alpha process the fusion of three helium nuclei (alpha particles) into a carbon nucleus. This process is sometimes called helium burning, and it occurs in many old stars.

tuning-fork diagram a diagram devised by Hubble to classify the various forms of spiral, elliptical, and irregular galaxies. The diagram is named for its shape.

turnoff point the location on the main sequence where a star's evolution causes it to move away from the main sequence toward the red giant region. The location of the turnoff point can be used to deduce the age of a star cluster.

T Tauri star a type of extremely young star that varies erratically in its light output.

21-cm line a spectrum line at radio wavelengths produced by un-ionized (neutral) hydrogen.

U

ultraviolet a portion of the electromagnetic spectrum with wavelengths shorter than that of visible light but longer than that of X rays. By convention the ultraviolet region extends from about 10 to 300 nm.

Universal time the time kept at Greenwich, England. Universal time is the same as Greenwich mean time. Most local times (in the United States these are Eastern, Central, Mountain, Pacific, and so forth) differ from it by an even number of hours.

Universe the largest astronomical structure we know of. The Universe contains all matter and radiation and encompasses all space.

V

Van Allen radiation belts doughnut-shaped regions surrounding the Earth containing charged particles trapped by the Earth's magnetic field.

variable star a star whose luminosity changes in time.

vernal equinox the spring equinox in the northern hemisphere. Spring begins on the vernal equinox, which is on or near March 21.

visible spectrum the part of the electromagnetic spectrum that we can see with our eyes. It consists of the familiar colors red, orange, yellow, green, blue, and violet.

visual binary star a pair of stars held together by their mutual gravity and in orbit about each other, and which can be seen with a telescope as separate objects.

visual double star two stars that appear to lie very close together on the sky but which in reality are at greatly different distances.

W

wave-particle duality the theory that electromagnetic radiation may be treated as either a particle (photon) or an electromagnetic wave.

wavelength the distance between wavecrests. It determines the color of visible light and is generally denoted by the Greek letter λ.

weak force the force responsible for radioactive decay of atoms. Now known to be linked to electric and magnetic forces and therefore called the electroweak force.

white dwarf a dense star whose radius is approximately the same as the Earth's but whose mass is comparable with the Sun's. White dwarfs burn no nuclear fuel and shine by residual heat. They are the end stage of stellar evolution for stars like the Sun.

white light visible light exhibiting no color of its own but composed of a mix of all colors. Sunlight and many artificial light sources are "white."

Wien's law a relation between a body's temperature and the wavelength at which it emits radiation most intensely. Hotter bodies radiate more intensely at shorter wavelengths. Mathematically, the law states that $\lambda_m = 3 \times 10^6 / T$, where λ_m is the wavelength of maximum emission in nanometers and T is the body's temperature on the Kelvin scale.

X

x-ray binaries a binary star system in which one of the stars, or the gas associated with a star, emits x-rays intensely. Such systems generally contain a collapsed object such as a neutron star or a black hole. *See also* x-ray burster.

x-ray burster a stellar system producing repetitive outbursts of x-ray radiation. x-ray bursters are thought to consist of a neutron star and a normal star in a close binary system. Mass from the normal star spills onto the neutron star, where it slowly accumulates and heats. Eventually the temperature becomes large enough to initiate nuclear fusion. The released energy heats the material further, causing more fusion, and leading to an explosion that we observe as the x-ray burst.

Y

year the time that it takes the Earth to complete its orbit around the Sun; that is, the period of the Earth's orbit.

Z

Zeeman effect the splitting of a single spectrum line into two or three lines by a magnetic field. A method for detecting magnetic fields in objects from their radiation.

zenith the point on the celestial sphere that lies directly overhead at your location.

zodiac a band running around the celestial sphere in which the planets move.

zone of avoidance a band running around the sky in which few galaxies are visible. It coincides with the Milky Way and is caused by dust that is within our galaxy. This dust blocks the light from distant galaxies.

Index

Northern Horizon

Eastern Horizon

Western Horizon

Southern Horizon

The Night Sky in Autumn

Northern Horizon

Eastern Horizon

Western Horizon

Southern Horizon

The Night Sky in Winter